Lecture Notes in Economics and Mathematical Systems 502

Springer
Berlin
Heidelberg
New York
Barcelona
Hong Kong
London
Milan
Paris
Singapore
Tokyo

Nicolas Hadjisavvas
Juan Enrique Martínez-Legaz
Jean-Paul Penot (Eds.)

Generalized Convexity and Generalized Monotonicity

Proceedings of the 6th International Symposium on Generalized Convexity/Monotonicity, Samos, September 1999

Springer

Editors

Prof. N. Hadjisavvas
Department of Mathematics
University of the Aegean
83200 Karlovassi/Samos, Greece

Prof. J.E. Martínez-Legaz
CODE and Department of Economics
Universitat Autonoma de Barcelona
08193 Bellaterra, Spain

Prof. J.-P. Penot
Department of Mathematics
Faculty of Sciences
Université de Pau
64000 Pau, France

Cataloging-in-Publication data applied for

Die Deutsche Bibliothek - CIP-Einheitsaufnahme

Generalized convexity and generalized monotonicity : proceedings of the 6th International Symposium on Generalized Convexity/Monotonicity, Samos, September 1999 / Nicolas Hadjisavvas ... (ed.). - Berlin ; Heidelberg ; New York ; Barcelona ; Hong Kong ; London ; Milan ; Paris ; Singapore ; Tokyo : Springer, 2001
(Lecture notes in economics and mathematical systems ; 502)
ISBN 3-540-41806-7

ISSN 0075-8450
ISBN 3-540-41806-7 Springer-Verlag Berlin Heidelberg New York

Springer-Verlag Berlin Heidelberg New York
a member of BertelsmannSpringer Science+Business Media GmbH

http://www.springer.de

Typesetting: Camera ready by author
Printed on acid-free paper SPIN: 10833049 55/3142/du 5 4 3 2 1 0

Preface

A famous saying (due to Herriot) defines culture as "what remains when everything is forgotten". One could paraphrase this definition in stating that generalized convexity is what remains when convexity has been dropped. Of course, one expects that some convexity features remain. For functions, convexity of epigraphs (what is above the graph) is a simple but strong assumption. It leads to beautiful properties and to a field in itself called convex analysis. In several models, convexity is not present and introducing genuine convexity assumptions would not be realistic.

A simple extension of the notion of convexity consists in requiring that the sublevel sets of the functions are convex (recall that a sublevel set of a function is the portion of the source space on which the function takes values below a certain level). Its first use is usually attributed to de Finetti, in 1949. This property defines the class of quasiconvex functions, which is much larger than the class of convex functions: a nondecreasing or nonincreasing one-variable function is quasiconvex, as well as any one-variable function which is nonincreasing on some interval $(-\infty, a]$ or $(-\infty, a)$ and nondecreasing on its complement. Many other classes of generalized convex functions have been introduced, often for the needs of various applications: algorithms, economics, engineering, management science, multicriteria optimization, optimal control, statistics. Thus, they play an important role in several applied sciences.

A monotone mapping F from a Hilbert space to itself is a mapping for which the angle between $F(x) - F(y)$ and $x - y$ is acute for any x, y. It is well-known that the gradient of a differentiable convex function is monotone. The class of monotone mappings (and the class of multivalued monotone operators) has remarkable properties. This class has been generalized in various directions, with applications to partial differential equations, variational inequalities, complementarity problems and more generally, equilibrium problems. The classes of generalized monotone mappings are more or less related to the classes of generalized functions via differentiation or subdifferentiation procedures. They are also linked via several other means.

It is the purpose of this volume to present some aspects of the recent evolution of these two fields, since it constitutes the proceedings of the Sixth International Symposium on Generalized Convexity and Monotonicity (August 30 - September 3, 1999). Previous symposia were held in Vancouver (Canada) in 1980, Canton (U.S.A.) in 1986, Pisa (Italy) in 1988, Pécs (Hungary) in 1992, and Marseille (France), in 1996. The sixth symposium took place in the town of Karlovassi, on the island of Samos (Greece), the native birthplace of Pythagoras and Aristarchus. It was preceded by a summer school which made accessible the field to young researchers and which treated questions of wider scope or of more basic nature than most of the research papers presented in the symposium.

In addition to new contributions in well established topics such as fractional programming, multicriteria optimization, nonsmooth analysis, d.c. optimization, optimality conditions, stochastic convexity, variational inequalities and equilibrium problems, several new tracks in emerging areas are explored in this volume. Also a number of applications are presented. The first part of the volume (invited papers) contains the written versions of the invited talks given by H. Konno, P. Pardalos and A. Prekopa on fractional programming, nonconvex quadratic programming and probability theory, respectively; according to the tradition of the generalized convexity symposia, the invited talks are on neighboring topics (rather than on generalized convexity itself) and try to emphasize the applications of generalized convexity and its connections with other fields.

Each contribution to these proceedings has been submitted to two referees or more. The editors are most grateful to the anonymous referees who often made a difficult and patient job.

Special thanks are due to our colleague George Tsapogas who spent a lot of efforts in making the whole set in a unified style and in solving a number of technical questions, and to Thea Vigli-Papadaki, Evagelia Anifanti and Manto Katsiani who assisted the organizers: N. Hadjisavvas (Samos, Greece, chair), R. Cambini (Pisa, Italy), A. Daniilidis (Pau, France), J.B.G. Frenk (Rotterdam, The Netherland) and S. Schaible (Riverside, USA).

The Symposium was financially supported by the Department of Mathematics of the University of the Aegean, the Ministry of Education, the Ministry of Culture, the Ministry of the Aegean, the General Secretariat of Research and Technology, the Municipality of Karlovassi, the Prefecture of Samos, the European Union (through the program EPEAEK) and many private companies and individuals, to whom we express our gratitude.

The research group on Generalized Convexity (http://www.ec.unipi.it/) has proved to be a valuable link between the researchers in this field. Its role is the future is likely to be still more important than in the past. Hopefully, the same can be said for the whole field.

March 2001

Nicolas Hadjisavvas
Juan Enrique Martínez-Legaz
Jean-Paul Penot

Contents

Invited Papers

Minimization of the Sum of Several Linear Fractional Functions

Hiroshi Konno

Department of Industrial Engineering and Management,
Tokyo Institute of Technology

Abstract. This paper is concerned with global minimization algorithms for solving rank-p linear fractional programming problems, namely the minimization of the sum of $p(\geq 2)$ linear fractional functions over a polytope. Algorithms to be discussed are : parametric simplex algorithm for rank-2 problems, convergent approximate algorithm for rank-3 problems, generalized convex multiplicative programming approach and branch and bound algorithm using piecewise convex underestimating function. We will show that we are now able to obtain a globally optimal solution for up to rank-10 problems in a practical amount of time.

2000 Mathematics Subject Classification. 90C32.

Keywords and phrases. linear fractional functions, global optimization, parametric simplex method, convex multiplicative programming problem, branch and bound algorithm.

1 Introduction

Minimization of a linear and nonlinear fractional function over a polytope attracted attention of researchers since the pioneering works of Charnes and Cooper [5] and Dinkelbach [7]. It has been demonstrated in [5] that a global minimum of linear fractional function can be calculated by a variant of simplex algorithm. Also, a global minima of convex - concave type nonlinear fractional programming problems can be obtained by applying Dinkelbach's procedure. A number of real world problems have been formulated as linear/nonlinear fractional programming problems and solved in a successful manner using these algorithms. Naturally, many researchers tried to extend these algorithms to yet more difficult class of problems, but without a remarkable progress, at least until recently.

The main purpose of this paper is to review our efforts in constructing practical algorithms for minimizing the sum of $p(\geq 2)$ linear fractional functions :

$$(LF_p) \quad \left| \begin{array}{l} \text{minimize} \quad \displaystyle\sum_{i=1}^{p} \frac{d_i^T x + d_{i0}}{c_i^T x + c_{i0}} \\ \text{subject to } x \in X, \end{array} \right. \qquad (1)$$

where $c_i, d_i \in R^n, c_{i0}, d_{i0} \in R^1, i = 1, \ldots, p$, and X is a non-empty and bounded polyhedral set :

$$X = \{x \in R^n | Ax \leq b, x \geq 0\}, \tag{2}$$

where $A \in R^{m \times n}, b \in R^m$.

The problem (1) will be called a rank-p linear fractional programming problem. To avoid technical complications, we assume throughout that

$$c_i^T x + c_{i0} > 0, \quad i = 1, \ldots, p, \quad \forall x \in X, \tag{3}$$

which is valid for all practical applications. For applications of (LF_p), readers are referred to [2,12,14,18,21].

It is well known that a positively valued linear fractional function is quasi-convex. However, the sum of linear fractional functions are no longer quasi-convex even if it is positively valued. Therefore, we need to employ some kind of global optimization techniques to compute a globally optimal solution.

In section 2, we will introduce the well-known Charnes-Cooper transformation [5] to reduce the sum of p linear fractional functions to the sum of $p-1$ linear fractional functions and a linear function. This procedure plays an important role to construct efficient algorithms. Section 3 will be devoted to a primal-dual parametric simplex algorithm [19] for solving rank-2 problems. This algorithm can solve large scale problems in less than twice as much computation time as that required to solve a linear programming problem of the same size. We next discuss an efficient algorithm for calculating an approximate optimal solution for rank-3 problems using the algorithm for rank-2 problems.

The next two sections will be devoted to practical algorithms for minimizing up to rank-10 problems. In Section 4, we will present an algorithm [20] using a generalized convex multiplicative programming approach. This method is based upon parametrization/outer approximation scheme [16].

Next, in Section 5, we will present a branch and bound algorithm [13]. We first convert the linear fractional programming problem to a minimization of a linear function under constraints containing products of two variables. We then apply a branch and bound method using a piecewise linear concave envelope of nonlinear terms. This algorithm is an application of a general algorithm developed in [3] for solving a problem containing products of two variables. To our knowledge, this is the most efficient algorithm for solving higher rank linear fractional programming problems up to $p = 10$. For more complete survey of fractional programming, readers are referred to [6,22].

2 Charnes - Cooper transformation

Let us consider a rank-p linear fractional programming problem :

$$(LF_p)\quad \left|\begin{array}{l} \text{minimize } \displaystyle\sum_{i=1}^{p} \frac{d_i^T x + d_{i0}}{c_i^T x + c_{i0}} \\ \\ \text{subject to } Ax \leq b, x \geq 0, \end{array}\right. \tag{4}$$

where $A \in R^{m\times n}, b \in R^m$, $c_i, d_i \in R^n, c_{i0}, d_{i0} \in R^1, i = 1, \ldots, p$.We will assume throughout this paper that the feasible region

$$X = \{x \in R^n | Ax \leq b, x \geq 0\}, \tag{5}$$

is non-empty and bounded. Also, we will assume

$$c_i^T x + c_{i0} > 0, \quad i = 1, \ldots, p, \quad \forall x \in X. \tag{6}$$

Let us define

$$y_0 = 1/(c_p^T x + c_{p0}). \tag{7}$$

Note that $y_0 > 0$ for all $x \in X$. Thus we have an alternative representation of (LF_p) :

$$\left|\begin{array}{l} \text{minimize } \displaystyle\sum_{i=1}^{p-1} \frac{d_i^T x + d_{i0}}{c_i^T x + c_{i0}} + (d_p^T x + d_{p0})y_0 \\ \\ \text{subject to } Ax \leq b, x \geq 0, \\ \qquad\qquad (c_p^T x + c_{p0})y_0 = 1. \end{array}\right. \tag{8}$$

Let us now define a new problem

$$(Q_p)\quad \left|\begin{array}{l} \text{minimize } \displaystyle\sum_{i=1}^{p-1} \frac{d_i^T y + d_{i0}y_0}{c_i^T y + c_{i0}y_0} + d_p^T y + d_{p0}y_0 \\ \\ \text{subject to } Ay - by_0 \leq 0, \quad y \geq 0, y_0 \geq 0, \\ \qquad\qquad c_p^T y + c_{p0}y_0 = 1. \end{array}\right. \tag{9}$$

Proposition 1. *Any feasible solution* (y, y_0) *of (9) satisfies* $y_0 > 0$.

Proof. Suppose that there exists a feasible solution $(y, 0)$ such that $y \neq 0$. Then y satisfies $Ay \leq 0, y \geq 0$, which implies that the set X is unbounded, a contradiction to our assumption. □

Theorem 1. *Let* (y^*, y_0^*) *be an optimal solution of (9). Then* $x^* = y^*/y_0^*$ *is an optimal solution of (4).*

Proof. Follows directly from proposition 1. □

The procedure to convert (4) to (9) is called the Charnes -Cooper transformation [5]. All algorithms for (LF_p) to be presented in this paper depend upon this transformation.

3 Parametric simplex algorithm for rank-two and three linear fractional programming problems

Let us consider a rank-2 linear fractional programming problem after the Charnes-Cooper transformation :

$$(Q_2)\quad \left|\begin{array}{l} \text{minimize} \;\; \dfrac{d_1^T y + d_{10} y_0}{c_1^T y + c_{10} y_0} + d_2^T y + d_{20} y_0 \\[2ex] \text{subject to } Ay - by_0 \le 0, \;\; y \ge 0, y_0 \ge 0, \\ \qquad\qquad c_2^T y + c_{20} y_0 = 1. \end{array}\right. \tag{10}$$

Let

$$Y = \{(y, y_0) | Ay - by_0 \le 0, \; c_2^T y + c_{20} y_0 = 1, \;\; y \ge 0, \; y_0 \ge 0\} \tag{11}$$

Also let

$$\xi = c_1^T y + c_{10} y_0, \tag{12}$$

and

$$\xi_{min} = \min\{c_1^T y + c_{10} y_0 \mid (y, y_0) \in Y\}, \tag{13}$$
$$\xi_{max} = \max\{c_1^T y + c_{10} y_0 \mid (y, y_0) \in Y\}. \tag{14}$$

Then the problem (10) is equivelant to:

$$\left|\begin{array}{l} \text{minimize} \;\; \dfrac{1}{\xi}(d_1^T y + d_{10} y_0) + d_2^T y + d_{20} y_0 \\ \text{subject to } c_1^T y + c_{10} y_0 = \xi, \\ \qquad\qquad (y, y_0) \in Y, \\ \qquad\qquad \xi_{\text{min}} \le \xi \le \xi_{\text{max}}. \end{array}\right. \tag{15}$$

For fixed value of $\xi \in [\xi_{min}, \xi_{max}]$, the problem

$$(Q_2(\xi))\quad \left|\begin{array}{l} \text{minimize} \;\; \dfrac{1}{\xi}(d_1^T y + d_{10} y_0) + d_2^T y + d_{20} y_0 \\[2ex] \text{subject to } c_1^T y + c_{10} y_0 = \xi, \\ \qquad\qquad (y, y_0) \in Y. \end{array}\right. \tag{16}$$

is a linear programming problem. Therefore, we can solve (15) by using a primal-dual parametric simplex algorithm. It is straightforward to see that the objective function can be represented as $\alpha\xi + \beta/\xi + \gamma$ in each subinterval of the parametric simplex procedure. Therefore, we can calculate the minimal value of the function in each subinterval by an analytical method, from which we can recover an optimal value of ξ and the associated value of (y, y_0). By Theorem 2, we can recover the optimal solution of (Q_2).

It has been demonstrated in [19] that this algorithm is extremely efficient. Most of the randomly generated and/or structured problems can be solved in computation time less than twice as much as that required to calculate ξ_{min}. See Hirsche [9] for the improvement of this algorithm. Also, Falk and Palocsay [8] and Cambini et al. [4] develop alternative algorithms for rank-2 problems.

Let us next consider a rank-3 problem :

$$(Q_3) \quad \left| \begin{array}{l} \text{minimize} \quad \dfrac{d_1^T y + d_{10} y_0}{c_1^T y + c_{10} y_0} + \dfrac{d_2^T y + d_{20} y_0}{c_2^T y + c_{20} y_0} + d_3^T y + d_{30} y_0 \\ \\ \text{subject to } (y, y_0) \in Y. \end{array} \right. \tag{17}$$

Let

$$\eta = c_2^T y + c_{20} y_0,$$

and let η_{min}, η_{max} be, respectively the minimal and maximal values of η over the feasible region. Then the problem can be represented in a parametric form:

$$\left| \begin{array}{l} \text{minimize} \quad \dfrac{d_1^T y + d_{10} y_0}{c_1^T y + c_{10} y_0} + \dfrac{1}{\eta}(d_2^T y + d_{20} y_0) + (d_3^T y + d_{30} y_0) \\ \\ \text{subject to } c_2^T y + c_{20} y_0 = \eta, \\ \qquad\qquad (y, y_0) \in Y \\ \qquad\qquad \eta_{\min} \le \eta \le \eta_{\max}. \end{array} \right. \tag{18}$$

When η is fixed, the associated subproblem $P_3(\eta)$ reduces to a rank-2 problem, so that it can be solved by a primal-dual parametric simplex algorithm. In a recent paper [12], we proposed the following procedure to calculate an approximate optimal solution of (Q_3)

Algorithm PS. *Primal-Dual Parametric Simplex Subdivision Algorithm*

Step 1. Calculate η_{min} and η_{max} by solving a linear programming programs generate a finite number of grid points η_k, $k = 0, 1, \ldots, K$ satisfying the condition

$$\eta_{min} = \eta_0 \ < \eta_1 \ < \eta_2 \ < \ldots \ < \eta_K \ = \eta_{max}.$$

Step 2. For $k = 0, 1, \ldots, K$, solve $P(\eta_k)$ by primal dual simplex method. Let f_k be the minimal value of the objective function of $P(\eta_k)$.
Let $f_{k^*} = min\,\{f_k | k = 0, 1, \ldots, K\}$ and let (y^*, y_0^*) be an optimal solution of the problem $P(\eta_{k^*})$.

Step 3. Stop. $x^* = y^*/y_0^*$ is an approximate optimal solution of (4). □

We tested the primal-dual parametric simplex subdivision algorithm PS and compared it with the convex multiplicative programming algorithm CM to be discussed in the next section. We generated test problems (LF_3) where the elements of $A \in R^{m\times n}$, $c_j, d_j \in R^n$, $c_{j0}, d_{j0} \in R^1$, $j = 1,2,3$ are randomly generated in the unit interval $[0,1]$. Also, the elements of b are randomly generated in the interval $[0.1, 1.0]$ to satisfy the condition (6). Table 1 shows the computation time for CM algorithm and PS algorithm.The convergence condition $\varepsilon = 10^{-5}$ was chosen for CM algorithm. Also, we divided the interval $[\eta_{min}, \eta_{max}]$ into 100 subintervals of equal length in PS algorithm. We solved ten test problems of each size and listed the average computation time and its standard deviation. All test problems satisfy the condition $\eta_{max} - \eta_{min} \le 1.0$, so that $\varepsilon < 0.01$ for 100 subdivisions.

Table 1. Computation Time

		(m,n)			
		(5,10)	(10,20)	(15,30)	(20,40)
CM	Ave. CPU time (sec)	6.76	47.01	221.58	(535.14)
	Std. Dev.(sec)	5.54	38.93	102.22	(245.98)
PS	Ave. CPU time (sec)	1.65	11.27	46.21	152.81
	Std. Dev.(sec)	0.25	1.96	6.98	23.24

Comparison was terminated at (m,n) = (20,40) since some test problems could not be solved within 10 minutes by CM algorithm. (The data for (20,40) in the bracket show the average and standard deviation of eight test problems which were solved in 10 minutes). We see from this that PS algorithm is about 5 times faster than CM algorithm. This is due to the fact that the number of pivots of primal-dual parametric simplex algorithm is relatively stable. The remarkable difference is that the standard deviation of PS is at least 10 times less than CM. We conclude from this that PS algorithm depends primarily on the size of the problem and the subdivision parameter K, while CM algorithm depends on both the size and the data. The quality of the solutions of PS and CM algorithms are more or less the same. Let us note that test problems are all dense. Sparse problems can be solved much faster. For alternative subdivision schemes, readers are referred to [12].

4 Generalized convex multiplicative programming algorithm

In this section, we will discuss an application of a generalized convex multiplicative programming approach [20] to rank-p linear fractional programming problem :

$$(Q_p)\quad \left|\begin{array}{l}\text{minimize } \sum_{i=1}^{p-1} \dfrac{d_i^T y + d_{i0} y_0}{c_i^T y + c_{i0} y_0} + d_p^T y + d_{p0} y_0 \\ \text{subject to } (y, y_0) \in Y.\end{array}\right. \tag{19}$$

In this section, we assume

$$d_i^T y + d_{i0} y_0 \geq 0,\ i = 1, \ldots, p,\ \ \forall\ (y, y_0) \in Y, \tag{20}$$

which is equivelant to

$$d_i^T x + d_{i0} \geq 0,\ i = 1, \ldots, p,\ \forall x \in X,$$

by noting $y_0 > 0$ for all $(y, y_0) \in Y$. Let us introduce a pair of $p-1$ dimensional auxiliary variables:

$$\xi = (\xi_1, \ldots, \xi_{p-1})^T,\quad \eta = (\eta_1, \ldots, \eta_{p-1})^T. \tag{21}$$

and consider the following problem in $n + 2(p-1)$ dimensional space:

$$\left|\begin{array}{l}\text{minimize } \dfrac{1}{2}\sum_{i=1}^{p-1} \left\{ \xi_i (d_i^T y + d_{i0} y_0)^2 + \eta_i \dfrac{1}{(c_i^T y + c_{i0} y_0)^2} \right\} + d_p^T y + d_{p0} y_0 \\ \text{subject to } (y, y_0) \in Y, \\ \qquad\qquad \xi_i \eta_i \geq 1,\ \xi_i \geq 0,\ \eta_i \geq 0,\ \ i = 1, \ldots, p-1.\end{array}\right. \tag{22}$$

Theorem 2. *Let $(x^*, \xi_1^*, \cdots, \xi_{p-1}^*, \eta_1^*, \cdots, \eta_{p-1}^*)$ be an optimal solution of (22). Then (y^*, y_0^*) is an optimal solution of (19).*

Proof. Follows from the following simple observation:

$$\min\{\alpha^2 u + \beta^2 v \mid uv \geq 1,\ u \geq 0,\ v \geq 0\} = 2|\alpha\beta|$$

□

Note that the problem (22) is a convex minimization problem for any fixed value of (ξ, η) since $(d_i^T y + d_{i0} y_0)^2$ and $1/(c_i^T y + c_{i0} y_0)^2$ are both convex. Let

$$G(\xi, \eta) = \min \left\{ \frac{1}{2} \sum_{i=1}^{p-1} \left\{ \xi_i (d_i^T y + d_{i0} y_0)^2 + \eta_i \frac{1}{(c_i^T y + c_{i0} y_0)^2} \right\} + \right.$$
$$\left. + d_p^T y + d_{p0} y_0 \;\middle|\; (y, y_0) \in Y \right\} \tag{23}$$

By Theorem 3, the problem (19) can be solved by minimizing $G(\xi,\eta)$ over the convex region:

$$\Psi = \{(\xi,\eta) \mid \xi_i\eta_i \geq 1,\ \xi_i \geq 0,\ \eta_i \geq 0,\ \ i=1,\ldots,p-1\} \tag{24}$$

Proposition 2. *The function $G(\xi,\eta)$ is a non-decreasing concave function on Ψ.*

Proof. Follows from the fact that ξ_i's and η_i's appear linearly in the minimand of (22). □

Let (ξ^*,η^*) be the global minimizer of $G(\xi,\eta)$ over Ψ.

Proposition 3. *Let*

$$\alpha_i = \min\{(c_i^T y + c_{i0}y_0)(d_i^T y + d_{i0}y_0) \mid (y,y_0) \in Y\}, \tag{25}$$
$$\beta_i = \max\{(c_i^T y + c_{i0}y_0)(d_i^T y + d_{i0}y_0) \mid (y,y_0) \in Y\}. \tag{26}$$

Then

$$1/\beta_i \leq \xi_i^* \leq 1/\alpha_i,\ \alpha_i \leq \eta_i^* \leq \beta_i,\ \ i=1,\ldots,p-1.$$

Proof. For $(y,y_0) \in Y$, the minimum of the function

$$G_i(\xi,\eta) = \min\left\{\xi_i(d_i^T y + d_{i0}y_0)^2 + \eta_i \frac{1}{(c_i^T y + c_{i0}y_0)^2}\right\}$$

over Ψ is attained by
$\xi_i = 1/(c_i^T y + c_{i0}y_0)(d_i^T y + d_{i0}y_0), \eta_i = (c_i^T y + c_{i0}y_0)(d_i^T y + d_{i0}y_0).$ □

Let us note that α_i's and β_i's can be calculated in an efficient way by applying parametric simplex algorithm [15].

We are now ready to apply the outer approximation method proposed in [16] to the following problem:

$$\left|\begin{array}{l}\text{minimize } G(\xi,\eta)\\ \text{subject to } (\xi,\eta) \in \Psi.\end{array}\right. \tag{27}$$

For completeness, we will present the outline of the algorithm. Let

$$\Omega_0 = \{(\xi,\eta) \mid 1/\beta_i \leq \xi_i^* \leq 1/\alpha_i,\ \alpha_i \leq \eta_i^* \leq \beta_i,\ \ i=1,\ldots,p-1\}.$$

We will generate a sequence of polytopes:

$$\Omega_0 \supset \Omega_1 \supset \Omega_2 \supset \cdots \supset \Omega_k \supset \cdots \supset \Psi,$$

where Ω_{k+1} is generated by adding a cut $l_k(\xi,\eta) \leq 0$ to Ω_k. Let

$$\Psi_i = \{(\xi_i,\eta_i) \mid \xi_i\eta_i \geq 1,\ (\xi_i,\eta_i) \geq 0\},$$
$$\Omega_{i0} = \{(\xi_i,\eta_i) \mid 1/\beta_i \leq \xi_i \leq 1/\alpha_i,\ \alpha_i \leq \eta_i \leq \beta_i\}.$$

The feasible region $\Psi \cap \Omega_0$ can be represented as follows;

$$\Psi \cap \Omega_0 = (\Psi_1 \cap \Omega_{10}) \times \cdots \times (\Psi_{p-1} \cap \Omega_{p-1,0}).$$

Let (ξ^k, η^k) be an optimal solution of k-th approximation problem:

$$\left| \begin{array}{l} \text{minimize} \;\; G(\xi,\eta) \\ \text{subject to } (\xi,\eta) \in \Omega_k. \end{array} \right. \tag{28}$$

Since $G(\xi,\eta)$ is concave and Ω_k is a polytope in $p-1$ dimensional space, (ξ^k, η^k) can be obtained by enumerating the vertices of Ω_k and solving subproblems:

$$P(\xi,\eta) \left| \begin{array}{l} \text{minimize} \;\; \frac{1}{2}\sum_{i=1}^{p-1}\left\{\xi_i(d_i^T y + d_{i0}y_0)^2 + \eta_i \frac{1}{(c_i^T y + c_{i0}y_0)^2}\right\} + \\ \qquad\qquad\qquad\qquad\qquad\qquad + d_p^T y + d_{p0}y_0 \\ \text{subject to } (y, y_0) \in Y, \end{array} \right. \tag{29}$$

for each vertex. In case $\xi_i^k \eta_i^k \geq 1-\epsilon$ $(i = 1,\ldots,p-1)$ for small enough $\epsilon > 0$, then we can terminate the computation. In case $\xi_i^k \eta_i^k < 1-\epsilon$ for some i, let

$$\xi_t^k \eta_t^k = \min_{1\leq i\leq p-1} \xi_i^k \eta_i^k$$

and define

$$l_k(\xi,\eta) \equiv 2 - \xi_t\sqrt{\eta_t^k/\xi_t^k} - \eta_t\sqrt{\xi_t^k/\eta_t^k}. \tag{30}$$

Proposition 4. *Assume that $(\xi^k, \eta^k) \notin \Psi$, then*

$$l_k(\xi,\eta) \leq 0, \quad \forall(\xi,\eta) \in \Psi \text{ and } l_k(\xi^k,\eta^k) > 0.$$

Proof. See [20] for details. □

Algorithm OA. Outer Approximation Algorithm.

Step 0. $k = 0$.

Step 1. Solve (P_k) to obtain an optimal solution (ξ^k, η^k) by enumerating all (newly generated) vertices of Ω_k and solving convex subproblems for each of them.

Step 2. If $\xi_t^k \eta_t^k > 1-\epsilon$ for all t, then terminate. Otherwise generate a cut $l_k(\xi,\eta) \leq 0$ and define

$$\Omega_{k+1} = \Omega_k \cap \{(\xi,\eta) \mid l_k(\xi,\eta) \leq 0\}.$$

Step 3. $k = k+1$ and return to *Step 1.* □

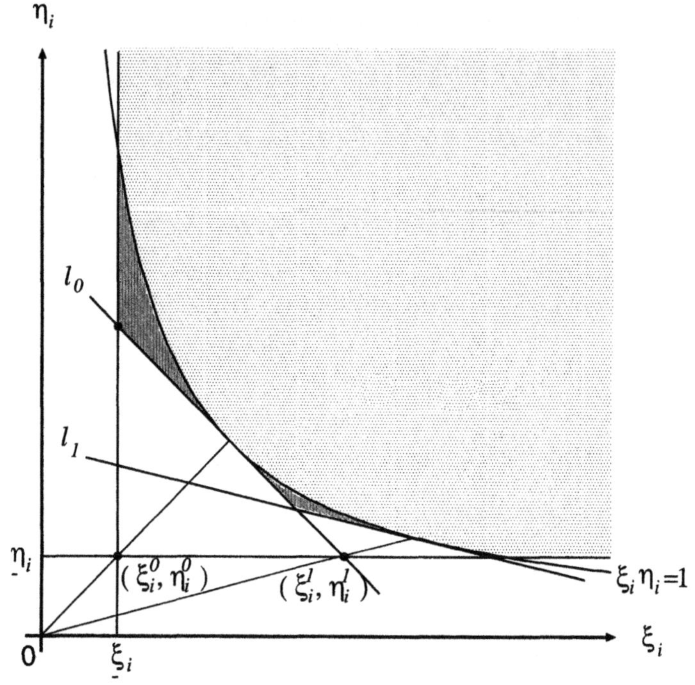

Fig. 1. Illustration of the cutting plane for $k = 1$.

It has been established in [20] that the sequence (ξ^k, η^k) converges to the global minimum of $G(\xi, \eta)$ over Ω.

We conducted computational experiments on the Algorithm OA for two classes of test problems:

Class 1. Randomly generated problems: where all components of the matrix A and vectors c_i, d_i are randomly generated in the unit interval [0, 1] and the components of the vector b are randomly generated in the interval [0.1, 1].

Class 2. Structured problems: where the feasible region is the unit hypercube and the objective function is generated in such a way that the components of c_i's and d_i's are either ± 1 or 0 and c_{i0}'s and d_{i0}'s are $c_{i0}/d_{i0} \gg 1$ for some i and $c_{i0}/d_{i0} \approx 1$ for other i's.

Table 2 and Table 3 show, respectively the results of computational experiments on these problems. We generated ten test problems for each class and calculated the average CPU time, the average number of cuts and the average number of vertices generated by adding these cuts. We also listed the standard deviation of these data in the bracket. The number of vertices corresponds to the number of subproblems to be solved before termination where tolerance ϵ is set at 10^{-5}. Each convex subproblem $P(\xi, \eta)$ was solved by the software package ASNOP. As expected, the CPU time sharply increases as p increases since the number of newly added vertices and therefore the number of subproblems increases. In contrast, the number of cuts and vertices are less sensitive to the size of the problem m and n.

Let us note that the CPU time for structured problems (*Class 2*) is significantly less than randomly generated problems (*Class 1*). The computation time should be much less if we use an alternative algorithm for solving subproblems by exploiting their special structures.

For more details of the computational experiments, readers are referred to [20].

Table 2. Randomly Generated Problem (19)

p	2	3	3	3	4	4	5
m	80	30	80	80	10	30	10
n	100	50	50	100	10	50	10
Avg.CPU time (sec.)	175.2	381.0	206.6	1849.1	51.6	2456.1	656.2
(Std. Dev.)	(0.92)	(206.9)	(65.6)	(1299.7)	(56.5)	(1926.0)	(1065.9)
Avg. # of cuts	6.4	17.5	12.9	15.3	24.0	24.1	28.2
(Std. Dev.)	(0.9)	(2.7)	(1.5)	(4.0)	(6.7)	(4.7)	(6.5)
Avg. # of vertices	13.8	190.2	109.2	154.8	1669.4	1547.0	10063.0
(Std. Dev.)	(1.8)	(49.5)	(25.8)	(75.1)	(1138.9)	(770.3)	(8307.0)

Table 3. Structured Problem (20) ($p = 3$)

p	3	3	3	3
n	30	50	70	80
Avg. CPU time (sec)	20.2	58.2	170.0	303.7
(Std. Dev.)	(10.9)	(32.8)	(64.1)	(129.2)
Avg. # of cuts	6.5	6.7	6.8	7.5
(Std. Dev.)	(1.3)	(1.5)	(0.6)	(0.9)
Avg. # of vertices	56.6	59.8	59.8	71.8
(Std. Dev.)	(18.1)	(23.1)	(9.6)	(17.0)

5 A branch and bound algorithm using a concave overestimating function

Let us consider a rank-p linear fractional programming problem

$$(Q_p)\quad \left|\begin{array}{l} \text{minimize } \displaystyle\sum_{i=1}^{p-1} \frac{d_i^T y + d_{i0} y_0}{c_i^T y + c_{i0} y_0} + d_p^T y + d_{p0} y_0 \\ \text{subject to } (y, y_0) \in Y. \end{array}\right. \tag{31}$$

As in Section 4, we assume the following :

$$c_i^T y + c_{i0} y_0 > 0, \quad i = 1, \ldots, p, \ \forall (y, y_0) \in Y. \tag{32}$$

Under this assumption, the problem (31) is equivalent to

$$(P_0)\left|\begin{array}{lll} \text{minimize g(z)} = \displaystyle\sum_{i=1}^{p-1} w_i + d_p^T y + d_{p0} y_0 & & \\ \text{subject to} \quad v_i = c_i^t y + c_{i0} y_0, & i = 1, \ldots, p-1, \\ \qquad\qquad\quad u_i = d_i^t y + d_{i0} y_0, & i = 1, \ldots, p-1, \\ \qquad\qquad\quad v_i w_i - u_i \geq 0, & i = 1, \ldots, p-1, \\ \qquad\qquad\quad (y, y_0) \in Y. & \end{array}\right. \tag{33}$$

where

$$z = (y, y_0, u_1 \ldots, u_{p-1}, v_1 \ldots, v_{p-1}, w_1 \ldots, w_{p-1}). \tag{34}$$

The equivalence of (33) and (31) can be checked by noting the conditions $w_i \geq u_i / v_i$ and that this inequality holds with equalities at the optimal solution.

Let us now consider a piecewise linear concave overestimating function of $v_i w_i$ over a rectangle $D_i^0 = [v_i^0, V_i^0] \times [w_i^0, W_i^0]$, where v_i^0, V_i^0 and w_i^0, W_i^0 are, respectively minimal and maximal values of v_i and w_i over the feasible region.

Proposition 5. *Let*

$$f_i(v_i, w_i) = min[v_i^0 w_i + w_i^0 v_i + v_i^0 w_i^0, \ V_i^0 w_i + W_i^0 v_i + V_i^0 W_i^0] \tag{35}$$

Then

$$f_i(v_i, w_i) \geq v_i w_i, \quad \forall (v_i, w_i) \in D_i^0 \tag{36}$$

Also it satisfies the following conditions

$$f_i(v_i, w_i) = v_i w_i, \quad (v_i, w_i) \in \partial D_i^0, \tag{37}$$

$$|f_i(v_i, w_i) - v_i w_i| \leq \frac{(V_i^0 - v_i^0)(W_i^0 - w_i^0)}{4} \tag{38}$$

Proof. Follows from elementary arithmetics. □

Proposition 6. *The function $f_i(v_i, w_i)$ is a concave envelope of a quadratic function $v_i w_i$*

Proof. See [1] or [14] (Corollary 3.6). □

We see from (38) that the maximal discrepancy between $v_i w_i$ and its concave envelope is proportional to the area of the rectangle D_i^0. Let us now define a convex minimization problem

$$\left| \begin{array}{lll} \text{minimize g(z)} = \sum_{i=1}^{p-1} w_i + d_p^T y + d_{p0} y_0 & & \\ \text{subject to} \quad v_i = c_i^t y + c_{i0} y_0, & i = 1, \ldots, p-1, & \\ \qquad u_i = d_i^t y + d_{i0} y_0, & i = 1, \ldots, p-1, & \\ \qquad f_i(v_i, w_i) - u_i \geq 0, & i = 1, \ldots, p-1, & \\ \qquad (y, y_0) \in Y. & & \end{array} \right. \tag{39}$$

Proposition 7. *If (39) is infeasible, then (33) is also feasible.*

Let $\hat{z}$ be an optimal solution of (39), Then

(*a*) $g(\hat{z})$ gives the lower bound of the minimal value of $g(z)$ of (33),

(*b*) $\hat{z}$ is an optimal solution of (33) if it is a feasible solution of (33).

Let us note that the problem (38) is a linear program since $f_i(v_i, w_i)$'s are piecewise linear concave functions. Therefore we can construct a branch and bound algorithm [13] if we can provide a convergent branching procedure. One standard method is the following bisection of the hyper-rectangle $D^0 = \prod_{i=1}^{p-1} [v_i^0, V_i^0] \times [w_i^0, W_i^0]$ into two subrectangles by using the longest edge. Let us assume without loss of generality that $V_s^0 - v_s^0$ is the longest edge of D^0. Then we partition D^0 into two subrectangles D^1 and D^2 at the mid-point of the interval $[v_s^0, V_s^0]$. An alternative and yet more efficient scheme is an ω - subdivision scheme, where we partition the hyperrectangle at the point $\hat{v}_s$, rather than the mid-point of v_s^0 and V_s^0.

Algorithm BB. Branch and Bound Algorithm.

Step 0. Let $\varepsilon_c > 0, \varepsilon_f > 0, L = -\infty, \mathcal{P} = \{(P_0)\}$. Calculate a feasible solution $\hat{z}$ of (P_0) by applying appropriate local search algorithm. Let $L_0 = U = g(\hat{z})$.

Step 1. If $\mathcal{P} = \emptyset$, then go to Step 8. Otherwise go to Step 2.

Step 2. Choose a problem $(P_s) \in \mathcal{P}$ such that $L_s = \min\{L_j | P_j \in \mathcal{P}\}$:

$$(P_s) \quad \left| \begin{array}{lll} \text{minimize} & g(z) & \\ \text{subject to} & v_i = c_i^t y + c_{i0} y_0, & i = 1, \ldots, p-1, \\ & u_i = d_i^t y + d_{i0} y_0, & i = 1, \ldots, p-1, \\ & v_i w_i - u_i \geq 0, & i = 1, \ldots, p-1, \\ & v_i^s \leq v_i \leq V_i^s, & i = 1, \ldots, p-1, \\ & w_i^s \leq w_i \leq W_i^s, & i = 1, \ldots, p-1, \\ & (y, y_0) \in Y. & \end{array} \right. \tag{40}$$

$\mathcal{P} = \mathcal{P} \backslash \{(P_s)\}$.

Step 3. Generate $(\bar{P}_s)$ by replacing $v_i w_i$ by a concave overestimating function $f_i(v_i, w_i)$ and solve $(\bar{P}_s)$.

Step 4. If $(\bar{P}_s)$ is infeasible, go to Step 1. Otherwise let z^s be an optimal solution of $(\bar{P}_s)$. If $\sum_{i=1}^{p-1} |f_i(v_i^s, w_i^s) - v_i^s w_i^s| \leq \varepsilon_f$, then $U_s = g(z^s)$ and go to Step 5. Otherwise $L_s = g(z^s)$ and go to Step 7.

Step 5. If $U_s < U$, then $\tilde{z} = z^s$, $U = g(z^s)$ and delete all subproblems (P_t) from $\mathcal{P}$ such that $L_t > U$, and go to Step 7.

Step 6. Generate two subproblems (P_{l+1}) and (P_{l+2}), by subdividing the hyper-rectangular $\prod_{i=1}^{p-1} [v_i^s, V_i^s] \times [w_i^s, W_i^s]$ using the longest edge.

Step 7. $L = minL_s$. If $U - L < \varepsilon_c$, go to Step 8. Otherwise goto Step 1.

Step 8. Stop. (ε_c - convergence has been reached.) Optimal solution z^* is given by $\tilde{z}$ □

Theorem 3. *The branch and bound algorithm proposed above terminates in finitely many steps.*

Proof. The bisection scheme is exhaustive as proved in Konno-Thach-Tuy [17] (Corollary 6.3). □

We tested this algorithm on rank-p linear fractional programming problem

$$\left| \begin{array}{l} \text{minimize} \ \sum_{i=1}^{p} \frac{d_i^T x + d_{i0}}{c_i^T x + c_{i0}} \\ \\ \text{subject to} \ Ax \leq b, x \geq 0, \end{array} \right.$$

where $c_j, d_j, y \in R^n$, $c_{i0}, d_{i0} \in R^1$ $(j = 1, \ldots, p)$, $A = (a_{ij}) \in R^{m \times n}$, $b \in R^m$. All elements of the matrix A and vector b, c_i's, d_i's are randomly generated from the uniform distribution in certain intervals: $A \in [0.01, 1.0]$, $b \in [0.1, 10.0]$, so that the feasible region is always non-empty and bounded. Also,

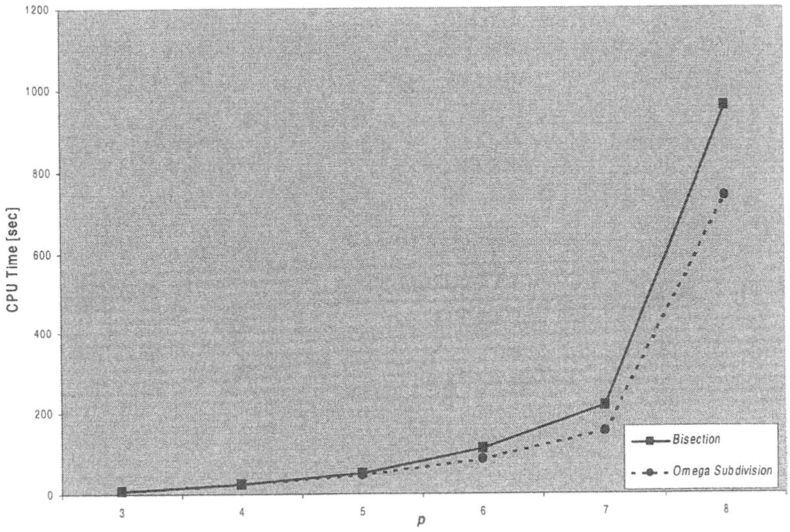

Fig. 2. Computation Time for LFP's ($n = 10, m = 30$)

we chose $c_i \in [-0.1, 0.1]$, $d_i \in [-0.1, 0.1]$, $\forall i$. Also, c_{i0}, d_{i0} are chosen in such a way that the condition (6) is satisfied. Ten test problems are generated for each size of the parameter (p, n, m).

Figure 2 shows the computational results for the case $(p = 4, n = 10, m = 30)$. Table 4 and 5 show the results for larger test problems. We see from Figure 2 that both bisection and ω - subdivision perform more of less equally.

6 Conclusion

We reviewed some of the recent algorithmic efforts for solving up to rank-8 linear fractional programming problems. We showed that lower rank problems can be solved in an efficient way by special purpose algorithms. On the other hand, higher rank problems can only be solved by a less efficient branch and bound algorithm BB. According to Figure 2, average computation time of the BB algorithm jumps at $p = 8$. However, most of the test problems can be solved fast and only one "hard" problem contribute to this jump. Large variance of computation time is common to all branch and bound type algorithms. Also, nearly optimal solutions are generated at the earlier stage of computation even for a few hard problems. Therefore, we can in fact solve almost all problems up to rank-10 problems from the practical point

Table 4. Computation Time of the Standard Bisection Algorithm .

p	(n, m)	Average CPU time (s.d.)	Average # of iteration (s.d.)
3	(10,30)	8.1(1.2)	5.7(2.0)
	(30,10)	34.4(7.2)	9.8(6.0)
	(30,50)	35.1(4.6)	6.9(2.9)
	(50,30)	77.8(9.3)	6.9(3.1)
4	(10,30)	23.7(5.2)	16.1(7.2)
	(30,10)	87.8(30.6)	26.5(16.7)
	(30,50)	78.2(24.7)	16.4(10.6)
	(50,30)	161.4(29.8)	16.8(7.4)
5	(10,30)	52.2(16.8)	32.2(16.3)
	(30,10)	257.7(172.2)	77.1(66.3)
6	(10,30)	111.5(33.4)	49.5(24.9)
7	(10,30)	217.2(70.4)	85.8(33.1)
8	(10,30)	958.4(845.2)	276.4(215.7)

Table 5. Computation Time of ω - subdivision Algorithm.

p	(n, m)	Average CPU time (s.d)	Average # of iteration (s.d)
3	(10,30)	8.1(0.9)	5.7(1.7)
	(30,10)	37.8(7.0)	10.2(6.0)
	(30,50)	36.3(6.2)	7.7(4.3)
	(50,30)	77.7(9.9)	6.9(3.2)
4	(10,30)	23.3(6.4)	16.8(8.4)
	(30,10)	87.3(29.4)	26.4(15.9)
	(30,50)	79.8(30.0)	16.8(13.0)
	(50,30)	151.6(28.2)	14.6(7.4)
5	(10,30)	46.3(13.5)	26.5(13.5)
	(30,10)	238.71(165.9)	70.9(65.5)
6	(10,30)	84.9(27.3)	39.4(19.6)
7	(10,30)	154.9(52.9)	55.7(25.6)
8	(10,30)	736.9(917.9)	200.6(237.15)

of view, particularly if we employ more elaborate scheme in a branch and bound algorithm.

We understand that most of the practical applications of linear fractional programming are of single digit rank. However, some applications such as stochastic shipping problems discussed in [8] require the solution of higher rank problems, say up to 50, which is still beyond the reach of our ability. However, we believe that the remarkable progress in global optimization will soon enable us to solve these difficult problems.

Acknowledgements. This research was supported in part by the Grant-in Aid for Scientific Research of the Ministry of Education, Science and Culture, Grant No. B(2) 10450041.

References

1. Al-Khayyal, F.A. and Falk, J.E., "Jointly Constrained Biconvex Programming", *Mathematics of Operations Research,* 8 (1983) 273-281.
2. Almogy, Y. and Levin, O., "Parametric Analysis of a Multi-Stage Stochastic Shipping Problem", *Proc. of the Fifth IFORS Conference,* Venice, 359-370, 1969.
3. Androulakis, I.P., Maranas,C.D. and Floudas,C.A.,"αBB: A Global Optimization Method for General Constrained Nonconvex Problems", *Journal of Global Optimization,* 7 (1995) 337-363.
4. Cambini, A., Martein, L. and Schaible, S., "On Maximizing a Sum of Ratios", *J. of Information and Optimization Science,* 10 (1989) 141-151.
5. Charnes, A. and Cooper, W. W., "Programming with Linear Fractional Functions", *Naval Research Logistics,* 9 (1962) 181-186.
6. Craven, B. D., *Fractional Programming,* Heldermann-Verlag Berlin, 1988.
7. Dinkelbach, W., "On Nonlinear Fractional Programming", *Management Science,* 13 (1967) 492-498.
8. Falk, J. E. and Palocsay, S. W., "Optimizing the Sum of Linear Fractional Functions", in *Recent Advances in Global Optimization* (Floudas, C. A. *et al.*(eds.)), 221-258, Princeton University Press, 1992.
9. Hirsche, J.,"Optimizing of Sum and Products of Linear Fractional Function under Linear Constraints",Technical Report 3,Department of Computer Science and Scientific Computing,Martin-Luther-Universitat, 1995.
10. Horst, R. and Pardalos, P., *Handbook of Global Optimization,* Kluwer Academic Publishers, 1995.
11. Horst, R. and Tuy, H., *Global Optimization: Deterministic Approaches(3rd edition),* Springer Verlag, 1995.
12. Konno, H. and Abe, N.,"Minimization of the Sum of Three Linear Fractional Functions", *Journal of Global Optimization,* 15 (1999) 419-432.
13. Konno, H. and Fukaishi, K., "A Branch and Bound Algorithm for Solving Low Rank Linear Multiplicative and Fractional Programming Problems",Technical Report of Department of IE and Management, Tokyo Institute of Technology, 1999, (to appear in *Journal of Global Optimization*).
14. Konno, H. and Inori, M., "Bond Portfolio Optimization by Bilinear Fractional Programming", *J. of the Operations Research Society of Japan,* 32 (1989) 143-158.
15. Konno, H. and Kuno, T., "Linear Multiplicative Programming", *Mathematical Programming,* 56 (1992) 51-64.
16. Konno, H., Kuno, T. and Yajima, Y., "Global Minimization of a Generalized Convex Multiplicative Function", *J. of Global Optimization,* 4 (1994) 47-62.
17. Konno, H., Thach, P. T. and Tuy, H., *Optimization on Low Rank Nonconvex Structures,* Kluwer Academic Publishers, 1997.
18. Konno, H. and Watanabe, H., "Bond Portfolio Optimization Problems and Their Applications to Index Tracking: A Partial Optimization Approach", *J. of the Operations Research Society of Japan,* 39 (1996) 295-306.

19. Konno, H., Yajima, Y. and Matsui, T., "Parametric Simplex Algorithms for Solving a Special Class of Nonconvex Minimization Problems", *J. of Global Optimization,* 1 (1991) 65-81.
20. Konno, H. and Yamashita, H., "Minimaization of the Sum and the Product of Several Linear Fractional Functions", *Naval Research Logistics,* 46 (1999) 583-591.
21. Schaible, S., "A Note on the Sum of a Linear and Linear-Fractional Function", *Naval Research Logistics,* 24 (1977) 691-693.
22. Schaible, S., "Fractional Programming", *Handbook of Global Optimization,* (Horst, R. and Pardalos, P. eds.,) Kluwer Academic Publishers, 1995, 495-608.
23. Tuy, H.,*Convex Analysis and Global Optimization*, Kluwer Academic Publishers,1998.

Discrete Higher Order Convex Functions and their Applications

András Prékopa

RUTCOR, Rutgers Center for Operations Research
640 Bartholomew Road Piscataway, NJ 08854-8003
prekopa@rutcor.rutgers.edu

Abstract. In this paper we present an overview about the recently developed theory of discrete moment problems, i.e., moment problems where the supports of the random variables involved are discrete. We look for the minimum or maximum of a linear functional acting on an unknown probability distribution subject to a finite number of moment constraints. Using linear programming methodology, we present structural theorems, in both the univariate and multivariate cases, for the dual feasible bases and show how the relevant problems can be solved by suitable adaptations of the dual method. The condition on the objective function is a kind of higher order convexity, expressed in terms of divided differences. A variant of the above, the discrete binomial moment problem, as well as generalization for discrete variable Chebyshev systems are also discussed. Finally, we present novel applications to valuations of financial instruments.

2000 Mathematics Subject Classification. 26B25, 30E05, 65C50, 90C05.

Keywords and phrases. Discrete moment problem, Linear programming, Higher order convex functions, Probability bounds, Option pricing.

1 Introduction

Discrete moment problems came to prominence by the discovery (Samuels and Studden (1989), Prékopa (1988,1990a,1990b) that the sharp Bonferroni bounds can be obtained as optimum values of discrete moment problems.

The first sharp bound for the probability of the union of n events was obtained by Dawson and Sankoff (1967). They assumed the knowledge of the first two binomial moments of the number of events which occur. Kwerel (1975a,b) reformulated and extended the problem using linear programming methodology, where the first three binomial moments of the occurencies are supposed to be known. The general problems, where there is no limitation regarding the numbers of binomial moments have been formulated and studied by Prékopa (1988,1990a).

The multivariate discrete moment problem has been introduced and studied by Prékopa (1992,1999).

By the use of the methodology of discrete moment problems a number of other known probability bounds could be derived (see the above cited papers by Samuels and Studden, Prékopa, furthermore the paper by Boros and Prékopa (1989)).

While Samuels and Studden follow the guidelines of the general moment problems in their discussion, Prékopa uses linear programming methodology, which enables him to come up with simple algorithmic solution to the problem wherever the size of the problem prevents us to present the bounds in closed forms.

The simplest discrete moment problem, where power moments are used, is closely connected with divided differences, higher order convex functions (defined in terms of divided differences) and Lagrange interpolation.

Let $f(z)$, $z \in Z = \{z_0, \dots, z_n\}$ be a discrete variable function, where $z_0 < \dots < z_n$. Its first order divided differences are designated and defined as:

$$[z_{i_1}, z_{i_2}; f] = \frac{f(z_{i_2}) - f(z_{i_1})}{z_{i_2} - z_{i_1}}, \ i_1 \neq i_2. \tag{1. 1}$$

The kth order divided difference is designated and defined as:

$$[z_{i_1}, z_{i_2}, \dots, z_{i_{k+1}}; f] = \frac{[z_{i_2}, \dots, z_{i_{k+1}}; f] - [z_{i_1}, \dots, z_{i_k}; f]}{z_{i_{k+1}} - z_{i_1}}, \tag{1. 2}$$

where $z_{i_1}, z_{i_2}, \dots, z_{i_{k+1}}$ are $k+1$ distinct elements of Z. Let $[z; f] = f(z)$, by definition.

The function f is said to be (strictly) convex of order k on Z, if all of its kth order divided differences are (positive) nonnegative.

A sufficient condition for that is the following: f is defined in $[z_0, z_n]$ and has (positive) nonnegative kth order derivatives in (z_0, z_n).

First order convexity means that the function is nondecreasing, second order convexity means that the function, obtained by connecting all neighbors of the points $(z_i, f(z_i)), i = 0, 1, \dots, n$ by straight lines, is convex in the classical sense.

We have the determinental formula

$$[z_{i_1}, z_{i_2}, \dots, z_{i_{k+1}}; f] = \frac{\begin{vmatrix} 1 & 1 & \dots & 1 \\ z_{i_1} & z_{i_2} & \dots & z_{i_{k+1}} \\ \vdots & \vdots & & \vdots \\ z_{i_1}^{k-1} & z_{i_2}^{k-1} & \dots & z_{i_{k+1}}^{k-1} \\ f(z_{i_1}) & f(z_{i_2}) & \dots & f(z_{i_{k+1}}) \end{vmatrix}}{\begin{vmatrix} 1 & 1 & \dots & 1 \\ z_{i_1} & z_{i_2} & \dots & z_{i_{k+1}} \\ \vdots & \vdots & & \vdots \\ z_{i_1}^{k-1} & z_{i_2}^{k-1} & \dots & z_{i_{k+1}}^{k-1} \\ z_{i_1}^{k} & z_{i_2}^{k} & \dots & z_{i_{k+1}}^{k} \end{vmatrix}}. \qquad (1.\ 3)$$

A theorem (see, e.g., Prékopa (1995)) asserts that if all kth order divided differences of f, corresponding to consecutive points, are (positive) nonnegative, then all kth oder divided fifferences of f are (positive) nonnegative. This fact is also a simple consequence of a general theorem of Fekete (1912). Let $I \subset \{0, 1, \dots, n\}$ with $|I| = m + 1 \quad (m \leq n)$. The Lagrange polynomial corresponding to the set of points $\{z_i, i \in I\}$ is defined as

$$L_I(z) = \sum_{h \in I} f(z_h) L_{I,h}(z), \qquad (1.\ 4)$$

where

$$L_{I,h}(z) = \prod_{i \in I \setminus \{h\}} \frac{z - z_h}{z_i - z_h}, \qquad h \in I \qquad (1.\ 5)$$

is the hth fundamental polynomial. We will need $L_I(z)$ at the points in Z but it can be defined for all real z values. The polynomial can be written in Newton's form:

$$L_I(z) = \sum_{h=0}^{m} [z_j, \, j \in I^h; f] \prod_{i \in I^{h-1}} (z - z_i), \qquad (1.\ 6)$$

where I^h is the set of the first $h + 1$ points in I. and $\prod_{i \in I^{-1}} (z - z_i) = 1$ by definition. An important formula in Lagrange interpolation is the following:

$$f(z) - L_I(z) = [z_j, j \in I, \, z; f] \prod_{i \in I} (z - z_i). \qquad (1.\ 7)$$

In the s-variate case first we define $Z_j = \{z_{j0}, \dots, z_{jn_j}\}$, $z_{j0} < \dots < z_{jn_j}$, $j = 1, \dots, s$. Let $Z = Z_1 \times \cdots \times Z_s$ and consider a function $f(z)$, $z \in Z$. The divided difference correponding to a subset

$$Z_{I_1 \dots I_s} = \{z_{1i}, i \in I_1\} \times \dots \times \{z_{si}, i \in I_s\} = Z_{1I_1} \times \dots \times = Z_{sI_s}$$

of the set Z can be defined in an iterative manner in such a way that first we take k_1-th order divided difference of f with respect to z_1, where $k_1 = |I_1|-1$, then the k_2-th order divided difference of that with respect to z_2, where $k_2 = |I_2| - 1$, etc. This can be executed in a mixed manner, the result will always be the same.

Let $[z_{1i}, i \in I_1; \ldots; z_{si}, i \in I_s; f]$ designate this divided difference and call it of order $(k_1, \ldots, k_s)$. The sum $k_1 + \ldots + k_s$ will be called the total order of the divided difference.

The set on which the above divided difference is defined is the Cartesian product of sets on the real line. Let us term such sets *rectangular*. Divided differences on non-rectangular sets have also been defined in the literature (see, e.g., Karlin, Micchelli and Rinott (1986)). These require, however, smooth functions while ours are defined on discrete sets.

A Lagrange interpolation polynomial corresponding to the points in $\{z_{1i}, i \in I_1\} \times \ldots \times \{z_{si}, i \in I_s\}$ is defined by the equation

$$L_{I_1 \cdots I_s}(z_1, \ldots, z_s) = \sum_{i_1 \in I_1} \cdots \sum_{i_s \in I_s} f(z_{1i_1}, \ldots, z_{si_s}) L_{I_1 i_1}(z_1) \cdots L_{I_s i_s}(z_s), \tag{1. 8}$$

where

$$L_{I_j i_j}(z_j) = \prod_{h \in I_j \setminus \{i_j\}} \frac{z_j - z_{ji_j}}{z_{jh} - z_{ji_j}}, \quad j = 1, \ldots, s. \tag{1. 9}$$

The polynomial (1.8) coincides with the function f at every point of the set $Z_{I_1 \ldots I_s}$ and is of degree $m_1 \ldots m_s$.

Newton's form of the Lagarange polynomial (1.8) can be given as follows.

Let $I_j^{(k_j)}$ designate the set of the first $k_j + 1$ elements of $I_j, 0 \le k_j \le m_j, j = 1, \ldots, s$. Then the required form is

$$L_{I_1 \ldots I_s}(z_1, \ldots, z_s) = \sum_{k_1=0}^{m_1} \cdots \sum_{k_s=0}^{m_s} [z_{1h}, h \in I_1^{(k_1)}; \ldots; z_{sh}, h \in I_s^{(k_s)}; f] * \prod_{j=1}^{s} \prod_{h \in I_j^{(k_j-1)}} (z_j - z_{jh}) \tag{1. 10}$$

2 The univariate discrete power moment problem

The problem is defined as the following LP:

$$\begin{aligned} &\text{Min (Max)} \sum_{i=0}^{n} f(z_i)x_i \\ &\text{subject to} \\ &\sum_{i=0}^{n} z_i^k x_i = \mu_k, \quad k = 0, 1, \ldots, m \\ &x_i \geq 0, \quad i = 0, 1, \ldots, n. \end{aligned} \tag{2. 1}$$

Here known are $z_0, \ldots, z_n$; $f(z_0), \ldots, f(z_n)$ and $\mu_1, \ldots, \mu_m; \mu_0 = 1$, by definition. The unknowns are $x_0, x_1, \ldots, x_n$ which form a probability distribution with support $Z = \{z_0, z_1, \ldots, z_n\}$. If X is a random variable with support Z and we know $\mu_k = E(X^k), k = 1, \ldots, m$, than the optimum values of problem (2.1) provide us with sharp lower and upper bounds for $E[f(X)]$. The term "sharp" refers to the fact that knowing only $\mu_1, \ldots, \mu_m$, no better bounds can be given to $E[f(X)]$, than the optimum values of problems (2.1).

If we introduce the notations:

$$\mathbf{a_i} = \begin{pmatrix} 1 \\ z_i \\ \vdots \\ z_i^m \end{pmatrix}, \ f_i = f(z_i), i = 0, \ldots, n; \ \mathbf{b} = \begin{pmatrix} 1 \\ \mu_1 \\ \mu_2 \\ \vdots \\ \mu_m \end{pmatrix},$$

then problem (2.1) can be written in the form:

$$\begin{aligned} &\text{Min (Max)} \sum_{i=0}^{n} f_i x_i \\ &\text{subject to} \\ &\sum_{i=0}^{n} \mathbf{a_i} x_i = \mathbf{b} \\ &x_i \geq 0, \quad i = 0, \ldots, n. \end{aligned} \tag{2. 2}$$

The matrix $A = (\mathbf{a_0}, \ldots, \mathbf{a_n})$ is an $(m+1) \times (n+1)$ Vandermonde matrix, hence every collection of $m+1$ vectors of A forms a basis in this LP.

The basis B is said to be feasible if

$$B^{-1}\mathbf{b} \geq 0 \tag{2. 3}$$

and dual feasible in the minimization (maximization) problem if

$$f_i - f_B^T B^{-1} \mathbf{a_i} \geq (\leq)\, 0, \quad i = 0, \dots, n, \tag{2. 4}$$

where f_B is the vector of basic components of $f = (f_0, f_1, \dots, f_n)^T$. An inequality in (2.4) holds with equality sign if i is the subscript of a basic vector. In what follows I_B designates the set of subscripts of the basic vectors.

For fixed B, consider the m-degree polynomial

$$L_{I_B}(z) = f_B^T B^{-1} \begin{pmatrix} 1 \\ z \\ \vdots \\ z^m \end{pmatrix}. \tag{2. 5}$$

By the remark made above, we have the relations

$$L_{I_B}(z_i) = f_i = f(z_i), i \in I_B \tag{2. 6}$$

and this implies that $L_{I_B}(z)$ is the m-degree Lagarange polynomial corresponding to the points $\{z_i, i \in I_B\}$. The application of formula (1.7) gives

$$\begin{aligned} & f(z) - L_{I_B}(z) \\ &= [z_i, i \in I_B, z; f] \prod_{i \in I_B} (z - z_i). \end{aligned} \tag{2. 7}$$

Assume now that all $m+1$st order divided differences of f are positive. Then, from (2.7) we see that $f(z) - L_{I_B}(z) \neq 0$ for $z \notin \{z_i, i \in I_B\}$. Since the left hand side of (2.4) is the same as that of (2.7) for $z = z_i$, it follows that each inequality in (2.4) holds strictly if $i \notin I_B$. This means that all bases in problems (2.1)-(2.2) are dual non-degenerate.

If we look again at (2.7), we see that the first factor on the right hand side is positive for every $z \notin \{z_i, i \in I_B\}$. Thus, the basis B is dual feasible in the minimization (maximization) problem iff

$$\prod_{i \in I_B} (z - z_i) > (<)\, 0 \quad \textit{for } z \notin \{z_i, i \in I_B\}. \tag{2. 8}$$

This immediately implies

Theorem 2.1. *If the function f has positive divided differences of order $m+1$, then B is a dual feasible basis to problem (2.1)-(2.2) iff the subscript set I_B has the following structure:*

	$m+1$ *even*	$m+1$ *odd*
min	$i, i+1, \dots, j, j+1$	$0, i, i+1, \dots, j, j+1$
max	$0, i, i+1, \dots, j, j+1, n$	$i, i+1, \dots, j, j+1, n$

If B_1 (B_2) is a dual feasible basis in the minimization (maximization) problem (2.1)-(2.2), then we have the inequalities

$$L_{B_1}(z) \le f(z) \le L_{B_2}(z) \quad z \in Z \tag{2. 9}$$

and

$$E[L_{B_1}(X)] \le E[f(X)] \le E[L_{B_2}(X)]. \tag{2. 10}$$

Since $L_{B_1}(z), L_{B_2}(z)$ are m-degree polynomials, inequality (2.10) provides us with bounds for $E[f(X)]$, based on the knowledge of the moments $\mu_1, \ldots, \mu_m$. The sharp bounds correspond to bases B_1, B_2 which are optimal in the linear programs (2.1)-(2.2).

Remark: If the function f has only nonnegative divided differences of order $m+1$, then only the one way assertion holds: if I_B has the structure in Theorem 2.1, then B is a dual feasible basis in problem (2.1)-(2.2).

One can immediately derive bounding formulas for the cases where the number of utilized moments is small. If we only know μ_1, then any dual feasible basis in the minimization problem has subscripts $j, j+1$ and the only dual feasible basis in the maximization problem has subscripts $0, n$. The latter one is optimal in the maximization problem since any LP that has feasible solution and finite optimum, has at least one primal-dual feasible basis. The former one is optimal in the minimization problem if the basis is also primal feasible, the condition for which is: $z_j \le \mu \le z_{j+1}$. With this j we have

$$\begin{aligned} \tfrac{z_{j+1}-\mu_1}{z_{j+1}-z_j} f(z_j) + \tfrac{\mu_1 - z_j}{z_{j+1}-z_j} f(z_{j+1}) \\ \le E[f(X)] \\ \le \tfrac{z_n-\mu_1}{z_n-z_0} f(z_0) + \tfrac{\mu_1-z_0}{z_n-z_0} f(z_n). \end{aligned} \tag{2. 11}$$

Note that if $f(z)$ is defined and is convex in the entire interval $[z_0, z_n]$ and the random variable may take any value in this interval, then for $E[f(X)]$ we have Jensen's inequality, as a lower bound and the Edmundson-Madansky inequality as an upper bound. The latter one coincides with the upper bound in (2.11) while the former one is equal to $f(\mu_1)$ which is different from the lower bound in (2.11). In fact, it is a weaker bound than the one in (2.11).

One may formulate the general statment that for a discrete random variable X the discrete moment bounds are always better than the general moment bounds, i.e., the bounds that we can obtain as optimum values of the problems

$$\begin{aligned} &\text{Min (Max)} \int_{z_0}^{z_n} f(z) dF(z) \\ &\text{subject to} \\ &\int_{z_0}^{z_n} z^k dF(z) = \mu_k, \quad k = 0, \ldots, m, \end{aligned} \tag{2. 12}$$

where $F(z)$ is a probability distribution function with support $[z_0, z_n]$.

Bounding formulas for the cases when μ_1, μ_2 or μ_1, μ_2, μ_3 are known, are presented in Section 5.9 of Prékopa (1995), for the cases of nonnegative integer valued random variables. An upper bound can be obtained (along the lines the upper bound in section 6.2.5 of the same book is obtained) if $\mu_1, \mu_2, \mu_3, \mu_4$ are known, otherwise we have to use algorithms to solve the problems.

3 Dual algorithm for the solution of problem (2.1)

We adapt the dual algorithm of Lemke (1954) to solve the problem. Assume that f has positive divided differences of order $m+1$.

The algorithm starts by picking a dual feasible basis. After that, at each iteration, one vector leaves the basis and another one enters, while the dual feasibility is always preserved.

In the simplex algorithm first we determine the incoming vector and then the outgoing vector. In the dual algorithm it is the other way around. Once the outgoing vector has been determined, the determination of the incoming vector is very simple. In fact, if we remove a vector from the basis, then, by Theorem 2.1, there is one and only one way to restore dual feasibility, by an incoming vector, and it can be found by a simple search procedure. In view of this, we have to concentrate on the problem of identifying an outgoing vector.

Let $I_B = \{i_0, i_1, \ldots, i_m\}$, where $i_0 < i_1 < \ldots < i_m$. If

$$(B^{-1}\mathbf{b})_k < 0 \tag{3. 1}$$

then $\mathbf{a}_{\mathbf{i_k}}$ may be an outgoing vector. Since we have that

$$\mathbf{b} = E\left[\begin{pmatrix} 1 \\ X \\ \vdots \\ X^m \end{pmatrix}\right]$$

$$B^{-1}\mathbf{b} = E\left[B^{-1}\begin{pmatrix} 1 \\ X \\ \vdots \\ X^m \end{pmatrix}\right]$$

$$L_I(z) = f_B^T B^{-1} \begin{pmatrix} 1 \\ z \\ \vdots \\ z^m \end{pmatrix},$$

it follows that

$$(B^{-1}\mathbf{b})_k = E[L_{I,k}(X)], \tag{3. 2}$$

where $L_{I,k}(z)$ is the kth Lagrange fundamental polynomial

$$L_{I,k}(z) = \prod_{j \in I \setminus \{i_k\}} \frac{z - z_{i_k}}{z_{i_j} - z_{i_k}}. \tag{3. 3}$$

The sign of the denominator in (3.3) equals $(-1)^{m-k}$. Thus, to determine the sign of (3.3) we have to look at the polynomial

$$\prod_{j \in I \setminus \{i_k\}} (z - z_{i_j}),$$

replace X for z and take expectation. The obtained value equals

$$\begin{aligned} &\mu_m - \left(\textstyle\sum_{j \in I \setminus \{i_k\}} z_{i_j}\right) \mu_{m-1} \\ &+ \cdots + (-1)^m \textstyle\prod_{j \in I \setminus \{i_k\}} z_{i_j}. \end{aligned} \tag{3. 4}$$

If the value in (3.4), multiplied by $(-1)^{m-k}$, is negative, then the kth vector of the basis can be chosen as the outgoing vector.

The algorithm to solve problem (2.1) can be summarized as follows.

Step 1. Pick any dual feasible basis in agreement with Theorem 2.1. Let $I = \{i_0, i_1, \ldots, i_m\}$ designate the set of subscripts of the basis vectors.

Determination of the outgoing vector.
Step 2. Take any element $i_k \in I$ and compute the coefficients of $\mu_m, \mu_{m-1}, \ldots, \mu_1$ in (3.4).

Step 3. Compute the value in (3.4) and multply it by $(-1)^{m-i_k}$. If it is negative, then $\mathbf{a}_{i_k}$ may be an outgoing vector. Otherwise take another element of I. Repeat until an outgoing vector is identified. Otherwise go to step 5.

Determination of the incoming vector.
Step 4. If the outgoing vector is identified, then find that vector which restores dual feasibility of the basis. Choose it as the incoming vector. Go to Step 3.

Step 5. Stop, we have $B^{-1}\mathbf{b} \geq \mathbf{0}$, hence the basis is primal feasible too, i.e., it is optimal. The value $f_B^T B^{-1}\mathbf{b}$ is a lower or upper bound for $E[f(X)]$, depending on which type of dual feasible bases we have used in the algorithm.

Since all bases are dual non-degenerate, it follows that cycling cannot occur and the algorithm terminates in a finite number of steps.

Discrete (as well as general) moment problems are frequently solved for large m, n values, e.g., $m \geq 20$ and n several hundred. Since large size Vandermonde matrices are numerically unstable, problem (2.1) cannot always be solved by general purpose LP packages. This is the reason why we need the above described algorithm to efficiently solve the problem. In this algorithm we work with $(m+1) \times (m+1)$ matrices, at each iteration and the calculation of the values (3.4) can be carried out in a stable manner, in a reasonable time, if $m \approx 20$, say. At the final step, however, when we compute the optimal value: $f_B^T B^{-1}\mathbf{b}$ (and not only check the signs of some values), special care has to be taken and the use of special algorithms to solve Vandermonde systems of equations (see, e.g., Chun and Kailath (1991) and the references there) are advisable.

If the divided differences of order $m+1$ of the function f are only nonnegative, then the above described algorithm needs some modification: as long as the dual feasible bases are dual non-degenerate we do the same as before but whenever dual degeneracy occurs, some anti-cycling rule should be applied (see, e.g., Prékopa (1995), Chapter 1).

4 Discrete binomial moment problems and the use of Chebyshev systems

The discrete binomial moment problem can be stated as

$$\begin{aligned} &\text{Min (Max)} \sum_{i=0}^{n} f(z_i)x_i \\ &\text{subject to} \\ &\sum_{i=0}^{n} \binom{z_i}{k} x_i = S_k, \quad k = 0,\dots,m \\ &x_i \geq 0, \quad i = 0,\dots,n, \end{aligned} \tag{4. 1}$$

where

$$S_k = E\left[\binom{X}{k}\right] \tag{4. 2}$$

is the kth binomial moment of the random variable with support set $\{z_0, z_1, \ldots, z_n\}$. We have the relations: $S_1 = \mu_1$, $S_2 = \frac{1}{2}(\mu_2 - \mu_1)$, $S_3 = \frac{1}{6}(\mu_3 - 3\mu_2 + 2\mu_1)$ etc. In view of these and similar relations between the binomial coefficients and powers, the equality constraints of problem (4.1) can be transformed into those of problem (2.1) (and vice versa) by simple linear transformation. This implies that if the $m+1$st order divided differences of the function f are positive, then Theorem 2.1 holds true, further, B is dual (primal) feasible in problem (4.1) iff it enjoys the same property in problem (2.1). Thus, we can find an optimal basis to problem (4.1) by finding one to problem (2.1). However, there are important problems of the type (4.1), where the above condition, for the function f, does not hold.

The binomial moment problems are connected with n events $A_1, \ldots, A_n$ and the random variable

$$X = X_1 + \ldots + X_n,$$

where

$$X_i = \begin{cases} 1, \text{ if } A_i \text{ occurs} \\ 0, \text{ otherwise}, \ i = 1, \ldots, n. \end{cases}$$

A frequently asked question is: what is the probability that at least one out of the n events occur. One answer to this question is given by the inclusion-exclusion formula:

$$\begin{aligned} &P(A_1 \cup \cdots \cup A_n) \\ &= \sum_{k=1}^{n} (-1)^{k-1} \sum_{1 \le i_1 < \cdots < i_k \le n} P(A_{i_1} \cap \cdots \cap A_{i_k}). \end{aligned} \tag{4. 3}$$

However, terms corresponding to large k values are freqently impossible to compute and in such cases we have to be satisfied with bounds. We make use of the following:

Theorem 4.1. *The following equations hold true:*

$$\begin{aligned} &S_k - E\left[\binom{X}{k}\right] \\ &= \sum_{1 \le i_1 < \cdots < i_k \le n} P(A_{i_1} \cap \cdots \cap A_{i_k}). \end{aligned} \tag{4. 4}$$

To obtain bounds for the probability of the union $A_1 \cup \cdots \cup A_n$ we may proceed in the following way. We formulate the linear programs

$$\begin{aligned} &\text{Min (Max)} \sum_{i=1}^{n} x_i \\ &\text{subject to} \\ &\sum_{i=1}^{n} \binom{i}{k} x_i = S_k, \ k = 1, \ldots, m \\ &x_i \ge 0, \quad i = 1, \ldots, n. \end{aligned} \tag{4. 5}$$

If we compare problems (4.1) and (4.5), where in problem (4.1) we have $z_i = i, f(i) = 1, i = 1, \ldots, n$, $f(0) = 0$, we see that the difference is that in problem (4.5) the constraint of S_0 and the variable x_0 are missing. On the other hand, we can establish relationships between the optimum values of the two problems.

Let $V_{min}(V_{max})$ designate the optimum value of the minimization (maximization) problem in (4.5). It is easy to see that the optimum value of the minimization problem (4.1) is the same $V_{\min}$, while the optimum value of the maximization problem (4.1) is $\min(V_{\max}, 1)$. In view of this relationship we work with problem (4.5) rather than with problem (4.1). The sharp bounds for the union are given by the inequalities:

$$V_{\min} \leq P(A_1 \cup \cdots \cup A_n) \leq \min(V_{\max}, 1). \tag{4. 6}$$

Discrete moment problems, more general than the power or binomial moment problems can be formulated by the use of the concept of a discrete Chebyshev system.

Definition 4.1. We say that the functions $g_0(z), g_1(z), \ldots, g_k(z)$, $z \in Z$ form a (discrete variable) Chebyshev system if for every $z_0 < z_1 < \cdots < z_k$ we have

$$\begin{vmatrix} g_0(z_0) & g_0(z_1) & \ldots & g_0(z_k) \\ g_1(z_0) & g_1(z_1) & \ldots & g_1(z_k) \\ \vdots & \vdots & & \vdots \\ g_k(z_0) & g_k(z_1) & \ldots & g_k(z_k) \end{vmatrix} > 0. \tag{4. 7}$$

If only the $\geq$ sign holds in (4.7), then we call it a weak Chebyshev system. For more details about Chebyshev systems the reader is referred to Karlin-Studden (1996) or Krein- Nudelman (1977).

Definition 4.2. Let $g_0(z), g_1(z), \ldots, g_{k-1}(z)$, $z \in Z$ be a Chebyshev system. We say that the function $f(z), z \in Z$ is convex of order k with respect to this Chebyshev system if for every $z_{i_0} < z_{i_1} < \cdots < z_{i_k}$ we have

$$\begin{vmatrix} g_0(z_{i_0}) & g_0(z_{i_1}) & \ldots & g_0(z_{i_k}) \\ g_1(z_{i_0}) & g_1(z_{i_1}) & \ldots & g_1(z_{i_k}) \\ \vdots & \vdots & & \vdots \\ g_{k-1}(z_{i_0}) & g_{k-1}(z_{i_1}) & \ldots & g_{k-1}(z_{i_k}) \\ f(z_{i_0}) & f(z_{i_1}) & \ldots & f(z_{i_k}) \end{vmatrix} \geq 0. \tag{4. 8}$$

If always the strict inequality holds, then f is said to be strictly convex of order k with respect to the system $g_0(z), g_1(z), \ldots, g_{k-1}(z)$.

Now we formulate the discrete moment problem:

$$\begin{aligned}
&\text{Min (Max)} \sum_{i=0}^{n} f(z_i)x_i \\
&\text{subject to} \\
&\sum_{i=0}^{n} g_k(z_i)x_i = \mu_k, \quad k = 0, 1, \ldots, m \\
&x_i \geq 0, \quad i = 0, 1, \ldots, n,
\end{aligned} \tag{4. 9}$$

where $g_0(z), g_1(z), \ldots, g_m(z)$, $z \in Z$ form a Chebyshev system and the function $(-1)^{m+1} f(z)$, $z \in Z$ is strictly convex of order $m+1$ with respect to this system. This means that if in (4.8), written up for $k = m+1$, we put $f(z_{i_0}), f(z_{i_1}), \ldots, f(z_{i_m})$ in the first row, then the resulting determinant is always strictly positive.

A dual feasible basis structure theorem, similar to Theorem 2.1, can be derived for problem (4.9). We will state it at the end of the derivation.

In what follows, we use problem (2.2) as another form of problem (4.8). Let B be a basis in problem (4.8). In view of the formula

$$\begin{pmatrix} 1 & f_B^T \\ 0 & B \end{pmatrix}^{-1} = \begin{pmatrix} 1 & -f_B^T B^{-1} \\ 0 & B^{-1} \end{pmatrix},$$

we may write

$$\begin{pmatrix} 1 & f_B^T \\ 0 & B \end{pmatrix}^{-1} \begin{pmatrix} f_i \\ \mathbf{a_i} \end{pmatrix} = \begin{pmatrix} f_i - f_B^T B^{-1} \mathbf{a_i} \\ B^{-1} \mathbf{a_i} \end{pmatrix},$$

or

$$\begin{pmatrix} 1 & f_B^T \\ 0 & B \end{pmatrix} \begin{pmatrix} f_i - f_B^T B^{-1} \mathbf{a_i} \\ B^{-1} \mathbf{a_i} \end{pmatrix} = \begin{pmatrix} f_i \\ \mathbf{a_i} \end{pmatrix}.$$

If we use Cramer's rule, we obtain the relation for $f_i - f_B^T B^{-1} \mathbf{a_i}$, the first component of the solution of the above system of linear equations:

$$f_i - f_B^T B^{-1} \mathbf{a_i} = \frac{1}{|B|} \begin{vmatrix} f_i & f_B^T \\ \mathbf{a_i} & B \end{vmatrix}, \tag{4. 10}$$

where $|B|$ designates the determinant of the matrix B.

Let I_B designate the subscript set of the basic vectors. If $i \in I_B$, then the value in (4.10) is 0. If $i \notin I_B$, then, by assumption, $|B| > 0$ and since the columns of the second determinant are columns of a positive determinant, if suitably rearranged, it follows that the right hand side of (4.10) is different from 0, and thus, all bases are dual nondegenerate.

If we look at a minimization problem, then B is dual feasible iff (4.10) is positive for every $i \notin I_B$. This immediately implies that if $m+1$ is even, then the basic vectors form consecutive pairs and if m is odd, then any basis consists of consecutive pairs and $\mathbf{a_n}$. Similar statement holds for the maximization problem. Both cases are summarized in the following

Theorem 4.2. *The basis B is dual feasible in problem (4.8) iff the subcript set I_B of the basic vectors has the following structure:*

	$m+1$ *even*	$m+1$ *odd*
min	$i, i+1, \ldots, j, j+1$	$0, i, i+1, \ldots, j, j+1$
max	$0, i, i+1, \ldots, j, j+1, n$	$i, i+1, \ldots, j, j+1, n$

Before presenting the application of this theorem, we state another one on binomial coefficients.

Theorem 4.3. *(Gessel and Viennot (1985), Prékopa (1988)). Consider the matrix*

$$D = \begin{pmatrix} 1 & 1 & 1 & 1 & 1 & \cdots & 1 \\ & 1 & 2 & 3 & 4 & \cdots & n \\ & & 1 & \binom{3}{2} & \binom{4}{2} & \cdots & \binom{n}{2} \\ & & & 1 & \binom{4}{3} & \cdots & \binom{n}{3} \\ & & & & & \ddots & \\ & & & & & & \binom{n}{n} \end{pmatrix}. \qquad (4.\ 11)$$

All minors (determinants of square submatrices) of D that have all positive entries in the main diagonal, are positive.

Corollary 4.1. *Consider the following submatrix of D:*

$$E = \begin{pmatrix} 1 & 1 & 1 & 1 & \cdots & 1 & 1 & \cdots & 1 \\ 1 & 2 & 3 & 4 & \cdots & m & m+1 & \cdots & n \\ & 1 & \binom{3}{2} & \binom{4}{2} & \cdots & \binom{m}{2} & \binom{m+}{2} & \cdots & \binom{n}{2} \\ & \cdot & \cdot & \cdot & \cdots & \cdot & \cdot & \cdots & \cdot \\ & & & & & 1 & \binom{m+1}{m} & \cdots & \binom{n}{m} \end{pmatrix}.$$

Theorem 4.3 implies that all minors of order $m+1$ from E and all minors of order m from the last m rows of E are positive.

The above Corollary can be applied to discover the collection of dual feasible bases of problems (4.5). In fact, the Corollary implies that if $Z = \{1, 2, \ldots, n\}$, then the functions $g_k(i) = \binom{i}{k}, i \in Z$, $k = 1, \ldots, m$ form a Chebyshev system and the constant function $(-1)^m$ is strictly convex of order m with respect to this system. Thus, Theorem 4.2 implies

Theorem 4.4. *The basis B is dual feasible in problem (4.5) iff the subscript set I_B has the structure in Theorem 4.2, where $m+1$ is replaced by m and 0 is replaced by 1.*

As special cases we can easily derive the classical bounding formulas for the probability of the union, if $m = 2$. In the minimization problem any dual feasible basis consists of two consicutive vectors. In order to find out which one of them is primal feasible, i.e., optimal, we have to look for that i, for which the solution of the equations:

$$ix_i + (i+1)x_{i+1} = S_1$$
$$\binom{i}{2}x_i + \binom{i+1}{2}x_{i+1} = S_2$$

satisfies $x_i \geq 0$, $x_{i+1} \geq 0$. It turns out that this i equals

$$i = 1 + \left\lfloor \frac{2S_2}{S_1} \right\rfloor . \tag{4. 12}$$

The lower bound, i.e., the sum of x_1 and x_2 is

$$\frac{2}{i+1}S_1 - \frac{2}{i(i+1)}S_2. \tag{4. 13}$$

This result was first established by Dawson and Sankoff (1967).

As regards the maximization problem, there is just one dual feasible basis and it corresponds to the subscript set $\{1, n\}$. If we solve the equation

$$x_1 + nx_n = S_1$$
$$\binom{n}{2}x_n = S_2$$

and add x_1 and x_2, then we obtain the upper bound for the union:

$$S_1 - \frac{2}{n}S_2. \tag{4. 14}$$

It is interesting to remark that the classical Bonferroni inequalities

$$P(A_1 \cup \cdots \cup A_n) \geq S_1 - S_2 + \cdots + S_{m-1} - S_m \tag{4. 15}$$

if m is even and

$$P(A_1 \cup \cdots \cup A_n) \leq S_1 - S_2 + \cdots + S_{m-2} - S_{m-1} + S_m\,, \tag{4. 16}$$

if m is odd, can also be derived from problems (4.1). In fact, the basis corresponding to $I_B = \{1, 2, \ldots, m\}$ is dual feasible in the minimization problem, if m is even and dual feasible in the maximization problem, if m is odd and the right hand side values in (4.15) and (4.16) are the corresponding objective function values.

For further closed form inequalities, derived from problem (3.1), see Kwerel (1975a,b), Boros and Prékopa (1989), Prékopa (1995).

If we look for the probability that at least r or exactly r out of n events occur, then to obtain bounds for them we have to use problem (2.1) or (4.1), for the cases of $Z = \{0, 1, \ldots, n\}$ and the functions

$$f(i) = \begin{cases} 1, \text{ if } i \geq r \\ 0, \text{ if } i < r \end{cases} \tag{4. 17}$$

and

$$f(i) = \begin{cases} 1, \text{ if } i = r \\ 0, \text{ if } i \neq r. \end{cases} \tag{4. 18}$$

However, these functions do not have all positive or all negative divided differences on Z, hence the above theory cannot directly be applied to these cases. Still, the collection of dual feasible bases is known in these cases too and dual methods for the solution of the relevant moment problems have been developed (see Prékopa (1990a,b, 1995)).

5 Multivariate discrete moment problems

Let $X_1, \ldots, X_s$ be random variables with supports $Z_1, \ldots, Z_s$, respectively, and introduce the notations

$$\begin{gathered} x_{i_1 \cdots i_s} = P(X_1 = z_{1i_1}, \cdots, X_s = z_{si_s}) \\ 0 \leq i_j \leq n_j,\ j = 1, \ldots, s \end{gathered} \tag{5. 1}$$

$$\mu_{\alpha_1 \cdots \alpha_s} = \sum_{i_1=0}^{n_1} \cdots \sum_{i_s=0}^{n_s} z_{1i_1}^{\alpha_1} \cdots z_{si_s}^{\alpha_s} x_{i_1 \cdots i_s}, \tag{5. 2}$$

where $\alpha_1, \cdots, \alpha_s$ are nonnegative integers. The number $\mu_{\alpha_1 \cdots \alpha_s}$ is called the $(\alpha_1, \cdots, \alpha_s)$-order moment of the random vector $(X_1, \cdots, X_s)$ and $\alpha_1 + \cdots + \alpha_s$ the total order of the moment.

We are looking for lower and upper bounds on

$$E\left[f(X_1, \ldots, X_s)\right], \tag{5. 3}$$

where f is some function defined on the discrete set Z. In most cases f is supposed to be some multivariate higher order convex function or one of the following types:

$$f(z_1, \ldots, z_s) = \begin{cases} 1, \text{ if } z_i \geq z_{r_i},\ i = 1, \ldots, s \\ 0, \text{ if otherwise} \end{cases} \tag{5. 4}$$

and

$$f(z_1, \ldots, z_s) = \begin{cases} 1, \text{ if } z_i = z_{r_i} \\ 0, \text{ if otherwise.} \end{cases} \tag{5. 5}$$

The choice of the function (5.4) enables us to create bounds for $P(X_1 \geq z_{r_1}, \ldots, X_s \geq z_{r_s})$, while the function (5.5) is used for bounding $P(X_1 = z_{r_1}, \ldots, X_s = z_{r_s})$.

We formulate two variants for the multivariate discrete power moment problem. For the sake of simplicity we introduce the notation $f_{i_1 \cdots i_s} = f(z_{1i_1}, \cdots, z_{si_s})$. The first problem is

$$\begin{aligned} &\text{Min (Max) } \sum_{i_1=0}^{n_1} \cdots \sum_{i_s=0}^{n_s} f_{i_1 \cdots i_s} x_{i_1 \cdots i_s} \\ &\text{subject to} \\ &\sum_{i_1=0}^{n_1} \cdots \sum_{i_s=0}^{n_s} z_{1i_1}^{\alpha_1} \cdots z_{si_s}^{\alpha_s} x_{i_1 \cdots i_s} = \mu_{\alpha_1 \cdots \alpha_s} \\ &\text{for } 0 \leq \alpha_j \leq m_j, \ j = 1, \cdots, s \\ &x_{i_1 \cdots i_s} \geq 0, \quad \text{all } i_1, \cdots, i_s. \end{aligned} \tag{5. 6}$$

The second problem is

$$\begin{aligned} &\text{Min (Max) } \sum_{i_1=0}^{n_1} \cdots \sum_{i_s=0}^{n_s} f_{i_1 \cdots i_s} x_{i_1 \cdots i_s} \\ &\text{subject to} \\ &\sum_{i_1=0}^{n_1} \cdots \sum_{i_s=0}^{n_s} z_{1i_1}^{\alpha_1} \cdots z_{si_s}^{\alpha_s} x_{i_1 \cdots i_s} = \mu_{\alpha_1 \cdots \alpha_s} \\ &\text{for } \alpha_j \geq 0, \ j = 1, \cdots, s; \ \alpha_1 + \cdots \alpha_s \leq m \\ &x_{i_1 \cdots i_s} \geq 0, \quad \text{all } i_1, \cdots, i_s. \end{aligned} \tag{5. 7}$$

In these problems the moments $\mu_{\alpha_1 \cdots \alpha_s}$ are supposed to be known and the decision variables are the unknown probabilities $x_{i_1 \cdots i_s}$. The two problems differ only in the assumptions regarding $\alpha_1, \ldots, \alpha_s$. The assumption in problem (5.7) is more natural than the one in (5.6). In fact, if, e.g., $s = 2$, then it is natural to assume that we know all moments up to the second order: $E(X_1), E(X_2), E(X_1^2), E(X_2^2), E(X_1 X_2)$, as required in problem (5.7), rather than the collection of moments that appears in problem (5.6) for $m_1 = 2, m_2 = 2$.

However, it is much easier to create dual feasible bases for problem (5.6), than for problem (5.7). The reason is that we can easily create bases for problem (5.6) by forming the tensor product of univariate bases and if the function f is the product of s suitably defined univariate functions, then we may derive dual feasible basis structure theorems, by the use of the univariate counterparts. Bases of problem (5.6) which are tensor products of univariate bases will be called rectangular.

Assume now that the function f is of the form

$$f(z_1, \ldots, z_s) = f_1(z_1) \cdots f_s(z_s). \tag{5. 8}$$

Under this assumption the following two theorems (see Prékopa (1998)) provide us with simple methods to construct lower and upper bounds for the value (5.3). Before formulating the statements we write up the univariate discrete moment problems corresponding to the components of the random vector $(X_1, \ldots, X_s)$:

$$\begin{aligned} &\text{Min (Max)} \quad \sum_{i=0}^{n_j} f_j(z_{ji}) x_i \\ &\text{subject to} \\ &\sum_{i=0}^{n_j} z_{ji}^{\alpha} x_i = \mu_\alpha^{(j)}, \alpha = 0, 1, \ldots, m_j \\ &x_i \geq 0, \, i = 0, 1, \ldots, n_j, \end{aligned} \tag{5. 9}$$

where $\mu_\alpha^{(j)} = E(X_j^\alpha)$, $\alpha = 0, 1, \ldots, m_j$, $j = 1, \ldots, s$.

Theorem 5.1. *Suppose that $f_j(z) \geq 0$ for all $z \in Z_j$, $j = 1, \ldots, s$. If for each $j\,(1 \leq j \leq s)$ we are given a B_j that is dual feasible relative to the maximization problem (5.9), then their tensor product $B = B_1 \otimes \cdots \otimes B_s$ is a dual feasible basis relative to the maximization problem (5.6).*

Moreover, if I_j is the set of subscripts of B_j and $L_{I_j}(z)$ is the corresponding Lagrange polinomial, then we have the inequality

$$E[f(X_1, \ldots, X_s)] \leq E[L_{I_1}(X_1) \cdots L_{I_s}(X_s)] \tag{5. 10}$$

Theorem 5.2. *If we replace the assumptions $f_j(z) \geq 0$, $z \in Z_j$ by the assumptions $L_{I_j}(z) \geq 0$, $z \in Z_j$, in Theorem 5.1, and write minimization instead of maximization, then the assertion of the theorem holds true with reversed inequality in (5.10).*

For the cases of problem (5.7) we have only a few types of dual feasible bases. We briefly mention what are these but disregard their detailed presentation. The interested reader may consult with the paper Prékopa (1998).

If we write up problem (5.7) in a more compact form, then each column of the matrix of the equality constraints can be represented by a subscript vector $(i_1, \ldots, i_s)$. Let I designate the subscript set of a basis.

Assumption. All divided differences of $f(z_1, \ldots, z_s)$, $(z_1, \ldots, z_s) \in Z_1 \times \cdots \times Z_s$ of total order $m+1$ are nonnegative and all moments $\mu_{\alpha_1 \ldots \alpha_s}$ of total order m are known.

Under this assumption we have the following dual feasible basis structure:

$$I = \{(i_1, \ldots, i_s) | i_j \geq 0, \, j = 1, \ldots, s; \, i_1 + \cdots + i_s \leq m\}$$

is a dual feasible basis subscript set for the minimization problem (5.7);

$$I = \{(i_1, \ldots, i_s) | 0 \leq i_j \leq n_j, \, j = 1, \ldots, s; \, n_1 i_1 + \cdots + n_s - i_s \leq m\}$$

is a dual feasible basis subscript set in the minimization (maximization) problem (5.7), if $m+1$ is even (odd).

If we write up the Lagrange polynomial corresponding to the points

$$(z_{1i_1}, \ldots, z_{si_s}), (i_1, \ldots, i_s) \in I,$$

replace X_j for $z_j, j = 1, \ldots, s$ and take expectation, then we obtain a lower (upper) bound for the value (5.3) if I corresponds to a minimization (maximization) problem.

The bounds, corresponding to these bases, are not sharp, however. The sharp bounds can be obtained if we carry out the dual method starting form these bases. An algorithm, similar to that presented in Section 3, can be designed for these cases too. More dual feasible bases are presented in Nagy and Prékopa (2000).

We can combine the two problems (5.6), (5.7) in such a way that we rely on the simpler problem (5.6) to find good dual feasible bases but we use moments of total order m. We illustrate the procedure in case of $s = 2$, for the sake of simplicity. We assume that all divided differences of $f(z_1, z_2)$, of total order at most $m+1$, are nonnegative.

First we pick some functions $f_0(z_1), f_1(z_1), \ldots, f_m(z_1)$, $g_0(z_2), g_1(z_2), \ldots, g_m(z_2)$, where we assume that $f_i(z_1)$ and $g_i(z_2)$ have nonnegative divided differences of order $i+1$, as single-variable functions. We may choose these functions, e.g., in such a way that we pick $m+1$ values for $z_2 : z_{20}, z_{21}, \ldots, z_{2m}$

and form $f_i(z_1) = f(z_1, z_{2i}), i = 0, 1, \ldots, m.$ and do the same with the first variable to obtain $g_i(z_2) = f(z_{1i}, z_2)$, $i = 0, 1, \ldots, m$.

Assume that all functions $f_i(z_1)g_{m-i}(z_2)$ and their lower bounding polynomials are nonnegative.

The next step is to write up the linear programming problem

$$\text{Min} \sum_{i=0}^{m} E[f_i(X_1)g_{m-i}(X_2)]y_i$$

subject to

$$\sum_{i=0}^{m} f_i(z_1)g_{m-i}(z_2)y_i \geq f(z_1, z_2),\ (z_1, z_2) \in Z_1 \times Z_2 \tag{5. 11}$$

$$\sum_{i=0}^{m} y_i = 1$$

$$y_i \geq 0,\ i = 0, 1, \ldots, m,$$

if we want to find upper bound for $E[f(X_1, X_2)]$ or

$$\text{Max} \sum_{i=0}^{m} E[f_i(X_1)g_{m-i}(X_2)]y_i$$

subject to

$$\sum_{i=0}^{m} f_i(z_1)g_{m-i}(z_2)y_i \leq f(z_1, z_2),\ (z_1, z_2) \in Z_1 \times Z_2 \tag{5. 12}$$

$$\sum_{i=0}^{m} y_i = 1$$

$$y_i \geq 0,\ i = 0, 1, \ldots, m,$$

if we want to find lower bound.

Each term $f_i(z_1)g_{m-i}(z_2)$ in problems (5.11),((5.12)) can be bounded from above (below) by the product of two polynomials with degrees i and $m-i$, respectively. If we replace the terms $f_i(z_1)g_{m-i}(z_2)$ by the upper bounding polynomials in the constraints and lower bounding polynomials in the objective function and then solve problem (5.11), then the optimum value is a lower bound for $E[f(X_1, X_2)]$. An upper bound for $E[f(X_1, X_2)]$ can be obtained from Problem (5.12) in a similar way.

The multivariate binomial moment problems are very similar to problems (5.6), (5.7), we only have to replace $z_{ki_k}^{\alpha_k}$ by $\binom{z_{ki_k}}{\alpha_k}$. In the simplest case we take $Z_i = \{0, 1, \ldots, n_i\}$. These problems are suitable to create bounds for

Boolean functions of finite sequences of events $A_{i1}, \ldots, A_{in_i}$, $i = 1, \ldots, s$.

We can improve on the univariate discrete moment bounds, that can be obtained for a single sequence of events $A_1, \ldots, A_n$, in such a way that we subdivide this sequence into s subsequences and create the multivariate bounds. If each subsequence consists of one single event., then we obtain the Boolean probability bounding scheme (see Prékopa, Vizvári, Regös (1997)).

6 Applications

Probability bounds based on discrete moment problems have been applied to a large number of practical problems. Among those which use LP formulation and higher order moments we mention the reliability bounds for transportation systems (Prékopa and Boros (1991)), communication systems (Prékopa, Boros and Lih (1991)), PERT problem (Prékopa and Long (1992)) and the use of probability bounds for probabilistic constrained stochastic programming problems (Prékopa (1999)). In this section we present some applications in economics and finance.

A utility function $u(z)$ usually indicates a function with the property that $u'(z) > 0$ and $u''(z) < 0$. Some authors (see, e.g., Konno et al. (1993)) argue, however, that in a standard situation the higher order derivatives of the function $u(z)$ also satisfy $u^{(k)} > 0$ for k odd and $u^{(k)} < 0$ for k even. Having this property and a discrete random variable X, we can create lower and upper bounds for $E[u(X)]$ for any number m for which the moments $\mu_k = E(X^k), k = 1, \ldots, m$ are known. For an odd m we use the discrete moment expectation bounds in their original forms while for an even m we create the bounds for $E[-u(X)]$ and then take the negatives of the bounds. If $m \leq 4$, then we can present the formulas in closed forms, otherwise we obtain them by executing the algorithm presented in Section 2.

An interesting, new field of application is the valuation of financial instruments, e.g., options.

Bounding option prices (values) has a past history. Without aiming completeness, we mention the papers by Ritchken (1985), Lo (1987), Grundy (1991), Zhang (1994), and Dulá (1995). The bounds, presented in these papers, are based on continuous moment problems, where mostly the knowledge of the first two moments is assumed. The paper by Dulá uses multivariate distribution and assumes the knowledge of the expectations and the covariance matrix. The paper by Grundy (1991) presents results where higher order moment is used but it is the nth moment alone the knowledge of which is assumed.

Bounding option prices is important because the assumption that the price of the asset follows geometric Brownian motion proves to be invalid in some cases and no other analytic form may be available to replace it. However, from past history data we may obtain moment information and, working with discrete random variables, we can obtain quite good bounds, even if the number of utilized moments is small.

We illustrate the moment bound construction on the European call option. Let us introduce the notations:

t = time now
T = future time, expiration of option
$S(t)$ = price of underlying asset now
$S(T)$ = random future price of asset
K = striking price
r = rate of interest, we assume continuous compounding
c = price of the option.

If we use "risk neutral valuation", then the price of the option is given by the equation:

$$c = e^{-r(T-t)} E([S(T) - K]_+ | S(t) = s). \tag{6. 1}$$

The Black-Scholes formula gives the value of c for the case where $S(\tau)$, $\tau \geq 0$ has the form

$$S(\tau) = S(0) e^{\sigma Z(\tau) + \mu\tau}, \tag{6. 2}$$

where $Z(\tau)$, $\tau \geq 0$ is the standard Brownion motion process, i.e., (a) $Z(0) = 0$, (b) the process has independent increments, (c) $Z(\tau)$ has the distribution $N(0, \tau)$, and $\sigma > 0$, μ are constants (see Black and Scholes (1973)). The process (6.2) is called multiplicative Brownion motion process.

Now we drop the assumption that $S(\tau)$, $\tau \geq 0$ is a multiplicative Brownion motion process. Let $t = 0$, for the sake of simplicity and assume that $S(T)$ has the form

$$S(T) = e^{\alpha Z + \beta}, \tag{6. 3}$$

where Z is a random variable with support $\{0, \Delta, 2\Delta, \ldots, n\Delta\}$ with $\Delta > 0$. We also assume that there exists an h such that

$$e^{\alpha h \Delta + \beta} - K = 0. \tag{6. 4}$$

Thus, the payoff is 0, if $Z \leq h\Delta$ and is $e^{\alpha k \Delta} - K$, if $Z = k \geq h$. We present two methods for bounding the option price c.

In the first method we assume that the first m conditional moments of Z, given that $e^{\alpha Z+\beta} > K$, are known. Let $\nu_1, \ldots, \nu_m$ designate them. Here too, $\nu_0 = 1$, by definition. Since we have the equation

$$\begin{aligned} c &= e^{-rT} E([S(T) - K]_+) \\ &= e^{-rT} E(e^{\alpha Z+\beta} - K | e^{\alpha Z+\beta} > K) P(e^{\alpha Z+\beta} > K), \end{aligned} \tag{6. 5}$$

we derive lower and upper bounds for the factors in the second line of (6.5), for c. The bounds for $E(e^{\alpha Z+\beta} - K | e^{\alpha Z+\beta} > K)$ are the optimum values of the LP's:

$$\begin{gathered} \text{Min (Max)} \sum_{i=h}^{n} (e^{\alpha i \Delta+\beta} - K) x_i \\ \text{subject to} \\ \sum_{i=h}^{n} (i\Delta)^k x_i = \nu_k,\ k = 0, 1, \ldots, m \\ x_i \geq 0,\ i = h, \ldots, n. \end{gathered} \tag{6. 6}$$

Note that the function

$$f(i) = e^{\alpha i \Delta+\beta} - K,\ i = h, \ldots, n$$

has positive divided differences of all orders that can be defined on the support set $\{h, \ldots, n\}$. Thus, there is no limitation, from this point of view, regarding the number of moments that we can use in the bounding LP.

The lower and upper bounds for

$$P(e^{\alpha i \Delta+\beta} > K) = P(\frac{1}{\Delta} Z \geq h + 1)$$

can be based on the moments

$$\mu_k = \frac{1}{\Delta^k} E(Z^k),\ k = 0, 1, \ldots, m,$$

using the discrete moment problems (2.1) with the function (4.17) and $r = h + 1$.

The bounds, obtained this way for the option price c, may not be tight enough because we get them as products of bounds obtained by the use of two separate optimization problems.

In the second method we assume that we know the first m moments of Z: $\mu_1, \ldots, \mu_m$. To obtain bounds for c we solve the LP's:

$$\text{Min (Max)} \sum_{i=0}^{n} [e^{\alpha i\Delta+\beta} - K]_+ x_i$$

$$\begin{aligned} &\text{subject to} \\ &\sum_{i=0}^{n} (i\Delta)^k x_i = \mu_k,\ k = 0, 1, \ldots, m \\ &x_i \geq 0,\ i = 0, 1, \ldots, n. \end{aligned} \tag{6. 7}$$

Our discrete moment problem methodology, however, requires that the coefficients in the objective function of problem (6.7) should have positive or at least nonnegative divided differences of order $m+1$. We will prove that, under some condition, this function has nonnegative divided differences of order $m+1$.

We keep the assumption, mentioned in the first method, that for some positive integer $h\,(0 < h < n)$ we have the equality $e^{\alpha h\Delta+\beta} = K$.

Theorem 6.1. *The function*

$$f(i\Delta) = [e^{\alpha i\Delta+\beta} - K]_+,\ i = 0, 1, \ldots, n \tag{6. 8}$$

has positive divided differences of order up to m, if

$$\alpha \geq \frac{\ln m}{\Delta}.$$

Proof. Since the support set $\{0, \Delta, 2\Delta, \ldots, n\Delta\}$ has equidistant points, it is enough to show the nonnegativity of the differences (rather than the divided differences) of order up to m. In addition, we may look at the function

$$g(i) = e^{-\beta} f(i\Delta),\ i = 0, 1, \ldots, n,$$

instead of the function (6.8), and prove the assertion for this, where the support set is $\{0, 1, \ldots, n\}$.

It is well-known that the kth order differences of any function g (defined on integers), corresponding to the points $x, x+1, ..., x+k$, is given by the formula

$$\Delta^k g(x) = \sum_{i=0}^{k} (-1)^k \binom{k}{i} g(x+k-i). \tag{6. 9}$$

Using (6.9) we can derive the following results for our function g. If $x \geq h$, then $g(i) = e^{\alpha i\Delta} - e^{\alpha h\Delta}$, $i = x, x+1, \ldots, x+k$, hence we have $\Delta^k g(x) > 0$.

If, on the other hand, $x + k \le h$, then trivially $\Delta^k g(x) = 0$; if $x < h$ and $x + k > h$, then

$$\Delta^k g(x) = \sum_{i=0}^{x+k-h} (-1)^i \binom{k}{i} g(x+k-i)$$

$$= \sum_{i=0}^{x+k-h} (-1)^i \binom{k}{i} \left(e^{\alpha(x+k-i)\Delta} - e^{\alpha h \Delta}\right) \qquad (6.\ 10)$$

$$= e^{\alpha h \Delta} \sum_{i=0}^{x+k-h} (-1)^i \binom{k}{i} \left(e^{\alpha(x+k-h-i)\Delta} - 1\right).$$

The last term in the above sum is zero but we keep it, if $x + k - h$ is odd, otherwise we drop it. It is well-known that

$$-\sum_{i=0}^{j} (-1)^i \binom{k}{i} > 0$$

for any odd, $j < k$. This implies that if $x + k - h$ is odd then

$$\Delta^k g(x) > e^{\alpha h \Delta} \sum_{i=0}^{x+k-h} (-1)^i \binom{k}{i} e^{\alpha(x+k-h-i)\Delta},$$

and if $x + k - h$ is even, then

$$\Delta^k g(x) > e^{\alpha h \Delta} \sum_{i=0}^{x+k-h-1} (-1)^i \binom{k}{i} e^{\alpha(x+k-h-i)\Delta}.$$

In both cases we combine each term, corresponding to an even i, with the next term. Any combined terms are nonnegative if

$$\frac{k-i}{i+1} \le e^{\alpha\Delta}. \qquad (6.\ 11)$$

Since $(k-i)/(i+1)$ is decreasing in i, its greatest value corresponds to $i = 0$, it follows that the difference (6.10) is positive, if

$$\alpha \ge \frac{\ln k}{\Delta}. \qquad (6.\ 12)$$

Since we have assumed (6.12) to hold for $k = m$, it follows that it holds for every $k \le m$.

References

1. Black, M. and M. Scholes (1973). The Pricing of Corporate Liabilities. Journal of Political Economy 81, 637-659.

2. Boros, E. and A. Prékopa (1989). Closed Form Two-Sided Bounds for Probabilities that Exactly r and at Least r out of n Events Occur. Mathematics of Operations Research 14, 317-342.
3. Chun, J. and T. Kailath (1991). Displacement Structure for Hankel, Vandermonde, and Related (Derived) Matrices. Linear Algebra and its Applications 151, 199-227.
4. Dawson, D. and A. Sankoff (1967). An Inequality for Probabilities. Proceedings of the American Mathematical Society 18, 504-507.
5. Dulá, J.H. (1995). A Multidimensional Semi-Parametric Upper Bound for Pricng a General Class of Complex Call Options. University of Mississippi, MS 38677.
6. Fekete, M. and G, Pólya (1912). Über ein Problem von Laguerre. Rendiconti del Circolo Matematico di Palermo 23, 89-120.
7. Galambos, J. and J. Simonelli (1996). Bonferroni-Type Inequalities with Applications, Springer, New York.
8. Gessel, I. and G. Viennot (1985). Binomial Determinants, Paths, and Hook Length Formulae. Advances in Mathematics 58, 300-321.
9. Grundy, B.D. (1991). Option Prices and the Underlying Asset's Return. The Journal of Finance 46, 1045-1069.
10. Karlin, S. and W.J. Studden (1966). Tchebycheff Systems: With Applications in Analysis and Statistics. Interscience, New York.
11. Karlin,S., C.A, Micchelli and Y. Rinott (1986). Multivariate Splines: A Probabilistic Perspective. Journal of Multivariate Analysis 20, 69-90.
12. Konno, H., H., Shirakawa and H. Yamazaki (1993). A Mean-Absolute Deviation-Skewness Portfolio Optimization Model. Annals of Operations Research 45, 205-220.
13. Krein, M. and A. Nudelman (1977). The Markov Moment Problem and Extremal Problems. Translations of Mathematical Monographs 50, American Mathematical Society, Providece, R.I.
14. Kwerel, S.M. (1975a). Most Stringent Bounds on Aggregated Probabilities of Partially Specified Dependent Probability Systems. Journal of the American Statistical Association 70, 472-479.
15. Kwerel, S.M. (1975b) Bounds on the Probability of a Union and Intersection of m Events. Advances of Applied Probability 7, 431-448.
16. Lemke, C.E. (1854) The Dual Method for Solving the Linear Programming Problem. Naval Research Logistic Quarterly 1, 36-47.
17. Lo, A. W. (1987). Semi-Parametric Upper Bounds for Option Prices and Expected Payoffs. Journal of Financial Economics 19, 373-387.
18. Nagy, G. and A. Prékopa (2000). On Multivariate Discrete Moment Problems and their Applications to Bounding Expectations and Probabilities. RUTCOR Research Report 41-2000.
19. Popoviciu, T. (1945). Les Functions Convex. Actualitiés Scientifique et Industrielles 992, Hermann, Paris.
20. Prékopa, A. (1988). Boole-Bonferroni Inequalities and Linear Programming. Operations Research 36, 145-162.
21. Prékopa, A. (1990a). Sharp Bounds on Probabilities Using Linear Programming. Operations Research 38, 227-239.
22. Prékopa, A. (1990b). The Discrete Moment Problem and Linear Programming. Discrete Applied Mathematics 27, 235-254.

23. Prékopa, A. (1992). Inequalities on Expectations Based on the Knowledge of Multivariate Moments. In: Stochastic Inequalities (M. Shaked and Y.Z. Tong eds.), Institute of Mathematical Statistics, Lecture Notes- Monograph Series 22, 309-331.
24. Prékopa, A. (1995). Stochastic Programming. Kluwer Academic Publishers, Dodrecht, Boston.
25. Prékopa, A. (1998). Bounds on Probabilities and Expectations Using Multivariate Moments of Discrete Distributions. Studia Scientiarum Mathematiearum Hungarica 34, 349-378.
26. Prékopa, A. (1999). The Use of Discrete Moment Bounds in Probabilistic Constrained Stochastic Programming Models. Annals of Operations Research 85, 21-38.
27. Prékopa, A. and E. Boros (1991). On the Existence of a Feasible Flow in a Stochastic Transportation Network. Operations Research 39, 119-129.
28. Prékopa, A., E. Boros and Keh-Wei Lih (1991). The Use of Binomial Moments for Bounding Network Reliability. DIMACS Series in Discrete Mathematics and Theoretical Computer Science 5, 197-212.
29. Prékopa, A. and J. Long (1992). New Bounds and Approximations for the Probability Distribution of the Length of the Critical Path. RUTCOR Research Report 16-92.
30. Prékopa, A., B. Vizvári and G. Regös (1997). Lower and Upper Bounds on Probabilities of the Boolean Functions of Events. RUTCOR Research Report 21-1997.
31. Ritchken, P. (1985). On Option Pricing Bounds. The Journal of Finance 40, 1219-1233.
32. Zhang, P. G. (1994). Bounds for Option Prices and Expected Payoffs. Review of Quantitative Finance and Accounting, 179-197.

Cuts and Semidefinite Relaxations for Nonconvex Quadratic Problems

Yasutoshi Yajima[1], Motakuri V. Ramana[2], and Panos M. Pardalos[2]

[1] Tokyo Institute of Technology, Meguro-ku, Tokyo 152-8552, Japan
[2] University of Florida, Gainesville, FL 32611-6595, USA

Abstract. In this paper, we propose a positive semidefinite relaxation for indefinite quadratic problems with box constraints. We show that a cutting plane procedure can be applied for generating tighter positive semidefinite relaxations. Computational results indicate that the relative gap obtained by the relaxations is less than 1% when the number of variables is up to 100. We also show a branch and bound procedure for obtaining an exact solution of the indefinite problems. Our branching scheme exploits the special structure of box constraints as well as the bounding procedure in which a pair of primal and dual solutions of the associated positive semidefinite relaxation is incorporated.

2000 Mathematics Subject Classification. Primary 90C26, 90C20; Secondary 90C22, 90C57, 49M20.

Keywords and phrases. Nonconvex quadratic programs, cut polytope, semidefinite programming, cutting plane method, branch and bound.

1 Introduction

Let us consider the following quadratic problem with box constraints:

$$(QP)\left|\begin{array}{l}\text{Maximize } \ q(y) = y^T Py + 2b^T y \\ \text{Subject to } -e \le y \le e,\end{array}\right. \tag{1}$$

where y is an n dimensional variable vector, P is an arbitrary $n \times n$ real symmetric matrix, b is an n dimensional vector and e is a vector of all ones.

If the objective function $q(\cdot)$ is concave, i.e, the matrix P is negative semidefinite, the problem can be solved in polynomial time using the ellipsoid method [5] or interior point methods e.g. [29,13]. When the objective function $q(\cdot)$ is convex, the problem (QP) can be reduced into the following quadratic $(-1, 1)$-problem:

$$(CQP)\left|\begin{array}{l}\text{Maximize } \ y^T Py + 2b^T y \\ \text{Subject to } y \in \{-1, +1\}^n,\end{array}\right. \tag{2}$$

which is known to be equivalent to the *max-cut problem*. Many algorithms have been proposed for solving the max-cut problem. Among those are eigenvalue methods [7,23], cutting plane methods which are associated with the *cut polytope* [2,20,4], branch and bound methods [21], and positive semidefinite (SDP) relaxations [12,18], see also [11,22] for extensions and the recent survey of semidefinite programming.

Several methods have been proposed for the problems with indefinite P. Among those are a branch and bound method by Hansen *et. al.* [16], and polynomial time algorithms [25,27,14] and an active set method [6] for obtaining approximate solutions. Other algorithms which include the indefinite problem as a special case can be found in the recent survey [1] and references therein.

Recently, Yajima and Fujie [26] have proposed a new cutting plane approach which generates a quite good approximate solution of the problem (QP) with a general indefinite quadratic objective function by using the *linearization approach*. Similar ideas of the linearization are well-known from the literature [15,20,24], and has been extensively studied [8] as relaxations of the quadratic $(-1,1)$-problem (CQP). In [26], Yajima and Fujie derive an equivalent problem with a linear objective function over a convex hull of a nonconvex region which is closely related to the cut polytope. They show that some classes of facets defining inequalities of the cut polytope can be valid for the convex hull.

Motivated by this result, we are going to show a generalized result which gives not only a tighter relaxation of the general indefinite problem (QP) but also an insight that any linearization approaches for the quadratic $(-1,1)$-problem (CQP) could contribute to the problem with continuous variables. We also show a branch and bound procedure for obtaining an exact solution of the indefinite problems. Our branching scheme exploits the special structure of box constraints as well as the bounding procedure in which a pair of primal and dual solutions of the associated SDP relaxation is incorporated.

The paper is organized as follows. Section 2 generalizes the results given in [26] and shows a SDP relaxation for the general indefinite problem (QP). Section 3 describes details of the branch and bound algorithm, which includes a powerful procedure for fixing variables. In Section 4, we show the results of our computational experiments.

2 Positive semidefinite relaxations

Let us consider the following nonconvex quadratic problem with box constraints:

$$(QP)\left|\begin{array}{l}\text{Maximize } \; q(y) = y^T P y + 2b^T y \\ \text{Subject to } -e \le y \le e,\end{array}\right. \tag{3}$$

where $y \in R^n$, $P \in R^{n\times n}$, $b \in R^n$, and e is a vector of all ones whose dimension is given by context.

Let $\mathcal{F}_n$ be the feasible region of (QP), i.e.,

$$\mathcal{F}_n = \{\, y \in R^n \mid -e \le y \le e \,\}.$$

It is well known that the problem can be homogenized by introducing a new variable y_{n+1} in the following way:

$$(QP)' \left| \begin{array}{l} \text{Maximize } x^T Q x \\ \text{Subject to } x \in \mathcal{F}_{n+1}, \end{array} \right. \tag{4}$$

where $x = \left(y^T \; y_{n+1}\right)^T \in R^{n+1}$ and $Q = \begin{bmatrix} P & b \\ b^T & 0 \end{bmatrix}$.

Let us introduce an equivalent formulation with matrix variables. Let S_{n+1} be a set of real symmetric matrices of order $n+1$, and let us define the convex hull of the subset of S_{n+1}:

$$FCut_{n+1} = Conv\{X \in S_{n+1} \mid X = xx^T, \; x \in \mathcal{F}_{n+1}\}.$$

Then, we have the following convex maximization problem:

$$(P_{FCut}) \left| \begin{array}{l} \text{Maximize } \langle Q, X \rangle \\ \text{Subject to } X \in FCut_{n+1}, \end{array} \right. \tag{5}$$

where $X \in S_{n+1}$, and $\langle A, B \rangle$ denotes the inner product of symmetric matrices A and B. The problem (P_{FCut}) is equivalent to the problem $(QP)'$ (or (QP)) in the following sense:

Lemma 1. *There exists an optimal solution X^* of (P_{FCut}) such that $X^* = x^* {x^*}^T$ where x^* is an optimal solution of $(QP)'$.*

Proof. Obvious. □

Let us introduce the set of positive semidefinite matrices whose diagonal elements are less than or equal to one, i.e.,

$$F\mathcal{E}_{n+1} = \{X \in S_{n+1} \mid d(X) \leq e, \; X \succeq 0\},$$

where $d(X)$ is a vector generated by the diagonal elements of X, and $X \succeq 0$ means X is positive semidefinite.

Obviously, we have the inclusion:

$$FCut_{n+1} \subseteq F\mathcal{E}_{n+1}.$$

Thus, the following semidefinite programming problem

$$(P_{SDP}) \left| \begin{array}{l} \text{Maximize } \langle Q, X \rangle \\ \text{Subject to } X \in F\mathcal{E}_{n+1} \end{array} \right. \tag{6}$$

gives an upper bound of the problem $(QP)'$. Recently, Ye [28] shows that this semidefinite relaxation generates a 4/7 approximate solution of (QP) by extending results by Goemans and Williamson [12] and Nesterov [19]. Also, Fujie and Kojima [9] propose a semidefinite relaxation for general nonconvex quadratic problems.

Similar approaches have also been applied for the case of the quadratic optimization over $(-1, 1)$-binary vectors. Let

$$\mathcal{B}_{n+1} = \{x \in R^{n+1} \mid x_i = -1 \text{ or } 1,\ i = 1, \ldots, n+1\}$$

be a set of $(-1, 1)$-binary vectors, and let us consider the following quadratic $(-1, 1)$-programming problem:

$$\left|\begin{array}{l}\text{Maximize } x^T Q x \\ \text{Subject to } x \in \mathcal{B}_{n+1}.\end{array}\right. \tag{7}$$

Let us define the following convex subset of S_{n+1}:

$$Cut_{n+1} = Conv\{X \in S_{n+1} \mid X = xx^T,\ x \in \mathcal{B}_{n+1}\},$$

which is called the *cut polytope*. Then, the problem (7) can be rewritten by the following matrix form:

$$\left|\begin{array}{l}\text{Maximize } \langle Q, X\rangle \\ \text{Subject to } X \in Cut_{n+1}.\end{array}\right. \tag{8}$$

The cut polytope has been extensively studied, and many classes of valid inequalities have been proposed. See [8] for a recent survey of theoretical results. One of the easiest inequalities are the *triangle inequalities*, which define the following *metric polytope*,

$$\begin{aligned} Met_{n+1} = \{X \in S_{n+1} \mid d(X) = e,& \\ X_{ij} + X_{ik} + X_{jk} &\geq -1,\quad 1 \leq i, j, k \leq n+1, \\ X_{ij} - X_{ik} - X_{jk} &\geq -1,\quad 1 \leq i, j, k \leq n+1\}, \end{aligned}$$

where X_{ij} denotes the i-j element of X. Moreover, it has been shown that the following class of inequalities are also valid for the cut polytope.

$$\sum_{i=1}^{n} \sum_{j=i+1}^{n+1} \delta_i \delta_j X_{ij} \geq \left(1 - \sum_{i=1}^{n+1} \delta_i^2\right) / 2, \tag{9}$$

where $\delta_i \in \{-1, 0, 1\},\ i = 1, 2, \ldots, n+1$ and $\sum_{i=1}^{n+1} \delta_i$ is odd. We note that the triangle inequalities are the special case of (9). Later, we will refer to the number of nonzero δ_i's, i.e., $\sum_{i=1}^{n+1} \delta_i^2$, as *the size* of inequalities (9).

We are going to demonstrate that the polytope Cut_{n+1} plays an important role in generating a tight relaxation of $FCut_{n+1}$. At first, let us introduce a simple notation. For any matrix $A \in S_n$, we define a matrix $A^{\#} \in S_n$ in the following way:

$$\{A^{\#}\}_{ij} = \begin{cases} 1 & \text{if } i = j, \\ \{A\}_{ij} & \text{otherwise} \end{cases}.$$

Now let us consider the set defined below:

$$SCut_{n+1} = \{X \in S_{n+1} \mid d(X) \leq e,\ X^{\#} \in Cut_{n+1},\ X \succeq 0\}. \tag{10}$$

We have the theorem.

Theorem 1.

$$FCut_{n+1} \subseteq SCut_{n+1}.$$

We note that this result is *not obvious* since $FCut_{n+1}$ *includes* Cut_{n+1}. The proof of the theorem will be given later.

The above theorem indicates that any valid linear inequality (including facets defining inequalities) for Cut_{n+1} is also valid for $FCut_n$. For example, since the triangle inequalities are valid for Cut_{n+1}, one can get the following inclusion:

$$\begin{aligned} FCut_{n+1} \subseteq \{X \in S_{n+1} \mid d(X) \leq e, \\ & X_{ij} + X_{ik} + X_{jk} \geq -1, \quad 1 \leq i, j, k \leq n+1, \\ & X_{ij} - X_{ik} - X_{jk} \geq -1, \quad 1 \leq i, j, k \leq n+1, \\ & X \succeq 0\}, \end{aligned}$$

which results in a tighter relaxation than $F\mathcal{E}_{n+1}$.

Recently, in [26], Yajima and Fujie have shown that certain classes of inequalities, which include the triangle inequalities, are still valid for $FCut_n$. They also give numerical results which show that the following problem

$$\left| \begin{array}{l} \text{Maximize } \langle Q, X \rangle \\ \text{Subject to } d(X) \leq e, \\ \qquad X_{ij} + X_{ik} + X_{jk} \geq -1, \quad 1 \leq i, j, k \leq n+1, \\ \qquad X_{ij} - X_{ik} - X_{jk} \geq -1, \quad 1 \leq i, j, k \leq n+1, \\ \qquad X \succeq 0 \end{array} \right. \tag{11}$$

gives a quite good approximate solution of (QP). Theorem 1 generalizes the result to arbitrary valid inequalities of Cut_{n+1}.

We first show the following lemma to prove the theorem.

Lemma 2. *For any $x \in \mathcal{F}_{n+1}$, the matrix $(xx^T)^\#$ can be expressed as the convex combination of matrices zz^T such that $z \in \mathcal{B}_{n+1}$. More precisely, for each $z \in \mathcal{B}_{n+1}$ let*

$$\lambda(z, x) = \prod_{z_i = 1} \left(\frac{1 + x_i}{2} \right) \prod_{z_i = -1} \left(\frac{1 - x_i}{2} \right), \tag{12}$$

be the coefficient associated to zz^T, then the matrix $(xx^T)^\#$ can be expressed as the following convex combination:

$$(xx^T)^\# = \sum_{z \in \mathcal{B}_{n+1}} \lambda(z, x) zz^T, \tag{13}$$

$$\sum_{z \in \mathcal{B}_{n+1}} \lambda(z, x) = 1 \textit{ and } \lambda(z, x) \geq 0, \tag{14}$$

where x_i and z_i denote the i-th elements of x and z, respectively. Moreover, we have

$$x = \sum_{z \in \mathcal{B}_{n+1}} \lambda(z,x) z. \tag{15}$$

Proof. We prove the results by induction. When $n = 0$, the results are obvious.

Now suppose that (13),(14) and (15) hold for n. We show that they also hold for $n+1$. From the induction hypothesis, for any $\hat{x} \in \mathcal{F}_n$, the matrix $(\hat{x}\hat{x}^T)^\#$ can be expressed as

$$(\hat{x}\hat{x}^T)^\# = \sum_{\hat{z} \in \mathcal{B}_n} \hat{\lambda}(\hat{z},\hat{x}) \hat{z}\hat{z}^T,$$

where

$$\sum_{\hat{z} \in \mathcal{B}_n} \hat{\lambda}(\hat{z},\hat{x}) = 1, \text{ and } \hat{\lambda}(\hat{z},\hat{x}) \geq 0.$$

Then, for each $z = \left(\hat{z}^T \; z_{n+1}\right)^T \in \mathcal{B}_{n+1}$, let us define

$$\lambda(z,x) = \begin{cases} \hat{\lambda}(\hat{z},\hat{x}) \cdot (1 + x_{n+1})/2 & \text{if } z_{n+1} = 1 \\ \hat{\lambda}(\hat{z},\hat{x}) \cdot (1 - x_{n+1})/2 & \text{if } z_{n+1} = -1, \end{cases}$$

which satisfies (12). It follows that

$$\begin{aligned} \sum_{z \in \mathcal{B}_{n+1}} \lambda(z,x) &= \sum_{\hat{z} \in \mathcal{B}_n} \hat{\lambda}(\hat{z},\hat{x}) \cdot \frac{1 + x_{n+1}}{2} + \sum_{\hat{z} \in \mathcal{B}_n} \hat{\lambda}(\hat{z},\hat{x}) \cdot \frac{1 - x_{n+1}}{2} \\ &= 1. \end{aligned}$$

Obviously, $\lambda(z,x) \geq 0$, $\forall z \in \mathcal{B}_{n+1}$. Then, (14) holds for $n+1$.

Similarly, for any $x = \left(\hat{x}^T \; x_{n+1}\right)^T \in \mathcal{F}_{n+1}$, we get

$$\begin{aligned} &\sum_{z \in \mathcal{B}_{n+1}} \lambda(z,x) z \\ &= \sum_{\hat{z} \in \mathcal{B}_n} \hat{\lambda}(\hat{z},\hat{x}) \cdot \frac{1 + x_{n+1}}{2} \begin{pmatrix} \hat{z} \\ 1 \end{pmatrix} + \sum_{\hat{z} \in \mathcal{B}_n} \hat{\lambda}(\hat{z},\hat{x}) \cdot \frac{1 - x_{n+1}}{2} \begin{pmatrix} \hat{z} \\ -1 \end{pmatrix} \\ &= \frac{1 + x_{n+1}}{2} \begin{pmatrix} \sum_{\hat{z} \in \mathcal{B}_n} \hat{\lambda}(\hat{z},\hat{x}) \hat{z} \\ 1 \end{pmatrix} + \frac{1 - x_{n+1}}{2} \begin{pmatrix} \sum_{\hat{z} \in \mathcal{B}_n} \hat{\lambda}(\hat{z},\hat{x}) \hat{z} \\ -1 \end{pmatrix} \\ &= \begin{pmatrix} \hat{x} \\ x_{n+1} \end{pmatrix}, \end{aligned}$$

and also

$$
\begin{aligned}
&\sum_{z\in\mathcal{B}_{n+1}} \lambda(z,x)zz^T \\
&= \sum_{\hat{z}\in\mathcal{B}_n} \hat{\lambda}(\hat{z},\hat{x})\cdot\frac{1+x_{n+1}}{2}\begin{pmatrix}\hat{z}\\1\end{pmatrix}\left(\hat{z}^T\ 1\right) \\
&\quad+\sum_{\hat{z}\in\mathcal{B}_n} \hat{\lambda}(\hat{z},\hat{x})\cdot\frac{1-x_{n+1}}{2}\begin{pmatrix}\hat{z}\\-1\end{pmatrix}\left(\hat{z}^T\ -1\right) \\
&= \frac{1+x_{n+1}}{2}\begin{bmatrix}\sum_{\hat{z}\in\mathcal{B}_n}\hat{\lambda}(\hat{z},\hat{x})\hat{z}\hat{z}^T & \sum_{\hat{z}\in\mathcal{B}_n}\hat{\lambda}(\hat{z},\hat{x})\hat{z}\\ \sum_{\hat{z}\in\mathcal{B}_n}\hat{\lambda}(\hat{z},\hat{x})\hat{z}^T & 1\end{bmatrix} \\
&\quad+\frac{1-x_{n+1}}{2}\begin{bmatrix}\sum_{\hat{z}\in\mathcal{B}_n}\hat{\lambda}(\hat{z},\hat{x})\hat{z}\hat{z}^T & -\sum_{\hat{z}\in\mathcal{B}_n}\hat{\lambda}(\hat{z},\hat{x})\hat{z}\\ -\sum_{\hat{z}\in\mathcal{B}_n}\hat{\lambda}(\hat{z},\hat{x})\hat{z}^T & 1\end{bmatrix} \\
&= \begin{bmatrix}\sum_{\hat{z}\in\mathcal{B}_n}\hat{\lambda}(\hat{z},\hat{x})\hat{z}\hat{z}^T & x_{n+1}\sum_{\hat{z}\in\mathcal{B}_n}\hat{\lambda}(\hat{z},\hat{x})\hat{z}\\ x_{n+1}\sum_{\hat{z}\in\mathcal{B}_n}\hat{\lambda}(\hat{z},\hat{x})\hat{z}^T & 1\end{bmatrix} = \begin{bmatrix}(\hat{x}\hat{x}^T)^{\#} & x_{n+1}\hat{x}\\ x_{n+1}\hat{x}^T & 1\end{bmatrix} \\
&= (xx^T)^{\#}.
\end{aligned}
$$

Thus, (13) and (15) also hold true for $n+1$. □

Now, we are ready to show the proof of Theorem 1 easily.

Proof (of Theorem 1). For any $x \in \mathcal{F}_{n+1}$, we have shown that $(xx^T)^{\#} \in Cut_{n+1}$. Also, it is obvious to see $xx^T \succeq 0$, and $d(xx^T) \leq e$, which implies that xx^T belongs to $SCut_{n+1}$. By virtue of the convexity of the positive semidefinite constraint, any convex combinations of points such that $xx^T \in FCut_{n+1}$ belong to $SCut_{n+1}$. □

At this moment, the equivalence between $FCut_{n+1}$ and $SCut_{n+1}$ remains open. The following problem, however, gives an optimal value of $(QP)'$ in the two extreme cases.

$$
\left|\begin{array}{l}\text{Maximize } \langle Q, X\rangle \\ \text{Subject to } X \in SCut_{n+1}.\end{array}\right. \tag{16}
$$

Lemma 3. *Let X^* be an optimal solution of (16). There exists an optimal solution x^* of $(QP)'$ such that ${x^*}^TQx^* = \langle Q, X^*\rangle$ if $d(P) \geq 0$ or $-P \succeq 0$.*

Proof. There exists an optimal solution such that $X^*_{ii} = 1$ if $Q_{ii} \geq 0$. Therefore, when $d(P) \geq 0$, we can assume all the diagonal elements of X^* is 1, which implies $X^{*\#} = X^*$. Then it is trivial to see that (16) is coincident with (8).

We now show the case when $-P \succeq 0$. Let

$$X^* = \begin{bmatrix} Y & y \\ y^T & 1 \end{bmatrix} \quad \text{and} \quad x = \begin{pmatrix} y \\ 1 \end{pmatrix},$$

where $Y \in S_n$, $y \in R^n$ and $x \in R^{n+1}$. We have

$$\begin{aligned} x^T Q x &= 2b^T y + y^T P y \\ &= 2b^T y + \langle P, Y \rangle - \langle P, (Y - yy^T) \rangle \\ &= \langle Q, X^* \rangle - \langle P, (Y - yy^T) \rangle \end{aligned}$$

Since $X^* \succeq 0$ and $d(X^*) \le e$, we have $Y - yy^T \succeq 0$ and $y \in \mathcal{F}_n$. This implies that $x \in \mathcal{F}_{n+1}$. Also, it follows from $-P \succeq 0$ that $-\langle P, (Y - yy^T) \rangle \geq 0$. Therefore,

$$x^T Q x \geq \langle Q, X^* \rangle.$$

Since (16) gives an upper bound of $(QP)'$, x is an optimal solution of $(QP)'$, which completes the proof. □

3 Branch and bound

The purpose of this section is to describe the branch and bound algorithm in which the SDP relaxation is involved to generate upper bounds. The branching techniques used in this section are also discussed in [16].

3.1 Branching tree

Here, let us recall the original problem:

$$(QP) \left| \begin{array}{l} \text{Maximize } \; q(y) = y^T P y + 2b^T y \\ \text{Subject to } \; y \in \mathcal{F}_n. \end{array} \right. \tag{17}$$

For simplicity, let us consider the branching on the nth variable y_n. Let

$$P = \begin{bmatrix} \hat{P} & \hat{p} \\ \hat{p}^T & P_{nn} \end{bmatrix}, \quad b = \begin{pmatrix} \hat{b} \\ b_n \end{pmatrix} \quad \text{and} \quad y = \begin{pmatrix} \hat{y} \\ y_n \end{pmatrix} \tag{18}$$

be the partition of P, b and y, where $\hat{P} \in S_{n-1}, \hat{p}, \hat{b}, \hat{y} \in R^{n-1}$ and $P_{nn}, b_n, y_n \in R^1$. The branching operations rely on the following facts [16]. If P_{nn} is non-negative, then there exists an optimal solution y^* which satisfies either $y_n^* = +1$ or $y_n^* = -1$. Otherwise, if P_{nn} is negative, y^* satisfies either $y_n^* = +1$, $y_n^* = -1$, or $\dfrac{\partial q(y)}{\partial y_n} = 0$ which is equivalent to the following equation:

$$y_n = -\frac{\hat{p}^T \hat{y} + b_n}{P_{nn}}. \tag{19}$$

Therefore, We can eliminate the variable y_n in any of these cases, and generate subproblems with exactly $n-1$ variables in the following way.

- When y_n is fixed to $\delta = \pm 1$:

$$\left| \begin{array}{l} \text{Maximize } \hat{y}^T \hat{P} \hat{y} + 2(\hat{b} + \delta \hat{p})^T \hat{y} + P_{nn} + 2\delta b_n \\ \text{Subject to } \hat{y} \in \mathcal{F}_{n-1}. \end{array} \right. \tag{20}$$

- When $y_n = -(\hat{p}^T \hat{y} + b_n)/P_{nn}$:

$$\left| \begin{array}{l} \text{Maximize } \hat{y}^T \left(\hat{P} - \dfrac{\hat{p}\hat{p}^T}{P_{nn}} \right) \hat{y} + 2 \left(\hat{b} - \dfrac{b_n}{P_{nn}} \hat{p} \right)^T \hat{y} - \dfrac{b_n^2}{P_{nn}} \\ \text{Subject to } \hat{y} \in \mathcal{F}_{n-1}. \end{array} \right. \tag{21}$$

Remark 1. We note that the inequality constraints,

$$-1 \leq y_n = -\frac{\hat{p}^T \hat{y} + b_n}{P_{nn}} \leq 1, \tag{22}$$

should be added to (21). However, it is obvious to see that (21), or its SDP relaxation, still gives an upper bound of the original problem if the subproblem contains an optimal solution of (QP). We ignore these inequalities to keep the subproblem simple. The inequalities (22) will be considered again later.

It is also worth noting that the quadratic matrix of the objective function of (21) can be obtained by Gauss-Jordan elementary operations. Using the last column of the matrix P of (18), one can reduce P to

$$\begin{bmatrix} \hat{P}' & \hat{p} \\ 0 & P_{nn} \end{bmatrix} \quad \text{where } \hat{P}' = \hat{P} - \frac{\hat{p}\hat{p}^T}{P_{nn}}.$$

□

Since these subproblems have exactly the same structure as the original one, i.e., a quadratic problem with box constraints, we can perform the same branching operations, and generate the *branching tree*. Let $N = \{i \mid i = 1, 2, \ldots, n\}$. For each node of the branching tree, let J^+, J^- and J° denote the sets of indices such that

$$\begin{aligned} J^+ &= \{\, i \in N \mid y_i \text{ is set to } +1 \,\}, \\ J^- &= \{\, i \in N \mid y_i \text{ is set to } -1 \,\}, \\ J^\circ &= \{\, i \in N \mid y_i \text{ is eliminated by (19) } \}. \end{aligned}$$

Also, let $I = N \setminus (J^+ \cup J^- \cup J^\circ)$. Each node of the branching tree is characterized by the disjoint subsets J^+, J^- and J°.

Now, suppose that the node characterized by (J^+, J^-, J°) is generated. For simplicity, let $J^\pm = J^+ \cup J^-$ and assume that

$$I = \{1, 2, \ldots, n'\}, \; J^\circ = \{n'+1, n'+2, \ldots, n''\}$$

and

$$J^{\pm} = \{n'' + 1, n'' + 2, \ldots, n\}.$$

Let

$$P = \begin{bmatrix} P(I) & P(I, J^{\circ}) & P(I, J^{\pm}) \\ P(J^{\circ}, I) & P(J^{\circ}) & P(J^{\circ}, J^{\pm}) \\ P(J^{\pm}, I) & P(J^{\pm}, J^{\circ}) & P(J^{\pm}) \end{bmatrix}, \quad b = \begin{pmatrix} b(I) \\ b(J^{\circ}) \\ b(J^{\pm}) \end{pmatrix}, \quad y = \begin{pmatrix} y(I) \\ y(J^{\circ}) \\ y(J^{\pm}) \end{pmatrix}$$

be the partition of P, b and y corresponding to (J^+, J^-, J°). Also, let us assume that the quadratic objective function of the subproblem associated with the node (J^+, J^-, J°) is given by

$$y(I)^T \bar{P} y(I) + 2\bar{b}^T y(I),$$

where $\bar{P} \in S_{n'}$ and $\bar{b} \in R^{n'}$.

Here, we need to investigate the matrix $\bar{P}$ carefully. The matrix $\bar{P}$ can be generated in the following way. First, deleting rows and columns corresponding to the indices which are set to ± 1, one can obtain the matrix

$$\begin{bmatrix} P(I) & P(I, J^{\circ}) \\ P(J^{\circ}, I) & P(J^{\circ}) \end{bmatrix}.$$

Then, a series of Gauss-Jordan elementary operations result in

$$\left[\begin{array}{c|cccc} \bar{P} & & & * & \\ \hline & \bar{p}_{n'+1} & & & \\ & & \bar{p}_{n'+2} & & * \\ 0 & & & \ddots & \\ & & 0 & & \bar{p}_{n''} \end{array}\right].$$

By the construction of J°, the matrix $P(J^{\circ})$ has been reduced to an upper triangle matrix whose diagonal elements are strictly negative. Thus, the following lemma holds.

Lemma 4. *At each node of the branching tree, the matrix $P(J^{\circ})$ is negative definite.*

Therefore, $P(J^{\circ})$ is nonsingular, and one can also obtain the objective function of each subproblem by substituting

$$y_j = 1, \text{ for all } j \in J^+, \quad \text{and} \quad y_j = -1, \text{ for all } j \in J^- \tag{23}$$

and

$$y(J^{\circ}) = -P(J^{\circ})^{-1} \left\{ b(J^{\circ}) + P(J^{\circ}, I) y(I) + P(J^{\circ}, J^{\pm}) y(J^{\pm}) \right\}. \tag{24}$$

We note that the system (24) corresponds to

$$\frac{\partial q(y)}{\partial y_j} = 0, \;\; j \in J^\circ.$$

Now, let us consider the leaf nodes of the branching tree where the variable y is uniquely determined. It is trivial to see that there are 2^n leaf nodes (J^+, J^-, J°) such that $J^+ \cup J^- = N$ and $J^\circ = \emptyset$. Moreover, let $P(K)$ $K \subseteq N$ be an arbitrary negative definite principal sub-matrix of P. For each $P(K)$, there exist $2^{|N \setminus K|}$ leaf nodes (J^+, J^-, J°) such that $J^\circ = K$, i.e., the variable y is determined uniquely as follows:

$$y_j = 1, j \in J^+, \;\; y_j = -1, j \in J^-,$$

and

$$y(J^\circ) = -P(J^\circ)^{-1}\left\{ b(J^\circ) + P(J^\circ, J^\pm) y(J^\pm) \right\},$$

where (J^+, J^-) is an arbitrary partition of $N \setminus K$. In particular, the branching tree has 3^n leaf nodes if and only if the matrix P is negative definite, and has 2^n leaf nodes if and only if P is positive definite.

So far, ignoring the inequalities (22), we have built the branching tree. However, the following lemma ensures that the branching operations does not exclude an optimal solution of (QP).

Lemma 5. *Let y^* be an optimal solution of (QP) such that the number of binding constraints is maximum. Then, y^* exists among the leaf nodes of the branching tree.*

Proof. Without loss of generality, we can assume that

$$-1 < y_j^* < 1, j \in K = \{1, 2, \ldots, n'\} \tag{25}$$

and

$$y_j^* = \pm 1, j \in \bar{K} = \{n' + 1, n' + 2, \ldots, n\} \tag{26}$$

is the solution which satisfies the assumption of the lemma. Let P, b and y be partitioned

$$P = \begin{bmatrix} P(K) & P(K, \overline{K}) \\ P^T(K, \overline{K}) & P(\overline{K}) \end{bmatrix}, \;\; b = \begin{pmatrix} b(K) \\ b(\overline{K}) \end{pmatrix}, \;\; y = \begin{pmatrix} y(K) \\ y(\overline{K}) \end{pmatrix}.$$

Let

$$\left| \begin{array}{l} \text{Maximize } \; y(K)^T P(K) y(K) + 2\{b(K) + P(K, \overline{K}) y^*(\overline{K})\}^T y(K) + const. \\ \text{Subject to } -1 \le y_j \le 1, \;\; j \in K, \end{array} \right. \tag{27}$$

be the problem obtained by substituting (26) into (QP). Since the problem has the optimal solution (25) in its interior, the matrix $P(K)$ should be negative definite. Otherwise, the problem has an optimal solution in its boundary, which contradicts the assumption. Thus, there exists a node (J^+, J^-, J°) such that $J^\circ = K$. This completes the proof. □

3.2 Bounding

At each node (J^+, J^-, J°) of the branching tree, we obtain the following subproblem by (23) and (24):

$$\left|\begin{array}{l} \text{Maximize } y^T P y + 2b^T y \\ \text{Subject to } y_j = 1,\ j \in J^+, \\ \qquad y_j = -1,\ j \in J^-, \\ \qquad y(J^\circ) = -P(J^\circ)^{-1}\left\{ b(J^\circ) + P(J^\circ, I)y(I) + P(J^\circ, J^\pm)y(J^\pm) \right\}, \\ \qquad -1 \le y_j \le 1,\ j \in J^\circ, \\ \qquad -1 \le y_j \le 1,\ j \in I. \end{array}\right. \tag{28}$$

Eliminating variables $y_j,\ j \in \{J^\pm \cup J^\circ\}$, we can rewrite (28) compactly as follows:

$$\left|\begin{array}{l} \text{Maximize } y(I)^T \bar{P} y(I) + 2\bar{b}^T y(I) \\ \text{Subject to } -1 \le y_j \le 1,\ j \in I, \\ \qquad -1 \le a_j^T y(I) + c_j \le 1,\ j = 1, 2, \ldots, |J^\circ|, \end{array}\right. \tag{29}$$

where

$$\begin{bmatrix} a_1^T \\ a_2^T \\ \vdots \\ a_{|J^\circ|}^T \end{bmatrix} = -P(J^\circ)^{-1} P(J^\circ, I), \begin{bmatrix} c_1 \\ c_2 \\ \vdots \\ c_{|J^\circ|} \end{bmatrix} = -P(J^\circ)^{-1}\left\{ b(J^\circ) + P(J^\circ, J^\pm) \right\}.$$

For simplicity, let us redefine y, b and P, and let us rewrite the problem (29) as follows:

$$\left|\begin{array}{l} \text{Maximize } y^T P y + 2b^T y \\ \text{Subject to } -1 \le y_j \le 1,\ j = 1, 2, \ldots, k, \\ \qquad -1 \le a_j^T y + c_j \le 1,\ j = 1, 2, \ldots, l, \end{array}\right. \tag{30}$$

where $k = |I|$, $l = |J^\circ|$, $y, b \in R^k$, and $P \in S_k$.

It is obvious that the node need not to be explored if the feasible region:

$$\left\{ y \in R^k \;\middle|\; \begin{array}{l} -1 \le y_j \le 1,\ j = 1, 2, \ldots, k, \\ -1 \le a_j^T y + c_j \le 1,\ j = 1, 2, \ldots, l, \end{array} \right\} \tag{31}$$

is empty, i.e., the node is infeasible, which can be checked by solving a linear programming problem.

Otherwise, we can define the following relaxation which gives an upper bound.

$$\left|\begin{array}{l} \text{Maximize } \langle Q_0, X \rangle \\ \text{Subject to } d(X) \le e,\ X^\# \in Cut_{k+1},\ X \succeq 0, \\ \qquad \langle Q_j, X \rangle \le 1,\ j = 1, 2, \ldots, l, \\ \qquad X_{k+1,k+1} = 1, \end{array}\right. \tag{32}$$

where $X \in S_{k+1}$,

$$Q_0 = \begin{bmatrix} P & b \\ b^T & 0 \end{bmatrix}, \quad \text{and} \quad Q_j = \begin{bmatrix} a_j a_j^T & c_j a_j \\ c_j a_j^T & c_j^2 \end{bmatrix}, \; j = 1, 2, \ldots, l. \tag{33}$$

In our computational experiments described in Section 4, we will use the inequalities (9) as cutting planes. Recall that the inequalities (9) are given by

$$\sum_{i=1}^{n} \sum_{j=i+1}^{n+1} \delta_i \delta_j X_{ij} \geq \left(1 - \sum_{i=1}^{n+1} \delta_i^2\right) /2,$$

where $\delta_i \in \{-1, 0, 1\}$, $i = 1, 2, \ldots, n+1$ and $\sum_{i=1}^{n+1} \delta_i$ is odd, and that the number of nonzero δ_i's is called the size of inequalities (9). Then, we have the following cutting plane procedure. First, we solve:

$$\left| \begin{array}{l} \text{Maximize } \langle Q_0, X \rangle \\ \text{Subject to } d(X) \leq e, \; X \succeq 0, \\ \qquad\qquad \langle Q_j, X \rangle \leq 1, \; j = 1, 2, \ldots, l, \\ \qquad\qquad X_{k+1,k+1} = 1. \end{array} \right. \tag{34}$$

Let X^* be an optimal solution. In general, X^* does not satisfy (9). We first enumerate all triangle inequalities, and detect those which violate X^*. For each violating inequality, we try to extend the size of it. In particular, enumerating every pair (i, j) of distinct indices such that δ_i and δ_j are 0, and setting them to ± 1, we can generate a pair $(\delta_i, \delta_j) \in \{-1, +1\}^2$ which maximizes the amount of violation. We iteratively increase the size by two and detect larger size inequalities until no violating inequalities are generated. Then, we sort those in violation with respect to the distance between X^* and a hyperplane which defines those inequalities, and add a number of maximal distance (i.e., deepest) inequalities to (34), which could generate a tighter SDP relaxation.

3.3 Fixing variables

For simplicity, let us write the SDP relaxation generated by adding a number of inequalities in the cutting plane procedure as follows:

$$\left| \begin{array}{l} \text{Maximize } \langle Q_0, X \rangle \\ \text{Subject to } \langle A_0, X \rangle = r_0 = 1, \\ \qquad\qquad \langle A_i, X \rangle \leq r_i, \; i = 1, \ldots, m, \\ \qquad\qquad X \succeq 0. \end{array} \right. \tag{35}$$

We note that the equality constraint $\langle A_0, X \rangle = 1$ corresponds to $X_{k+1,k+1} = 1$ of (34), and the inequality constraints of (35) include those of (34) as well as those which have been added during the cutting plane procedure. We can define the dual of (35) as follows:

$$\left|\begin{array}{ll} \text{Minimize} & \sum_{i=0}^{m} r_i t_i \\ \text{Subject to} & \sum_{i=0}^{m} A_i t_i - Q_0 \succeq 0, \\ & t_i \geq 0,\ i = 1, \ldots, m. \end{array}\right. \tag{36}$$

Let us consider the Lagrangian function

$$\langle Q_0, X\rangle - \sum_{i=0}^{m} t_i(\langle A_i, X\rangle - r_i).$$

It is obvious to see that if X satisfies the constraints of (35) and $t_i \geq 0$ for all $i = 1, 2, \ldots, m$, we have

$$\langle Q_0, X\rangle \leq \langle Q_0, X\rangle - \sum_{i=1}^{m} t_i(\langle A_i, X\rangle - r_i),$$

which also holds true when $X = xx^T$, where $x^T = (y^T, 1)$ and y satisfies (31).

Thus, let us define a quadratic function,

$$L(y,t) = (y^T, 1)\left\{Q_0 - \sum_{i=1}^{m} t_i A_i\right\}\begin{pmatrix} y \\ 1 \end{pmatrix} + \sum_{i=1}^{m} t_i r_i,$$

which can be an overestimator of the objective function, i.e.,

$$y^T P y + 2b^T y \leq L(y,t)$$

for all y satisfying (31). Moreover, let t^* be an optimal solution of the dual problem (36), we can expect that $L(y, t^*)$ would be a fairly tight concave overestimator of the objective function over (31).

Let $\underline{q}$ be the lower bound obtained so far. Also, let i be an index such that the i-th diagonal elements of matrix P of (30) is positive or zero, which implies that the current node should be branched either $y_i = +1$ or $y_i = -1$. Then, let us consider the following problems:

$$\left|\begin{array}{l} \text{Maximize } L(y,t^*) \\ \text{Subject to } y \text{ satisfies (31)}, \quad y_i = 1. \end{array}\right. \tag{37}$$

and

$$\left|\begin{array}{l} \text{Maximize } L(y,t^*) \\ \text{Subject to } y \text{ satisfies (31)}, \quad y_i = -1. \end{array}\right. \tag{38}$$

Let L^*_{+i} and L^*_{-1} be optimal values of (37) and (38), respectively, which can be generated by standard convex optimization techniques since these are concave maximization problems, and which also give upper bounds of the original problem. Thus, if $L^*_{+i} < \underline{q}$ ($L^*_{-i} < \underline{q}$), the variable y_i should be fixed to -1 (+1). Obviously, theses problems (37) and (38) can be optimized far easier than the SDP relaxation generated in the cutting plane procedure.

3.4 Algorithm

Now, we summarize the branch and bound algorithm. We explore the branching tree by the best-first search strategy. We perform the following steps at each node (J^+, J^-, J°) of the branching tree. Let $\overline{q}$ be the upper bound obtained so far.

Step 1. Define the subproblem (30) and check the feasibility of (31). If (31) is empty, the current node is fathomed.

Step 2. (Cutting Plane Procedure)

- **2-1.** Solve the SDP relaxation (34). Let X^* be an optimal solution.
- **2-2.** Generate inequalities (9) which cut off X^*, and sort those with respect to the distance between X^* and hyperplanes which define the associated inequalities.
  Add M maximal distance inequalities to (34) and define the SDP relaxation (35).
- **2-3.** Solve (35). If the optimal value is less than $\underline{q}$, the current node is fathomed.

Remark 2. We note that, we can iterate **Step 2** until no violating inequality is detected. However, it would require a huge amount of computations to solve the associated SDP relaxation. In our implementation, we set $M = |I|$, and iterate **Step 2** six times.

Step 3. (Fixing Variables)

- **3-1.** For each $i \in I = N \setminus (J^+ \cup J^- \cup J^\circ)$, solve (37) and (38).
- **3-2.** If $L^*_{+i} < \underline{q}$, then $J^- = J^- \cup \{i\}$.
  If $L^*_{-i} < \underline{q}$, then $J^+ = J^+ \cup \{i\}$.

We iterate these steps until no variables are fixed at **Step 3**, or the node is fathomed.

If the node is not fathomed, then we choose a branching variable by the following rule. Let $\langle Q_0, X\rangle$ be the objective function of the the finally obtained SDP relaxation. We note that Q_0 is a $(k+1) \times (k+1)$ symmetric matrix defined by (33), where $k = |I|$. For each $i \in I$, we compute $\sum_{j=1}^{k+1} |\{Q_0\}_{ij}|$, and select a variable which attains a maximum, and branch the node on that variable.

We update the lower bound when all variables are fixed at a feasible point. Furthermore, we use a local search procedure to update the bound whenever we obtain an optimal solution X^* of the SDP relaxation. Let $(y^{*T}, 1)$ be the last row of X^*. We note that $-e \le y^* \le e$. Thus, we solve (29) by the steepest descent search starting with the initial solution y^* until a local optimum is found. If a better solution is generated, we update the lower bound.

We terminate the branch and bound when

$$\overline{q} \le \underline{q} + \epsilon|\underline{q}|$$

is satisfied, where ϵ is a given tolerance.

4 Computational experiments

We will show the results of our computational experiments.

We implemented the algorithms in C language, and tested on a Digital AlphaServer 8400 (437MHz CPU, 1GB memory) workstation with no parallel options. We used SDPA 4.10 [10] as the subroutine for the SDPs.

We generated the following problems with $(0, 1)$ box constraints:

$$\left| \begin{array}{l} \text{Minimize} \quad y^T P y + b^T y \\ \text{Subject to } 0 \le y \le e, \end{array} \right. \tag{39}$$

where $y \in R^n$, $P \in S_n$, and $b \in R^n$. Two different types of test problems were generated.

The coefficients of the first type were generated as

$$P = \lambda A_1 A_1^T - (1 - \lambda) A_2 A_2^T, \quad \text{and} \quad b = 0,$$

where the elements of $A_1, A_2 \in S_n$ were assigned randomly between -50 and 50. We note that λ is a parameter which controls the convexity of the objective function. We rewrite (39) into the $(-1, +1)$ box constraints form of (QP).

We first show the performance of the cutting plane procedure with inequalities (9). To this end, we iterated **Step 2.** of the algorithm given in Section 3.4 until the relative gap given by

$$\frac{\overline{q} - \underline{q}}{|\underline{q}|} \times 100 \tag{40}$$

reaches less than 1%. We solved ten problems for each size and λ.

In Table 1, we show the average of CPU time (in seconds) and the number of inequalities involved in the relaxation when the procedure terminates. The rows labeled "size3" and "size5$^+$" correspond to the number of inequalities whose size, i.e., $\sum_{i=1}^{n+1} \delta_i^2$ of (9), are three and five or more, respectively. We also show the average of the relative gap obtained by SDP relaxations without cuts. The table demonstrates that regardless of the convexity, i.e., λ, we can obtain fairly strong bounds when the size of the problems are up

Table 1. Average Results for Different Types of Objective Function

		Number of Variables (n)			
		60	80	100	120
$\lambda = 0.1$	CPU	30.3	77.7	260.9	832.4
	# of cuts				
	size3	114.7	177.1	274.0	386.5
	size5$^+$	7.5	9.2	30.2	77.9
	SDP gap	7.3	7.9	7.2	6.8
$\lambda = 0.5$	CPU	47.2	179.0	3922.7	4794.2
	# of cuts				
	size3	168.3	265.7	428.7	486.0
	size5$^+$	11.5	48.5	221.1	223.9
	SDP gap	7.84	8.30	10.56	8.97
$\lambda = 0.9$	CPU	27.0	188.8	389.5	722.7
	# of cuts				
	size3	126.5	221.8	330.2	433.0
	size5$^+$	2.5	36.5	59.9	70.5
	SDP gap	4.68	5.84	6.75	6.44

Table 2. Computational Results When $\lambda = 0.5$

		Number of Variables (n)			
		60	80	100	120
nodes	ave.	1.2	1.8	4.3	4.1
	s.d.	0.6	1.2	4.2	3.2
CPU	ave.	55.7	217.4	952.6	1519.8
	s.d.	33.7	144.1	907.9	1175.7

to $n = 120$. As we have expected, the SDP based relaxation with cutting planes generates tight bounds if the objective functions are almost convex ($\lambda = 0.9$). More interestingly, it also works well for the problems with almost concave ($\lambda = 0.1$) objectives. We note that the inequalities included in these relaxations mostly consist of the simple triangle inequalities. On the other hand, we have found that the gap of the SDP bounds without cuts reaches almost 10% when the convex and concave terms are balanced ($\lambda = 0.5$), and that a larger number of size five, or more, inequalities (9) are required to improve bounds. We would like to emphasize that, even for those difficult problems, a larger number of size five, or more, inequalities contribute to yield quite tight bounds when n is up to 120.

We also implemented the branch and bound procedure as described in Section 3 with $\epsilon = 0.01\%$, and solved the hardest problems with $\lambda = 0.5$. We solved the same problems as in Table 1. We give the average (ave.) and standard derivation (s.d.) of the number of branching nodes and CPU time (in seconds) in Table 2. When the size n of the problems is up to 80, most problems have been solved optimally at the root node of the branching tree, while larger size problems require branchings.

Table 3. Computational Results When $n = 60$

		Density of P				
		0.1	0.3	0.5	0.7	0.9
nodes	ave.	2.2	1.8	3.4	2.4	2.8
	s.d.	1.0	1.0	0.8	1.2	1.5
CPU	ave.	188.5	197.4	181.3	136.6	555.0
	s.d.	137.1	144.2	32.8	67.5	416.0

Table 4. Computational Results When $n = 80$

		Density of P				
		0.1	0.3	0.5	0.7	0.9
nodes	ave.	2.8	3.6	3.8	6.4	8.2
	s.d.	1.0	2.9	1.6	4.0	4.7
CPU	ave.	437.5	1631.5	1570.3	2559.7	3362.6
	s.d.	200.7	1586.9	585.9	1632.4	1918.5

Table 5. Computational Results When $d = 0.1$

		Number of Variables (n)				
		40	60	80	100	120
nodes	ave.	1.4	2.2	2.8	3.6	3.8
	s.d.	0.8	1.0	1.0	0.8	1.6
CPU	ave.	16.5	188.5	437.5	1688.7	4778.5
	s.d.	5.1	137.1	200.7	1467.9	2577.2

In Table 3 $\sim$ Table 8, we give the results of the second type problems whose coefficients were generated in the same way of Hansen *et al.* [16], i.e., the elements of P and b were assigned randomly between -50 and 50 with various density (d). We note that these matrices have almost the same number of positive and negative eigenvalues, which would be considered to generate hard problems.

We solved five problems for each size and density by the branch and bound procedure with $\epsilon = 0.01\%$, and show the average (ave.) and standard derivation (s.d.) of branching nodes and CPU time (in seconds).

Table 6. Results by Hansen *et al.*

Density (d)	Number of variables (n) 30	40	50	60	70	80
0.1	0.13	0.36	1.14	92.32	81.13	488.06
0.2	0.70	8.90	132.30			
0.3	3.39	143.08				
0.4	9.12					
0.5	50.85					
0.6	116.26					
0.7	160.94					

Table 3 and Table 4 show the results for $n = 60$ and $n = 80$ when the density varies, while Table 5 shows those when P is sparse. Also, Table 6 shows the results of the branch and bound algorithm by Hansen *et al.* presented in [16]. They also show the average CPU time (in seconds) of ten problems in each case.

Since the computation environment used in [16] is different, we can not compare them against ours directly. However, we can see from Table 6 that the computational effort increases dramatically as the density and the size of the matrix P increase. It is obvious to see that moderate density problems with 60 or 80 variables would be out of reach by their algorithm.

On the other hand, Table 3 and Table 4 show that the number of nodes generated by our branch and bound grows quite slow and remains moderate even when the matrix is dense. Our branch and bound is less sensitive to the increase of the density of P. Moreover, when P is sparse, we can see from Table 5 that the average number of branching nodes increases slowly as the size of n gets larger, and that the standard derivations are fairly small. This is due to the fact that the quality of the bounds generated by the semidefinite relaxation with the cutting planes is very tight and less sensitive to the size of n.

Table 7. Quality of Bounds When $d = 0.1$

n	type	gap (%)	# of size3	# of size 5^+	CPU
150		1.09	808	1526	>48H
	TRI	2.52	2520		55457.3
	SDP	15.7			62.0
200		5.08	1054	2037	>48H
	TRI	6.64	2885		83785.5
	SDP	17.7			153.4
250		6.93	1536	1923	>48H
	TRI	8.03	3847		>48H
	SDP	19.3			389.3

Table 8. Quality of Bounds When $d = 0.9$

n	type	gap (%)	# of size3	# of size 5^+	CPU
100		2.08	573	1374	>48H
	TRI	4.34	1307		5014.3
	SDP	17.2			16.7
150		8.31	809	1716	>48H
	TRI	9.98	1734		13128.9
	SDP	21.4			62.0
200		8.86	1035	1894	>48H
	TRI	10.2	2348		39293.0
	SDP	21.4			163.2
250		9.27	1364	1867	>48H
	TRI	10.1	3065		78380.3
	SDP	19.0			377.1

Finally, we give some results of the second type problems with over 100 variables, which can not be solved optimally in a practical amount of CPU time by the branch and bound procedure, and consider the limit of our approach. We show the quality of the bounds obtained by the cutting plane procedure with inequalities (9). We iterated **Step 2.** of the algorithm until the CPU time exceeded 48 hours or no cutting planes were generated. We solved one instance for each size and density.

In Table 7 and 8, we show the relative gap (40) and the number of inequalities when the procedure terminates. The column labeled "# of size3" ("# of size5^+") corresponds to the number of added inequalities whose size is three (five or more). We also solved the same problems by employing only the inequalities whose size is three, i.e., the class of triangle inequalities. These results are shown in the row labeled "TRI". To compare the performance of the cutting plane procedure we show the results obtained by the SDP relaxation without cuts, which are labeled "SDP". CPU times (in seconds) are shown if the procedure terminates with no violating inequalities.

It is obvious to see that the cutting plane procedure significantly improves SDP bounds especially for the problems where n is relatively small. When n is over 200, however, the gap obtained by the triangle inequalities have reached more than 10%. It seems that triangle inequalities are not strong enough to describe tight relaxations for larger size problems. In fact, within the same amount of CPU time limits, we can always obtain tighter bounds by employing inequalities whose size are five or more. Also, we should note that after iterating the procedure for 48 hours, we still detected a huge number of violating inequalities which could result in further improvements. However, the primal–dual interior point method used in SDPA would require considerable amount of CPU time and thus there is no hope to get further improvements by incorporating these inequalities within a practical amount of CPU time limit. Several different algorithms such as a dual potential-reduction method [3], or a spectral bundle method [17] could contribute to overcome the limitation of the size of the problems.

5 Conclusion

In this paper, we have described new SDP relaxations for indefinite quadratic problems with box constraints. We have shown that the simple SDP relaxation (6) can be tightened by incorporating inequalities derived from valid inequalities for the cut polytope.

Our computational results show that the branch and bound approach practically solves the problem with up to $n = 120$ when P is sparse and $n = 100$ when P is dense.

We found that semidefinite relaxations with the simple triangle inequalities generate fairly tight bounds when the objective functions are almost concave or convex, or when the matrices of the objective functions are sparse. Also, even if the convex and concave parts of the objective functions are balanced, one can obtain tighter bounds by including more complicated inequalities derived from the cut polytope.

References

1. ANGELIS, P. L., PARDALOS, P. M., AND TORALDO, G. Quadratic programming with box constraints. In *Developments in Global Optimization*, vol. 18 of *Nonconvex Optimization and Its Applications*. Kluwer Acad. Publ., Dordrecht, 1997, pp. 73–93.
2. BARAHONA, F., AND MAHJOUB, A. R. On the cut polytope. *Math. Programming 36*, 2 (1986), 157–173.
3. BENSON, S., YE, Y., AND ZHANG, X. Solving large-scale sparse semidefinite programs for combinatorial optimization. Tech. rep., Technical report, Department of Management Sciences, The University of Iowa, 1997.
4. BOROS, E., AND HAMMER, P. L. Cut-polytopes, Boolean quadratic polytopes and nonnegative quadratic pseudo-Boolean functions. *Math. Oper. Res. 18*, 1 (1993), 245–253.
5. CHUNG, S. J., AND MURTY, K. G. Polynomially bounded ellipsoid algorithms for convex quadratic programming. In *Nonlinear Programming, 4*, O. L. Mangasarian, R. R. Meyer, and S. M. Robinson, Eds. Academic Press, New York, 1981, pp. 439–485.
6. COLEMAN, T. F., AND HULBERT, L. A. A direct active set algorithm for large sparse quadratic programs with simple bounds. *Math. Programming 45*, 3 (Ser. B) (1989), 373–406.
7. DELORME, C., AND POLJAK, S. Laplacian eigenvalues and the maximum cut problem. *Math. Programming 62*, 3, Ser. A (1993), 557–574.
8. DEZA, M. M., AND LAURENT, M. *Geometry of cuts and metrics*, vol. 15 of *Algorithms and Combinatorics*. Springer-Verlag, Berlin, 1997.

9. Fujie, T., and Kojima, M. Semidefinite programming relaxation for nonconvex quadratic programs. *J. Global Optim. 10*, 4 (1997), 367–380.
10. Fujisawa, K., Kojima, M., and Nakata, K. Sdpa (semi-definite programming algorithm) – user's manual version 4.10. Dept. of Math. and Comp. Sci. B-308, Tokyo Institute of Technology, 1995.
11. Goemans, M. X. Semidefinite programming in combinatorial optimization. *Math. Programming 79*, 1-3, Ser. B (1997), 143–161.
12. Goemans, M. X., and Williamson, D. P. Improved approximation algorithms for maximum cut and satisfiability problems using semidefinite programming. *J. Assoc. Comput. Mach. 42*, 6 (1995), 1115–1145.
13. Han, C.-G., Pardalos, P. M., and Ye, Y. Computational aspects of an interior point algorithm for quadratic programming problems with box constraints. In *Large-scale numerical optimization*, T. F. Coleman and Y. Li, Eds. SIAM, Philadelphia, PA, 1990, pp. 92–112.
14. Han, C.-G., Pardalos, P. M., and Ye, Y. On the solution of indefinite quadratic problems using an interior-point algorithm. *Informatica 3*, 4 (1992), 474–496, 594, 602.
15. Hansen, P. Methods of nonlinear $0-1$ programming. *Ann. Discrete Math. 5* (1979), 53–70.
16. Hansen, P., Jaumard, B., Ruiz, M., and Xiong, J. Global minimization of indefinite quadratic functions subject to box constraints. *Naval Res. Logist. 40*, 3 (1993), 373–392.
17. Helmberg, C., and Rendl, F. A spectral bundle method for semidefinite programming. Tech. rep., Technical Report SC 97-37, Konrad-Zuse-Zentrum für Informationstechnik Berlin, 1997.
18. Helmberg, C., and Rendl, F. Solving quadratic (0,1)-problems by semidefinite programs and cutting planes. *Math. Programming 82*, 3 (1998), 291–316.
19. Nesterov, Y. E. Quality of semidefinite relaxation for nonconvex quadratic optimization. Tech. rep., CORE, Universite Catholique de Louvain, Belgium, 1997.
20. Padberg, M. The Boolean quadric polytope: some characteristics, facets and relatives. *Math. Programming 45*, 1 (Ser. B) (1989), 139–172.
21. Pardalos, P. M., and Rodgers, G. P. Computational aspects of a branch and bound algorithm for quadratic zero-one programming. *Computing 45*, 2 (1990), 131–144.
22. Pardalos, P. M., and Wolkowicz, H., Eds. *Topics in semidefinite and interior-point methods*, vol. 18 of *Fields Insitute Communications.* Americam Mathematical Society, 1998.
23. Poljak, S., and Rendl, F. Solving the max-cut problem using eigenvalues. *Discrete Appl. Math. 62*, 1-3 (1995), 249–278.
24. Sherali, H. D., and Tuncbilek, C. H. A reformulation-convexification approach for solving nonconvex quadratic programming problems. *J. Global Optim. 7*, 1 (1995), 1–31.
25. Vavasis, S. A. Approximation algorithms for indefinite quadratic programming. *Math. Programming 57*, 2, Ser. B (1992), 279–311.
26. Yajima, Y., and Fujie, T. A polyhedral approach for nonconvex quadratic programming problems with box constraints. Dept. of Math. and Comp. Sci. B-323, Tokyo Institute of Technology, 1996.
27. Ye, Y. On affine scaling algorithms for nonconvex quadratic programming. *Math. Programming 56*, 3, Ser. A (1992), 285–300.

28. Ye, Y. Approximating quadratic programming with quadratic constraints. Tech. rep., Department of Management Sciences The University of Iowa, 1997.
29. Ye, Y., and Tse, E. An extension of Karmarkar's projective algorithm for convex quadratic programming. *Math. Programming 44*, 2 (Ser. A) (1989), 157–179.

Contributed papers

The Steiner Ratio of L_p^3

Jens Albrecht and Dietmar Cieslik

University of Greifswald
Germany

Abstract. We consider Steiner's Problem in $\mathcal{L}_p^3$, which is a three-dimensional space equipped with p-norm. Steiner's Problem is the "Problem of shortest connectivity", that means, given a finite set N of points in the plane, search for a network interconnecting these points with minimal length. This shortest network must be a tree and is called a Steiner Minimal Tree (SMT). It may contain vertices different from the points which are to be connected. Such points are called Steiner points.
If we do not allow Steiner points, that means, we only connect certain pairs of the given points, we get a tree which is called a Minimum Spanning Tree (MST) for N. Steiner's Problem is very hard as well in combinatorial as in computational sense, but on the other hand, the determination of an MST is simple. Consequently, we are interested in the greatest lower bound for the ratio between the lengths of these both trees:

$$m(3,p) := \inf \left\{ \frac{L(\text{SMT for } N)}{L(\text{MST for } N)} : N \subseteq \mathcal{L}_p^3 \text{ is a finite set} \right\},$$

which is called the Steiner ratio (of $\mathcal{L}_p^3$).
We look for estimates for $m(3,p)$, depending on the parameter p, and determine general upper bounds for the Steiner ratio of $\mathcal{L}_p^3$.

2000 Mathematics Subject Classification. 05C05 , 52A21

Keywords and phrases. Steiner Trees, Steiner ratio.

1 Introduction

Starting with the famous book "What is Mathematics" by Courant and Robbins the following problem has been popularized under the name of Steiner: For a given finite set of points in a metric space find a network which connects all points of the set with minimal length. Such a network must be a tree, which is called a Steiner Minimal Tree (SMT). It may contain vertices other than the points which are to be connected. Such points are called Steiner points.[1] A classical survey of this problem in the Euclidean plane was given by Gilbert and Pollak [11]. An updated one can be found in [15].

[1] The history of Steiner's Problem started with P.Fermat early in the 17th century and C.F.Gauß in 1836. At first perhaps with the famous book *What is Mathematics* by R.Courant and H.Robbins in 1941, this problem became popularized under the name of Steiner.

If we don't allow Steiner points, that is if we connect the given set of points only, this is a Minimum Spanning Tree (MST). Starting with Boruvka in 1926 and Kruskal in 1956 Minimum Spanning Trees have a well-documented history [12] and effective constructions [2], [16].

Obviously, these problems depend essentially on the way distances in the plane are determined. We consider finite-dimensional spaces with p-norm, defined in the following way: Let A_d be the d-dimensional affine space. For the points $X = (x_1, ..., x_d)$ and $Y = (y_1, ..., y_d)$ of the space we define the distance by

$$\rho(X,Y) = \rho_p(X,Y) = \left(\sum_{i=1}^{d} |x_i - y_i|^p\right)^{1/p}$$

where $1 \leq p < \infty$ is a real number. If p runs to infinity we get the so-called Maximum distance

$$\rho_\infty(X,Y) = \max\{|x_i - y_i| : 0 \leq i \leq d\}$$

In each case we obtain a finite-dimensional Banach space written by $\mathcal{L}_p^d$.

A (finite) graph $G = (V, E)$ with the set V of vertices and the set E of edges is embedded in this space in the sense that

- V is a finite set of points in the space;
- Each edge $\underline{XY} \in E$ is a segment $\{tX + (1-t)Y : 0 \leq t \leq 1\}$, $X, Y \in V$; and
- The length of G is defined by

$$L(G) = L_p(G) = \sum_{\underline{XY} \in E} \rho_p(X,Y).$$

Now, **Steiner's Problem of Minimal Trees** is the following:
Given a finite set N of points in the space $\mathcal{L}_p^d$. Find a connected graph $G = (V, E)$ embedded in the space such that $N \subseteq V$ and $L_p(G)$ is minimal as possible.

A solution of Steiner's Problem must be a tree and is called a Steiner Minimal Tree (SMT) for N in $\mathcal{L}_p^d$. The vertices in the set $V \setminus N$ are called Steiner points. We may assume that for any SMT $T = (V, E)$ for N the following holds: The degree of each Steiner point is at least 3 and $|V \setminus N| \leq |N| - 2$. For these facts and more results about the combinatorial structure of SMT's compare [5].

We have algorithms to construct SMT's in the plane $\mathcal{L}_p^2$, see [6], but they need exponential time. A method to find an SMT in the higher-dimensional spaces $\mathcal{L}_p^d$, $d > 2$, is still unknown, except the cases of Euclidean spaces $\mathcal{L}_2^d$, see [18], and the spaces with rectilinear norm $\mathcal{L}_1^d$, see [8] and [20]. On the other hand, we said above that it is simple to find an MST.

The relative defect, which describes the length of an SMT divided by the length of an MST, is given in the **Steiner Ratio**:

$$m(d,p) = \inf \left\{ \frac{L(\text{SMT for } N)}{L(\text{MST for } N)} : N \subset \mathcal{L}_p^d \text{ finite set} \right\}.$$

$m(d,p)$ is a measure of how good an MST as an approximation of Steiner's Problem in the space $\mathcal{L}_p^d$ is. It is not hard to see that we have $1 \geq m(d,p) \geq 1/2$ for every dimension d and any value of p, see [11]. In general, the exact value of the Steiner ratio $m(d,p)$ is still unknown.

2 Several special cases

As an introductory example consider in the d-dimensional affine space A_d the unit ball $B(1)$ of ρ_1. $B(1)$ is the convex hull of

$$N = \{\pm(0,...,0,1,0,...,0) : \text{the i'th component is equal to } 1, i = 1,\dots,d\}. \tag{1}$$

Then the set N contains $2d$ points. The distance of any two different points in N equals 2. Hence, an MST for N has the length $2(2d-1)$. Conversely, an SMT for N with the Steiner point $(0,...,0)$ has the length $2d$.[2] This implies

$$m(d,1) \leq \frac{d}{2d-1}. \tag{2}$$

In view of (2) we see that the lower bound 1/2 is the best possible for the Steiner ratio over the class of all spaces $\mathcal{L}_p^d$.[3]
Graham and Hwang [13] conjectured that equality holds, which is true in the planar case: $m(2,1) = \frac{2}{3} = 0.6666\ldots$, see [14].[4]

$\mathcal{L}_2^d$ is the d-dimensional Euclidean space. It is easy to see that

$$m(d,2) \leq \sqrt{\frac{1}{2} + \frac{1}{2d}}. \tag{3}$$

The Euclidean Steiner ratio in the planar case is well-known, since Du and Hwang [7] showed

$$m(2,2) = \frac{\sqrt{3}}{2} = 0.86602... \tag{4}$$

[2] For the fact that this tree is indeed an SMT compare [8] and [20].
[3] And also of all finite-dimensional Banach spaces.
[4] The methods given by Hwang in [14] do not seem to be applicable to proving the conjecture in the higher dimensional case.

Better estimates than (3) for the Euclidean Steiner ratio in higher dimensions, based on investigations of finite sets, can be found by Chung, Gilbert [3], Smith [18] and Du, Smith [9]:

dimension	upper bound by Chung, Gilbert	upper bound by Smith	upper bound by Du, Smith
$=2$	0.86602...		
$=3$	0.81305...	0.81119...	0.78419...
$=4$	0.78374...	0.76871...	0.74398...
$=5$	0.76456...	0.74574...	0.72181...
$=6$	0.75142...	0.73199...	0.70853...
$=7$	0.74126...	0.72247...	0.70012...
$=8$	0.73376...	0.71550...	0.69455...
$=9$	0.72743...	0.71112...	0.69076...
$=10$	0.72250...		0.68812...
$=11$	0.71811...		0.68624...
$=20$	0.69839...		
$=40$	0.68499...		
$=80$	0.67775...		
$=160$	0.67392...		
$\to\infty$	0.66984...		

The first column was computed by Chung and Gilbert considering regular simplices, the second by Smith investigating regular octahedra, respectively cross polytopes. The third column used the ratio of sausages.
For each dimension $d > 2$, at present, the exact value of the quantity $m(d, 2)$ is not yet known. In particular, this is true for $d = 3$. Smith and McGregor Smith [19] conjectured

$$\begin{aligned} m(3,2) &= \sqrt{\frac{283}{700} - \frac{3\sqrt{21}}{700} + \frac{9\sqrt{11-\sqrt{21}}\sqrt{2}}{140}} \\ &= 0.78419\ldots. \end{aligned}$$

If we consider the planes with p-norm we have for the Steiner ratio the inequalities

$$\frac{2}{3} \leq m(2,p) \leq \frac{\sqrt{3}}{2}. \tag{5}$$

The first inequality becomes an equality if $p = 1$ or $p = \infty$. Equality in the second inequality we have if and only if $p = 2$. For a proof see [17] and [10]; further estimates for the Steiner ratio of $\mathcal{L}_p^2$, depending on the value of p, we find in [1], [4] and [17].

Another approach to investigate the Steiner ratios of $\mathcal{L}_p^d$ is to use an isometric embedding from $\mathcal{L}_2^{d'}$ into $\mathcal{L}_p^d$, if it exists. Then, we have $m(d,p) \leq$

$m(d',2)$. For instance, by this way we find
$m(d,4) \leq m(3,2) \leq 0.81119\ldots$ for $d > 5$;
$m(d,4) \leq m(4,2) \leq 0.76871\ldots$ for $d > 10$;
$m(d,4) \leq m(7,2) \leq 0.72247\ldots$ for $d > 28$;
$m(d,4) \leq m(23,2) \leq 0.69839\ldots$ for $d > 275$.
For more results in this sense see [5].

3 Using tetrahedrons

Our goal is to estimate the quantities $m(3,p)$ with help of investigations for configurations of points in a regular situation in the space $\mathcal{L}_p^3$. To do this we start with the consideration of tetrahedrons.
Let $1 < p < \infty$. We consider the four points

$$\begin{aligned} P_1 &= (1,0,0), \\ P_2 &= (0,1,0), \\ P_3 &= (0,0,1) \quad \text{and} \\ P_4 &= (1,1,1) \end{aligned}$$

in the three-dimensional space, which build an equilateral set.
We are interested in an SMT for these points; then we will find an upper bound for $m(3,p)$.[5]

An SMT T for four points has at most two Steiner points:

$$\begin{aligned} S_1 &= (x_1, y_1, z_1) \quad \text{and} \\ S_2 &= (x_2, y_2, z_2). \end{aligned}$$

We assume, that S_1 is adjacent to P_1, P_2 and S_2 to P_3, P_4. Moreover, we may assume that S_1 and S_2 are within the tetrahedra and consequently in the (open) cube $(0,1)^3$. The minimality of T means that the function $F : \mathbb{R}^6 \longrightarrow \mathbb{R}$:

$$\begin{aligned} &F(x_1, y_1, z_1, x_2, y_2, z_2) = \\ &\quad f_1(x_1, y_1, z_1) + f_2(x_1, y_1, z_1, x_2, y_2, z_2) + f_3(x_2, y_2, z_2), \end{aligned}$$

whereby

$$\begin{aligned} f_1(x_1, y_1, z_1) &= \rho(S_1, P_1) + \rho(S_1, P_2) \\ &= ((1-x_1)^p + y_1^p + z_1^p)^{1/p} + (x_1^p + (1-y_1)^p + z_1^p)^{1/p}, \end{aligned}$$

$$\begin{aligned} f_2(x_1, y_1, z_1, x_2, y_2, z_2) &= \rho(S_1, S_2) \\ &= (|x_1 - x_2|^p + |y_1 - y_2|^p + |z_1 - z_2|^p)^{1/p} \end{aligned}$$

[5] Additionally we will find several facts about the degrees of Steiner points; see the last section.

and

$$\begin{aligned} f_3(x_2, y_2, z_2) &= \rho(S_2, P_3) + \rho(S_2, P_4) \\ &= (x_2^p + y_2^p + (1 - z_2)^p)^{1/p} + ((1 - x_2)^p + (1 - y_2)^p + (1 - z_2)^p)^{1/p} \end{aligned}$$

is minimal.

To investigate the functions f_i, $i = 1, 2, 3$, we show

Lemma 1. *Let $1 < p < \infty$, $0 < z < 1$ and let f and g be two functions over the open range $D(f) = D(g) = (0, 1)^2$ with*

$$f(x, y) = ((1 - x)^p + y^p + z^p)^{1/p} + (x^p + (1 - y)^p + z^p)^{1/p}$$

and

$$g(x, y) = (x^p + y^p + (1 - z)^p)^{1/p} + ((1 - x)^p + (1 - y)^p + (1 - z)^p)^{1/p}.$$

Then it holds

$$f\left(\frac{1}{2}, \frac{1}{2}\right) < f(x, y)$$
$$g\left(\frac{1}{2}, \frac{1}{2}\right) < g(x, y)$$

for all $(x, y) \in D(f) \backslash \{(\frac{1}{2}, \frac{1}{2})\}$ and for all $(x, y) \in D(g) \backslash \{(\frac{1}{2}, \frac{1}{2})\}$.

Proof. We consider the function f and use the following abbreviations

$$\begin{aligned} H_1 &= H_1(x, y) := (1 - x)^p + y^p + z^p \\ H_2 &= H_2(x, y) := x^p + (1 - y)^p + z^p. \end{aligned}$$

The first and second partial derivatives exit overall and we have

$$\begin{aligned}
f_x(x,y) &= -H_1^{1/p-1}(1-x)^{p-1} + H_2^{1/p-1}x^{p-1}\\
f_y(x,y) &= H_1^{1/p-1}y^{p-1} - H_2^{1/p-1}(1-y)^{p-1}\\
f_{xx}(x,y) &= \left(\frac{1}{p}-1\right)pH_1^{1/p-2}(1-x)^{2p-2} + H_1^{1/p-1}(p-1)(1-x)^{p-2}\\
&\quad + \left(\frac{1}{p}-1\right)pH_2^{1/p-2}x^{2p-2} + H_2^{1/p-1}(p-1)x^{p-2}\\
&= (p-1)H_1^{1/p-2}(1-x)^{p-2}(y^p+z^p)\\
&\quad +(p-1)H_2^{1/p-2}x^{p-2}((1-y)^p+z^p)\\
f_{yy}(x,y) &= \left(\frac{1}{p}-1\right)pH_1^{1/p-2}y^{2p-2} + H_1^{1/p-1}(p-1)y^{p-2}\\
&\quad + \left(\frac{1}{p}-1\right)pH_2^{1/p-2}(1-y)^{2p-2} + H_2^{1/p-1}(p-1)(1-y)^{p-2}\\
&= (p-1)H_1^{1/p-2}y^{p-2}((1-x)^p+z^p)\\
&\quad +(p-1)H_2^{1/p-2}(1-y)^{p-2}(x^p+z^p)\\
f_{xy}(x,y) &= (p-1)H_1^{1/p-2}y^{p-1}(1-x)^{p-1} + (p-1)H_2^{1/p-2}x^{p-1}(1-y)^{p-1}.
\end{aligned}$$

Particularly,

$$f_x\left(\frac{1}{2},\frac{1}{2}\right) = f_y\left(\frac{1}{2},\frac{1}{2}\right) = 0 \tag{6}$$

$$f_{xx}(x,y) > 0 \quad \text{for all } (x,y) \in D(f). \tag{7}$$

The set $D(f)$ is the neighborhood of the point $\left(\frac{1}{2},\frac{1}{2}\right)$. Using Taylors theorem for this neighborhood and stop after the first step we get for arbitrary points $(x,y) = \left(\frac{1}{2}+\Delta x, \frac{1}{2}+\Delta y\right) \in D(f)\setminus\left\{\left(\frac{1}{2},\frac{1}{2}\right)\right\}$ the equality

$$\begin{aligned}
&f(x,y) - f\left(\frac{1}{2},\frac{1}{2}\right) =\\
&f_x\left(\frac{1}{2},\frac{1}{2}\right)\Delta x + f_y\left(\frac{1}{2},\frac{1}{2}\right)\Delta y + \frac{1}{2}\Bigg(\Delta x^2 f_{xx}\left(\frac{1}{2}+\theta\Delta x, \frac{1}{2}+\theta\Delta y\right)\\
&+2\Delta x\Delta y f_{xy}\left(\frac{1}{2}+\theta\Delta x,\frac{1}{2}+\theta\Delta y\right) + \Delta y^2 f_{yy}\left(\frac{1}{2}+\theta\Delta x,\frac{1}{2}+\theta\Delta y\right)\Bigg)
\end{aligned}$$

whereby $0 < \theta < 1$.

Paying attention to (6) and (7) we get

$$2\left(f(x,y)-f\left(\frac{1}{2},\frac{1}{2}\right)\right)=\underbrace{\frac{1}{f_{xx}}}_{>0}\left[\underbrace{(\Delta x f_{xx}+\Delta y f_{xy})^2}_{>0}+\underbrace{\Delta y^2}_{\geq 0}(f_{xx}f_{yy}-f_{xy}^2)\right]$$

and, moreover using

$$\begin{aligned}
f_{xx}f_{yy}-f_{xy}^2 &= (p-1)^2\big[(H_1H_2)^{(1/p)-2}[(1-x)(1-y)]^{p-2}(y^p+z^p)(x^p+z^p)\\
&\quad +H_1^{(2/p)-4}[(1-x)y]^{p-2}(y^p+z^p)((1-x)^p+z^p)\\
&\quad +H_2^{(2/p)-4}[x(1-y)]^{p-2}((1-y)^p+z^p)(x^p+z^p)\\
&\quad +(H_1H_2)^{(1/p)-2}(xy)^{p-2}((1-y)^p+z^p)((1-x)^p+z^p)\big]\\
&\quad -(p-1)^2\big[H_1^{(2/p)-4}[y(1-x)]^{2p-2}+H_2^{(2/p)-4}[x(1-y)]^{2p-2}\\
&\quad +2(H_1H_2)^{(1/p)-2}[xy(1-x)(1-y)]^{p-1}\big]\\
&> (p-1)^2(H_1H_2)^{(1/p)-2}[xy(1-x)(1-y)]^{p-2}\\
&\quad \cdot\Big[[xy]^2+[(1-x)(1-y)]^2-2xy(1-x)(1-y)\Big]\\
&= (p-1)^2(H_1H_2)^{(1/p)-2}[xy(1-x)(1-y)]^{p-2}\\
&\quad \cdot[xy-(1-x)(1-y)]^2\\
&\geq 0
\end{aligned}$$

we find

$$f(x,y)-f\left(\frac{1}{2},\frac{1}{2}\right)>0.$$

Similarly we find the inequality for the function g. □

Furthermore, we have for arbitrary values $z_1, z_2 \in (0,1)$

$$\begin{aligned}
f_2(x_1,y_1,z_1,x_2,y_2,z_2) &= (|x_1-x_2|^p+|y_1-y_2|^p+|z_1-z_2|^p)^{1/p}\\
&\geq f_2\left(\frac{1}{2},\frac{1}{2},z_1,\frac{1}{2},\frac{1}{2},z_2\right).
\end{aligned} \tag{8}$$

In view of lemma 1 and the inequality (8) we find

$$\begin{aligned}
F(x_1,y_1,z_1,x_2,y_2,z_2) &> f_1\left(\tfrac{1}{2},\tfrac{1}{2},z_1\right)+f_2\left(\tfrac{1}{2},\tfrac{1}{2},z_1,\tfrac{1}{2},\tfrac{1}{2},z_2\right)+\\
&\qquad f_3\left(\tfrac{1}{2},\tfrac{1}{2},z_2\right)\\
&= F\left(\frac{1}{2},\frac{1}{2},z_1,\frac{1}{2},\frac{1}{2},z_2\right)
\end{aligned}$$

for all $x_1, y_1, z_1, x_2, y_2, z_2 \in (0,1)$ with $(x_1, y_1, x_2, y_2) \neq (\frac{1}{2}, \frac{1}{2}, \frac{1}{2}, \frac{1}{2})$. Hence, we see that the function F, for arbitrary choice of z_1 and z_2, find their unique determined minimum in

$$x_1 = y_1 = x_2 = y_2 = \frac{1}{2}.$$

In other terms, the Steiner points lie on the line $x = y = \frac{1}{2}$. Moreover, paying attention to that we look for a minimum of the length of the tree T, we obtain $z_2 \geq z_1$. Then the function F is simplified to the function G

$$G(z_1, z_2) = 2\left(2\left(\frac{1}{2}\right)^p + z_1^p\right)^{1/p} + (z_2 - z_1) + 2\left(2\left(\frac{1}{2}\right)^p + (1 - z_2)^p\right)^{1/p}.$$

The partial derivatives

$$\frac{\partial G}{\partial z_1}(z_1, \cdot) = 2\left(2^{1-p} + z_1^p\right)^{(1/p)-1} z_1^{p-1} - 1 \tag{9}$$

$$\frac{\partial G}{\partial z_2}(\cdot, z_2) = 2\left(2^{1-p} + (1 - z_2)^p\right)^{(1/p)-1} (1 - z_2)^{p-1} + 1 \tag{10}$$

are strictly monotone in z_1 respectively z_2 and have at most one zero in $(0,1)$. For the zero a of (9), the value $1 - a$ is an unique zero of (10). Consequently, for the Steiner points it holds $z_2 = 1 - z_1$.

Now we have seen that the function G can be simplified to the function H with

$$H(z) = 4\left(2^{1-p} + z^p\right)^{1/p} + 1 - 2z$$

which is to minimize in the range $(0, \frac{1}{2}]$. Hence, we are interested in the zeros of

$$H'(z) = 4\left(2^{1-p} + z^p\right)^{(1/p)-1} z^{p-1} - 2.$$

Lemma 2. *Let* $1 < p < \infty$ *and* q *conjugated to* p [6]. *Then the function* $f : \mathbb{R}^+ \to \mathbb{R}$ *with*

$$f(x) = 4\left(2^{1-p} + x^p\right)^{(1/p)-1} x^{p-1} - 2$$

has an uniquely determined zero

$$x_0 = 2^{-1/q}(2^q - 1)^{-1/p}.$$

Moreover,

$$x_0 \begin{cases} \leq \frac{1}{2} & : \text{ if } \quad 1 < p \leq \frac{\log 3}{\log 3 - \log 2} = 2.709511... \\ > \frac{1}{2} & : \text{ if } \quad \frac{\log 3}{\log 3 - \log 2} < p < \infty \end{cases}$$

[6] that means $1/p + 1/q = 1$

Proof. We consider the derivation of f. Then it holds

$$\begin{aligned} f'(x) &= 4\left[\left(\frac{1}{p}-1\right)(2^{1-p}+x^p)^{(1/p)-2}px^{2p-2}+(2^{1-p}+x^p)^{(1/p)-1} \right. \\ &\qquad \left. \cdot(p-1)x^{p-2}\right] \\ &= 4(p-1)(2^{1-p}+x^p)^{(1/p)-2}x^{p-2}\left[-x^p+(2^{1-p}+x^p)\right] \\ &> 0 \end{aligned}$$

for $x > 0$.
Hence, f is strictly monotone increasing and can have at most one zero. Then we find the assertion with help of several elementary operations. □

Lemma 2 implies that for $p \geq \frac{\log 3}{\log 3 - \log 2}$ the both Steiner points S_1, S_2 coincide to the Steiner point $S = \left(\frac{1}{2}, \frac{1}{2}, \frac{1}{2}\right)$ of degree 4.

All our investigations give the following

Theorem 1. *Let $1 < p < \infty$ and let q be conjugated to p. Then we have for the Steiner ratio of $\mathcal{L}_p^3$*

$$m(3,p) \leq \begin{cases} \frac{1}{3}\left(2^{-1/p}+(2^q-1)^{1/q}\right) & : \quad \text{if} \quad 1 < p \leq \frac{\log 3}{\log 3 - \log 2} \\ \left(\frac{2}{3}\right)^{1/q} & : \quad \text{otherwise} \end{cases}$$

Proof. It is

$$L(T) = \begin{cases} H(x_0) = 2^{1/p}(2^q-1)^{1/q}+1 & : \quad \text{if} \quad 1 < p \leq \frac{\log 3}{\log 3 - \log 2} \\ 4\left(3\left(\frac{1}{2}\right)^p\right)^{1/p} = 2\sqrt[p]{3} & : \quad \text{otherwise} \end{cases}$$

In the same situation an MST for the given points has the length

$$L_M := L(\text{MST for } \{P_1, \ldots, P_4\}) = 3\sqrt[p]{2}.$$

□

4 Using cross-polytopes

We consider a cross-polytope created by the set of nodes $N = \{P_1, \ldots, P_6\}$ whereby

$$\begin{aligned} P_1 &= (1,0,0), \\ P_2 &= (0,1,0), \\ P_3 &= (0,0,1), \\ P_4 &= (-1,0,0), \\ P_5 &= (0,-1,0) \quad \text{and} \\ P_6 &= (0,0,-1). \end{aligned}$$

The points P_i and P_j, $i \neq j$, have the distance

$$\rho(P_i, P_j) = \begin{cases} 2 & : \text{ if } |i-j| = 3 \\ 2^{1/p} \leq 2 & : \text{ otherwise} \end{cases}$$

and consequently an MST for N has the length

$$L_p(\text{MST for } N) = 5 \cdot 2^{1/p}. \tag{11}$$

On the other hand, using the fact that it holds $\rho(P_i, o) = 1$ for $i = 1, \ldots, 6$, there is a tree for $N \cup \{o\}$ of the length 6. Hence,

Theorem 2. *For the Steiner ratio of a three-dimensional $\mathcal{L}_p$-space it holds*

$$m(3,p) \leq \frac{6}{5} \cdot \left(\frac{1}{2}\right)^{1/p}.$$

In the tree above we use only one Steiner point, namely o, and it is to assume that we will find better estimates if we introduce more of such points.[7] We have to expect at most four Steiner points:

$$\begin{aligned} S_1 &= (x, x, 0), \\ S_2 &= (-x, 0, x), \\ S_3 &= (0, -x, -x) \quad \text{and} \\ S_4 &= (0, 0, 0) \end{aligned}$$

whereby $0 \leq x < 1$. Our tree T consits of the edges $\underline{S_i, P_{2i-1}}$, $\underline{S_i, P_{2i}}$, $\underline{S_i, S_4}$ with $i = 1, 2, 3$. Consequently,

$$L(T) = f(x) = 6 \cdot (x^p + (1-x)^p)^{1/p} + 3 \cdot 2^{1/p} x.$$

Now, we minimize the function f and consider the derivation f':

$$f'(x) = 6 \cdot [x^p + (1-x)^p]^{1/p - 1} (x^{p-1} - (1-x)^{p-1}) + 3 \cdot 2^{1/p}.$$

In the case of the rectilinear space, that means $p = 1$, we have $f'(x) = 6 > 0$ overall. Hence, f is minimal in $x = 0$. All Steiner points coincide with o, and we find no better bounds than before.

Now we assume $p > 1$, and paying attention to the second derivation of f, we see f' strictly monotone increasing, since

$$\begin{aligned} f''(x) &= 6 \cdot (((1/p) - 1)(x^p + (1-x)^p)^{1/p-2} p (x^{p-1} - (1-x)^{p-1})^2 \\ &\qquad (x^p + (1-x)^p)^{1/p-1} (p-1)(x^{p-2} + (1-x)^{p-2})) \\ &= 6(p-1)(x^p + (1-x)^p)^{1/p-2} (-(x^{p-1} - (1-x)^{p-1})^2 \\ &\qquad + (x^p + (1-x)^p)(x^{p-2} + (1-x)^{p-2})) \\ &= \ldots (-x^{2p-2} + 2x^{p-1}(1-x)^{p-1} - (1-x)^{2p-2} \\ &\qquad + x^{2p-2} + x^p (1-x)^{p-2} + (1-x)^p x^{p-2} + (1-x)^{2p-2}) \\ &> 0. \end{aligned}$$

[7] This must be true, since a Steiner point for an SMT in the space $\mathcal{L}_p^3$ has at most a degree less than six.

Moreover, f' is continuous, and we have in view of

$$f'(0) = 3 \cdot 2^{1/p} - 6 < 0 \text{ and}$$
$$f'\left(\frac{1}{2}\right) = 3 \cdot 2^{1/p} > 0$$

that there is a uniquely determined value x_0 with $0 < x_0 < 1$ and $f'(x_0) = 0$, which is a minimum of f, since $f''(x_0) > 0$. Particularly, $f(x_0) < f(0)$, that means we find a strictly better upper bound for the Steiner ratio. Using the equation (11) we have

Corollary 1. *Let $1 \leq p \leq \infty$ and x_0 how described above. Then*

$$m(3,p) \leq \frac{6}{5}\left(\frac{1}{2}\right)^{1/p} ({x_0}^p + (1-x_0)^p)^{1/p} + \frac{3}{5}x_0.$$

Now, we will consider a cross-polytope in another way which will give us better bounds for the Steiner ratio if $p > 2$.[8]
At first we assume that $p \neq \infty$ and consider the set $N = \{P_1, \ldots, P_6\}$ of given points with

$$\begin{aligned} P_1 &= (x, x-1, 1-x), \\ P_2 &= (x, x, 2-x), \\ P_3 &= (1,0,1), \\ P_4 &= (0,0,0), \\ P_5 &= (0,1,1) \quad \text{and} \\ P_6 &= (x-1, x, 1-x), \end{aligned}$$

whereby x denotes the unique zero of the function f with

$$f(x) = x^p + 2(x-1)^p - 2$$

over the range $(1, 2)$.
The mutual distances between the points are

$$\rho(P_i, P_j) = \begin{cases} (2x^p + (2-x)^p)^{1/p} > 2^{1/p} & : \text{ if } \{i,j\} = \{1,5\}, \{2,4\}, \{3,6\} \\ 2^{1/p} & : \text{ otherwise} \end{cases}$$

Hence, an MST for N has the length

$$L(\text{MST for } N) = 5 \cdot 2^{1/p}. \tag{12}$$

[8] But between $p = 2$ and $p \approx 2.0619508$ the bound will be greater than the bounds given in the section before.

We consider the tree $T = (N \cup \{S_1, \ldots, S_4\}, E)$ for N, where

$$\begin{aligned}
S_1 &= \left(x - a, x - \frac{1}{2}, \frac{3}{2} - x\right), \\
S_2 &= \left(b, \frac{1}{2}, \frac{1}{2}\right), \\
S_3 &= \left(a, \frac{1}{2}, \frac{1}{2}\right), \text{ and} \\
S_4 &= \left(x - b, x - \frac{1}{2}, \frac{3}{2} - x\right)
\end{aligned}$$

are the Steiner points and

$$E = \{\underline{S_1P_1}, \underline{S_1P_2}, \underline{S_1S_4}, \underline{S_4S_2}, \underline{S_4P_6}, \underline{S_2P_3}, \underline{S_2S_3}, \underline{S_3P_4}, \underline{S_3P_5}\}$$

is the set of edges.
With respect of

$$0 < a \leq b \leq \frac{x}{2} \tag{13}$$

we get

$$\begin{aligned}
L_p(T) &= g(a, b) \\
&= 4(a^p + 2^{1-p})^{1/p} + 2(b - a) + 2((1 - b)^p + 2^{1-p})^{1/p} \\
&\qquad +((x - 2b)^p + 2(x - 1)^p)^{1/p}.
\end{aligned}$$

To search the minimal positions of the function g we investigate the partially derivations.

$$\begin{aligned}
\frac{\partial g}{\partial a}(a, \cdot) &= 4(a^p + 2^{1-p})^{1/p-1} a^{p-1} - 2, \\
\frac{\partial g}{\partial b}(\cdot, b) &= 2 - 2((1 - b)^p + 2^{1-p})^{1/p-1}(1 - b)^{p-1} \\
&\qquad -2((x - 2b)^p + 2(x - 1)^p)^{1/p-1}(x - 2b)^{p-1}.
\end{aligned}$$

We cannot describe a zero of $\partial g/\partial b$ in an explicit manner, but $\frac{1}{2}$ is a good approximation. On the other hand, the function $\partial g/\partial a$ was investigated in lemma 2. In view of (13) we define

$$a = \begin{cases} 2^{-1/q}(2^q - 1)^{-1/p} & : \text{ if } 1 < p \leq \frac{\log 3}{\log 3 - \log 2} = 2.709511\ldots \\ \frac{1}{2} & : \text{ if } \frac{\log 3}{\log 3 - \log 2} < p < \infty \end{cases}$$

and consequently with $1/p + 1/q = 1$ we get

$$
\begin{aligned}
L(T) &= 4\Big(\Big[2^{-1/q}(2^q-1)^{-1/p}\Big]^p + 2^{1-p}\Big)^{1/p} + 1 - 2\Big[2^{-1/q}(2^q-1)^{-1/p}\Big] \\
&\qquad + 3^{1/p}x \\
&= 4\Big(2^{1-p}(2^q-1)^{-1} + 2^{1-p}\Big)^{1/p} - 2^{1/p}(2^q-1)^{-1/p} + 1 + 3^{1/p}x \\
&= 4\Big(2^{1-p}(2^q-1)^{-1}\left[1+(2^q-1)\right]\Big)^{1/p} - 2^{1/p}(2^q-1)^{-1/p} + 1 + 3^{1/p}x \\
&= 2^{2-1/q+q-1}(2^q-1)^{-1/p} - 2^{1/p}(2^q-1)^{-1/p} + 1 + 3^{1/p}x \\
&= (2^q-1)^{-1/p}2^{1/p}[2^q-1] + 1 + 3^{1/p}x \\
&= (2^q-1)^{1/q}2^{1/p} + 1 + 3^{1/p}x
\end{aligned}
$$

for $1 < p \leq \frac{\log 3}{\log 3 - \log 2}$ and

$$
\begin{aligned}
L(T) &= 4(2^{-p} + 2^{1-p})^{1/p} + 3^{1/p}x \\
&= 3^{1/p}(x+2)
\end{aligned}
$$

for $\frac{\log 3}{\log 3 - \log 2} < p < \infty$.
In view of (12) we have proved

Theorem 3. *Let $1 < p < \infty$, $1/p + 1/q = 1$ and let x_0 be the unique determined zero of the function f with $f(x) = x^p + 2(x-1)^p - 2$ in the range $(1,2)$. Then the Steiner ratio of $\mathcal{L}_p^3$ can be estimated by*

$$
m(3,p) \leq \begin{cases} \frac{1}{5}\left((2^q-1)^{1/q} + \left(\frac{1}{2}\right)^{1/p} + \left(\frac{3}{2}\right)^{1/p}x_0\right) & : \text{ if } 1 < p \leq \frac{\log 3}{\log 3 - \log 2} \\ \frac{1}{5}\left(\frac{3}{2}\right)^{1/p}(x_0+2) & : \text{ if } \frac{\log 3}{\log 3 - \log 2} < p < \infty \end{cases}
$$

Using theorem 3 for $p = \infty$ with the quantities $x = 1$ and $a = b = 1/2$ give the value 3/5 for the Steiner ratio, but here we have with help of another considerations, see [5], the better bound

$$
m(3,\infty) \leq \frac{4}{7}.
$$

5 Concluding remarks

Our considerations give also the following interesting fact: We have seen that an SMT in a space $\mathcal{L}_p^3$ for several values p has a Steiner point of degree 4. On the other hand, we know by [5] and [17] that for arbitrary Euclidean spaces $\mathcal{L}_2^2$ and two-dimensional L_p-spaces, with $1 < p < \infty$, the degree of any Steiner point in each SMT is exactly 3. For forthcoming facts about the degree of Steiner points see [21] and [22].

References

1. J.Albrecht and D.Cieslik, The Steiner Ratio of $\mathcal{L}_p$-planes, to appear in *Approximation and Complexity in Numerical Optimization: Continuous and Discrete Problems*, Kluwer Academic Publishers.
2. D. Cheriton and R.E. Tarjan, Finding Minimum Spanning Trees, *SIAM J. Comp.* 5(1976) pp. 724-742.
3. F.R.K. Chung and E.N. Gilbert, Steiner Trees for the Regular Simplex, *Bull. Inst. Math. Ac. Sinica*, 4(1976), pp. 313-325.
4. D. Cieslik, The Steiner-ratio in Banach-Minkowski planes. In R. Bodendieck, editor, *Contemporary Methods in Graph Theory*, Bibliographisches Institut (BI), Mannheim, 1990, 231-247.
5. D. Cieslik, *Steiner Minimal Trees*, Kluwer Academic Publishers, 1998.
6. D. Cieslik and J. Linhart, Steiner Minimal Trees in L_p^2, *Discrete Mathematics*, (155)1996, pp. 39-48.
7. D.-Z. Du and F.K.Hwang, A Proof of the Gilbert-Pollak Conjecture on the Steiner Ratio, *Algorithmica*, 7(1992), pp. 121-136.
8. D.Z.Du and F.K. Hwang, Reducing the Steiner Problem in a normed space, *SIAM J. Computing*, 21(1992), pp. 1001–1007.
9. D.Z.Du and W.D.Smith, Disproofs of Generalized Gilbert-Pollak Conjecture on the Steiner Ratio in three or more Dimensions, *J. of Combinatorial Theory*, Series A, 74(1996), 115-130.
10. B. Gao, D.Z. Du and R.L. Graham, A Tight Lower Bound for the Steiner Ratio in Minkowski Planes, *Discrete Mathematics*, 142(1993), 49-63.
11. E.N. Gilbert and H.O. Pollak, Steiner Minimal Trees, *SIAM J. Appl. Math.*, 16(1968), pp. 1-29.
12. R.L. Graham and P. Hell, On the History of the Minimum Spanning Tree Problem, *Ann. Hist. Comp.*, 7(1985), pp. 43-57.
13. R.L. Graham and F.K. Hwang, A Remark on Steiner Minimal Trees, *Bull. of the Inst. of Math. Ac. Sinica*, (4)1976, pp. 177-182.
14. F.K. Hwang, On Steiner Minimal Trees with rectilinear distance; *SIAM J. Appl. Math.*, (30)1976, pp. 104-114.
15. F.K. Hwang, D.S. Richards, and P. Winter, *The Steiner Tree Problem*, Ann. Discr. Math., (53)1992.
16. J.B. Kruskal,On the shortest spanning subtree of a graph and the travelling salesman problem, *Proc. of the Am. Math. Soc.*, 7(1956), pp. 48-50.
17. Z.C. Liu and D.-Z. Du, On Steiner Minimal Trees with $\mathrm{L_p}$ Distance, *Algorithmica*, (7)1992, pp. 179-192.
18. W.D. Smith, How To Find Steiner Minimal Trees in Euclidean d-Space, *Algorithmica*, 7(1992), pp. 137-178.
19. W.D. Smith and J.M. Smith, On the Steiner Ratio in 3-Space, *J. of Combinatorial Theory, A*, 65(1995), pp. 301-332.
20. T.L. Snyder, On the exact Location of Steiner Points in General Dimension, *SIAM J. Computing*, 21(1992), pp.163-180.
21. K. Swanepoel, Extremal Problems in Minkowski Space related to Minimal Networks, *Proc. of the Am. Math. Soc.*, (124)1996, pp. 2513–2518.
22. K. Swanepoel, *Combinatorial Geometry of Minkowski Spaces*, PhD, University of Pretoria, 1997.

Normal Cones to Sublevel Sets: An Axiomatic Approach

Applications in Quasiconvexity and Pseudoconvexity

Didier Aussel[1] and Aris Daniilidis[2]

[1] Département de Mathématiques, Université de Perpignan,
66860 Perpignan Cedex, France
E-mail : aussel@univ-perp.fr

[2] Laboratoire de Mathématiques Appliquées, CNRS ERS 2055, Université de Pau et des Pays de l'Adour, Avenue de l'Université, 64000 PAU, France.
E-mail: aris.daniilidis@univ-pau.fr

Abstract. An axiomatic approach of normal operators to sublevel sets is given. Considering the Clarke-Rockafellar subdifferential (resp. quasiconvex functions), the definition given in [4] (resp. [5]) is recovered. Moreover, the results obtained in [4] are extended in this more general setting. Under mild assumptions, quasiconvex continuous functions are classified, establishing an equivalence relation between functions with the same normal operator. Applications in pseudoconvexity are also discussed.

2000 Mathematics Subject Classification. Primary 52A01; Secondary 49J52, 26E25

Keywords and phrases. Normal cone, quasiconvexity, pseudoconvexity.

1 Introduction

The notion of a "normal cone to sublevel sets", i.e. a multivalued operator associating with every function f and every point x of its domain the normal cone to the sublevel set $S_{f(x)}$ has first been introduced and studied in [5], where the authors discussed continuity properties of this operator (or variants of it) when applied to quasiconvex functions. Subsequently, several authors used this notion (see [13], [10], [11] e.g.) for dealing with quasiconvex optimization problems.

In [4], a modification on the original definition ([5]) of the normal operator has been proposed, consisting in considering for every x the polar cone of the Clarke tangent cone of $S_{f(x)}$ at x. This new definition coincides with the previous one whenever the function f is quasiconvex, whereas it has the advantage to allow simple characterizations of various types of quasiconvexity in terms of corresponding types of quasimonotonicity of the normal operator.

In this work, following the lines of [4], we give an axiomatic formulation for the concept of normal operator, based on an abstract notion of subdifferential, see Section 2. Subsequently, we present some applications in quasiconvexity (Sections 3 and 5) and in pseudoconvexity (Section 4).

Throughout this paper, X will be a Banach space with dual X^*, and f a lower semicontinuous (lsc) function on X with values in $\mathbb{R} \cup \{+\infty\}$. For any $x \in X$ and any $x^* \in X^*$ we denote by $\langle x^*, x \rangle$ the value of the functional x^* at the point x. We also use the standard notation: $B_\delta(x)$ for the closed ball centered at x with radius $\delta > 0$, $\operatorname{dom} f := \{x \in X : f(x) \neq +\infty\}$ for the domain of the function f and $S_{f(x)} := \{x' \in X : f(x') \leq f(x)\}$ (resp. $S^-_{f(x)} = \{x' \in X : f(x') < f(x)\}$) for the sublevel and the strict sublevel sets of f. For $x, y \in X$ we set $[x, y] = \{tx + (1-t)y : 0 \leq t \leq 1\}$ and we define the segments $]x, y]$, $[x, y[$ and $]x, y[$ analogously.

2 Abstract subdifferential and normal operator

Let us first recall from [2] the definition of an abstract subdifferential.

Definition 1. We call *subdifferential operator*, any operator ∂ associating to any Banach space X, any lower semicontinuous function $f : X \to \mathbb{R} \cup \{+\infty\}$ and any $x \in X$, a subset $\partial f(x)$ of X^*, and satisfying the following properties:

(P1) $\partial f(x) = \{x^* \in X^* \;:\; \langle x^*, y - x \rangle + f(x) \leq f(y), \;\; \forall y \in X\}$, whenever f is convex;

(P2) $0 \in \partial f(x)$, whenever f attains a local minimum at $x \in \mathit{dom} f$;

(P3) $\partial(f + g)(x) \subset \partial f(x) + \partial g(x)$, whenever g is real-valued convex continuous, and *∂-differentiable at x*,

where g *∂-differentiable at x* means that $\partial g(x)$ and $\partial(-g)(x)$ are nonempty.

In the sequel, we shall assume in addition that

$$\partial \subset \partial^{\uparrow} \quad \text{or} \quad \partial \subset \partial^{D+}$$

where $\partial^{\uparrow}$ is the Clarke-Rockafellar and ∂^{D+} the upper Dini subdifferential.
Let us recall that the definitions:

$$\partial^{\uparrow} f(x) = \{x^* \in X^* : \langle x^*, d \rangle \leq f^{\uparrow}(x, d), \text{ for all } d \in X\}$$

where

$$f^{\uparrow}(x, d) = \sup_{\varepsilon > 0} \limsup_{\substack{t \searrow 0 \\ y \to_f x}} \inf_{d' \in B_\varepsilon(d)} \frac{1}{t} \left(f(y + td') - f(y) \right).$$

and

$$\partial^{D+} f(x) = \{x^* \in X^* : \langle x^*, d \rangle \leq f^{D+}(x, d), \text{ for all } d \in X\}$$

where

$$f^{D+}(x, d) = \limsup_{t \searrow 0^+} \frac{1}{t} \left(f(x + td) - f(x) \right).$$

It is recalled that $t \searrow 0^+$ indicates the fact that $t > 0$ and $t \to 0$, while $x \to_f x_o$ means that both $x \to x_o$ and $f(x) \to f(x_o)$.
We further recall from [2] the following definition.

Definition 2. A norm $\|.\|$ on X is said to be *∂-smooth* if the functions of the following form are ∂-differentiable:

$$x \mapsto \Delta_2(x) := \sum_n \mu_n \|x - v_n\|^2,$$

where $\mu_n \geq 0$, the series $\sum_n \mu_n$ is convergent, and the sequence (v_n) converges in X.

Let us also introduce the notion of an "abstract" normal cone, based on the subdifferential ∂.

Definition 3. Let ∂ be a subdifferential operator. For any closed subset C of X and any point $x \in X$ we associate the normal cone to C at the point x defined by

$$N_C(x) = \begin{cases} \partial\psi_C(x) & \text{if } x \in C \\ \emptyset & \text{otherwise} \end{cases}$$

where ψ_C denotes the indicator function of C (i.e. $\psi_C(x) = 0$ if $x \in C$ and $+\infty$ if $x \notin C$).

For all classical subdifferentials (Clarke, lower and upper Hadamard, lower and upper Dini, Fréchet, proximal...) the subset $N_C(x)$ is effectively a cone. Although this property will not be used in the sequel, to be in accordance with the term "normal cone" of the above definition, we can assume that the abstract subdifferential fulfills the following property:

$$\text{For any function } f, \text{any } \lambda > 0 \text{ and any } x \in X, \quad \partial(\lambda f)(x) = \lambda \partial f(x).$$

Whenever the subdifferential operator is the lower Hadamard subdifferential ∂^{H-}, the corresponding normal cone is the classical Bouligand normal cone defined as follows

$$NK_C(x) = \{x^* \in X^* \;:\; \langle x^*, d\rangle \leq 0,\ \forall d \in K_C(x)\} \tag{1}$$

with

$$K_C(x) = \{y = \lim_{k\to\infty} y_k \;:\; \exists t_k \searrow 0 \text{ with } x + t_k y_k \in C,\ \ \forall k \in \mathbb{N}\}$$

On the other hand, if $\partial = \partial^{\uparrow}$, then we recover the Clarke normal cone

$$N_C^{\uparrow}(x) = \{x^* \in X^* \;:\; \langle x^*, d\rangle \leq 0,\ \forall d \in T_C(x)\} \tag{2}$$

with

$$d \in T_C(x) \Leftrightarrow \begin{cases} \forall \varepsilon > 0, \exists \delta > 0 \ \text{ such that} \\ \forall x' \in B_\delta(x) \cap C, \forall t \in]0, \delta[,\ (x' + tB_\varepsilon(d)) \cap C \neq \emptyset. \end{cases}$$

We are now in a position to define the normal operator associated with a lsc function.

Definition 4. Let ∂ be a subdifferential operator. For any lower semicontinuous function $f : X \to \mathbb{R} \cup \{+\infty\}$ we associate a multivalued operator $N_f : X \to 2^{X^*}$ - called normal operator - defined by

$$x \mapsto \begin{cases} N_{S_{f(x)}}(x) & \text{if } x \in \operatorname{dom} f \\ \emptyset & \text{otherwise} \end{cases}$$

Remark: 1) In the particular case $\partial = \partial^{\uparrow}$, we recover the definition used in [4] (see relation (2)).

2) Based on the strict sublevel sets (i.e. $S_\lambda^- = \{x \in X \ : \ f(x) < \lambda\}$) an analogous concept of normal operator (called strict normal operator) has been considered in [4] (extending the original definition of [5]) :

$$\tilde{N}_f(x) = \begin{cases} \emptyset & \text{if } x \notin \operatorname{dom} f \\ X^* & \text{if } x \in \operatorname{Argmin} f \\ N_{cl(S^-_{f(x)})}(x) & \text{if } x \in cl(S^-_{f(x)}) \\ \{0\} & \text{otherwise} \end{cases}$$

Since, as showed in [4], the operator N_f is more appropriate than $\tilde{N}_f$ for the normal characterization of the different types of quasiconvexity, the use of (large) sublevel sets has been preferred for the purpose of this paper.

A natural question immediately arises concerning the relation between the multivalued operators N_f and ∂f and in particular, the possible equality between $N_f(x)$ and $cone\,(\partial f(x)) := \{tx^* \ : t \geq 0 \text{ and } x^* \in \partial f(x)\}$. This equality is not true in general. In fact several counterexamples have been given in [4] for the case $\partial = \partial^{\uparrow}$. In the following proposition we shed more light on this topic.

Let us recall that a function $f : X \to \mathbb{R} \cup \{+\infty\}$ is called *quasiconvex* if its sublevel sets S_λ are convex subsets of X. Following [6], a locally Lipschitz function is said to be *regular* at a point x, if for any $d \in X$ the classical directional derivative $f'(x,d)$ exists and is equal to the Clarke directional derivative $f^\circ(x,d)$ defined as follows:

$$f^\circ(x,d) = \limsup_{t \searrow 0^+ \ y \to x} \frac{1}{t}\left(f(y+td) - f(y)\right)$$

Proposition 1. *Let $f : X \to \mathbb{R} \cup \{+\infty\}$ be lsc such that $0 \notin \partial f(X)$.*

i) If f is quasiconvex then, for any $x \in X$,

$$cone(\partial f(x)) \subset N_f(x).$$

ii) Let us suppose, in addition, that f is Lipschitz continuous and $\partial \subset \partial^{\uparrow}$. If f is quasiconvex or f is regular, then for any $x \in X$,

$$N_f(x) = cone(\partial f(x)).$$

Proof. For $i)$ let us suppose, for a contradiction, that $x \in \operatorname{dom} f$ is such that $\partial f(x) \not\subset N_f(x) = NK_{S_f(x)}$. Hence there exists $y \in S_f(x)$ and $x^* \in \partial f(x)$ verifying $\langle x^*, y - x\rangle > 0$. Let $\delta > 0$ be such that $\langle x^*, u - x\rangle > 0$ for all $u \in B_\delta(y)$. Since f is quasiconvex, it follows (see [3] e.g.) that $f(u) \geq f(x)$ for all $u \in B_\delta(y)$. But, since y is an element of $S_f(x)$, y is a local minimum of f and therefore $0 \in \partial f(y)$ which contradicts the hypothesis.

$ii)$ is a direct consequence of [6, Th. 2.4.7]. □

Remark 1. $a)$ As proved in [14, Lemma 5.3], if ∂ is the Fréchet subdifferential, then assertion $i)$ can be obtained without the assumption "$0 \notin \partial f(X)$".

$b)$ In assertion $ii)$ of the previous proposition, the Lipschitz assumption can not be dropped. Indeed, if we define the function $f : \mathbb{R} \to \mathbb{R}$ by $f(x) = \sqrt{x}$ if $x \geq 0$ and $f(x) = -\sqrt{-x}$ otherwise, then for any $x \neq 0$, $cone(\partial f(x)) = N_f(x)$, while for $x = 0$ we have $\partial f(0) = \emptyset$ and $N_f(0) = [0, +\infty[$.

3 Normal characterizations of quasiconvexity

In this section we establish 'normal' characterizations for quasiconvex and strictly (semistrictly) quasiconvex functions in terms of the abstract normal operator N_f. These characterizations have been derived in [4] in the particular case $\partial = \partial^\dagger$.

Let us first recall the relevant definitions. A function $f : X \to \mathbb{R} \cup \{+\infty\}$ is said to be *semistrictly quasiconvex* if f is quasiconvex and for any x, $y \in \operatorname{dom} f$ we have

$$f(x) < f(y) \Longrightarrow f(z) < f(y), \ \forall\, z \in [x, y[.$$

Similarly, f is called *strictly quasiconvex*, if it is quasiconvex and for any x, $y \in \operatorname{dom} f$ and $z \in]x, y[$ we have

$$f(z) < \max\{f(x), f(y)\}.$$

For any subset K of X, let us also recall that a multivalued operator $T : X \to 2^{X^*}$ is called *quasimonotone* on K if for all $x, y \in K$ we have

$$\exists x^* \in T(x),\ \langle x^*, y - x\rangle > 0 \Rightarrow \forall y^* \in T(y) \ : \ \langle y^*, y - x\rangle \geq 0.$$

Following [8] T is called *cyclically quasimonotone* (on K), if for every $x_1, x_2, \ldots, x_n \in X$ (resp. $x_1, x_2, \ldots, x_n \in K$), there exists $i \in \{1, 2, \ldots, n\}$ such that

$$\langle x_i^*, x_{i+1} - x_i\rangle \leq 0, \forall x_i^* \in T(x_i)$$

(where $x_{n+1} := x_1$).

Furthermore ([7]), the operator T is called *semistrictly quasimonotone* on K, if T is quasimonotone on K and for any x, $y \in K$ we have

$$\exists x^* \in T(x),\ \langle x^*, y - x\rangle > 0 \Longrightarrow$$
$$\exists z \in]\frac{x+y}{2}, y[,\ \exists z^* \in T(z) \ : \ \langle z^*, y - z\rangle > 0.$$

Finally T is called *strictly quasimonotone* if T is quasimonotone and for any x, $y \in K$ we have

$$\exists\, z \in]x,y[,\ \exists\, z^* \in T(z)\ :\ \langle z^*, y - x\rangle \neq 0.$$

Let us now recall from [3] the following characterization.

Proposition 2. *Let X be a Banach space admitting a ∂-smooth renorm and let $f : X \to \mathbb{R} \cup \{+\infty\}$ be a lsc function. Then f is quasiconvex iff ∂f is quasimonotone.*

For the forthcoming characterization we need the following lemmas:

Lemma 1. *Let C be a nonempty subset of X. The following statements are equivalent:*

i) C is closed and convex.
ii) The indicator function ψ_C is convex and lsc.
iii) The indicator function ψ_C is quasiconvex and lsc.

Proof. The proof is straightforward and will be omitted. □

Lemma 2. *For any lsc quasiconvex function f, and any $x \in dom(f)$ we have:*

$$N_f(x) = NK_{S_{f(x)}}(x)$$

Proof. For every $x \in dom f$, the set $C = S_{f(x)}$ is convex and closed, hence from Lemma 1 it follows that the function ψ_C is convex and lsc. Property (P1) of Definition 1 implies that $\partial\psi_C$ does not depend on the subdifferential operator. In particular $\partial\psi_C(x)$ coincides with the cones defined in (1) and (2) respectively. □

Theorem 1. *Let X be a Banach space admitting a ∂-smooth renorm and let $f : X \to \mathbb{R} \cup \{+\infty\}$ be a lsc function. Consider the following statements:*

i) f is a quasiconvex function.

ii) $\exists x^* \in N_f(x)$, $\langle x^*, y - x\rangle > 0 \Longrightarrow f(y) > f(x)$

iii) N_f is a (cyclically) quasimonotone operator.

Then we always have i) $\Rightarrow$ ii) $\Rightarrow$ iii). Moreover if, either $\partial^{H-} \subset \partial$ and X admits a Gâteaux-smooth renorm or, $\partial \subset \partial^{D+}$ and f is radially continuous or, $\partial \subset \partial^{\uparrow}$ and f is continuous, then iii) $\Rightarrow$ i), hence all these three conditions are equivalent.

Proof. *i) $\Rightarrow$ ii).* Let us suppose that for some $x^* \in N_f(x)$ we have $\langle x^*, y-x\rangle > 0$. It follows from Lemma 2 that $N_f(x) = NK_{S_{f(x)}}(x)$. Consequently $y - x$ is not an element of $K_{S_{f(x)}}(x) = cl(\cup_{\lambda>0}\lambda(S_{f(x)} - \{x\}))$. Hence, in particular, y is not an element of $S_{f(x)}$, i.e. $f(x) < f(y)$.

ii) $\Rightarrow$ *iii)*. Take any finite family $\{x_1, \dots, x_n\}$ of points of X and suppose that for $i \in \{1, \dots, n\}$, there exists $x_i^* \in N_f(x_i)$ such that $\langle x_i^*, x_{i+1} - x_i \rangle > 0$ where $x_{n+1} = x_1$. A contradiction immediately occurs since *ii)* yields $f(x_1) < f(x_2) < \cdots < f(x_{n+1}) = f(x_1)$.

iii) $\Rightarrow$ *i)*. Let us suppose, to a contradiction, that f is not quasiconvex. Then from Lemma 1 it follows that for some $x_0 \in \operatorname{dom} f$, the function $\psi_{x_0} := \psi_{S_{f(x_0)}}$ is not quasiconvex.

If $\partial^{H-} \subset \partial$ (and X admits a Gâteaux-smooth renorm) then, in view of Proposition 2, its lower Hadamard subdifferential $\partial^{H-}\psi_{x_0}$ is not quasimonotone. Hence there exist x, $y \in \operatorname{dom} \psi_{x_0} = S_{f(x_0)}$, $x^* \in \partial^{H-}\psi_{x_0}(x)$ and $y^* \in \partial^{H-}\psi_{x_0}(y)$ satisfying $\langle x^*, y - x \rangle > 0$ and $\langle y^*, x - y \rangle > 0$. Note now that $S_{f(x)} \subseteq S_{f(x_0)}$, from which it follows that $\psi_{x_0}(\cdot) \leq \psi_x(\cdot)$. We can easily conclude that $\psi_{x_0}^{H^-}(x, d) \leq \psi_x^{H^-}(x, d)$ for all d in X, hence $\partial^{H^-}\psi_{x_0}(x) \subseteq \partial^{H^-}\psi_x(x)$. Hence $x^* \in N_f(x)$ and (similarly) $y^* \in N_f(y)$ and we obtain the desired contradiction.

In both other cases, using again Proposition 2, we conclude to the existence of x, $y \in \operatorname{dom} \psi_{x_0} = S_{f(x_0)}$, $x^* \in \partial\psi_{x_0}(x)$ and $y^* \in \partial\psi_{x_0}(y)$ satisfying $\langle x^*, y - x \rangle > 0$ and $\langle y^*, x - y \rangle > 0$.

Now we claim that $f(x) = f(y) = f(x_0)$.

[We obviously have $f(x) \leq f(x_0)$. Let us now suppose that $f(x) < f(x_0)$.

If $\partial \subset \partial^{D+}$, then from the radial continuity of f we may find some $\delta > 0$ such that $f(u) < f(x_0)$ for any element u in the segment $(x - \delta(y-x), x + \delta(y-x))$. Then it follows that the function ψ_{x_0} is constant on this segment, which is not compatible with the inequality $\langle x^*, y - x \rangle > 0$. Hence $f(x) = f(x_0)$ and for the same reasons $f(y) = f(x_0)$.

If now $\partial \subset \partial^{\uparrow}$ (and the function f is continuous), then we may take a $\delta > 0$ such that $f(u) < f(x_0)$ for all $u \in B_\delta(x)$, hence the function ψ_{x_0} is locally constant on x, which contradicts the fact that $\langle x^*, y - x \rangle > 0$. Again we conclude that $f(x) = f(x_0) = f(y)$. The claim is proved.]

Now the proof is complete. Indeed $\psi_{x_0} = \psi_x = \psi_y$. Hence, in both cases x^* is an element of $\partial\psi_{x_0}(x) = \partial\psi_x(x) = N_{S_{f(x)}}(x) = N_f(x)$ and y^* is an element of $N_f(y)$ thus furnishing a contradiction with the quasimonotonicity of N_f. □

Using essentially the same proof as in [4] it is possible to obtain the following characterizations of semistrict and strict quasiconvexity in this more general framework. Let us thus state - without proof - these results.

Theorem 2. *Let X be a Banach space admitting a ∂-smooth renorm and let $f : X \to \mathbb{R} \cup \{+\infty\}$ be lsc and continuous on its domain $dom f$. Then the following statements are equivalent:*

i) *f is a semistrictly quasiconvex function.*

ii) $\exists x^* \in N_f(x) \; : \; \langle x^*, y - x \rangle > 0 \Longrightarrow f(y) > f(z), \; \forall z \in [x, y)$

iii) *N_f is a semistrictly quasimonotone operator on $dom f$.*

Theorem 3. *Let X be a Banach space admitting a ∂-smooth renorm and let $f : X \to \mathbb{R} \cup \{+\infty\}$ be lsc and continuous on $dom f$.*

Then f is strictly quasiconvex if and only if N_f is strictly quasimonotone on $dom f$.

4 Normal cones and pseudoconvexity

In this section we shall discuss relations between normal operators and pseudoconvexity. In [1], a *differentiable* function f was called *pseudoconvex*, if for every $x, y \in dom(f)$ the inequality $\langle df(x), y - x \rangle \geq 0$ ensures $f(y) \geq f(x)$. The notion of pseudoconvexity was subsequently extended into non-smooth functions, based on the concept of subdifferential (see [12], [3]). Let us further give the definition of pseudoconvexity in an even more abstract setting.

Definition 5. Given an operator $T : X \to 2^{X^*}$, a function $f : X \to \mathbb{R} \cup \{+\infty\}$ is called T-pseudoconvex, if for any $x, y \in dom(f)$ and $x^* \in T(x)$, the inequality $\langle x^*, y - x \rangle \geq 0$ implies $f(y) \geq f(x)$.

In case $T := \partial f$, we recover the definition given in [12] (see also [9] for a summary).

Since Definition 5 of $N_f \setminus \{0\}$-pseudoconvexity and Theorem 1 *ii)* are very similar, one may wonder whether quasiconvexity and $N_f \setminus \{0\}$-pseudoconvexity differ. It is shown below (Proposition 3) that for some particular case these concepts coincide. However this is not the case in general, as shows the example of the function $f : \mathbb{R} \to \mathbb{R}$, with

$$f(x) = \begin{cases} 1, \text{ if } \mid x \mid < 1 \\ 0, \text{ if } \mid x \mid \geq 1 \end{cases} \tag{3}$$

The above function is lower semicontinuous and T-pseudoconvex (for $T = N_f \setminus \{0\}$), without being quasiconvex.

A more general example of a lsc function satisfying for all $x, y \in \operatorname{dom} f$ the property:

$$\forall x^* \in T(x), \; \langle x^*, y - x \rangle \geq 0 \Longrightarrow f(y) \geq f(z), \; \text{for all } z \in [x, y] \tag{4}$$

without being quasiconvex is given below. (Relation (4) was taken as definition for T-pseudoconvexity in [9]).

Example: Let us consider the lsc function $f : \mathbb{R}^2 \to \mathbb{R}$ defined by

$$f(x,y) = \begin{cases} 1 \text{ , if} & x < 0 \text{ and } y > 0 \\ 0 \text{ , if} & xy \geq 0 \\ -x \text{ , if} & x > 0,\ y < 0 \text{ and } -y \geq x \\ y \text{ , if} & x > 0,\ y < 0 \text{ and } -y \leq x. \end{cases} \tag{5}$$

It is easily seen that f is $N_f \setminus \{0\}$-pseudoconvex, provided that $\partial \subset \partial^{\uparrow}$. On the other hand, since

$$S_{f(0,0)} = \mathbb{R}^2 \setminus \{(x,y) \ : \ x < 0,\ y > 0\}$$

the function f is not quasiconvex.

Proposition 3. *Let $f : X \to \mathbb{R} \cup \{+\infty\}$ be a lsc radially continuous function with convex domain. Then*

i) f quasiconvex $\Rightarrow f$ $N_f \setminus \{0\}$-pseudoconvex.

ii) if, moreover, $X = \mathbb{R}^n$ and $\partial = \partial^{\uparrow}$ then f is quasiconvex iff f is $N_f(x) \setminus \{0\}$-pseudoconvex.

Proof. i) Let us assume that x, $y \in \operatorname{dom} f$ and $x^* \in N_f(x) \setminus \{0\}$ are such that $\langle x^*, y - x\rangle \geq 0$. Since $x^* \neq 0$, there exists $d \in X$ such that $\langle x^*, d\rangle > 0$. Then for $y_n = y + \frac{1}{n}d$ (with $n \in \mathbb{N}$) we have $\langle x^*, y_n - x\rangle > 0$ which implies, by *ii)* of Theorem 1 that $f(y_n) > f(x)$. Since f is radially continuous this yields $f(y) \geq f(x)$ and f is $N_f \setminus \{0\}$-pseudoconvex.

ii) To prove the converse implication, let us suppose that f is $N_f(x) \setminus \{0\}$-pseudoconvex and (towards a contradiction) z is an element of $]x, y[$ verifying

$$f(z) > \max[f(x), f(y)].$$

Since f is radially continuous, we may assume that $f(x) > f(y)$ and that there exists $\tilde{z} \in]z, y[$ such that $f(x) < f(\tilde{z}) < f(z)$. It is also no loss of generality in assuming that $f(u) > f(\tilde{z})$ for all $u \in]z, \tilde{z}]$. Thus $\tilde{z}$ is on the boundary of the closed subset $S_{f(\tilde{z})}$ and consequently $N_f(\tilde{z})$ contains a nonzero element $\tilde{z}^*$ (see [6] e.g.). On the other hand, since $f(\tilde{z}) > f(x)$, we have $\langle \alpha^*, x - \tilde{z}\rangle < 0$ for any $\alpha^* \in N_f(\tilde{z}) \setminus \{0\}$. In particular, $\langle \tilde{z}^*, y - \tilde{z}\rangle > 0$ and, according to the $N_f \setminus \{0\}$-pseudoconvexity, $f(y) > f(\tilde{z})$ which is a contradiction. □

We also recall ([8]) that an operator T is called *cyclically pseudomonotone*, if for every $x_1, x_2, ..., x_n \in X$, the following implication holds:

$$\begin{aligned} &\exists i \in \{1,2,...,n\}, \exists x_i^* \in T(x_i) : \langle x_i^*, x_{i+1} - x_i\rangle > 0 \Longrightarrow \\ &\exists j \in \{1,2,...,n\}, \forall x_j^* \in T(x_j) : \langle x_j^*, x_{j+1} - x_j\rangle < 0 \end{aligned}$$

(where $x_{n+1} := x_1$).

Let us now state the following result, to be compared with Theorem 1.

Proposition 4. *Let X be a Banach space admitting a ∂-smooth renorm and $f : X \to \mathbb{R} \cup \{+\infty\}$ be a continuous function. The following statements are equivalent:*

i) *f is quasiconvex.*

ii) *$N_f \setminus \{0\}$ is (cyclically) pseudomonotone.*

Proof. *i)* $\Rightarrow$ *ii)*. Set $T(x) = N_f(x) \setminus \{0\}$ for all $x \in X$. Let any finite subset $\{x_1, x_2, ..., x_n\}$ of X and suppose (for a contradiction) that $\langle x_1^*, x_2 - x_1 \rangle > 0$ whereas for all $j \geq 2$, and all $x_j^* \in T(x_j)$, $\langle x_j^*, x_{j+1} - x_j \rangle \geq 0$ (where $x_{n+1} := x_1$). Since $x_j^* \neq 0$, using the same arguments as in part *i)* of the previous proof, we obtain $f(x_{j+1}) \geq f(x_j)$, for $j \geq 2$. On the other hand, since $\langle x_1^*, x_2 - x_1 \rangle > 0$ we infer by Theorem 1 *ii)* that $f(x_2) > f(x_1)$. The contradiction follows easily, since $x_{n+1} := x_1$. Hence T is cyclically pseudomonotone.

ii) $\Rightarrow$ *i)*. This implication follows from Theorem 1 (*iii)* $\Rightarrow$ *i)*), since the pseudomonotonicity of $N_f \setminus \{0\}$ obviously implies the quasimonotonicity of N_f. □

It is well known (see [8] e.g.) that every ∂f-pseudoconvex lsc function is quasiconvex. Combining with Proposition 3 *i)* and proposition 1 *i)* we thus recover easily the following known result:

Corollary 1. *Suppose that f is continuous and $0 \notin \partial f(X)$. Then*
f is quasiconvex $\Longleftrightarrow$ f is ∂-pseudoconvex

5 Normally equivalent functions

As observed in [4], two functions with the same normal operator may differ by more than an additive constant. Nevertheless, using the previous definition of T-pseudoconvexity (with $T = N_f \setminus \{0\}$), it is possible to characterize, under certain regularity assumptions, the set of quasiconvex functions having the same normal operator as a given quasiconvex function. This is the aim of Theorem 4.

Let us first define an equivalent relation on the set of all real-valued functions on X as follows:

$$f \sim g \Leftrightarrow N_f(x) = N_g(x), \quad \forall x \in X.$$

Remark: It follows directly from the definition that $f \sim \varphi \circ f$ for every $f : X \to \mathbb{R}$ and every strictly increasing function $\varphi : \mathbb{R} \to \mathbb{R}$, since the functions f and $\varphi \circ f$ have the same sublevel sets.

We now denote by $\mathcal{C}$ the class of continuous quasiconvex functions $f : X \to \mathbb{R}$ satisfying the following two regularity conditions:

(a) every local minimum is a global minimum

(b) the subset

$$Argmin\ f := \{x \in X : f(x) = \inf_X f\}$$

is included in a closed hyperplane of X.

Let us remark that assumption (a) can be rewritten as follows:

(a′) For every $\lambda \in f(X)$, $\lambda > \inf_X f$: $cl(S_\lambda^-) = S_\lambda$

and that, in finite dimensional spaces, (b) is equivalent to

(b′) the subset $Argmin\ f$ has an empty interior.

Hypothesis (a) has been used in [5] in order to obtain continuity results for the normal operator.

In the following theorem, we characterize the equivalent class, denoted by $\bar{f}$, of a given function f in $\mathcal{C}$.

Theorem 4. *The equivalent class $\bar{f}$ of a given function f in $\mathcal{C}$ is the set of all $N_f \setminus \{0\}$-pseudoconvex functions, that is*

$$\bar{f} = \{g \in \mathcal{C} \ : \ \exists x^* \in N_f(x) \setminus \{0\} \ : \ \langle x^*, y - x\rangle \geq 0 \Longrightarrow g(y) \geq g(x)\}.$$

Another way to express this result is to say that a function g of $\mathcal{C}$ has the same normal operator as a given function f of $\mathcal{C}$ if, and only if, g is $N_f \setminus \{0\}$-pseudoconvex.

Proof. Let us denote by $\mathcal{C}_f$ the subset of $\mathcal{C}$ defined by

$$\mathcal{C}_f = \{g \in \mathcal{C} \ : \ \exists x^* \in N_f(x) \setminus \{0\} \ : \ \langle x^*, y - x\rangle \geq 0 \Longrightarrow g(y) \geq g(x)\}.$$

(i) Let us first show $\bar{f} \subseteq \mathcal{C}_f$:
Suppose that $g \in \bar{f}$ and let x, $y \in X$ and $x^* \in N_f(x) \setminus \{0\} = N_g(x) \setminus \{0\}$ be such that

$$\langle x^*, y - x\rangle \geq 0. \tag{6}$$

If the inequality (6) is strict, then from Theorem 1 we conclude $g(y) > g(x)$.
In case where equality holds in (6), there exists a sequence $(y_n)_n \subset X$ converging to y such that $\langle x^*, y_n - x\rangle > 0$, for any $n \in \mathbb{N}$. It follows $g(y_n) \geq g(x)$, which together with the upper semicontinuity of g yields $g(y) \geq g(x)$.
(ii) We shall now show $\mathcal{C}_f \subseteq \bar{f}$:
Let any $g \in \mathcal{C}_f$.
<u>*Step 1:*</u> $N_f(x) \subseteq N_g(x)$, for all $x \in X$.
Assume, for a contradiction, that there exists $x \in X$ and $x^* \in N_f(x)$ such that $x^* \notin N_g(x)$.
Claim: $x \in Argmin\ g$

[Indeed, if x is not an element of *Argmin* g then, using assumption (a') and the fact that x^* is not an element of $N_g(x)$, we immediately obtain the existence of a point y of $S^-_{g(x)}$ satisfying $\langle x^*, y - x\rangle \geq 0$. A contradiction occurs since the definition of C_f now yields $g(y) \geq g(x)$. The claim is proved.]

Since $x^* \notin N_g(x)$, there exists $\bar{y} \in S_{g(x)} = Argmin\ g$ such that

$$\langle x^*, \bar{y} - x\rangle > 0. \tag{7}$$

Obviously

$$g(x) = g(\bar{y}) = \min g \tag{8}$$

On the other hand, x^* is an element of $N_f(x)$ and therefore, (7) implies, $f(\bar{y}) > f(x)$.

Pick now any λ in $]f(x), f(\bar{y})[$. Since f is continuous, there exists $\varepsilon > 0$ such that

$$B_\varepsilon(x) \subset S^-_\lambda(f). \tag{9}$$

Due to the closedness of $S_\lambda(f)$, one can find $t \in]0,1[$ such that

$$C_t \cap S_\lambda(f) = \emptyset \tag{10}$$

where $C_t = \{t\bar{y} + (1-t)u\ ;\ u \in B_\varepsilon(x)\}$. Since $int(C_t) \neq \emptyset$, assumption (b) implies the existence of a point $\tilde{x} \in B_\varepsilon(x)$ such that for $\tilde{y} = t\bar{y} + (1-t)\tilde{x}$ we have:

$$g(\tilde{y}) > g(\bar{y}) = g(x) \tag{11}$$

Thanks to (10), it is no loss of generality to assume that $f(z) > f(\tilde{x})$ for all z in $]\tilde{x}, \tilde{y}]$. Applying thus a separation argument to the disjoint convex sets $]\tilde{x}, \tilde{y}]$ and $S_{f(\tilde{x})}(f)$, we conclude that there exists $\tilde{x}^* \in N_f(\tilde{x}) \setminus \{0\}$ such that $\langle \tilde{x}^*, \tilde{y} - \tilde{x}\rangle \geq 0$.

The definition of C_f now yields $g(\tilde{y}) \geq g(\tilde{x})$. The contradiction is obtained, since, using (8) with the quasiconvexity of g we get $g(\tilde{y}) = g(\bar{y})$, which is not compatible with (11). Hence $N_f(x) \subseteq N_g(x)$, for all $x \in X$.

Step 2: $N_g(x) \subset N_f(x)$, for all $x \in X$.

We shall also proceed by contradiction. So let us suppose that there exist $x \in X$ and $x^* \in N_g(x)$ such that x^* is not an element of $N_f(x)$. This implies the existence of a point y of $S_{f(x)}(f)$ which is not in $S_{g(x)}(g)$, i.e. $g(y) > g(x)$.

Case 1: The interior of $S_{f(x)}(f)$ is nonempty.

In this case we claim that there exists $\bar{z}$ such that $f(\bar{z}) < f(x)$ and $g(\bar{z}) > g(x)$.

Indeed if $f(y) < f(x)$, then take $\bar{z} = y$. Otherwise we have $f(x) = f(y)$, and thanks to hypothesis (a') there exists a sequence $\{y_n\}_{n\geq 1}$ in $S^-_{f(x)}(f)$ converging to y. Since g is continuous and $g(y) > g(x)$, the claim follows for $\bar{z} = y_n$ and n sufficiently large.

Now one can separate (in a large sense) the subsets $S_{f(\bar{z})}(f)$ and $\{x\}$. Hence there exists $\bar{z}^* \in N_f(\bar{z}) \setminus \{0\}$ such that

$$\langle \bar{z}^*, x - \bar{z}\rangle \geq 0.$$

This immediately implies, from the definition of $\mathcal{C}_f$, that $g(\bar{z}) \leq g(x)$ which is impossible.

Case 2. The set $S_{f(x)}(f)$ has an empty interior.

In this case we have $f(x) = f(y) = \min f$. We shall conclude again to a contradiction. Indeed, by hypothesis (b) there exists $\alpha^* \in X^* \setminus \{0\}$ such that

$$Argmin\ f \subseteq H_{\alpha^*} = \{u \in X \ : \ \langle \alpha^*, u - y \rangle = 0\}$$

Thus $\alpha^* \in N_f(y) \setminus \{0\}$, hence according to the definition of $\mathcal{C}_f$, $g(x) \geq g(y)$ which is impossible.

Consequently N_f coincides with N_g and the proof is complete. □

Example: If $X = \mathbb{R}$, the class $\mathcal{C}$ consists of the equivalent classes determined by the functions $\bar{f}_1(x) = x$, $\bar{f}_2(x) = -x$ and $\bar{f}_{3,\alpha}(x) = |x - \alpha|$ (for $\alpha \in \mathbb{R}$). For example, the function defined in Remark 1(b), after Proposition 1, is an element of $\bar{f}_1$.

Remarks: 1. Two equivalent functions $f, g \in \mathcal{C}$ do not necessarily have the same family of sublevel sets. Consider for instance the functions $f(x) = |x|$ and $g(x) = \max\{x, -2x\}$. Note that both functions belong to the class defined by $\bar{f}_{3,0}$ (see the previous example).

2. It is possible to consider quasiconvex functions taking the value $+\infty$. In this case one can obtain a result similar to Theorem 4 under the assumption that all functions have the same domain. Without this assumption, the fore mentioned result is not true, as can be shown by easy counterexamples.

Acknowledgement. The authors are grateful to N. Hadjisavvas for quoting the possibility of extending Theorem 1 in some particular cases.

The research of the second author was supported by the TMR post-doctoral grant ERBFMBI CT 983381.

References

1. Avriel, M., Diewert, W.T., Schaible, S. & Zang, I., *Generalized Concavity*, Plenum Publishing Corporation, New York, 1988.
2. Aussel, D., Corvellec, J.-N. & Lassonde, M., Nonsmooth constrained optimization and multidirectional mean value inequalities, *SIAM J. Optim.* **9** (1999), 690-706.
3. Aussel D., Subdifferential Properties of Quasiconvex and Pseudoconvex Functions: A Unified Approach, *J. Optim. Th. Appl.* **97** (1998), 29-45.
4. Aussel, D. & Daniilidis, A., Normal characterization of the main classes of quasiconvex functions, *Set-Valued Anal.*, **8** (2000), 219-236.
5. Borde, J. & Crouzeix, J.-P., Continuity Properties of the Normal Cone to the Level Sets of a Quasiconvex Function, *J. Optimization Theory Appl.* **66** (1990), 415-429.

6. CLARKE, F., Optimization and nonsmooth analysis, Wiley Interscience, New York 1983 (Republished in 1990: Vol. 5, Classics in Applied Mathematics, Society for Industrial and Applied Mathematics, Philadelphia, Pa.).
7. DANIILIDIS, A. & HADJISAVVAS, N., Characterization of nonsmooth semistrictly quasiconvex and strictly quasiconvex functions, *J. Optimization Theory Appl.* **102** (1999), 525-536.
8. DANIILIDIS, A. & HADJISAVVAS, N., On the Subdifferentials of Quasiconvex and Pseudoconvex Functions and Cyclic Monotonicity, *J. Math. Anal. Appl.* **237** (1999), 30-42.
9. HADJISAVVAS, N. & SCHAIBLE, S., "Generalized Monotone Multivalued Maps", in: *Encyclopedia of Optimization*, (Pardalos, P & Floudas, G eds), Kluwer (to appear).
10. HASSOUNI, A. Quasimonotone multifunctions; applications to optimality conditions in quasiconvex programming, *Numer. Funct. Anal. and Optimiz.* **13** (1992), 267-275.
11. HASSOUNI, A. & JADDAR, A., On quasiconvex functions and applications to optimality conditions in nonlinear programming, *Applied Mathematics Letters* (to appear).
12. PENOT, J.-P. & QUANG, P.H., Generalized convexity of functions and generalized monotonicity of set-valued maps, *J. Optimization Theory Appl.* **92** (1997), 343-356.
13. PENOT, J.-P., Are generalized derivatives useful for generalized convex functions?, in J.-P. Crouzeix, J.-E. Martinez-Legaz and M. Volle (eds), *Generalized Convexity, Generalized Monotonicity*, Kluwer Academic Publishers, Dordrecht, pp. 3-59, 1998.
14. VOLLE, M., Duality for the level sum of quasiconvex functions and applications, *ESAIM: Control, Optimization and Calculus of Variations* **3** (1998), 329-343.

Multiobjective Programming with ρ-convex Functions

Antonio Beato-Moreno, Rafaela Osuna-Gómez, Antonio Rufián-Lizana, and Pascual Ruiz-Canales

Facultad de Matemáticas, Universidad de Sevilla, Sevilla, 41012, Spain

Abstract. During the last years numerous results about optimality in mathematical programming problems with generalized convex functions have been obtained, with special attention to multiobjetive problems.

In this paper, we consider multiobjective problems with ρ- convex functions and some additional hypothesis. The objective is to deduce theoretical properties that permit to develop algorithms in order to derive efficient points for the analyzed problems. In particular, we study the application of properties to certain biobjective problems.

2000 Mathematics Subject Classification. Primary 90C29; Secondary 90C20, 90C25.

Keywords and phrases. generalized convexity, multiobjective programming, efficiency.

1 Introduction

In the last decades several generalizations for the concept of convexity have been proposed. The concept of ρ-convexity was introduced by Vial [31] who also gave some applications to mathematical programming and geometry.

The aim of this paper is to establish some procedures to obtain efficient points in multicriteria programming problems with ρ-convex objective functions. In particular, we develop results to give a search procedure for efficient points in quadratic bicriteria programming problems.

In the sequel $B(x, r)$ is the open ball of radius r and center x, and $\partial B(x,r)$ is its boundary. Let $X \subset I\!R^n$ be a nonempty convex set and $f = [f_1, \ldots, f_m]$ a vectorial function with $m \geq 2$ components f_i, $i \in \{1, \ldots, m\}$, defined on X. The multicriteria programming problem (MP) can be written as

$$\text{(MP)} \quad \text{“minimize”} \quad f(x), \qquad \text{subject to } x \in X.$$

The decision making processes with multiple criteria appear frequently in many areas of research. As examples we can cite scheduling (Ref. [2], Ref. [22]), nutrition planning (Ref. [6], Ref. [11]), financial planning (Ref. [22]) and many others (Ref. [9], Ref. [14], Ref. [15], Ref. [28] and Ref. [33]).

In problem (MP) each scalar function $f_i(x)$ measures the effect produced in objective i when the decisor selects action x. When the criterion to discriminate between all the possible decisions $x \in X$ is that the effect in each objective will be smallest possible, the concept of solution for the problem (MP) that arises in a natural way is the concept of (locally) efficient solution.

Definition 1. We say that a point $x_0 \in X$ is an (locally) efficient point for (MP) if there does not exist $x \in X$ ($x \in B(x_0, r)$, for some $r > 0$) such that $f(x) \leq f(x_0), f(x) \neq f(x_0)$.

Let $Eff(MP)$ ($Leff(MP)$) be the set of (locally) efficient points for problem (MP). We say that a point $x \in X$ dominate to $x' \in X$ in (MP) if $f(x) \leq f(x'), f(x) \neq f(x')$.

From the theoretical point of view, one of the principal difficulties in the problem (MP) is the characterization of efficient points. A frequently used technique is to obtain local conditions that all efficient points $x_0 \in X$ must fulfill. In this way we obtain a subset that contains the efficient set. Thereafter additional conditions are found under which the obtained points are efficient. Therefore, the first phase of the procedure of detecting efficient points is the determination of locally efficient points for problem (MP). In the second phase we give conditions under which locally efficient points are efficient.

There exist several results that characterize efficient points, weakly efficient points and properly efficient points under generalized convexity and differentiability hypothesis through first order conditions. Results of this type appear collected in Ref. [1], Ref. [9], Ref. [7], Ref. [9], Ref. [12], Ref. [20] or Ref. [26]. If we eliminate the generalized convexity hypothesis, it is only possible to characterize the locally efficient points, locally weakly efficient points and locally properly efficient points. We will analyze how to obtain efficient points from locally efficient points in (MP) with a class of ρ-convex objective functions.

2 Unconstrained multicriteria problems with ρ convex functions

Let $f = [f_1, \ldots, f_m]$ be a vectorial function with $m \geq 2$ components f_i, $i = 1, \ldots, m$ defined on $\mathbb{R}^n$. The unconstrained multicriteria problem (MPU) can be written as

$$\text{(MPU)} \quad \text{“minimize”} \quad f(x), \qquad \text{subject to } x \in \mathbb{R}^n.$$

In this paper we consider (MPU) with weakly convex functions. We propose a iteration procedure to obtain efficient points starting from locally efficient points. In particular we characterize locally efficient points in bicriteria quadratic programming problems and then we apply the proposed procedure.

Definition 2. Let $g : D \longrightarrow \mathbb{R}$ be a real-valued function defined on a convex subset D of $\mathbb{R}^n$. f is said to be ρ-convex if there exists some real number ρ such that $\forall x_1, x_2 \in D$, $\forall \lambda \in [0, 1]$

$$g(\lambda x_1 + (1 - \lambda)x_2) \leq \lambda g(x_1) + (1 - \lambda)g(x_2) - \rho\lambda(1 - \lambda)\|x_1 - x_2\|^2.$$

If $\rho > 0$, the function is said to be strongly convex.
If $\rho = 0$, the function is convex.
If $\rho < 0$, the function is said to be weakly convex.

Proposition 1. (**Vial,[31]**) *If g_1 and g_2 are, respectively, ρ_1-convex and ρ_2-convex, and α_1 and α_2 are positive real numbers, then $\alpha_1 g_1 + \alpha_2 g_2$ is $(\alpha_1\rho_1 + \alpha_2\rho_2)$-convex.*

Proposition 2. (**Vial,[31]**) *$g : I\!R^n \longrightarrow I\!R$ is ρ-convex if and only if there exists a convex function $h : I\!R^n \longrightarrow I\!R$ such that $g(x) = h(x) + \rho\|x\|^2$.*

Proposition 3. (**Vial,[31]**) *A twice differentiable function is ρ-convex if and only if there exists some real ρ such that for any $x \in I\!R^n$ and $v \in I\!R^n$*

$$\langle \nabla^2 g(x)v, v\rangle \geq 2\rho\|v\|^2.$$

In particular, a quadratic form is ρ-convex with respect to $\rho = e/2$, where e is the smallest eigenvalue of the Hessian matrix.

Remark 1. Let $g : I\!R^n \longrightarrow I\!R$ be a real-valued quadratic function with $g(x) = x^t Ax + b^t x$, where A is a matrix $n \times n$, $b \in I\!R^n$. As consequence of the previous proposition, $g(x)$ is $e/2$-convex, because it is the sum of an affine function (0-convex) and a quadratic form ($e/2$-convex), where e is the smallest eigenvalue of the Hessian matrix.

In this paper, we will delimit the spectrum of functions for which we will obtain results to weakly convex quadratic functions. In future works, we intend to study the extension of these results to other weakly convex functions.

3 Unconstrained multicriteria problem with weakly convex quadratic functions

We will call quadratic multicriteria programming problem to

$$\text{(QMP)} \qquad \begin{array}{l} \text{Min} \quad f(x) = (f_1(x), \ldots, f_m(x)), \\ x \in I\!R^n \end{array}$$

where $f_i(x) = 1/2\ x^t A_i x + b_i^t x$, $x \in I\!R^n$, A_i is a $n \times n$ matrix, $b_i \in I\!R$, $i = 1, \ldots, m$.

Problem (QMP) has been widely studied by several authors, for example in Ref. [4], Ref. [14] and Refs. [3]-[27].

Let the function $H : \partial B_n(0,1) \longrightarrow I\!R$ defined by

$$H(d) = \max_{i \in \{1,\ldots,m\}} d^t A_i d, \qquad d \in \partial B_n(0,1).$$

There are several works about the function $H(d)$. For references and results about the function $H(d)$, the reader is referred to the paper [10].

We will consider the sets D^+, D^- and D^0 as $H^{-1}(R_+)$, $H^{-1}(R_-)$ and $H^{-1}(0)$ respectively.

Remark 2. Obviously $\partial B_n(0,1) = D^+ \cup D^- \cup D^0$.

Lemma 1. *Consider the problem (QMP). If $D^- \neq \emptyset$ then $Eff(QMP) = \emptyset$.*

Proof. Suppose that $D^- \neq \emptyset$. Let $d \in D^-, d \neq 0$. Then

$$d^t A_i d < 0, \quad \forall i \in \{1, \ldots, m\}.$$

Let $x_0 \in I\!R^n$. Then

$$f_i(x_0 + \lambda d) = f_i(x_0) + \lambda \nabla f_i(x_0)^t d + 1/2\ \lambda^2 d^t A_i d.$$

We can obtain points with smaller value than x_0 in all the objective functions by taking $\lambda \longrightarrow +\infty$. Consequently $Eff(QMP) = \emptyset$. □

Assumption: In the sequel we will suppose that $D^+ = \partial B_n(0,1)$.

Let $x_0 \in Leff(QMP)$, $i \in \{1, \ldots, m\}$ and $d \in I\!R^n$. The following instances may occur:

1. $d^t A_i d < 0$, $\nabla f_i(x_0)^t d > 0$. The function $g^i_{x_0,d}(\lambda) = f_i(x_0 + \lambda d)$ is weakly convex. Therefore we can define the α-extension of a point $x \in I\!R^n$ as follows.

   Definition 3. (Ref. [17]) Let $h : I\!R^n \longrightarrow I\!R$ be a concave function, $x \in I\!R^n$, and let α be a real number satisfying $\alpha \leq h(x)$. A point $y \in I\!R^n$ is called α-extension of x in direction $d \in I\!R^n \setminus \{0\}$ (with respect to h) if

   $$y = x + \theta d \quad with \quad \theta \ = \ \sup\{t : h(x + td) \geq \alpha\}.$$

   As $x_0 \in Leff(QMP)$, we derive from the figure 1 that there is no point $x' = x_0 + \lambda d$ such that $f_i(x') \leq f_i(x_0)$ for $\lambda < \theta$ and θ an $f_i(x_0)$-extension of x_0.
2. $d^t A_i d > 0$, $\nabla f_i(x_0)^t d < 0$. The function $g^i_{x_0,d}(\lambda) = f_i(x_0 + \lambda d)$ is strongly convex. Therefore we can define the α-extension of a point $x \in I\!R^n$ as follows.

   Definition 4. Let $h : I\!R^n \longrightarrow I\!R$ be a convex function, $x \in I\!R^n$, and let α be a real number satisfying $\alpha \geq h(x)$. A point $y \in I\!R^n$ is called α-extension of x in direction $d \in I\!R^n \setminus \{0\}$ (with respect to f) if

   $$y = x + \theta d \quad with \quad \theta \ = \ \sup\{t : h(x + td) \leq \alpha\}.$$

   As $x_0 \in Leff(QMP)$, we derive from the figure 2 that there is no point $x' = x_0 + \lambda d$ such that $f_i(x') \leq f_i(x_0)$ for $\lambda > \theta$ and θ an $f_i(x_0)$-extension of x_0.

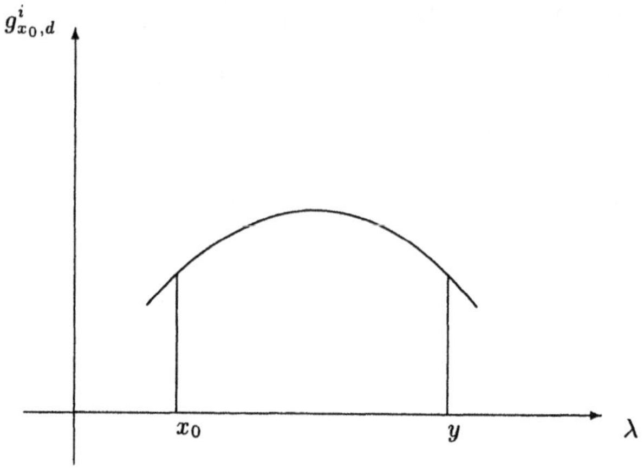

Fig. 1. $f_i(x_0) - $ *extension* of a weakly convex quadratic function

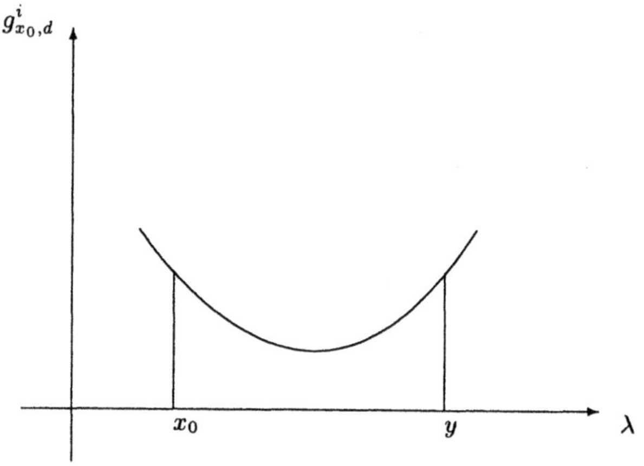

Fig. 2. $f_i(x_0) - $ *extension* of a strongly convex quadratic function

3. $d^t A_i d \geq 0$, $\nabla f_i(x_0)^t d \geq 0$. There is no point $x' = x_0 + \lambda d$ such that $f_i(x') \leq f_i(x_0)$ for $\lambda > 0$, except if $d^t A_i d = \nabla f_i(x_0)^t d = 0$.
4. $d^t A_i d \leq 0$, $\nabla f_i(x_0)^t d \leq 0$. Then $f_i(x') \leq f_i(x_0)$, where $x' = x_0 + \lambda d$, $\forall \lambda \geq 0$.

For each $d \in \partial B_n(0,1)$, define

$$g^i_{(x_0,d)}(\lambda) = f_i(x_0 + \lambda d)$$

$$X_1(d) \quad = \{i \in \{1,\ldots,m\} \ / d^t A_i d < 0\}$$

$$X_2(d) \quad = \{i \in \{1,\ldots,m\} \ / d^t A_i d \geq 0\}$$

$$\overline{g(d)} \quad = \max_{i \in X_1(d)} \quad \{\|y - x_0\| : y \text{ is a } f(x_0) - \text{extension of } g^i_{(x_0,d)}(\lambda)\}$$

$$\underline{g(d)} \quad = \min_{i \in X_2(d)} \quad \{\|y - x_0\| : y \text{ is a } f(x_0) - \text{extension of } g^i_{(x_0,d)}(\lambda)\}$$

$$L \quad = \{d \in \partial B_n(0,1) \ / \text{ there is not } i \text{ for which } f_i(x_0 + \lambda d) > f_i(x_0), \ \forall \lambda > 0\}.$$

Lemma 2. *Let $x_0 \in Leff(QMP)$.If*

$$\sup_{d \in L} \left(\overline{g(d)} - \underline{g(d)}\right) < 0,$$

then $x_0 \in Eff(QMP)$.

Proof. By contradiction. Given $x_0 \in Leff(QMP)$, suppose that the assumption of the lemma is verified but $x_0 \notin Eff(QMP)$. This is equivalent to saying that there exists a value $\lambda_0 > 0$ and a vector $d_0 \in \partial B(0,1)$ such that the point $y = x_0 + \lambda_0 d_0$ dominates the point x_0, that is,

$$f_i(x_0 + \lambda_0 d_0) \leq f_i(x_0), \qquad i = 1,\ldots,m, \tag{1}$$

with at least one strict inequality.

Obviously $d_0 \in L$. Consider the sets $X_1(d_0)$ and $X_2(d_0)$, which verify

$$\{1,\ldots,m\} \subseteq X_1(d_0) \cup X_2(d_0).$$

From the definition of $\overline{g(d_0)}$ we have

$$\lambda_0 \leq \overline{g(d_0)},$$

and from the definition of $\underline{g(d_0)}$ we get

$$\lambda_0 \geq \underline{g(d_0)}.$$

Therefore,

$$\underline{g(d_0)} \leq \lambda_0 \leq \overline{g(d_0)}$$

and this contradicts the hypothesis of lemma. □

Corollary 1. *Let x_0 be a locally efficient point for the problem (QMP). Let $\beta = \min_{d \in \partial B_n(0,1)} \; g(d)$. Then there is no $x' \in B(x_0, \beta)$ such that $f(x') \leq f(x_0)$.*

Proof. It follow from the proof of lemma 2. □

Evidently this result gets greater usefulness when the type of objective function permits an easy computation of the value m.

Lemma 3. *We consider the problem (QMP). Let $x_0 \in Leff(QMP)$ and*

$$F(d) = \max_{i \in X_1(d)} \quad 2\nabla f_i(x_0)^t d.$$

Suppose that there exists a number ρ such that $F(d) \geq \rho$, $\forall d \in \partial B_n(0,1)$. Then there is no $x' \in B(x_0, \frac{\rho}{-\gamma})$ such that $f(x') \leq f(x_0)$, where $\gamma < 0$ is the smallest negative eigenvalue of the matrices $A_i, i \in \{1, \ldots, m\}$.

Proof. Let $d \in \partial B_n(0,1)$. By the proof of the lemma 2, the value β of the previous corollary must fulfill:

$$\beta \geq \frac{-2\nabla f_i(x_0)^t d}{d^t A_i d}, \quad \forall i \in X_1(d).$$

Let γ_i be the smallest eigenvalue of the matrix A_i, $i \in \{1, \ldots, m\}$. Then $d^t A_i d \geq \gamma_i$ and therefore

$$\beta \geq \min_{d \in \partial B_n(0,1)} \quad \max_{i \in X_1(d)} \quad \frac{-2\nabla f_i(x_0)^t d}{d^t A_i d} \geq$$

$$\geq \min_{d \in \partial B_n(0,1)} \quad \max_{i \in X_1(d)} \quad \frac{-2\nabla f_i(x_0)^t d}{\gamma_i} =$$

$$= \min_{d \in \partial B_n(0,1)} \quad \frac{F(d)}{-\gamma_i} \geq \max_{d \in \partial B_n(0,1)} \quad \frac{F(d)}{-\gamma},$$

where $\gamma = \min_{i \in \{1, \ldots, m\}} \; \gamma_i$. As by hypothesis $F(d) \geq \rho$, $\forall d \in \partial B_n(0,1)$, we obtain $\beta \geq \rho / -\gamma$. □

The proof of the lemma 2 provides also a condition to decide if a locally efficient point is efficient in (QMP).

Theorem 1. *Let $x_0 \in Leff(QMP)$. Suppose that*

$$\max_{i \in \{1, \ldots, m\}} \quad d^t A_i d > \varepsilon, \quad \forall d \in \partial B_n(0,1), \quad \varepsilon > 0.$$

If there is no $x' \in B(x_0, M)$ such that $f(x') \leq f(x_0)$, where

$$M = \max_{i \in \{1, \ldots, m\}} \quad \frac{2\|\nabla f_i(x_0)\|}{\varepsilon},$$

then $x_0 \in Eff(QMP)$.

Proof. As it is deduced from the proof of lemma 2, if there exists a point x that dominates x_0, then $\|x - x_0\| \leq M$, with

$$M = \max_{d \in \partial B_n(0,1)} \underline{g(d)},$$

from where, according to the assumption,

$$\begin{aligned}\underline{g(d)} &= \min_{i \in X_2(d)} \frac{-2\nabla f_i(x_0)^t d}{d^t A_i d} \leq \\ &\leq \min_{i \in X_2(d)} \frac{|-2\nabla f_i(x_0)^t d|}{\varepsilon} \leq \\ &\leq \min_{i \in X_2(d)} \frac{\|-2\nabla f_i(x_0)^t\|}{\varepsilon}.\end{aligned}$$

Then,

$$\begin{aligned}M &\leq \max_{d \in \partial B_n(0,1)} \min_{i \in X_2(d)} \frac{\|-2\nabla f_i(x_0)^t\|}{\varepsilon} \leq \\ &\leq \max_{i \in \{1,\dots,m\}} \frac{\|-2\nabla f_i(x_0)^t\|}{\varepsilon}.\end{aligned}$$

Therefore if there does not exist a point that dominates x_0 in $B(x_0, M)$, then there does not exist a point in $\mathbb{R}^n$ with this property and $x_0 \in Eff(QMP)$. □

With the previous theorem, we can reduce an unconstrained problem to a constrained problem that can be studied by branch and bound techniques.

4 Efficient points in (QMP)

In this section we will give a procedure to obtain efficient point from locally efficient points in (QMP).

The procedure will start from locally efficient points, so we give a characterization of the locally efficient points in the problem (QMP).

4.1 Characterization of locally efficient points

We will use the following notation:

$$\begin{aligned}E(x_0) &\equiv \{d \in \partial B_n(0,1) \ / \ \max_{i \in \{1,\dots,m\}} \nabla f_i(x_0)^t d = 0\} \\ I(d) &\equiv \{i \in \{1,\dots,m\} \ / \ \nabla f_i(x_0)^t d = 0\}.\end{aligned}$$

Theorem 2. *We consider the problem (QMP). Then $x_0 \in Leff(QMP)$ if and only if there exists $\lambda_1, \dots, \lambda_m \geq 0$ not all zero such that*

$$\lambda_1 \nabla f_1(x_0) + \dots + \lambda_m \nabla f_m(x_0) = 0 \tag{2}$$

and $\forall d \in E(x_0)$ we obtain $d^t A_j d > 0$ for some $j \in I(d)$ or $\nabla f_i(x_0)^t d = d^t A_i d = 0, \ \forall i \in I(d)$.

Proof. We will prove that (2) is sufficient. The condition (2) implies (see 4.2.3 of Ref. [21]) that the system

$$\nabla f_i(x_0)^t d < 0, \quad \forall i \in \{1,\ldots,m\},$$

does not have a solution $d \in I\!R$.

Therefore if there exists $d \in I\!R^n$ and a index $i \in \{1,\ldots,m\}$ such that $\nabla f_i(x_0)^t d < 0$, there must exist $j \neq i$ such that $\nabla f_j(x_0)^t d \geq 0$.

1. If it is $\nabla f_j(x_0)^t d > 0$ for some $j \in \{1,\ldots,m\}$, evidently in some neighborhood of x_0 the function $f_j(x)$ increases in direction d.
2. If $\max_{j\in\{1,\ldots,m\}} \nabla f_j(x_0)^t d = 0$, then by the Gordan theorem, $I(d) \neq \emptyset$ and $d^t A_j d > 0$ for some $j \in I(d)$ by hypothesis. Therefore the function $f_j(x)$ increases in direction d.
   When $\nabla f_i(x_0)^t d = 0$, $\forall i \in \{1,\ldots,m\}$, then by hypothesis $d^t A_i d = 0$, $\forall i \in \{1,\ldots,m\}$.

In any case, for each $d \in I\!R^n$ either there exists a function that increases in this direction or no function decreases in this direction. Then $x_0 \in Leff(QMP)$.

We will prove that the condition (2) is necessary.

We suppose that $x_0 \in Leff(QPM)$.

If (2) is not verified , we can find $d \in I\!R^n$ such that $\nabla f_i(x_0)^t d < 0$ $\forall i \in \{1,\ldots,m\}$, which contradicts that $x_0 \in Leff(QMP)$.

If (2) is verified, but the remaining hypothesis are not fulfilled, then there exists $d \in E(x_0)$ such that $\nabla f_i(x_0)^t d < 0$, $\forall i \notin I(d)$ and $\nabla f_i(x_0)^t d = 0$ $\forall i \in I(d)$, $d^t A_i d \leq 0$ $\forall i \in I(d)$ (but it can not occur $\nabla f_i(x_0)^t d = d^t A_i d = 0$ $\forall i \in I(d)$).

This contradicts that $x_0 \in Leff(QMP)$, since the functions $f_i(x)$ can decrease near x_0, $i \notin I_1(d)$, the functions $f_i(x)$ can decrease near x_0, $i \in I(d)$, $d^t A_i d < 0$ and the remaining functions $f_i(x)$, $i \in I(d)$, $d^t A_i d = 0$ do not increase. □

The following sufficient condition is deduced from the previous theorem.

Corollary 2. *We consider the problem* (QMP). $x_0 \in Leff(QMP)$ *if there exist* $\lambda_1,\ldots,\lambda_m \geq 0$*, not all zero, such that*

$$\lambda_1 \nabla f_1(x_0) + \ldots + \lambda_m \nabla f_m(x_0) = 0 \tag{3}$$

and $G(d) > 0$ *in* $R(x_0)$*, where*

$$G(d) = \max_{\{i \ / \ \lambda_i > 0\}} \quad d^t A_i d,$$
$$R(x_0) = \{d \in I\!R^n \ / \ \nabla f_i(x_0)^t d = 0, \ \forall i \text{ with } \lambda_i > 0\}.$$

Proof. Let $d \in I\!R^n$. If $\nabla f_i(x_0)^t d < 0$ for some $\lambda_i > 0$, then by (3), there must exist $j \neq i$ such that $\nabla f_j(x_0)^t d > 0$, and then the function $f_j(x)$ increases in direction d near x_0.

If $\nabla f_i(x_0)^t d = 0$, for all i such that $\lambda_i > 0$, then the hypothesis of the theorem assures that there exists an index j such that $\lambda_j > 0$ (and therefore $\nabla f_j(x_0)^t d = 0$) and $d^t A_j d > 0$. Then the function $f_j(x)$ increases in direction d near x_0.

If $\nabla f_i(x_0)^t d > 0$ then $f_i(x)$ increases in direction d near x_0.

In all three cases, there exists a increasing function in every direction $d \in I\!R^n$ near x_0. Thus $x_0 \in Leff(QMP)$. □

The characterization of locally efficient points obtained in the theorem 2 is not very easy to apply. Next we present a more suitable method to check if x_0 is a locally efficient point. To simplify the exposition, we will suppose that $\nabla f_i(x_0) \neq 0$, $\forall i \in \{1, \ldots, m\}$.

Algorithm 1: LOCAL EFFICIENCY TEST.

Step 0. Set $k = 1$, $\nabla f_i^1(x_0) = \nabla f_i(x_0)$, $L^1(x_0) = I\!R^n$.

Step 1 Find $\lambda_1^k, \ldots, \lambda_m^k \geq 0$ not all zero such that $\sum_{i=1}^m \lambda_i^k \nabla f_i^k(x_0) = 0$. If such λ_i^k do not exist, go to step 4.

Step 2 Set $L^{k+1}(x_0) \equiv L^k \cap \{d \in I\!R^n \ / \ \nabla f_i^k(x_0)^t d = 0, \forall i \text{ with } \lambda_i^j > 0, \ j \leq k\}$.

Step 3 Consider the subspace $L^{k+1}(x_0)$ instead of L^k and reiterate the process. In this step we should find the projection of $\nabla f_i^k(x_0)$ to $L^{k+1}(x_0)$, $\nabla f_i^{k+1}(x_0)$ and do $k := k + 1$. If it is possible to choose all the $\lambda_i^k > 0$, go to step 4. If not go to step 1.

Step 4 Check whether the system $d^t A_i d \leq 0$, $i = 1, \ldots, m$, $\lambda_i^j > 0$ for some $j \leq k$, does not have solution $d \in V$ with some strict inequality, where V is the set defined by:

$$V = \{d \in R^n \ / \ \nabla f_i(x_0)^t d = 0, \text{ if } \lambda_i^j > 0 \text{ for some } j \leq k,$$
$$\nabla f_i(x_0)^t d \leq 0, \text{ if } \lambda_i^j = 0 \text{ for all } j \leq k\}.$$

If there is no solution, then $x_0 \in Leff(QMP)$.

This local efficiency test can make $m - 1$ iterations at maximum before going to step 4. This step is the most complex one, since it implies to solve a quadratic inequalities system with linear constraints.

Remark 3. In the previous algorithm, when we conclude $x_0 \notin Leff(QMP)$, we derive a vector d_0 in the step 4 that can be used to find a point that dominates x_0 by a one dimensional search.

4.2 Obtaining efficient points

Let $x_0 \in Leff(QMP)$. The objective is to obtain an efficient point starting of x_0. For this we will follow the next algorithm:

Algorithm 2: OBTAINING EFFICIENT POINTS

Step 1 Check whether the current point is efficient.

Step 2 In negative case, find a direction and a distance for a new locally efficient point that dominates the current point.For this step we will use the results of section 3.

Reiterate the process with the new point until finding an efficient point.

The algorithm 2 provides an iterative method to find efficient points in the problem (QMP).

Next, we study in detail each step of the algorithm 2.

STEP 1

To explain this step, we suppose that x_0 is a locally efficient point. This implies that if there exists a vector $d \in \mathbb{R}^n$ such that $\nabla f_i(x_0)^t d = 0,\ \forall i$, then there exists an index i_0 such that $d^t A_{i_0} d > 0$ (since all the functions cannot decrease in that direction) or $d^t A_i d = 0,\ \forall i$. Therefore, there is no point that dominates x_0 in direction d.

From the previous reasoning we see that if a point y dominates x_0, we can find it advancing in a direction d such that

$$d \in \ L(x_0) \equiv \ \langle\ \nabla f_1(x_0), \ldots, \nabla f_m(x_0)\ \rangle.$$

Here, $\langle \cdot, \ldots, \cdot \rangle$ denote the subspace generated by its arguments.

Based on the results of section 3 and the previous paragraph, we set:

$$\begin{aligned} m(d) &= \max \quad \left\{ -\frac{2\nabla f_i(x_0)^t d}{d^t A_i d} : \ \nabla f_i(x_0)^t d > 0, d^t A_i d < 0 \right\} \\ M(d) &= \min \quad \left\{ \frac{2\nabla f_i(x_0)^t d}{d^t A_i d} : \ \nabla f_i(x_0)^t d < 0, d^t A_i d > 0 \right\}. \end{aligned}$$

To conclude that $x_0 \in Eff(QMP)$, we must calculate

$$\mu(x_0) = \max_{d \in L(x_0)} \quad (M(d) - m(d)). \tag{4}$$

When $\mu(x_0) \leq 0$ we will conclude that $x_0 \in Eff(QMP)$.

STEP 2

If $x_0 \notin Eff(QMP)$, we can use d_0, the solution of the problem outlined in (4), and advance by $M(x_0)$, arriving to the new point $x_0' = x_0 + M(x_0)d_0$. Thereafter, from x_0' we can obtain a new point $x_0'' \in Leff(QMP)$ that dominates x_0' through a local search method.

4.3 Numerical Example

We consider the following problem:

$$\begin{array}{ll} \min & (f_1(x), f_2(x)) \\ x \in I\!R^2 & \end{array}$$

where $f_1(x) = -2x_1^2 + 4x_2^2 + 4x_1 - 8x_2 + 2$ and $f_2(x) = 5x_1^2 - 4x_2^2 - 20x_1 + 32x_2 - 44$. If $x_0 \in Leff(QPM)$ there must exist $\lambda_1, \lambda_2 \geq 0$, not all zero, such that

$$\lambda_1 \nabla f_1(x_0) + \lambda_2 \nabla f_2(x_0) = 0$$

For $\lambda_2 = 0$ we obtain the point $(1,1)$, which is locally efficient. The remaining locally efficient points must verify

$$\lambda \nabla f_1(x_0) + \nabla f_2(x_0) = 0$$

with $\lambda \geq 0$.

The necessary condition in this problem is also sufficient, since if

$$\nabla f_1(x_0)^t d = 0$$

and furthermore $\nabla f_2(x_0)^t d = 0$, $d \in I\!R^2$, either $d^t A_1 d > 0$ or $d^t A_2 d > 0$.

Therefore the locally efficient points, excluding the point $(1,1)$, are those that verify:

$$\begin{cases} 4\lambda(x_1 - 1) - 10(x_1 - 2) & = 0 \\ -8\lambda(x_2 - 1) + 8(x_2 - 2) & = 0, \end{cases}$$

or rather,

$$\begin{cases} x_1 = \dfrac{2\lambda - 10}{2\lambda - 5} \\ \\ x_2 = \dfrac{-\lambda + 4}{-\lambda + 1} \end{cases} \tag{5}$$

with $\lambda \geq 0$.

The locally efficient points for this problem are represented in the figure 3. To apply the algorithm 2, we should start from a locally efficient point. For example, if we take the point that appears in the previous equations for $\lambda = 0.5$, that is $(x_1, x_2) = (2.25, 7)$, and applying algorithm 2 we obtain:

Step 1.- $x_0 = (2.25, 7)$ is not efficient, since for each direction (d_1, d_2),

$$f_1(2.25 + \lambda d_1, 7 + \lambda d_2) = -2(d_1^2 - 2d_2^2)\lambda^2 + (48d_2 - 5d_1)\lambda + 1127/8,$$
$$f_2(2.25 + \lambda d_1, 7 + \lambda d_2) = (5d_1^2 - 4d_2^2)\lambda^2 + ((5d_1 - 48d_2)/2)\lambda - 571/16,$$

and taking $d = (1, -1)$, it is M(d)=9.7797423, while $m(d) < M(d)$.

Step 2.- We take for example $d = (1, -1)$ and we advance from $x_0 = (2.25, 7)$ a quantity $\lambda(x_0) = 9.7797423$, obtaining the point

$$x_0' = (5.160575, -2.336590).$$

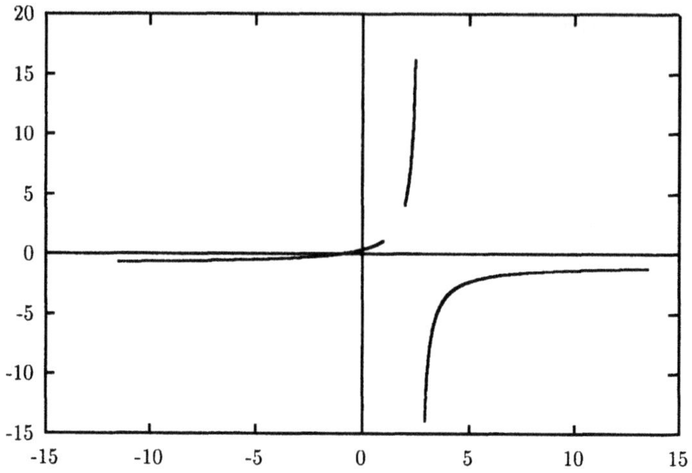

Fig. 3. Locally efficient points for the example.

This point is locally efficient, since it verifies the necessary first order conditions $\lambda = 1.899121479$ (in this problem this conditions are also sufficient for local efficiency). We return to the step 1.
Step 1.- We conclude that the point x_0' is efficient, since $\mu(x_0') \leq 0$.

5 Conclusions

In this paper we have studied the multicriteria programming problem with ρ-convex functions. We have delimited the spectrum of functions for which we will obtain results to weakly convex quadratic functions. The purpose of this paper is to establish some procedures to obtain efficient points in these problems.

We have characterized the locally efficient points for the problem (QMP) defined in section 3 and next we have proposed a iteration procedure to obtain efficient points starting from locally efficient points. The proposed method reduces the problem of finding efficient points in (QMP) to the resolution of one dimensional problems that can be solved by techniques of global optimization (see for example Ref. [13], Ref. [18], Ref. [17], Ref. [19], Ref. [24] or Ref. [32]).

The procedure could be adapted to more general functions through the use of α-extension defined in section 3.

References

1. M. Avriel, W.E. Diewert, S. Schaible and I. Zang (Eds.) (1988). Generalized Concavity, Plenum Press, New York.

2. U. Bagchi (1989). Simultaneous minimization of mean and variation of flow time and waiting time in Single Machine Systems, Oper. Res., **37**, 118–125.
3. A. Beato-Moreno (1995). Eficiencia en Programación Cuadrática Multiobjetivo PhD Thesis, Departamento de Estadística e Investigación Operativa, Universidad de Sevilla, Sevilla, Spain.
4. A. Beato-Moreno, P.L. Luque-Calvo, A. Rufián-Lizana and P. Ruiz-Canales (1997). Programación Cuadrática Bicriterio, XXIII Congreso Nacional de Estadística e Investigación Operativa, Valencia, Spain.
5. A. Beato-Moreno, P.L. Luque-Calvo, R. Blanquero Bravo and P. Ruiz-Canales (1997). Multiobjective Quadratic Problem: characterization of the efficient points, in Proceedings of the 5th Symposium on Generalized Convexity, Kluwer Academic Publisher.
6. H.P. Benson and T.L. Morin (1987). A bicriteria Mathematical Programming model for nutrition planning in developing nations, Management Sci., **53**, 1593–1601.
7. A. Ben-Israel, A. Ben-Tal and A. Charnes (1977). Necessary and sufficient conditions for a Pareto optimum in Convex Programming, Econometrica, **45**, 811–820.
8. M. Castellani and M. Pappalardo (1998) Characterizations of ρ-convex functions, in Generalized Convexity, Generalized Monotonicity, 219–233, edited by J.P. Crouzeix, J.E. Martinez-Legaz and M. Volle, Kluwer Academic Publishers, Dordrecht.
9. V. Chankong and Y.Y. Haimes (1983). Multiobjective Decision Making: Theory and Methodology, Elsevier Science Publishing Co. Inc. North-Holland.
10. X. Chen and Ya-xiang Yuan (1999). A note on quadratic forms, Math. Program., **86**, 187-197.
11. N. Demokan and A.H. Land (1981). A parametric Quadratic Program to solve a class of Bicriteria Decision Problems, J. Oper. Res. Soc., **32** 477-488.
12. R.R. Egudo (1989). Efficiency and generalized convex duality for Multiobjective Programs, J. Math. Anal. Appl., **138**, 84–94.
13. C.A. Floudas and P.M. Pardalos (1992). Recent advances in Global Optimization, Princeton Series in Computer Science, Princeton University Press, Princeton, New Jersey.
14. C.J. Goh and X.Q. Yang (1996). Analytic efficient solution set for Multicriteria Quadratic Programs, Eur. J. Oper. Res., **92**, 166–181.
15. A. Goicoechea, D.R. Hansen, and L. Duckstein (1982). Multiobjective Decision Analysis with engineering and business applications , Wiley, New York.
16. S. Helbig (1990). An algorithm for Quadratic Vector Optimization Problems, Z. Angew. Math. Mech., **70**, 751–753.
17. R. Horst, P.M. Pardalos and N.V. Thoai (1995). Introduction to Global Optimization, Kluwer Academic Publishers, Dordrecht.
18. R. Horst and H. Tuy (1990). Global Optimization: Deterministic Approaches , Springer-Verlag.
19. R. Horst and P.M.Pardalos (Eds.) (1995). Handbook of Global Optimization, Kluwer Academic Publishers, Dordrecht.
20. S. Komlósi, T. Rapksak and S. Schaible (Eds.) (1994). Generalized Convexity Springer-Verlag Heidelberg.
21. O.L. Mangasarian (1969). Nonlinear Programming McGraw Hill.
22. P.M. Pardalos, Y. Siskos and C. Zopounidis Ed. (1995). Advances in Multicriteria Analysis, Kluwer Academic Publishers, Dordrecht.

23. V. Preda (1992). On Efficiency and duality for Multiobjective Programs, J. Math. Anal. Appl., **166**, 365–377.
24. G.R. Reeves (1975). Global minimization in non convex all Quadratic Programming, Management Sci., **22**, 76–86.
25. Y. Sawaragi, H. Nakayama and T. Tanino (1985). Theory of Multiobjective Optimization Academic Press.
26. S. Schaible and W.T. Ziemba (Eds.), Generalized Concavity in Optimization and Economics Academic Press, New York, 1981.
27. S.D. Sharma and N. A. Sulayman (1993). A new method for Multi-objective Indefinite Quadratic Programming Problems, Acta Cienc. Indica Math., **19**, 23–26.
28. W. Stadler Ed. (1988). Multicriteria Optimization in Engineering and in the Sciences , Plenum, New York.
29. R.E. Steuer (1986). Multiple Criteria Optimization. John Wiley & Son, New York.
30. S. K. Suneja, C. R. Bector and C. Singh (1991). Duality in Multiobjective Fractional Programming involving strongly and weakly invex and related functions Opsearch, **28**, N. 3, 153–164.
31. J.P. Vial (1983). Strong and weak convexity of sets and functions, Math. Oper. Res., **8**, N. 2, 231–259.
32. V. Visweswaran and C.A. Floudas (1993). New properties and computational improvement of the GOP Algorithm for problems with quadratic objective functions and constraints, J. Glob. Optim., **3**, 439–462.
33. D.J. White (1990). A bibliography on the applications of Mathematical Programming Multiple-Objective methods, J. Oper. Res. Soc., **41**, 669–691.
34. D. J. White (1982). Optimality and Efficiency John Wiley & Sons, New York.

Vector Invex N-set Functions and Minmax Programming

Davinder Bhatia[1] and Promila Kumar[2]

[1] Department of Operational Research,
Faculty of Mathematical Sciences,
University of Delhi,
Delhi - 110 007, India.
[2] Department of Mathematics,
Gargi College, Khel Gaon Marg,
New Delhi - 110 049, India.

Abstract. Vector invexity and generalized vector invexity for n-set functions is introduced which is then utilized to establish sufficient optimality and duality results for a class of minmax programming problems involving n-set functions. Applications of these results to fractional programming problems are also presented.

2000 Mathematics Subject Classification. Primary ; Secondary .

Keywords and phrases. Optimality, duality, Minmax Programming, V-invexity.

1 Introduction

A class of minmax programming problems is an important class of programming problems. The significance of minmax models and methods is well known in many areas of decision making, engineering design and applied mathematics. A fairly extensive treatment of minmax theory and applications is given by Damyanov and Malozemov [10]. Bector and Chandra [4] extended duality for minmax programs involving pseudolinear functions. Weir and Mond [22] studied the same problems under pseudoconvexity assumptions. Jaykumar and Mond [12] introuduced the notion of V-invexity for a vector function and discussed its application to a class of constrained multiobjective optimization problem. Bector, Chandra and Kumar [5] applied this definition of V-invexity to establish sufficient optimality and duality results for minmax programs. All these authors worked for point functions.

Morris [17] defined the notion of local convexity, global convexity and differentiability for set functions and established optimality conditions and lagrangian duality for minmax programming problems involving set functions. The methods and results advanced by Morris have been used and further extended in [6], [8], [9], [11], [14], and [20]. A different approach to defining and characterizing the notion of convexity for set functions is followed in [15], [18], [19] and some optimality and duality results based on this approach are obtained in [15]. There always has been an urge to find the largest possible class of functions that satisfy the optimality conditions. For this reason, attempts have been made to weaken the concept of convexity. In the present paper we move a step further in this direction by introducing the notion of V-invexity and its generalization for n-set functions. Let us, now, consider the following minmax programming problem :

$$\text{(P)} \qquad \min \max_{1 \leqq j \leqq r} F_j(S)$$

$$\text{subject to } H_k(S) \leqq 0, \; k \in K = \{1,2,...,m\},$$

$$S = (S_1, S_2, ..., S_n) \in \mathcal{A}^n,$$

where $\mathcal{A}^n$ is the n-fold product of σ-algebra $\mathcal{A}$ of subsets of a given set X. F_j, $j \in J = \{1,...,r\}$ and $H_k, k \in K = \{1,...,m\}$ are real valued differentiable n-set functions defined on $\mathcal{A}^n$. For each $j \in J$, $F_j(S) \geqq 0$, for all $S \in \mathcal{A}^n$.

Sufficient optimality and duality results are established for the problem (P) under generalized V-invexity assumptions on the functions involved. These results are then applied to study fractional minmax programming problems.

2 Preliminaries

Throughout the paper we assume that $(X, \mathcal{A}, \mu)$ is a finite atomless measure space with $L_1(X, \mathcal{A}, \mu)$ separable, where μ is a measure defined on $\mathcal{A}$. Let 'd' be a pseudometric on $\mathcal{A}^n$, defined by

$$d(R,S) = \left[\sum_{i=1}^{n} \mu^2(R_i \triangle S_i)\right]^{1/2}, \quad R_i, S_i \in \mathcal{A}, i \in N.$$

Thus $(\mathcal{A}^n, d)$ is a pseudometric space which will serve as the domain for most of the functions used in this paper. For $h \in L_1(X, \mathcal{A}, \mu)$ and $Z \in \mathcal{A}$ with indicator function $I_Z \in L_\infty(X, \mathcal{A}, \mu)$ the integral $\int_Z h \, d\mu$ will be denoted by $\langle h, I_Z \rangle$.

We shall next define the notion of differentiability for n-set functions. As mentioned earlier, these notions were originally introduced by Morris [17] for set functions. The n-set counter part were defined by Corley [9]. Here we shall follow [9] as far as differentiability is concerned.

Definition 2.1. A function $P : \mathcal{A} \to R$ is said to be differentiable at $\bar{S}$ if there exists $DP(\bar{S}) \in L_1(X, \mathcal{A}, \mu)$ called the derivative of P at $\bar{S}$, such that for each $S \in \mathcal{A}$,

$$P(S) = P(\bar{S}) + \langle DP(S), I_S - I_{\bar{S}} \rangle + V_P(S, \bar{S})$$

where $V_P(S, \bar{S})$ is $O(d(S, \bar{S}))$, that is,

$$\lim_{d(S,\bar{S}) \to 0} V_P(S, \bar{S}) / d(S, \bar{S}) = 0$$

Definition 2.2. A function $Q : \mathcal{A}^n \to R$ is said to have partial derivative at $\bar{S} = (\bar{S}_1, \bar{S}_2, ..., \bar{S}_n)$ with respect to the ith argument if the function $P(S_i) = Q(\bar{S}_1, ..., \bar{S}_{i-1}, S_i, \bar{S}_{i+1}, ..., \bar{S}_n)$ has derivative $DP(\bar{S}_i), i \in N$. In that case the ith partial derivative of Q at $\bar{S}$ is defined to be $D_iQ(\bar{S}) = DP(\bar{S}_i)$, $i \in N$.

Definition 2.3. A function $Q : \mathcal{A}^n \to R$ is said to be differentiable at $\bar{S}$ if all the partial derivatives $D_iQ(\bar{S}), i \in N$, exist and satisfy the following condition:

$$Q(S) = Q(\bar{S}) + \sum_{i=1}^{n} \langle D_iQ(\bar{S}), I_{S_i} - I_{\bar{S}_i} \rangle + W_Q(S, \bar{S})$$

where $W_Q(S, \bar{S})$ is $O(d(S, \bar{S}))$, for all $S \in \mathcal{A}^n$.

Definition 2.4. A feasible solution $\bar{S}$ is said to be a regular feasible solution for (P) if there exists $\widehat{S} \in \mathcal{A}^n$ such that

$$H_k(\bar{S}) + \sum_{i=1}^{n} \langle D_iH_k(\bar{S}), I_{\widehat{S}_i} - I_{\bar{S}_i} \rangle < 0, \quad k \in K.$$

Now, we shall resort to the following equivalent problem :

$$\text{(EP)} \qquad \text{minimize } q$$

$$\text{subject to } F_j(S) \leqq q, \quad j \in J \tag{1}$$

$$H_k(S) \leqq 0, \quad k \in K \tag{2}$$

$$S \in \mathcal{A}^n, q \in \mathbb{R}_+ = \{x \in \mathbb{R} : x > 0\} \tag{3}$$

Lemma 2.1. [23] *If $\bar{S}$ is an optimal solution of (P) then $(\bar{S}, \bar{q})$ with $\bar{q} = \max_{1 \leqq j \leqq r} F_j(\bar{S})$ is an optimal solution of (EP). Conversely, if (S^0, q^0) is optimal for (EP), then S^0 is optimal for (P).*

Theorem 2.1. [23] *Let $\bar{S}$ be a regular optimal solution of (P). Assume that $F_j, j \in J$ and $H_k, k \in K$ are differentiable at $\bar{S}$. Then there exist $\bar{u} \in \mathbb{R}^r_+, \sum_{j=1}^{r} \bar{u}_j = 1, \bar{v} \in \mathbb{R}^m_+$, $\bar{q} \in \mathbb{R}_+$ with $\bar{q} = \max_{1 \leqq j \leqq r} F_j(\bar{S})$ such that*

$$\langle D_i(\bar{u}F)(\bar{S}) + D_i(\bar{v}H)(\bar{S}), I_{S_i} - I_{\bar{S}_i} \rangle \geqq 0, \quad S_i \in \mathcal{A}, i \in N \tag{4}$$

$$\bar{u}_j[F_j(\bar{S}) - q] = 0, \qquad j \in J \tag{5}$$

$$\bar{v}_k H_k(\bar{S}) = 0, \qquad k \in K \tag{6}$$

Definition 2.5. Given $\eta : \mathcal{A}^n \times \mathcal{A}^n \to \mathbb{R}^n$ and $\theta_j : \mathcal{A}^n \times \mathcal{A}^n \to \mathbb{R}^+ \backslash \{0\}, j \in J$, a vector function $F : \mathcal{A}^n \to \mathbb{R}^r$ is said to be V-invex with respect to η and $\theta_j, j \in J$, if for each $S, T \in \mathcal{A}^n$ and $j \in J$,

$$F_j(S) - F_j(T) \geqq \theta_j(S,T) \sum_{i=1}^{n} \eta_i(S,T) \langle D_iF_j(T), I_{S_i} - I_{T_i} \rangle$$

Definition 2.6. Given $\eta : \mathcal{A}^n \times \mathcal{A}^n \to \mathbb{R}^n$ and $\theta_j : \mathcal{A}^n \times \mathcal{A}^n \to \mathbb{R}^+ \backslash \{0\}, j \in J$, a vector function $F : \mathcal{A}^n \to R^r$ is said to be V-pseudoinvex with respect to η and $\theta_j, j \in J$, if for each $S, T \in \mathcal{A}^n$ and $j \in J$,

$$\sum_{j=1}^{r}\left[\sum_{i=1}^{n} \eta_i(S,T)\langle D_i F_j(T), I_{S_i} - I_{T_i}\rangle\right] \geqq 0$$

$$\Rightarrow \quad \sum_{j=1}^{r} \phi_j(S,T) F_j(S) \geqq \sum_{j=1}^{r} \phi_j(S,T) F_j(T)$$

or equivalently

$$\sum_{j=1}^{r} \phi_j(S,T) F_j(S) < \sum_{j=1}^{r} \phi_j(S,T) F_j(T)$$

$$\Rightarrow \quad \sum_{j=1}^{r}\left[\sum_{i=1}^{n} \eta_i(S,T)\langle D_i F_j(T), I_{S_i} - I_{T_i}\rangle\right] < 0$$

Definition 2.7. Given $\eta : \mathcal{A}^n \times \mathcal{A}^n \to \mathbb{R}^n$ and $\theta_j : \mathcal{A}^n \times \mathcal{A}^n \to R^+ \backslash \{0\}, j \in J$, a vector function $F : \mathcal{A}^n \to \mathbb{R}^r$ is said to be V-quasiinvex with respect to η and $\psi_j, j \in J$, if for each $S, T \in \mathcal{A}^n$ and $j \in J$,

$$\sum_{j=1}^{r} \psi_j F_j(S) \leqq \sum_{j=1}^{r} \psi_j F_j(T)$$

$$\Rightarrow \quad \sum_{j=1}^{r}\left[\sum_{i=1}^{n} \eta_i(S,T)\langle D_i F_j(T), I_{S_i} - I_{T_i}\rangle\right] \leqq 0$$

It is obvious that every V-invex function is both V-pseudoinvex (with $\phi_j = 1/\theta_j$) and V-quasiinvex (with $\psi_j = 1/\theta_j$). Also if we set $r = 1, \theta_j(S,T) = 1, \phi_j(S,T) = 1, \psi_j(S,T) = 1, j \in J$ and $\eta_i(S,T) = 1, i \in N$ then the above definition reduce to those of convexity, pseudoconvexity and quasiconvexity respectively. If F and G are point functions then the above definitions reduce to the definition of V-invexity given by Jaykumar and Mond [12].

3 Sufficient optimality conditions

Theorem 3.1. *Let* (T, q^0, u^0, v^0) *with* $T \in \mathcal{A}^n$, $q^0 \in \mathbb{R}, u^0 \in \mathbb{R}^r_+$, $v \in \mathbb{R}^m_+$ *satisfy*

$$\langle D_i(u^0 F)(T) + D_i(v^0 H)(T), I_{S_i} - I_{T_i}\rangle \geqq 0, \quad S_i \in A, i \in N \tag{7}$$

$$u_j^0(F_j(T) - q^0) = 0, \quad j \in J \tag{8}$$

$$v_k^0 H_k(T) = 0, \quad k \in K \tag{9}$$

$$\sum_{j=1}^{r} u_j^0 = 1 \tag{10}$$

Further, let

1. *$(u_1^0F_1, u_2^0F_2, ..., u_r^0F_r)$ be V-pseudoinvex with respect to $\eta : \mathcal{A}^n \times \mathcal{A}^n \to \mathbb{R}^n$ and $\phi_j : \mathcal{A}^n \times \mathcal{A}^n \to \mathbb{R}^+\backslash\{0\}, j \in J$*
2. *$(v_1^0H_1, v_2^0H_2, ..., v_m^0H_m)$ be V-quasiinvex with respect to the same η and $\psi_k : \mathcal{A}^n \times \mathcal{A}^n \to \mathbb{R}^+\backslash\{0\}, k \in K$*

with η satisfying $\eta^t\alpha < 0 \Rightarrow \alpha_i < 0$ for at least one $i \in N$.
Then T is optimal for (P) with the corresponding optimal objective value equal to q^0.

Proof. Let $S \in \mathcal{A}^n$ be any feasible solution to (P) or let (S, q) be any feasible solution to (EP). Let if possible

$$q < q^0. \tag{11}$$

Since $\phi_j(S,T) > 0, j \in J,\ S, T \in \mathcal{A}^n$, it follows from (1), (8), (10) and (11) that

$$\sum_{j=1}^{r} \phi_j(S,T)u_j^0F_j(S) < \sum_{j=1}^{r} \phi_j(S,T)u_j^0F_j(T) \tag{12}$$

V-pseudoconvexity of $(u_1^0F_1, u_2^0F_2, ..., u_r^0F_r)$ along with (12) yields that

$$\sum_{j=1}^{r}\sum_{i=1}^{n} \eta_i(S,T)\langle D_i(u_j^0F_j)(T), I_{S_i} - I_{T_i}\rangle < 0$$

$$\Rightarrow \quad \sum_{i=1}^{n} \eta_i(S,T)\langle D_i(u^0F)(T), I_{S_i} - I_{T_i}\rangle < 0 \tag{13}$$

Now (2) along with (9) yields

$$v_k^0H_k(S) \leqq v_k^0H_k(T), \quad k \in K \tag{14}$$

Now positivity of $\psi_k(S,T), k \in K, S, T \in \mathcal{A}^n$ along with (14) yields

$$\sum_{k=1}^{m} \psi_k(S,T)v_k^0H_k(S) \leqq \sum_{k=1}^{m} \psi_k(S,T)v_k^0H_k(T)$$

Now using V-quasiinvexity of $(v_1^0H_1, v_2^0H_2, ..., v_m^0H_m)$, we get

$$\sum_{k=1}^{m}\sum_{i=1}^{n} \eta_i(S,T)\langle D_i(v_k^0H_k)(T), I_{S_i} - I_{T_i}\rangle \leqq 0$$

$$\Rightarrow \quad \sum_{j=1}^{m} \eta_i(S,T)\langle D_i(v^0H)(T), I_{S_i} - I_{T_i}\rangle \leqq 0 \tag{15}$$

Adding (13) and (15) we get

$$\sum_{i=1}^{n} \eta_i(S,T)\langle D_i(u^0 F)(T) + D_i(v^0 H)(T), I_{S_i} - I_{T_i}\rangle < 0$$

$\Rightarrow \quad \langle D_i(u^0 F)(T) + D_i(v^0 H)(T), I_{S_i} - I_{T_i}\rangle < 0$, for at least one $i \in N$ which contradicts (7). Hence $q \geq q^0$, which ensures that (T^0, q^0) is optimal for (EP) proving thereby that T is optimal for (P) with optimal objective value q^0.□

4 Duality

In this section, we introduce a Mond-Weir [6] type dual (D) to the problem (P) and establish duality results under generalized V-invexity assumptions.

$$\begin{aligned}
\textbf{(D)} \qquad & \text{maximize } \xi \\
& \text{subject to } \langle D_i(uF)(T) + D_i(vH)(T), I_{S_i} - I_{T_i}\rangle \geqq 0 \qquad (16)\\
& \qquad i \in N, S, T \in \mathcal{A}^n \\
& u_j[F_j(T) - \xi] \geqq 0, \quad j \in J \qquad (17)\\
& v_k H_k(T) \geqq 0, \quad k \in K \qquad (18)\\
& T \in \mathcal{A}^n, u \in \mathbb{R}_+^r, \sum_{j=1}^{r} u_j = 1, \quad v \in \mathbb{R}_+^m, \xi \in \mathbb{R}_+ \qquad (19)
\end{aligned}$$

Theorem 4.1. (Weak Duality) *Let (S,q) be feasible for (EP) and let (T,ξ,u,v) be feasible for (D). Let $(u_1F_1, u_2F_2, ..., u_rF_r)$ be V-pseudoinvex with repect to functions η and $\phi_j, j \in J$ and $(v_1H_1, v_2H_2, ..., v_mH_m)$ be V-quasiinvex with respect to functions η and $\psi_k, k \in K$ such that $\eta^t\alpha < 0 \Rightarrow \alpha_i < 0$ for at least one $i \in N$. Then $q \geq \xi$.*

Proof. Let if possible $q < \xi$.
It follows from (1), (17) and (19) that

$$\sum_{j=1}^{r} \phi_j(S,T) u_j F_j(S) < \sum_{j=1}^{r} \phi_j(S,T) u_j F_j(T) \qquad (20)$$

where $\phi_j : A^n \times A^n \to \mathbb{R}^+ \backslash \{0\}$ i.e. $\phi_j(S,T) > 0$, $S, T \in \mathcal{A}^n$.
Now V-pseudoinvexity of $(u_1F_1, u_2F_2, ..., u_rF_r)$ along with (20) implies that

$$\begin{aligned}
& \textstyle\sum_{j=1}^{r} \sum_{i=1}^{n} \eta_i(S,T)\langle D_i(u_j^0 F_j)(T), I_{S_i} - I_{T_i}\rangle < 0 \\
\Rightarrow \quad & \textstyle\sum_{i=1}^{n} \eta_i(S,T)\langle D_i(uF)(T), I_{S_i} - I_{T_i}\rangle < 0 \qquad (21)
\end{aligned}$$

Now from (2) and (17) we have

$$v_k H_k(S) \leqq v_k H_k(T), \quad k \in K$$

for any function $\psi_k : A^n \times A^n \to \mathbb{R}^+\backslash\{0\}$

$$\psi_k(S,T)v_k H_k(S) \leqq \psi_k(S,T)v_k H_k(T) \tag{22}$$

Now V-quasiinvexity of $(v_1H_1, v_2H_2, ..., v_mH_m)$ along with (22) yields

$$\begin{aligned} & \textstyle\sum_{k=1}^{m}\sum_{j=1}^{n}\eta_i(S,T)\langle D_i(v_kH_k)(T), I_{S_i} - I_{T_i}\rangle \leqq 0 \\ \Rightarrow \quad & \textstyle\sum_{i=1}^{n}\eta_i(S,T)\langle D_i(vH)(T), I_{S_i} - I_{T_i}\rangle \leqq 0\,. \end{aligned} \tag{23}$$

Adding (21) and (23) we get

$$\sum_{i=1}^{n}\eta_i(S,T)\langle D_i(uF)(T) + D_i(vH)(T), I_{S_i} - I_{T_i}\rangle < 0$$

$$\Rightarrow \quad \langle D_i(uF)(T) + D_i(vH)(T), I_{S_i} - I_{T_i}\rangle < 0 \tag{24}$$

for at least one $i \in N$ (because $\eta^t\alpha < 0 \Rightarrow \alpha_i < 0$ for at least $i \in N$). Now (24) contradicts (16). Hence $q \geq \xi$. This completes the proof. □

Theorem 4.2. (Strong Duality) *Let $\bar{S}$ be a regular optimal solution of (P) and assume that $F_j, j \in J$ and $H_k, k \in K$ are differentiable at $\bar{S}$. Then there exist $\bar{u} \in \mathbb{R}^r_+$ $\sum_{j=1}^{r}\bar{u}_j = 1, \bar{v} \in \mathbb{R}^m_+, \bar{q} \in R_+$ such that $(\bar{S}, \bar{q}, \bar{u}, \bar{v})$ is feasible for (D). Further, if the conditions of weak duality theorem hold then $(\bar{S}, \bar{q}, \bar{u}, \bar{v})$ is an optimal solution of (D) and $\max\limits_{1\leqq j\leqq r} F_j(\bar{S}) = \bar{q}$.*

Proof. By Theorem 2.1 there exists $\bar{u}$, $\bar{v}$ and $\bar{q} = \max\limits_{1\leqq j\leqq r} F_j(\bar{S})$ such that $(\bar{S}, \bar{q}, \bar{u}, \bar{v})$ is a feasible solution of (D). Optimality of $(\bar{S}, \bar{q}, \bar{u}, \bar{v})$ follows from the weak duality theorem. □

Theorem 4.3. (Strict Converse Duality) *Let (S^0, q^0) be a regular optimal solution of (EP). Let (T^0, ξ^0, u^0, v^0) be optimal for (D) and let, at all (EP) feasible and (D) feasible solutions*

1. *$(u_1^0F_1, u_2^0F_2, ..., u_r^0F_r)$ be V-pseudoinvex with respect to $\eta : A^n \times A^n \to \mathbb{R}^n$ and $\phi_j : A^n \times A^n \to \mathbb{R}^+\backslash\{0\}, j \in J$,*
2. *$(v_1^0H_1, v_2^0H_2, ..., v_m^0H_m)$ be V-quasiinvex with respect to the same $\eta : A^n \times A^n \to \mathbb{R}^n$ and $\psi_k : A^k \times A^n \to \mathbb{R}^+\backslash\{0\}, k \in K$,*

with η satisfying $\eta^t\alpha < 0 \Rightarrow \alpha_i < 0$ for at least one $i \in N$.
Then $(S^0, q^0) = (T^0, \xi^0)$.

Proof. Since (S^0, q^0) is a regular optimal solution of (EP) at which the conditions (i) and (ii) stated above, are satisfied. Therefore, there exist $u^* \in \mathbb{R}^r_+, \sum_{j=1}^{r} u_j^* = 1, v^* \in \mathbb{R}^m_+$ such that (S^0, q^0, u^*, v^*) is optimal for (D).
Also, for $k \in K, v_k^0H_k(S^0) \leqq 0$ and $v_k^0H_k(T^0) \geqq 0$.

Therefore,

$$v_k^0 H_k(S^0) \leqq v_k^0 H_k(T^0)\,. \tag{25}$$

Now proceeding along the lines of Theorem 4.1, V-quasiinvexity of

$$(v_1^0 H_1, v_2^0 H_2, ..., v_m^0 H_m)$$

with respect to η and ψ_j along with (25) yields

$$\sum_{i=1}^{n} \eta_i(S^0, T^0)\langle D_i(v^0 H)(T^0), I_{S_i^0} - I_{T_i^0}\rangle \leqq 0 \tag{26}$$

Further, let if possible

$$q^0 < \xi^0 \tag{27}$$

Proceeding again along the lines of Theorem 4.1 and using V-pseudoinvexity of $(u_1^0 F_1, u_2^0 F_2, ..., u_r^0 F_r)$ along with (27) yields

$$\sum_{i=1}^{n} \eta_i(S^0, T^0)\langle D_i(u^0 F)(T^0), I_{S_i^0} - I_{T_i^0}\rangle \leqq 0\,. \tag{28}$$

Adding (26) and (28) we get

$$\sum_{i=1}^{n} \eta_i(S^0, T^0)\langle D_i(u^0 F)(T^0) + D_i(v^0 H)(T^0), I_{S_i^0} - I_{T_i^0}\rangle < 0$$

$\Rightarrow$ $\langle D_i(u^0 F)(T^0) + D_i(v^0 H)(T^0), I_{S_i^0} - I_{T_i^0}\rangle < 0$, for at least one $i \in N$ (because $\eta^t \alpha < 0 \Rightarrow \alpha_i < 0$ for at least one $i \in N$) which contradicts (16). Hence, $q^0 \geqq \xi^0$. If $q^0 > \xi^0$ we again get a contradiction to the fact that (S^0, q^0) is optimal for (EP).
Hence $q^0 = \xi^0$ i.e. $(S^0, q^0) = (T^0, \xi^0)$. □

5 Fractional minmax programs

In this section we consider the following discrete minmax fractional programming problem :

$$\text{(FP)} \qquad \begin{aligned} &\min \max\nolimits_{1 \leqq j \leqq r} \ (F_j(S)/G_j(S)) \\ &\text{subject to } H_k(S) \leqq 0, \quad k \in K, \\ &\qquad S = (S_1, S_2, ..., S_n) \in \mathcal{A}^n, \end{aligned}$$

where F_j, G_j, $j \in J$ and H_k, $k \in K$ are real valued differentiable n-set functions defined on $\mathcal{A}^n$. For each $j \in J$, $F_j(S) \geqq 0$, $G_j(S) > 0$, for all $S \in \mathcal{A}^n$. Minmax optimization problems of this type in which F_j, G_j, $j \in J$ and H_k, $k \in K$ are defined on $\mathbb{R}^n$ (n-dimensional Euclidean space) are called generalized fractional encountered in multiobjective programming [1], approximation theory [2], [3], goal programming [7], [13] and Economics [21]. Duality for a generalized linear fractional programming problem subject to only non-negativity constraints was originally considered by Von Neumann [21] in the process of modelling and analysis of an economic equilibrium problem.

Now, as in section 2, we shall resort to the following equivalent problem :

$$\text{(EFP)} \qquad \text{minimize } q$$
$$\text{subject to } F_j(S) - qG_j(S) \leqq 0, \quad j \in J, \tag{29}$$
$$H_k(S) \leqq 0, \quad k \in K, \tag{30}$$
$$S = (S_1, S_2, ..., S_n) \in \mathcal{A}^n. \tag{31}$$

Lemma 5.1. [6] *If $\bar{S}$ is an optimal solution of (FP), then $(\bar{S}, \bar{q})$ with $\bar{q} = \max_{1 \leqq j \leqq r} (F_j(\bar{S})/G_j(\bar{S}))$ is an optimal solution of (EFP). Conversely, if (S^0, q^0) is optimal for (EFP), then S^0 is optimal for (FP).*

Theorem 5.1 (Sufficient Optimality). *Let (T, q^0, u^0, v^0) with $T \in \mathcal{A}^n, q^0 \in R$, $u^0 \in R^r_+, v^0 \in R^m_+$ satisfy*

$$\langle \sum_{j=1}^{r} u_j^0 [D_i F_j(T) - q^0 D_i G_j(T)] + \sum_{k=1}^{m} v_k D_i H_k(T), I_{S_i} - I_{T_i} \rangle \geqq 0$$
$$\text{for all } S_i \in \mathcal{A}, \quad i \in N \tag{32}$$
$$u_j^0 [F_j(T) - q^0 G_j(T)] = 0, \quad j \in J \tag{33}$$
$$v_k^0 H_k(T) = 0, \quad k \in K \tag{34}$$
$$\sum_{j=1}^{r} u_j^0 = 1 \tag{35}$$

Further, let

1. *$(u_1^0(F_1 - q^0 G_1), u_2^0(F_2 - q^0 G_2), ..., u_r^0(F_r - q^0 G_r))$ be V-pseudoinvex with respect to $\eta : \mathcal{A}^n \times \mathcal{A}^n \to R^n$ and $\phi_j ; \mathcal{A}^n \times \mathcal{A}^n \to R^+ \backslash \{0\}, j \in K$*
2. *$(v_1^0 H_1, ..., v_m^0 H_m)$ be V-quasiinvex with respect to same η and $\eta^t \alpha < 0 \Rightarrow \alpha_i < 0$ for at least one $i \in N$. Then T is optimal for (FP) with the corresponding objective value equal to q^0.*

Proof. Let $S \in \mathcal{A}^n$ be any feasible solution to (FP) or let (S, q) be any feasible solution to (EFP).
Let if possible

$$q < q^0 . \tag{36}$$

It follows from (29) that

$$F_j(S)/G_j(S) \leqq q < q^0, \quad j \in J$$
$$\Rightarrow \quad F_j(S) - q^0 G_j(S) < 0, \quad j \in J \tag{37}$$

Since $\phi_j(S,T) > 0$, $j \in J$, $S,T \in \mathcal{A}^n$, it follows from (33), (35) and (37) that

$$\sum_{j=1}^{r} \phi_j(S,T)u_j^0[F_j(S) - q^0 G_j(S)] < \sum_{j=1}^{r} \phi_j(S,T)u_j^0[F_j(T) - q^0 G_j(T)]$$

Now, proceeding on the same lines as in Theorem 3.1 we gets contradiction to (36). Hence $q \geqq q^0$, which ensures that (T, q^0) is optimal for (EFP) proving thereby that T is optimal for (FP) with optimal objective value q^0. □

The following dual is proposed.

$$\textbf{(FP)} \quad \min_{1 \leqq j \leqq r} (F_j(S)/G_j(S))$$

$$\text{subject to } \langle D_i(u(F - \xi G))_T + D_i(vH)_T, I_{S_i} - I_{T_i} \rangle \geqq 0,$$

$$S_i \in A, i \in N \tag{38}$$

$$u_j(F_j(T) - \xi G_j(T)) \geqq 0, \quad j \in J \tag{39}$$

$$v_k H_k(T) \geqq 0, \quad k \in K \tag{40}$$

$$T \in A^n, u \in \mathbb{R}^r_+, \sum_{j=1}^{r} u_j = 1, \quad v \in \mathbb{R}^m_+, \xi \in \mathbb{R}_+. \tag{41}$$

Theorem 5.2 (Weak Duality). *Let (S, q) be feasible for (EFP) and (T, ξ, u, v) be feasible for (FD). Let $(u_1(F_1 - \xi G_1), ..., u_r(F_r - \xi G_r))$ be V-pseudoinvex with respect to functions η and ϕ_j, $j \in J$ and $(v_1 H_1, ..., v_k H_k)$ be V-quasiinvex with respect to η and ψ_k, $k \in K$ such that $\eta^t \alpha < 0 \Rightarrow \alpha_i < 0$ for at least one $i \in N$. Then $q \geqq \xi$.*

Proof. Let if possible

$$q < \xi . \tag{42}$$

It follows from (29) that

$$F_j(S)/G_j(S) \leqq q < \xi, \quad j \in J$$

$$\text{i.e.} \quad F_j(S) - \xi G_j(S) < 0, \quad j \in J \tag{43}$$

Since $\phi_j(S,T) > 0$, $j \in J$, $S,T \in \mathcal{A}^n$, it follows from (39), (41) and (43) that

$$\sum_{j=1}^{r} \phi_j(S,T)u_j[F_j(S) - \xi G_j(S)] < \sum_{j=1}^{r} \phi_j(S,T)u_j[F_j(T) - \xi G_j(T)]$$

Now, proceeding on the same lines as in Theorem 4.1 we arrive at a contradiction. Hence $q \geqq \xi$. This complete the proof. □

Strong duality results follow on the lines on Theorem 4.2 and is hence omitted.

Acknowledgment. The authors wish to express their deep gratitude to Professor R.N. Kaul (Retd.) Professor, Department of Mathematics, University of Delhi, Delhi for his inspiration throughout the preparation of this paper. Thanks are also due to the unknown referees for their useful suggestions.

References

1. D.J. Ashton and D.R. Atkins, 'Multicriteria programming for financial Planning', J. Op. Res. Soc.,30 (1979) 259-270.
2. I. Barrodale, 'Best Rational Approximation and Strict Qusiinvexity', SIAM J. Numer. Anal., 10 (1973), 8-12.
3. I. Barrodale, M.J.D. Powell and F.D.K Roberts, 'The Differential Correction Algorithm for Rational l_∞ Approximation', SIAM J. Numer. Anal. 9 (1972), 493-504.
4. C.R. Bector, and S.Chandra, 'Duality for Pseudolinear Minmax Programs', Asia Pacific J. Op.Res., Vol.3, (1993) pp. 86-94.
5. C.R. Bector, S. Chandra and V. Kumar, 'Duality for Minmax Programming Involving V-invex Functions', Optimization, (Vol.30), (1994) pp.93-103.
6. D. Bhatia and P. Kumar, 'A Note on Fractional Minmax Programs Containing n-Set Functions', J. Math. Anal. Appl., Vol.215, (1997) pp. 283-292.
7. A. Charnes and W.W. Cooper, Goal Programming and Multiobjective Optimization, European J. Op. Res. 1 (1977) 39-54.
8. J.H. Chou, W.S. Hsia and T.Y. Lee, 'On Multiple Objective Programming Problems with Set Functions', J. Math. Anal. Appl. 105 (1985) 383-394.
9. H.W. Corley, 'Optimization Theory for n-set Functions', J.Math. Anal. Appl., 127 (1987), 193-205.
10. V.F. Damyanov and V.N. Malozemov, Introduction to Minmax Programming, Wiley, New York (1974).
11. W.S. Hsia and T.Y. Lee, Proper D Solution of Multi-objective Programming Problems with Set Functions, J. Opt. Th. Appl. 53 (1987) 247-258.
12. V. Jeyakumar and B. Mond, 'On Generalized Convex Mathematical Programming', J. Aust. Math. Soc. (Ser.B), Vol.34, (1992) pp. 43-53.
13. J.S.H. Korubluth, 'A Survey of Goal Programming', Omega 1 (1973) 193-205.
14. H.C. Lai, S.S. Yang and G.R. Hwang, Duality in Mathematical Programming of Set Functions : On Fenchel Duality Theorem, J. Math. Anal. Appl., 95 (1983) 223-234.
15. P. Mazzolenni, On Constrained Optimization for convex set Functions In : A PREKOPA (Ed.) Survey of Mathematical Programming, Vol.1, North Holland, Amsterdam, New York, 1979, 273-290.
16. B. Mond and T. Weir, Generalized Concavity and Duality, in Generalized Cincavity in Optimization and Economics (Eds.) S. Schiable and W. Ziemba, Academic Press, New Delhi (1981).
17. R.J.T. Morris, 'Optimal Constrained Selection of a Measurable Subset', J. Math. Anal. Appl. 70 (1979) 546-562.
18. J.Rosenmuller, 'Some Properties of Convex Set Functions', Arch. Math. 22 (1971) 420-430.
19. J. Rosenmuller and H.G. Weidner, 'Extreme Convex Set Functions with finite Carries : General Theory', Discrete Math. 10 (1974) 343-382.
20. K. Tanaka and Y. Maruyama, 'The Multiobjective optimization Problem of Set Functions', J. Inform. Opt. Sci. 5 (1984) 293-306.
21. J. Von Neumann, 'A Model of General Economics Equilibrium',. Rev. Econ. Studies 13(1945) 1-9.
22. T. Weir and B. Mond, 'Sufficient Optimality Conditions and Duality for Pseudolinear Minmax Programs', Cahiers du C.E.R.O., (1991), Vol.33, pp. 123-128.

23. G.J. Zalmai, 'Optimality Conditions and Duality for Constrained Measurable Subset Selection Problem with Minmax Objective Functions', Optimization, Vol.20, pp. 377-395 (1989).

On the Supremum in Quadratic Fractional Programming

Alberto Cambini, Laura Carosi, and Laura Martein

Dept. of Statistics and Applied Mathematics,
University of Pisa.

Abstract. We consider the maximization of a ratio f of an arbitrary quadratic function and an affine function over a closed and unbounded set. The behavior of unbounded feasible sequences is studied in order to derive a) conditions under which f attains a finite supremum and b) conditions which guarantee that its supremum is finite. We first consider a function f where the quadratic form is semidefinite. We then obtain results for the case where the quadratic function is the product of two affine functions.

2000 Mathematics Subject Classification. Primary 90C32; Secondary 90C26.

Keywords and phrases. Fractional Programming, Generalized Convexity.

1 Introduction

Optimization problems with unbounded feasible regions have been exhaustively handled in recent literature. Theoretical and algorithmic aspects have been studied in order to find both constructive new methods and conditions under which the maximum value exists. See for example the survey proposed by Auslender [1] or the wide literature on fractional programming problems dealing with polyhedral feasible regions [8].

Even though we can find many results for optimality conditions, there is almost nothing known about the supremum of a function over an unbounded and closed feasible region. On the other hand, when a function does not attain its maximum value, it is important to know if its supremum is finite or not. For the remarkable role played both in optimization and in economic theory, we turn our attention to a nonlinear fractional problem where the objective function is a ratio between a quadratic and an affine function and the feasible region is a closed and unbounded set. We study the case where the quadratic function is convex or concave, or is the product between two affine functions. For these classes of problems, several necessary and/or sufficient conditions for a finite supremum are established by means of the asymptotic cone of the feasible set and suitable directions associated with such a cone.

To simplify notation and to save words, we take advantage of the fact that, for any bounded (unbounded) sequence, there exists a convergent (divergent in norm) subsequence. Namely, when we say that a bounded sequence $\{x_n\}$ converges to x, we mean that if this sequence is not convergent, then it is replaced by an appropriate subsequence and this subsequence is again denoted by $\{x_n\}$. Similar abuse of language and notation is applied to unbounded sequence, when we say that $\{x_n\}$ diverges to infinity.

2 On the supremum of a quadratic fractional problem

Consider the following quadratic fractional program

$$\sup_{x\in X}\left[f(x)=\frac{x^TQx+a^Tx+a_0}{b^Tx+b_0}\right] \tag{1}$$

where $a,b,x\in\Re^n$, $b\neq 0$, $a_0,b_0\in\Re$, Q is a symmetric matrix, X is a closed and unbounded set and $b^Tx+b_0>0$ for every $x\in X$.

In order to study the existence of finite supremum for Problem (1), it is necessary to analyze the behavior of f along unbounded feasible sequences.

With this aim, we recall the following definitions and properties [5], [7].

Definition 1. Let $X\subset\Re^n$ be a closed set. The asymptotic cone of X is the set

$$X_\infty=\{d\in\Re^n:\exists\{\alpha_n\}\subset\Re_+,\{x_n\}\subset X, such\ \ that\ \ \alpha_n\to 0^+,\alpha_nx_n\to d\}.$$

The following properties are well known:

Proposition 1. *Let X be a closed set and X_∞ its asymptotic cone.*
i) X_∞ is a closed cone.
ii) X is bounded if and only if $X_\infty=\{0\}$.

In what follows, without any loss of generality, any directions belonging to the asymptotic cone will be assume different from the zero vector.
The following Lemma holds.

Lemma 1. *Let $\{x_n\}$ be a feasible unbounded sequence for Problem (1) and set $d=\lim\limits_{n\to+\infty}\frac{x_n}{\|x_n\|}$. If $d^TQd\neq 0$, then $f(x_n)\to+\infty$ or $f(x_n)\to-\infty$ according to $d^TQd>0$ or $d^TQd<0$.*

Proof. It follows by computing the limit of

$$f(x_n)=\frac{\frac{x_n}{\|x_n\|}^TQ\frac{x_n}{\|x_n\|}\|x_n\|+a^T\frac{x_n}{\|x_n\|}+\frac{a_0}{\|x_n\|}}{b^T\frac{x_n}{\|x_n\|}+\frac{b_0}{\|x_n\|}}$$

□

As a direct consequence of the Lemma 1, we obtain a necessary and a sufficient condition for the supremum of Problem (1) to be finite.

Theorem 1. *i) If $\sup\limits_{x\in X}f(x)<+\infty$, then $d^TQd\leq 0$ for every $d\in X_\infty$.*
ii) If $d^TQd<0$ for every $d\in X_\infty$, then $\sup\limits_{x\in X}f(x)=\max\limits_{x\in X}f(x)$.

Proof. i) If there exists $d \in X_\infty$ such that $d^TQd > 0$, from Lemma 1 there exists a feasible unbounded sequence $\{x_n\}$ such that $\frac{x_n}{\|x_n\|} \to d$, $f(x_n) \to +\infty$ and this is absurd.

ii) Let $\{x_n\}$ be a feasible sequence such that $\lim_{n\to\infty} f(x_n) = \sup_{x\in X} f(x)$. If $\{x_n\}$ is unbounded, then $\frac{x_n}{\|x_n\|} \to d \in X_\infty$, so that, from Lemma 1, $f(x_n) \to -\infty$ and this is absurd. Consequently $\{x_n\}$ is bounded so that it converges to a point $x^* \in X$. From the continuity of the function f, we have

$$\lim_{n\to\infty} f(x_n) = f(x^*) = \sup_{x\in X} f(x).$$

□

Lemma 1 and Theorem 1 allow us to easily handle the definite case.

Corollary 1. *Consider Problem (1) .*

i) If Q is negative definite, then $\sup_{x\in X} f(x) = \max_{x\in X} f(x)$.

ii) If Q is positive definite and X is unbounded, then $\sup_{x\in X} f(x) = +\infty$.

Proof. i) follows directly from ii) of Theorem 1.

ii) follows directly from Lemma 1. □

Beside the definite case, the necessary condition stated in i) of Theorem 1 is not sufficient to reach finite supremum as it is shown by the following example.

Example 1. Consider Problem (1) where $f(x,y,z) = \dfrac{x^2+z^2}{x+y+1}$ and $X = \{(x,y,z) \in \Re^3 : x \geq 0, y = x^2, z = x\sqrt{x}\}$. It results $X_\infty = \{d = k(0,1,0),\ k \geq 0\}$, $d^TQd = 0$ and $\sup_{(x,y,z)\in X} f(x,y,z) = +\infty$ since $\lim_{x\to+\infty} f(x, x^2, x\sqrt{x}) = +\infty$.

The previous example points out the necessity to deepen the analysis in the case $d^TQd = 0$. With this aim, we will assume that Q is positive or negative semidefinite; as it is well known, for such matrices it results $d^TQd = 0$ if and only if $Qd = 0$.

Let $KerQ = \{x \in \Re^n : Qx = 0\}$ and denote with $(KerQ)^\perp$ its orthogonal subspace. Any element $z \in \Re^n$ can be written as $z = z^* + z^\perp$ where $z^* \in KerQ$ and $z^\perp \in (KerQ)^\perp$. Moreover let us define the following subsets of the asymptotic cone.

Definition 2. $X_\infty^Q = X_\infty \cap KerQ$

Definition 3. $X_\infty^+ = \{d \in X_\infty^Q : b^Td > 0\}$

Definition 4. $X_\infty^0 = \{d \in X_\infty^Q : b^Td = 0\}$

The following lemma holds.

Lemma 2. *Consider Problem (1) where Q is semidefinite and let $\{x_n\}$ be an unbounded feasible sequence with $\frac{x_n}{\|x_n\|} \to d$. Then i) and ii) hold.*

i) If $d \in X_\infty^+$ and $\frac{x_n^\perp}{\sqrt{\|x_n\|}} \to w$, then $\lim_{n\to+\infty} f(x_n) = \dfrac{w^TQw + a^Td}{b^Td}$

ii) If $d \in X_\infty^0$ and $\frac{x_n^\perp}{\sqrt{\|x_n\|}} \to w$ with $w^TQw + a^Td \neq 0$, then $\lim_{n\to+\infty} f(x_n) = +\infty$ or $-\infty$ according to $w^TQw + a^Td > 0$ or $w^TQw + a^Td < 0$.

iii) If $d \in X_\infty^Q$ and $\left\{\frac{x_n^\perp}{\sqrt{\|x_n\|}}\right\}$ is an unbounded sequence, then $f(x_n) \to +\infty$ or $f(x_n) \to -\infty$ according to Q is positive semidefinite or negative semidefinite.

Proof. i) ii) Taking into account that $x_n^TQx_n = x_n^{\perp T}Qx_n^\perp$ we have

$$f(x_n) = \frac{\frac{x_n^\perp}{\sqrt{\|x_n\|}}^T Q \frac{x_n^\perp}{\sqrt{\|x_n\|}} + a^T\frac{x_n}{\|x_n\|} + \frac{a_0}{\|x_n\|}}{b^T\frac{x_n}{\|x_n\|} + \frac{b_0}{\|x_n\|}}$$

and so i) and ii) hold.

iii) We have

$$f(x_n) = \frac{\frac{\|x_n^\perp\|^2}{\|x_n\|}\frac{x_n^\perp}{\|x_n^\perp\|}^T Q \frac{x_n^\perp}{\|x_n^\perp\|} + a^T\frac{x_n}{\|x_n\|} + \frac{a_0}{\|x_n\|}}{b^T\frac{x_n}{\|x_n\|} + \frac{b_0}{\|x_n\|}}$$

Since $\frac{x_n^\perp}{\|x_n^\perp\|} \to z \notin KerQ$, it results $z^TQz > 0$ or $z^TQz < 0$ according to Q is positive semidefinite or negative semidefinite. The thesis follows by noting that $\frac{\|x_n^\perp\|^2}{\|x_n\|} = \left(\frac{\|x_n^\perp\|}{\sqrt{\|x_n\|}}\right)^2 \to +\infty$ since $\left\{\frac{x_n^\perp}{\sqrt{\|x_n\|}}\right\}$ is an unbounded sequence.
□

As a consequence of Lemma 2, it remains to analyze the behavior of $f(x_n)$ with respect to feasible unbounded sequences belonging to the following set Y.

Definition 5. $Y = \{\{x_n\} \subset X : \frac{x_n}{\|x_n\|} \to d \in X_\infty^0,\ \frac{x_n^\perp}{\sqrt{\|x_n\|}} \to w$ *with* $w^TQw + a^Td = 0\}$.

Now we will assume $a \neq 0$ (for the case $a = 0$ see Remark 2.1).
Set $a^\perp = \{x \in \Re^n : a^Tx = 0\}$, $b^\perp = \{x \in \Re^n : b^Tx = 0\}$.

According with the previous definitions any element x_n of a feasible sequence can be written as follows:

$$\begin{aligned} &(1)\ x_n = a'_n + \alpha_n^* a \\ &(2)\ x_n = b'_n + \beta_n^* b \end{aligned} \qquad (2)$$

where $a'_n \in a^\perp, \beta'_n \in b^\perp$, α^*_n, β^*_n are scalars.

Hence we have

$$\begin{aligned}(1)\ & a^T x_n = \alpha^*_n \|a\|^2 \\ (2)\ & b^T x_n = \beta^*_n \|b\|^2\end{aligned} \tag{3}$$

Taking into account the above fact, we can define the following sequences

$$\begin{aligned}(1)\ & \alpha_n = \alpha^*_n + \tfrac{a_0}{\|a\|^2} \\ (2)\ & \beta_n = \beta^*_n + \tfrac{b_0}{\|b\|^2}\end{aligned} \tag{4}$$

Consequently, given any feasible unbounded sequence $\{x_n\}$, we can write the corresponding sequence $f(x_n)$ in the following way

$$f(x_n) = \frac{x_n^{\perp T} Q x_n^\perp + \alpha_n \|a\|^2}{\beta_n \|b\|^2} \tag{5}$$

Putting $u = \lim_{n\to+\infty} \frac{x_n^\perp}{\|x_n^\perp\|}$ and $\varepsilon_n = \dfrac{\alpha_n}{x_n^{\perp T} Q x_n^\perp} + \dfrac{1}{\|a\|^2}$, we are able to characterize the behavior of f along those feasible unbounded sequences $\{x_n\} \in Y$.

Lemma 3. *Consider Problem (1) where Q is semidefinite. Let $\{x_n\} \in Y$. Then, one of the following exhaustive cases occurs:*

Case a. $\dfrac{\|x_n^\perp\|^2}{\beta_n} \to 0.$

a.1) $f(x_n) \to \alpha_1 \dfrac{\|a\|^2}{\|b\|^2}$ *if* $\dfrac{\alpha_n}{\beta_n} \to \alpha_1$

a.2) $f(x_n) \to +\infty$ *if* $\dfrac{\alpha_n}{\beta_n} \to +\infty$

a.3) $f(x_n) \to -\infty$ *if* $\dfrac{\alpha_n}{\beta_n} \to -\infty$

Case b. $\dfrac{\|x_n^\perp\|^2}{\beta_n} \to \bar\beta \neq 0.$

b.1) $f(x_n) \to \dfrac{\bar\beta u^T Q u + \alpha_2 \|a\|^2}{\|b\|^2}$ *if* $\dfrac{\alpha_n}{\beta_n} \to \alpha_2$

b.2) $f(x_n) \to +\infty$ *if* $\dfrac{\alpha_n}{\beta_n} \to +\infty$

b.3) $f(x_n) \to -\infty$ *if* $\dfrac{\alpha_n}{\beta_n} \to -\infty$

Case c. $\dfrac{\|x_n^\perp\|^2}{\beta_n} \to +\infty.$

c.1) $f(x_n) \to +\infty$ *if* $\dfrac{\alpha_n}{\|x_n^\perp\|^2} \to +\infty$

c.2) $f(x_n) \to -\infty$ *if* $\dfrac{\alpha_n}{\|x_n^\perp\|^2} \to -\infty$

c.3) $f(x_n) \to +\infty$ *if* $\dfrac{\alpha_n}{\|x_n^\perp\|^2} \to \alpha_3$ *with* $u^T Q u + \alpha_3 \|a\|^2 > 0$

c.4) $f(x_n) \to -\infty$ *if* $\frac{\alpha_n}{\|x_n^\perp\|^2} \to \alpha_3$ *with* $u^T Qu + \alpha_3 \|a\|^2 < 0$

Case d. $\frac{\|x_n^\perp\|^2}{\beta_n} \to +\infty$, $\frac{\alpha_n}{\|x_n^\perp\|^2} \to \alpha_3$ *with* $u^T Qu + \alpha_3 \|a\|^2 = 0.$

Then $\varepsilon_n \to 0$ *and:*

d.1) $f(x_n) \to \alpha_4 u^T Qu \frac{\|a\|^2}{\|b\|^2}$ *if* $\varepsilon_n \frac{\|x_n^\perp\|^2}{\beta_n} \to \alpha_4$

d.2) $f(x_n) \to +\infty$ *if* $\frac{\varepsilon_n}{\beta_n} x_n^{\perp T} Q x_n^\perp \to +\infty$

d.3) $f(x_n) \to -\infty$ *if* $\frac{\varepsilon_n}{\beta_n} x_n^{\perp T} Q x_n^\perp \to -\infty$

Proof. Case a) and b). In these cases we can write

$$f(x_n) = \frac{\frac{\|x_n^\perp\|^2}{\beta_n} \frac{x_n^\perp}{\|x_n^\perp\|}^T Q \frac{x_n^\perp}{\|x_n^\perp\|} + \frac{\alpha_n}{\beta_n} \|a\|^2}{\|b\|^2}$$

and so we get the desired results, simply taking the limit of $f(x_n)$.

Case c). Dividing numerator and denominator by $\|x_n^\perp\|^2$, we obtain

$$f(x_n) = \frac{\frac{x_n^\perp}{\|x_n^\perp\|}^T Q \frac{x_n^\perp}{\|x_n^\perp\|} + \frac{\alpha_n}{\|x_n^\perp\|^2} \|a\|^2}{\frac{\beta_n}{\|x_n^\perp\|^2} \|b\|^2}$$

and again cases c.1),...,c.4) follow trivially by taking the limit.

Case d). It results $\lim_{n\to+\infty} \frac{\alpha_n}{x_n^{\perp T} Q x_n^\perp} = \lim_{n\to+\infty} \frac{\frac{\alpha_n}{\|x_n^\perp\|^2}}{\frac{x_n^\perp}{\|x_n^\perp\|}^T Q \frac{x_n^\perp}{\|x_n^\perp\|}} = \frac{\alpha_3}{u^T Qu} =$ $-\frac{1}{\|a\|^2}$. That implies $\varepsilon_n \to 0$. Since

$$f(x_n) = \frac{x_n^{\perp T} Q x_n^\perp + \alpha_n \|a\|^2}{\beta_n \|b\|^2} = \frac{\varepsilon_n x_n^{\perp T} Q x_n^\perp}{\beta_n} \frac{\|a\|^2}{\|b\|^2} =$$

$$\frac{\varepsilon_n \|x_n^\perp\|^2}{\beta_n} \frac{\frac{x_n^\perp}{\|x_n^\perp\|}^T Q \frac{x_n^\perp}{\|x_n^\perp\|} \|a\|^2}{\|b\|^2} \qquad (6)$$

we achieve the thesis. □

Lemma 2 and Lemma 3 allow us to obtain conditions which ensure that the supremum of Problem (1) is finite. With this regard, consider the following sets associated with an asymptotic direction:

$$
\begin{aligned}
Y^{+} &= \{\{x_n\} \subset X : \tfrac{x_n}{\|x_n\|} \to d \in X^0_{\infty}, \tfrac{x_n^{\perp}}{\sqrt{\|x_n\|}} \to w \text{ with } w^TQw + a^Td > 0\} \\
Y^{-} &= \{\{x_n\} \subset X : \tfrac{x_n}{\|x_n\|} \to d \in X^0_{\infty}, \tfrac{x_n^{\perp}}{\sqrt{\|x_n\|}} \to w \text{ with } w^TQw + a^Td < 0\} \\
W_1 &= \{(d,w) : \tfrac{x_n}{\|x_n\|} \to d \in X^{+}_{\infty}, \tfrac{x_n^{\perp}}{\sqrt{\|x_n\|}} \to w\} \\
\bar{z} &= \sup_{(d,w)\in W_1} \frac{w^TQw + a^Td}{b^Td} \\
Y_1^{+} &= \{\{x_n\} \in Y : \frac{\|x_n^{\perp}\|^2}{\beta_n} \to 0, \tfrac{\alpha_n}{\beta_n} \to +\infty\} \\
Y_1^{-} &= \{\{x_n\} \in Y : \frac{\|x_n^{\perp}\|^2}{\beta_n} \to 0, \tfrac{\alpha_n}{\beta_n} \to -\infty\} \\
\Gamma^1 &= \{\gamma \in \Re : \exists\{x_n\} \in Y, \text{ with } \frac{\|x_n^{\perp}\|^2}{\beta_n} \to 0, \tfrac{\alpha_n}{\beta_n} \to \gamma\} \\
\bar{y}_1 &= \sup_{\gamma\in\Gamma^1} \gamma \frac{\|a\|^2}{\|b\|^2}
\end{aligned} \tag{7}
$$

$$
\begin{aligned}
Y_2^{+} &= \{\{x_n\} \in Y : \frac{\|x_n^{\perp}\|^2}{\beta_n} \to \bar{\beta} \neq 0, \tfrac{\alpha_n}{\beta_n} \to +\infty\} \\
Y_2^{-} &= \{\{x_n\} \in Y : \frac{\|x_n^{\perp}\|^2}{\beta_n} \to \bar{\beta} \neq 0, \tfrac{\alpha_n}{\beta_n} \to -\infty\} \\
\Gamma^2 &= \{(u,\bar{\beta},\gamma) \in \Re^{n+2} : \exists\{x_n\} \in Y \ : \frac{\|x_n^{\perp}\|^2}{\beta_n} \to \bar{\beta} \neq 0, \tfrac{\alpha_n}{\beta_n} \to \gamma\ \} \\
\bar{y}_2 &= \sup_{(u,\bar{\beta},\gamma)\in\Gamma^2} \tfrac{\bar{\beta}u^TQu + \gamma\|a\|^2}{\|b\|^2}
\end{aligned} \tag{8}
$$

$$
\begin{aligned}
Y_3^{+} &= \{\{x_n\} \in Y : \frac{\|x_n^{\perp}\|^2}{\beta_n} \to +\infty, \tfrac{\alpha_n}{\|x_n^{\perp}\|^2} \to +\infty\} \\
Y_4^{+} &= \{\{x_n\} \in Y : \frac{\|x_n^{\perp}\|^2}{\beta_n} \to +\infty, \tfrac{\alpha_n}{\|x_n^{\perp}\|^2} \to \alpha_3, \text{ with } u^TQu + \alpha_3\|a\|^2 > 0\} \\
Y_3^{-} &= \{\{x_n\} \in Y : \frac{\|x_n^{\perp}\|^2}{\beta_n} \to +\infty, \tfrac{\alpha_n}{\|x_n^{\perp}\|^2} \to -\infty\} \\
Y_4^{-} &= \{\{x_n\} \in Y : \frac{\|x_n^{\perp}\|^2}{\beta_n} \to +\infty, \tfrac{\alpha_n}{\|x_n^{\perp}\|^2} \to \alpha_3, \text{ with } u^TQu + \alpha_3\|a\|^2 < 0\} \\
Y_5^{+} &= \{\{x_n\} \in Y : \frac{\|x_n^{\perp}\|^2}{\beta_n} \to +\infty, \tfrac{\alpha_n}{\|x_n\|^2} \to \alpha_3, u^TQu + \alpha_3\|a\|^2 = 0, \\
&\qquad \text{with } \varepsilon_n \frac{\|x_n^{\perp}\|^2}{\beta_n} \to +\infty\} \\
Y_5^{-} &= \{\{x_n\} \in Y : \frac{\|x_n^{\perp}\|^2}{\beta_n} \to +\infty, \tfrac{\alpha_n}{\|x_n\|^2} \to \alpha_3, u^TQu + \alpha_3\|a\|^2 = 0, \\
&\qquad \text{with } \varepsilon_n \frac{\|x_n^{\perp}\|^2}{\beta_n} \to -\infty\} \\
\Gamma^3 &= \{(u,\gamma) \in \Re^{n+1} : \exists\{x_n\} \in Y \text{ with } \tfrac{\alpha_n}{\|x_n\|^2} \to \alpha_3, u^TQu + \alpha_3\|a\|^2 = 0, \\
&\qquad \varepsilon_n \frac{\|x_n^{\perp}\|^2}{\beta_n} \to \gamma\} \\
\bar{y}_3 &= \sup_{(u,\gamma)\in\Gamma^3} \gamma \frac{u^TQu\,\|a\|^2}{\|b\|^2}
\end{aligned} \tag{9}
$$

According to 7, 8, and 9, we define the following sets which play an important role in deriving conditions for the supremum to be finite.

$$H^+ = Y^+ \cup Y_1^+ \cup Y_2^+ \cup Y_3^+ \cup Y_4^+ \cup Y_5^+$$
$$H^- = Y^- \cup Y_1^- \cup Y_2^- \cup Y_3^- \cup Y_4^- \cup Y_5^-$$

The following Theorem holds.

Theorem 2. *Consider Problem (1) where Q is negative semidefinite.*
i) The supremum is finite if and only if $H^+ = \emptyset$.
ii) $Supf(x) = +\infty$ if and only if $H^+ \neq \emptyset$.
iii) If $X_\infty^Q = \emptyset$, then $supf(x) = \max f(x)$.
iv) If $W_1 \cup \Gamma^1 \cup \Gamma^2 \cup \Gamma^3 \cup H^+ = \emptyset$, then $supf(x) = \max f(x)$.
v) If f has finite supremum, which is not attained as maximum value, then we obtain:

$$\sup f(x) = \sup \{\bar{z}, \bar{y}_1, \bar{y}_2, \bar{y}_3\}$$

Proof. i) $\Leftarrow$ Suppose on the contrary that $H^+ \neq \emptyset$. Hence there exists a feasible unbounded sequence $\{x_n\} \in Y^+ \cup Y_1^+ \cup Y_2^+ \cup Y_3^+ \cup Y_4^+ \cup Y_5^+$ and so at least one of the cases a.2, b.2, c.1, c.3, d.2. of Lemma 3 occurs. Then $f(x_n) \to +\infty$ and this can not be true.

$\Rightarrow$ Let $\{x_n\}$ be a feasible sequence such that $\lim_{n\to+\infty} f(x_n) = \sup_{x\in X} f(x)$. If $\{x_n\}$ is bounded, then the result trivially follows . If otherwise $\{x_n\}$ is unbounded, then $\frac{x_n}{\|x_n\|} \to d \in X_\infty$. Necessarily we have $d \in X_\infty^Q$ since otherwise $f(x_n) \to -\infty$ and this can not be true. Since $H^+ = \emptyset$, from Lemma 2 and Lemma 3 it can not be $f(x_n) \to +\infty$ and so we are done.

ii) follows directly from i).

iii) Since $X_\infty^Q = \emptyset$, the result follows directly from ii) of Theorem 1.

iv)Since $H^+ = \emptyset$, from i) we get a finite supremum. Consider an arbitrary unbounded feasible sequence $\{x_n\}$ with $\frac{x_n}{\|x_n\|} \to d \in X_\infty$. If $d \notin X_\infty^Q$, then from Lemma 1 we get $f(x_n) \to -\infty$. Suppose otherwise that $d \in X_\infty^Q$; since $W_1 \cup \Gamma^1 \cup \Gamma^2 \cup \Gamma^3 \cup H^+ = \emptyset$ from Lemma 2 and Lemma 3 we have $f(x_n) \to -\infty$. Then for every unbounded feasible sequence we have $f(x_n) \to -\infty$ and so there exists a bounded feasible sequence $\{x_n\}$ such that $f(x_n) \to \sup f(x)$.

v) Let $\{x_n\}$ be a feasible sequence such that $\lim_{n\to+\infty} f(x_n) = \sup_{x\in X} f(x)$. Since f does not attain its maximum value, $\{x_n\}$ is unbounded. The thesis follows from Lemma 3 cases a.1) b.1), d.1) and i) of Lemma 2. □

The following examples point out that when there exists $d \in X_\infty^Q$, with $W_1 \neq \emptyset$, it may happen that the supremum is attained or not.

Example 2. Consider Problem (1) where $f(x,y) = \dfrac{-2x^2+y+1}{x+y+3}$ and $X = \{(x,y) \in \Re^2 : y = x^2\}$. We have $X_\infty = \{d = k(0,1),\ k \geq 0\}$. Consider the feasible unbounded sequence $\{z_n\} = \{(x_n, x_n^2)\}$. It results $z_n^\perp = (x_n, 0)$ and

$\sqrt{\|z_n\|} = \|x_n\| \sqrt[4]{\frac{1}{x_n^2} + 1}$. Hence $\frac{z_n^{\perp}}{\sqrt{\|z_n\|}} \to w$, $w = (1,0)$ so that $W_1 \neq \emptyset$. In this case we have $\sup\limits_{(d,w)\in W_1} \frac{w^T Q w + a^T d}{b^T d} = -1$ while the maximum value of the function is $1/3$ attained at $(0,0)$.

Example 3. Consider Problem (1) where $f(x,y) = \dfrac{-x^2+y+1}{y+6}$ and
$X = \{(x,y) \in \Re^2 : y = 2x^2\}$. We have $X_\infty = \{d = k(0,1) \;\; k \geq 0\}$. Consider the feasible unbounded sequence $z_n = (x_n, 2x_n^2)$. It results $z_n^{\perp} = (x_n, 0)$ and $\sqrt{\|z_n\|} = \|x_n\| \sqrt[4]{\frac{1}{x_n^2} + 4}$. Hence $\frac{z_n^{\perp}}{\sqrt{\|z_n\|}} \to w$, $w = (\frac{1}{\sqrt{2}}, 0)$ so that $W_1 \neq \emptyset$. It easy to verify that $\sup\limits_{(d,w)\in W_1} \frac{w^T Q w + a^T d}{b^T d} = -\frac{1}{2} + 1 = \frac{1}{2} = \sup\limits_{x \in X} f(x)$ which is not attained as a maximum since it results $f(x, 2x^2) < 1/2$ for every $x \in \Re$.

Theorem 3. *Consider Problem (1) where Q is positive semidefinite. Set* $W = \{\{x_n\} \subset X : \frac{x_n}{\|x_n\|} \to d \in X_\infty^Q,\ \frac{\|x_n^{\perp}\|^2}{\|x_n\|} \to +\infty\}$.

i) If there exists $d \in X_\infty$ such that $d \notin KerQ$, then $\sup\limits_{x\in X} f(x) = +\infty$.

ii) If $X_\infty = X_\infty^Q$ and $W \cup H^+ \neq \emptyset$, then $\sup\limits_{x\in X} f(x) = +\infty$.

iii) If $X_\infty = X_\infty^Q$, $W \cup H^+ = \emptyset$, then the supremum of Problem (1) is finite. Moreover if the maximum is not reached on Problem (1), then it results

$$\sup_{x\in X} f(x) = \sup\{\bar{z}, \bar{y}_1, \bar{y}_2, \bar{y}_3\}$$

Proof. i) follows from Lemma 1.

ii) If $W \neq \emptyset$, the result follows from iii) of Lemma 2. If $H^+ \neq \emptyset$, at least one of the cases a.2), b.2) c.1), c.3) and d.2) of Lemma 3 occurs. Then $\sup\limits_{x\in X} f(x) = +\infty$.

iii) Let $\{x_n\}$ be a feasible sequence such that $\lim\limits_{n\to+\infty} f(x_n) = \sup\limits_{x\in X} f(x)$. From ii) the supremum is finite. Since f does not attain its maximum value, $\{x_n\}$ is unbounded. The thesis follows from Lemma 3 and i) of Lemma 2. □

The following examples show that when Q is positive semidefinite, $X_\infty = X_\infty^Q$ and $W \cup H^+ = \emptyset$, f may admit maximum value or not.

Example 4. Consider Problem (1) where $f(x,y) = \dfrac{y^2+1}{x+1}$ and
$X = \{(x,y) \in \Re^2 : x \geq 0,\ y = \frac{1}{x+1}\}$.
We have $X_\infty = \{d = k(1,0) \;\; k \geq 0\}$, $X_\infty = X_\infty^Q$ and $W \cup H^+ = \emptyset$. It results $\max f(x,y) = 1$ which is attained in $(0,1)$.

Example 5. Consider Problem (1) where $f(x,y) = \dfrac{x^2-4+y}{x+y+3}$ and $X = \{(x,y) \in \Re^2 : 0 \le x \le 1,\ y \ge 1\}$. We have $X_\infty = \{d = k(0,1),\ \ k \ge 0\}$, and $X_\infty = X_\infty^Q$. For every unbounded feasible sequence $\{(x_n, y_n)\}$ the corresponding sequence $\frac{(x_n^\perp, y_n^\perp)}{\sqrt{\|(x_n,y_n)\|}} = \left(\frac{x_n^\perp}{\sqrt{\|(x_n,y_n)\|}}, 0\right)$ is convergent to $(0,0)$, so that $W \cup H^+ = \emptyset$. Moreover the supremum

$$\sup f(x,y) = \sup_{(d,w)\in W_1} \frac{w^T Q w + 1}{1} = 1$$

which is not attained as maximum.

Remark 1. Let us note that all of the above results are established when $a \neq 0$. The case $a = 0$ can be obtained performing a similar analysis. With respect to the given results, it is sufficient to substitute in all the previous formulas $\|a\| = 1$ and $\alpha_n = a_0$.

In the particular case $a = 0,\ \ a_0 = 0$, we have the following corollary, whose proof can be obtained also by a direct analysis.

Corollary 2. *Consider problem (2.1) where* $f(x) = \frac{x^T Q x}{b^T x + b_0}$.
i) If Q is negative semidefinite, then the supremum is finite and if it is not reached as a maximum, then it results

$$supf(x) = \sup_{(u,\bar\beta)\in\tilde U} \frac{\bar\beta u^T Q u}{\|b\|^2} \tag{10}$$

where

$$\tilde U = \{(u,\bar\beta) : \exists\{x_n\} \in \tilde Y \ \ with \ \frac{\|x_n^\perp\|^2}{\beta_n} \to \bar\beta\}$$
$$\tilde Y = \{\{x_n\} \subset X : \frac{x_n}{\|x_n\|} \to d \in X_\infty^Q,\ \ \frac{\|x_n^\perp\|^2}{\beta_n} \to \bar\beta\}$$

ii) If Q is positive semidefinite, then the supremum is finite if and only if $\tilde Y_1 = \emptyset$, where

$$\tilde Y_1 = \{\{x_n\} \subset X : \frac{x_n}{\|x_n\|} \to d \in X_\infty^Q,\ \ \frac{\|x_n^\perp\|^2}{\beta_n} \to +\infty\}.$$

When the supremum is not reached as a maximum we have

$$supf(x) = \sup_{(u,\bar\beta)\in\tilde U} \frac{\bar\beta u^T Q u}{\|b\|^2} \tag{11}$$

3 On the supremum with a product affine functions as numerator

In this section we consider a particular quadratic fractional function for which we can study the behavior of any feasible unbounded sequence regardless of whether the quadratic form is semidefinite or not. In this sense we are able to provide necessary and sufficient conditions for the supremum to be finite in a more general context.

$$\sup_{x \in X} f(x) = \left[\frac{(a^T x + a_0)(c^T x + c_0)}{b^T x + b_0} \right] \tag{12}$$

where $a, b, c, x \in \Re^n$, $a_0, b_0, c_0 \in \Re$, $b^T x + b_0 > 0$ and X is any closed and unbounded set. We will assume $a \neq 0$, $c \neq 0$ (the cases $a = 0$ or $c = 0$ have been studied in [4]).

Specifying the results presented in Lemma 1 and Theorem 1, we obtain the following corollaries.

Corollary 3. *Let $\{x_n\}$ be a feasible unbounded sequence for Problem (12) and let d be such that $d = \lim_{n \to +\infty} \frac{x_n}{\|x_n\|}$ and $(a^T d)(c^T d) \neq 0$. Then $f(x_n) \to +\infty$ or $f(x_n) \to -\infty$ according to $(a^T d)(c^T d) > 0$ or $(a^T d)(c^T d) < 0$.*

Corollary 4. *i) If $\sup_{x \in X} f(x) < +\infty$, then $(a^T d)(c^T d) \leq 0$, $\forall\, d \in X_\infty$.*

ii) If $(a^T d)(c^T d) < 0$ for every $d \in X_\infty$, then $\sup_{x \in X} f(x) = \max_{x \in X} f(x)$.

The following example shows that the necessary condition stated in i) of Corollary 4 is not sufficient to have finite supremum.

Example 6. Consider Problem (12) where $f(x, y, z) = \frac{(x+1)(y+z+2)}{x+y+5}$ and $X = \{(x, y, z\} : x \geq 0, y = x^2, z = \sqrt{x}\}$. We get $X_\infty = \{d = k(0,1,0),\ k \geq 0\}$, $(a^T d)(c^T d) = 0$ and $\sup_{x \in X} f(x, y, z) = +\infty$.

Corollary 4 points out the necessity to deepen the case $(a^T d)(c^T d) = 0$; in order to study what happens in such a case, we consider $a^\perp$ and $b^\perp$ previously defined and $c^\perp = \{x \in \Re^n : c^T x = 0\}$. Set $D = a^\perp \cup c^\perp\}$ and

$$\begin{aligned} x_n &= c'_n + \gamma_n^* c \\ c^T x_n &= \gamma_n^* \|c\|^2 \\ \gamma_n &= \gamma_n^* + \tfrac{c_0}{\|c\|^2} \end{aligned} \tag{13}$$

where $c'_n \in c^\perp$ and γ_n, γ_n^* are scalars.

Recalling equations (1) and (2) of 4, we can write

$$f(x_n) = \frac{\left(\alpha_n^* \|a\|^2 + a_0\right)\left(\gamma_n^* \|c\|^2 + c_0\right)}{\beta_n^* \|b\|^2 + b_0} = \frac{\alpha_n \gamma_n}{\beta_n} \frac{\|a\|^2 \|c\|^2}{\|b\|^2} \tag{14}$$

The following lemma holds.

Lemma 4. *Consider Problem (12) and let $\{x_n\}$ be an unbounded feasible sequence with $\frac{x_n}{\|x_n\|} \to d$. Then i) and ii) hold.*

i) If $d \in D$ and $\frac{\alpha_n \gamma_n}{\beta_n} \to \delta_d$, then $\lim_{n \to +\infty} f(x_n) = \delta_d \frac{\|a\|^2 \|c\|^2}{\|b\|^2}$

ii) If $d \in D$ and $\frac{\alpha_n \gamma_n}{\beta_n} \to \infty$, then $f(x_n) \to +\infty$ or $f(x_n) \to -\infty$ according to $\frac{\alpha_n \gamma_n}{\beta_n} > 0$ or $\frac{\alpha_n \gamma_n}{\beta_n} < 0$ definitively.

Proof. The results follow immediately by taking the limit in (14). □

Lemma 4 allows us to obtain conditions which ensure that the supremum of Problem (12) is finite.

With this regard, for every $d \in X_\infty \cap D$, set

$$\Delta_d = \{\delta_d \in \Re : \exists \{x_n\} \subset X, \frac{x_n}{\|x_n\|} \to d \in X_\infty, \frac{\alpha_n \gamma_n}{\beta_n} \to \delta_d\}$$

and define Δ_d^* to be the set of all unbounded, definitively positive sequences $\left\{\frac{\alpha_n \gamma_n}{\beta_n}\right\}$.

The following theorem specifies when Problem (12) has maximum value and gives a characterization of the supremum when it is not attained as a maximum.

Theorem 4. *Consider Problem (12). Then the following statements hold.*

i) If there exists $d \in X_\infty$ such that $(a^T d)(c^T d) > 0$, then $\sup_{x \in X} f(x) = +\infty$.

ii) If there exists $d \in X_\infty \cap D$ such that $\Delta_d^* \neq \emptyset$, then $\sup_{x \in X} f(x) = +\infty$.

iii) Assume $(a^T d)(c^T d) \leq 0 \ \forall d \in X_\infty$. If $\Delta_d^* = \emptyset$ for $\forall d \in X_\infty \cap D$, then $\sup_{x \in X} f(x)$ is finite and if it is not attained as a maximum it results

$$\sup_{x \in X} f(x) = \frac{\|a\|^2 \|c\|^2}{\|b\|^2} \sup_{d \in X_\infty \cap D} \left(\sup_{\delta_d \in \Delta_d} \delta_d\right).$$

Furthermore if $\Delta_d = \emptyset \ \forall d \in X_\infty \cap D$, then $\sup_{x \in X} f(x) = \max_{x \in X} f(x)$.

Proof. i) follows from Corollary 3.

ii) follows from ii) of Lemma 4.

iii) Since X is closed, the supremum is finite if and only if $\lim_{n \to +\infty} f(x_n) \neq +\infty$ for every unbounded feasible sequence $\{x_n\}$. From Corollary 3, if

$$(a^T d)(c^T d) < 0,$$

then $\lim_{n\to+\infty} f(x_n) = -\infty$, while, from Lemma 4, $(a^T d)(c^T d) = 0$ and $\Delta_d^* = \emptyset$ imply $\lim_{n\to+\infty} f(x_n)$ is finite or $-\infty$. If in addition $\Delta_d = \emptyset\ \forall d \in X_\infty \cap D$, necessarily we have $\lim_{n\to+\infty} f(x_n) = -\infty$ for every unbounded sequence, so that the supremum is attained as a maximum. The proof is complete. □

The following examples point out that when there exists $d \in X_\infty \cap D$ with $\Delta_d^* = \emptyset$ and $\Delta_d \neq \emptyset$, we can have both maximum and finite supremum.

Example 7. Consider Problem (12) where $f(x,y,z) = \frac{(x+2)(y+z+2)}{y+2}$ and $X = \{(x,y,z) \in \Re^3 : 0 \le x \le 1, y \ge 0, z = \sqrt{y}\}$. It is easy to verify that $X_\infty = \{d = k(0,1,0),\ \ k \ge 0\}$ so that $d \in D$. Consider any unbounded feasible sequence $h_n = \{x_n, y_n, \sqrt{y_n}\}$. Since x_n is bounded, without any loss of generality we can assume $\lim_{n\to+\infty} x_n = \bar{x}$ with $\bar{x} \in [0,1]$. It results $\alpha_n = x_n + 2$, $\gamma_n = \frac{1}{2}\left(y_n + \sqrt{y_n}\right) + 1, \beta_n = y_n + 2, \frac{\alpha_n\gamma_n}{\beta_n} \to \frac{1}{2}\bar{x} + 1$, so that $\Delta_d^* = \emptyset$, $\Delta_d \neq \emptyset$ and $\frac{\|a\|^2\|c\|^2}{\|b\|^2} \sup_{d\in X_\infty\cap D}\left(\sup_{\delta_d\in\Delta_d} \delta_d\right) = 2\frac{3}{2} = 3$. Simple calculations show that the function f has maximum value $3 + \frac{3}{4}\sqrt{2}$ attained at the point $(1, 2, \sqrt{2})$.

Example 8. Consider Problem (12) where $f(x,y,z) = \frac{(3x+y+2)(z+1)}{x+y+10}$ and $X = \{(x,y,z) \in \Re^3 : x \ge 1, y = \sqrt{x}, 0 \le z \le 1\}$. It is easy to verify that $X_\infty = \{d = k(1,0,0),\ \ k \ge 0\}$ so that $d \in D$. Consider any unbounded feasible sequence $h_n = \{x_n, \sqrt{x_n}, z_n\}$. Since z_n is bounded, without any loss of generality, we can assume $\lim_{n\to+\infty} z_n = \bar{z}$ with $\bar{z} \in [0,1]$. It results $\alpha_n = \frac{1}{10}\left(3x_n + \sqrt{x_n}\right) + \frac{1}{5}$,
$\gamma_n = z_n + 1$, $\beta_n = \frac{1}{2}\left(x_n + \sqrt{x_n}\right) + 5$, $\frac{\alpha_n\gamma_n}{\beta_n} \to \frac{1}{5}3(\bar{z}+1)$ and $\delta_d = \frac{3}{5}(\bar{z}+1)$,
so that $\Delta_d^* = \emptyset$, $\Delta_d \neq \emptyset$ and $f(x_n, y_n, z_n) \to \frac{3}{5}(\bar{z}+1)$.
Simple calculations show that function f does not attain maximum value so that $\sup_{x\in X} f(x) = \frac{\|a\|^2\|c\|^2}{\|b\|^2} \sup_{\delta_d\in\Delta_d} \delta_d = 5\frac{6}{5} = 6$.

3.1 Particular cases

Consider the following problem

$$\sup_{x\in X} f(x) = \left(a^T x + a_0\right)\left(c^T x + c_0\right) \tag{15}$$

obtained from Problem (12) setting $b = 0$ and $b_0 = 1$.

With similar argument, for any unbounded sequence $\{x_n\}$, we have

$$f(x_n) = \alpha_n\gamma_n \|a\|^2 \|c\|^2 .$$

Setting

$$\Delta_d = \{\delta_d \in \Re : \exists\{x_n\} \subset X, \frac{x_n}{\|x_n\|} \to d \in X_\infty, \alpha_n\gamma_n \to \delta_d\}$$

and Δ_d^* the set of all unbounded definitively positive sequences $\{\alpha_n \gamma_n\}$, we have the following corollary.

Corollary 5. *Consider Problem (15). Then the following statements hold.*

i) If there exists $d \in X_\infty$ such that $(a^T d)(c^T d) > 0$, then $\sup_{x \in X} f(x) = +\infty$.

ii) If there exists $d \in X_\infty \cap D$ such that $\Delta_d^* \neq \emptyset$, then $\sup_{x \in X} f(x) = +\infty$.

iii) Assume $(a^T d)(c^T d) \leq 0$ for every $d \in X_\infty$. If $\Delta_d^* = \emptyset$, for every $d \in X_\infty \cap D$, then $\sup_{x \in X} f(x)$ is finite and if it is not attained, then it results

$$\sup_{x \in X} f(x) = \|a\|^2 \|c\|^2 \sup_{d \in X_\infty \cap D} \left(\sup_{\delta_d \in \Delta_d} \delta_d \right).$$

Furthermore if $\Delta_d = \emptyset$ for every $d \in X_\infty \cap D$, then $\sup_{x \in X} f(x) = \max_{x \in X} f(x)$.

As we have already seen in the general case, when $d \in X_\infty \cap D$ with $\Delta_d^* = \emptyset$ and $\Delta_d \neq \emptyset$, we can have both maximum and finite supremum not attained.

Example 9. Consider Problem (15) where $f(x, y) = (x + 1)y$ and $X = \{(x, y) \in \Re^2 : x \geq 1, y = \frac{1}{x}\}$. It is easy to verify that $X_\infty = \{d = k(1, 0),\ k \geq 0\}$ so that $d \in D$. Consider any unbounded feasible sequence $\{(x_n, y_n)\}$. It results $\alpha_n = x_n + 1$, $\gamma_n = \frac{1}{x_n}$, $\alpha_n \gamma_n = 1 + \frac{1}{x_n} \to 1$, $\delta_d = 1$, so that $\Delta_d^* = \emptyset$, $\Delta_d \neq \emptyset$ and $\|a\|^2 \|c\|^2 \sup_{d \in X_\infty \cap D} \left(\sup_{\delta_d \in \Delta_d} \delta_d \right) = 1$. Simple calculations show that the function f has maximum value 2 which is attained at $(1, 1)$.

Example 10. Consider Problem (15) where $f(x, y) = (x - 1)y$ and $X = \{(x, y) \in \Re^2 : x \geq 1, y = \frac{1}{x}\}$. It is easy to verify that $X_\infty = \{d = k(1, 0),\ k \geq 0\}$ so that $d \in D$. Consider any unbounded feasible sequence $\{(x_n, y_n)\}$. It results $\alpha_n = x_n - 1$, $\gamma_n = \frac{1}{x_n}$, $\alpha_n \gamma_n = 1 - \frac{1}{x_n} \to 1$, $\delta_d = 1$, so that $\Delta_d^* = \emptyset$, $\Delta_d \neq \emptyset$ and $\|a\|^2 \|c\|^2 \sup_{d \in X_\infty \cap D} \left(\sup_{\delta_d \in \Delta_d} \delta_d \right) = 1$. Simple calculations show that function f does not attain maximum value.

When $c = a$ and $c_0 = a_0$, Problem (12) becomes

$$\sup_{x \in X} \left[f(x) = \frac{(a^T x + a_0)^2}{b^T x + b_0} \right] \tag{16}$$

where $a, b, x \in \Re^n$, $a_0, b_0 \in \Re$ and X is any unbounded closed set, $b^T x + b_0 > 0$ for every $x \in X$.

It is well known that f is a convex function (see [3]) and specialyzing the previous results, we obtain the following corollary.

Corollary 6. *Consider Problem (16). The following statements hold.*

i) If $\exists d \in X_\infty : a^T d \neq 0$, *then* $\sup f(x) = +\infty$.

ii) If $\exists d \in X_\infty : a^T d = 0$ *and* $\Delta_d^* \neq \emptyset$, *then* $\sup f(x) = +\infty$.

iii) f *has finite supremum if and only if* $X_\infty \subset a^\perp$ *and* $\Delta_d^* = \emptyset$. *Moreover if* f *does not have maximum value, then*

$$\sup_{x \in X} f(x) = \frac{\|a\|^4}{\|b\|^2} \sup_{d \in X_\infty \cap D} \left(\sup_{\delta_d \in \Delta_d} \delta_d \right)$$

References

1. Auslender A., (1997), How to deal with the unbounded in optimization: Theory and Algorithms, *Mathematical Programming* **79** 3-18.
2. Avriel, M., Diewert, W.E., Schaible, S. and I. Zang, (1988), *Generalized Concavity, Mathematical Concepts and Methods in Science and Engineering*, vol.36, Plenum Press, New York.
3. Bazaraa M.S., Sheraly H. D., Shetty C.M, (1993), *Nonlinear programming*, Second edition John Wiley & Sons, Inc., New York.
4. Carosi L., On the supremum in linear fractional programming with respect to any closed unbounded feasible set, forthcoming in *Journal of Interdisciplinary Mathematics*..
5. Dedieu J.P., (1978), Cône asymptote d'un ensemble non convexe. Application à l'optimization, in *Comptes Rendus. Acádémie de Sciences Paris*, **287** 941-943, Série A- 941.
6. Martein L., (1988), Applicazione della programmazione frazionaria nel campo economico finanziario, Report 14, Dipartimento di Statistica e Matematica applicata all'Economia, Economia e Commercio, Università di Pisa, 1988.
7. Rockafellar R.T., (1970), *Convex Analysis*, Princeton University Press, Princeton, NJ.
8. Schaible S., (1995), Fractional Programming in *Handbook of Global Optimization* 495-608 Edited by R. Horst, P.M. Pardalos, Kluwer Academic Publishers, Dordrecht.

First and Second Order Characterizations of a Class of Pseudoconcave Vector Functions

Riccardo Cambini and Laura Martein

Dept. of Statistics and Applied Mathematics, University of Pisa,
Via Cosimo Ridolfi 10, 56124 Pisa, ITALY
E-mail: cambric@ec.unipi.it - lmartein@ec.unipi.it

Abstract. In this paper we deeply analyze a family of vector valued pseudoconcave functions which extends to the vector case both the concepts of scalar pseudoconcavity and scalar strict pseudoconcavity, as well as the optimality properties of these functions, such as the global optimality of local optima, of critical points and of points verifying Kuhn-Tucker conditions. The considered functions come out to be particularly relevant since it is possible for them to determine several first and second order characterizations; this offers a complete extension to the vector case of the well known concept of scalar pseudoconcavity and gives the chance to work in multiobjective optimization with all the properties of the scalar case.

2000 Mathematics Subject Classification. Primary 90C26; Secondary 90C29, 90C30.

Keywords and phrases. Vector Optimization, Optimality Conditions, Generalized Convexity.

1 Introduction

Pseudoconcave vector valued functions play a key role in multicriteria decision making and in multiobjective programming since their properties give the way to recognize the efficient points.

In these very last years, several classes of vector valued pseudoconcave functions have been introduced and studied with the aim of extending to the vector case some properties of scalar pseudoconcavity [3–12,14,16,17].

In this paper we deeply analyze a family of vector valued pseudoconcave functions, among all the proposed ones, which extends to the vector case both scalar pseudoconcavity and scalar strict pseudoconcavity as well as their optimality properties, such as the global optimality of local optima, of critical points and of points verifying Kuhn-Tucker conditions.

These functions, introduced in [6,8], come out to be particularly relevant, since it is possible to determine for them both first and second order characterizations.

The proposed class of vector valued pseudoconcave function thus offers a complete extension to the vector case of the well known concept of scalar pseudoconcavity and gives the chance to work in multiobjective optimization with all the properties of the scalar case.

2 Definitions and preliminary results

From now on we will consider a vector valued function f, a cone C and some of its subcones, verifying the following properties:

i) $f : A \to \Re^m$, defined on the open convex set $A \subseteq \Re^n$, is a differentiable vector valued function,

ii) $C \subset \Re^m$ is a closed convex pointed cone with nonempty interior,

iii) $C^* \subset \Re^m$ is any cone ([1]) such that $\text{Int}(C) \subseteq C^* \subseteq C$,

iv) $C^0 = C \setminus \{0\}$.

Some of the main results of this paper will be based on the following fundamental preliminary theorem which specifies the behaviour of function f when $J_f(x_0)(x - x_0) \in \text{Int}(C)$.

Theorem 1. *Let $x, x_0 \in A$, $x \neq x_0$. It follows that $J_f(x_0)(x - x_0) \in \text{Int}(C)$ if and only if the following condition (1) holds:*

$$\text{there exists a vector } \xi_{x,x_0} \in \text{Int}(C), \exists \lambda^* \in (0,1] \text{ such that: } \quad (1)$$
$$f(x_0 + \lambda(x - x_0)) \in f(x_0) + \lambda(1-\lambda)\xi_{x,x_0} + C \quad \forall \lambda \in (0, \lambda^*)$$

Proof. $\Rightarrow$) Suppose by contradiction that $\forall \xi_{x,x_0} \in \text{Int}(C)$, $\forall \lambda^* \in (0,1]$, $\exists \lambda \in (0, \lambda^*)$ such that $f(x_0 + \lambda(x - x_0)) \notin f(x_0) + \lambda(1-\lambda)\xi_{x,x_0} + C$. Let $c \in \text{Int}(C)$ and consider the sequences $\xi_{x,x_0} = \frac{1}{n}c \in \text{Int}(C)$ and $\lambda^* = \frac{1}{n} \in (0,1]$, $n = 1, 2, \ldots$, then there exists a sequence $\lambda_n \in (0, \frac{1}{n})$ such that:

$$\frac{f(x_0 + \lambda_n(x - x_0)) - f(x_0)}{\lambda_n} \notin \frac{1}{n}(1 - \lambda_n)c + C.$$

Since $\lim_{n\to+\infty} \lambda_n = 0$ we then have that:

$$J_f(x_0)(x - x_0) = \lim_{n\to+\infty} \frac{f(x_0 + \lambda_n(x - x_0)) - f(x_0)}{\lambda_n} \notin \text{Int}(C)$$

which is a contradiction.

$\Leftarrow$) By means of the hypothesis we have:

$$J_f(x_0)(x - x_0) = \lim_{\lambda\to 0^+} \frac{f(x_0 + \lambda(x - x_0)) - f(x_0)}{\lambda} \in \xi_{x,x_0} + C$$

so that the thesis follows since $\xi_{x,x_0} + C \subset \text{Int}(C)$ since C is a convex cone. □

[1] Note that, in order to have general results, we will consider also open cones and cones without the origin, that is to say sets K such that:

$$k \in K \quad \Longrightarrow \quad \lambda k \in K \ \forall \lambda > 0$$

Another useful result, still based on the behaviour of function f with respect to its Jacobian and Hessians matrices, is the following.

Theorem 2. *If the following condition holds:*

$$\exists \lambda^* \in (0,1) \textit{ such that:} \qquad (2)$$
$$f(x_0 + \lambda(x - x_0)) \in f(x_0) + C \quad \forall \lambda \in (0, \lambda^*)$$

then $J_f(x_0)(x - x_0) \in C$. *If* f *is also twice differentiable then whenever* $J_f(x_0)(x - x_0) = 0$ *it results* $(x - x_0)^T H_f(x_0)(x - x_0) \in C$.

Proof. By means of the hypothesis $\exists \lambda^* \in (0,1)$ such that:

$$\frac{f(x_0 + \lambda(x - x_0)) - f(x_0)}{\lambda} \in C \quad \forall \lambda \in (0, \lambda^*)$$

Given that C is a closed convex cone it follows that:

$$J_f(x_0)(x - x_0) = \lim_{\lambda \to 0} \frac{f(x_0 + \lambda(x - x_0)) - f(x_0)}{\lambda} \in C$$

and the thesis is proved.
Suppose now $J_f(x_0)(x - x_0) = 0$; by means of the hypothesis $\exists \lambda^* \in (0,1)$ such that:

$$\frac{f(x_0 + \lambda(x - x_0)) - f(x_0)}{\lambda^2} \in C \quad \forall \lambda \in (0, \lambda^*)$$

Let us consider the following second order Taylor expansion of f at x_0:

$$f(x_0 + \lambda(x - x_0)) = f(x_0) + \lambda J_f(x_0)(x - x_0) + \\ + \frac{\lambda^2}{2}(x - x_0)^T H_f(x_0)(x - x_0) + \lambda^2 \|x - x_0\|^2 \sigma(\lambda, 0)$$

where $\lim_{\lambda \to 0^+} \sigma(\lambda, 0) = 0$. Since $J_f(x_0)(x - x_0) = 0$ it follows:

$$\frac{f(x_0 + \lambda(x - x_0)) - f(x_0)}{\lambda^2} = \frac{1}{2}(x - x_0)^T H_f(x_0)(x - x_0) + \|x - x_0\|^2 \sigma(\lambda, 0)$$

so that, being C a closed cone, it results:

$$\frac{1}{2}(x - x_0)^T H_f(x_0)(x - x_0) = \lim_{\lambda \to 0} \frac{f(x_0 + \lambda(x - x_0)) - f(x_0)}{\lambda^2} \in C$$

and the theorem is proved. □

Note that the two previous preliminary results are not based on the generalized concavity of function f.

3 Pseudoconcavity and quasiconcavity

The classes of pseudoconcave functions, whose properties are going to be deepened on in this paper, are defined as follows (see [6,8]).

Definition 1. Let $C \subset \Re^m$ be a closed convex pointed cone with nonempty interior and let C^* be a cone such that $\text{Int}(C) \subseteq C^* \subseteq C$. A differentiable function $f : A \to \Re^m$, where $A \subseteq \Re^n$ is an open convex set, is said to be a $(C^*, \text{Int}(C))$-*pseudoconcave function* if the following implication holds $\forall x, x_0 \in A$, $x \neq x_0$:

$$f(x) \in f(x_0) + C^* \quad \Longrightarrow \quad J_f(x_0)(x - x_0) \in \text{Int}(C)$$

Remark 1. As we have pointed out in the previous note ([1]), C^* may coincide with $\text{Int}(C)$, C^0, C, or may be any cone contained in C and containing $\text{Int}(C)$, having or not the origin and not necessarily open or closed. Thus, in the previous Definition 1 we have defined a whole family of pseudoconcave functions, that is to say that for any chosen cone C^* we have the corresponding class of pseudoconcave function.

Note that in the scalar case, where $C = \Re_+$ and $\text{Int}(C) = \Re_{++}$, the classes of $(C^*, \text{Int}(C))$-pseudoconcave functions coincide with the well known pseudoconcave and strictly pseudoconcave functions when $C^* = \text{Int}(C) = \Re_{++}$ and $C^* = C = \Re_+$ respectively.

In the vector case the following concept of quasiconcavity has been also studied (see [6,8]).

Definition 2. Let $C \subset \Re^m$ be a closed convex pointed cone with nonempty interior and let C^1 and C^2 be cones such that $\text{Int}(C) \subseteq C^1 \subseteq C$ and $\text{Int}(C) \subseteq C^2 \subseteq C$. A function $f : A \to \Re^m$, where $A \subseteq \Re^n$ is a convex set, is said to be a (C^1, C^2)-*quasiconcave function* if the following implication holds $\forall x, x_0 \in A$, $x \neq x_0$:

$$f(x) \in f(x_0) + C^1 \quad \Longrightarrow \quad \begin{array}{c} f(x_0 + \lambda(x - x_0)) \in f(x_0) + C^2 \\ \forall \lambda \in (0, 1) \end{array}$$

Unlike the scalar case, the concept of vector pseudoconcavity does not imply the one of quasiconcavity introduced so far, as the next example shows (see [6,8]).

Example 1. Consider the cone $C = \Re^3_+$ and the following differentiable function $f : [0, 3] \to \Re^3$:

$$f(x) = \begin{cases} (-x^2 + 2x)[1/2, 1/2, 1]^T & \text{if } x \in [0, 1] \\ [1/2, 1/2, 1]^T + (-2x^3 + 9x^2 - 12x + 5)[1, -1, 0]^T & \text{if } x \in (1, 2) \\ [3/2, -1/2, 1]^T + (x - 2)^2[-5/6, 7/6, -1/3]^T & \text{if } x \in [2, 3] \end{cases}$$

This function is $(C^*, \text{Int}(C))$-pseudoconcave but it is not (C^1, C^2)-quasiconcave for any cones C^1 and C^2 such that $\text{Int}(C) \subseteq C^1 \subseteq C$ and $\text{Int}(C) \subseteq C^2 \subseteq C$ (see [6,8]) since $f(3) \in f(0) + \text{Int}(C)$ while $f(2) \notin f(0) + C$.

The relationship existing in the scalar case among pseudoconcave and quasiconcave functions can be restored in the vector case by means of the following new concept of vector local quasiconcavity. Note that local concepts of quasiconcavity and pseudoconcavity for scalar valued functions were introduced and investigated in [15,18,19] and proved to be useful in deepening generalized convexity.

Definition 3. Let $C \subset \Re^m$ be a closed convex pointed cone with nonempty interior and let C^1 and C^2 be cones such that $\text{Int}(C) \subseteq C^1 \subseteq C$ and $\text{Int}(C) \subseteq C^2 \subseteq C$. A function $f : A \to \Re^m$, with $A \subseteq \Re^n$ convex set, is said to be a *locally (C^1, C^2)-quasiconcave function* if the following implication holds $\forall x, x_0 \in A$, $x \neq x_0$:

$$f(x) \in f(x_0) + C^1 \quad \Longrightarrow \quad \begin{array}{c} \exists \lambda^* \in (0,1] \text{ such that } \forall \lambda \in (0, \lambda^*) \\ f(x_0 + \lambda(x - x_0)) \in f(x_0) + C^2 \end{array} \tag{3}$$

Theorem 3. *Let $f : A \to \Re^m$, with A open convex set, be a differentiable $(C^*, \text{Int}(C))$-pseudoconcave function, then it is also locally (C^1, C^2)-quasiconcave for any cones C^1 and C^2 such that* $\text{Int}(C) \subseteq C^1 \subseteq C^*$ *and* $\text{Int}(C) \subseteq C^2 \subseteq C$.

Proof. The inclusion relationship follows directly from the definitions and Theorem 1, being C a convex cone. □

Note that, as a particular case, any $(C^*, \text{Int}(C))$-pseudoconcave function is also locally (C^*, C^*)-quasiconcave and locally $(\text{Int}(C), C)$-quasiconcave. Note also that, by means of the definitions, the class of locally (C^1, C^2)-quasiconcave functions properly contains the class of (C^1, C^2)-quasiconcave functions studied in [6,8], as it is pointed out in Example 1.

In the scalar case the concepts of quasiconcavity and of local quasiconcavity coincide under continuity hypothesis.

Theorem 4. *A continuous scalar function $f : A \to \Re$, $A \subseteq \Re^n$ convex set, is quasiconcave if and only if it is locally quasiconcave.*

Proof. By means of the definitions every quasiconcave function is also locally quasiconcave. Suppose now by contradiction that f is locally quasiconcave but not quasiconcave, that is to say that $\exists x, y \in A$, $x \neq y$, $\exists \lambda_1 \in (0,1)$ such that $f(y) \geq f(x)$ and $f(x + \lambda_1(y - x)) < f(x)$; by means of the local quasiconcavity of f $\exists \lambda_2 \in (0,1]$ such that $\forall \lambda \in (0, \lambda_2)$ $f(x + \lambda(y - x)) \geq f(x)$, so that $\lambda_2 < \lambda_1$. Let us now define the following $\overline{\lambda} \in (\lambda_2, \lambda_1)$:

$$\overline{\lambda} = sup\{t \in [0,1] \text{ s.t. } f(x + \lambda(y - x)) \geq f(x) \quad \forall \lambda \in [0,t]\};$$

by means of the continuity of f we can easily prove that $f(x + \overline{\lambda}(y - x)) = f(x)$ and that $\exists \epsilon > 0$ such that:

$$f(x + \lambda(y - x)) < f(x + \overline{\lambda}(y - x)) = f(x) \quad \forall \lambda \in (\overline{\lambda}, \overline{\lambda} + \epsilon). \tag{4}$$

Being $f(y) \geq f(x + \overline{\lambda}(y - x)) = f(x)$ we have, by means of the local quasiconcavity of f, that $\exists \lambda^* \in (\overline{\lambda}, 1]$ such that $\forall \lambda \in (\overline{\lambda}, \lambda^*)$ $f(x + \lambda(y - x)) \geq f(x + \overline{\lambda}(y - x))$ and this contradicts (4). □

Note that if the scalar function is not continuous these classes do not coincide, as it is pointed out in the next example.

Example 2. Let us consider the following scalar function $f : \Re \to \Re$:

$$f(x) = \begin{cases} 2 & \text{if } x \neq 1 \\ 1 & \text{if } x = 1 \end{cases}$$

It comes out that f is locally quasiconcave even if it is not quasiconcave, since $f(2) = f(0)$ while $f(1) < f(0)$.

Note finally that Theorem 4 will allow us to show that the characterizations of pseudoconcave vector valued functions we are going to state in the forthcoming sections, are a generalization of the known results in the scalar case.

4 Pseudoconcavity and optimality

In this section we are going to show that the family of $(C^*, \mathrm{Int}(C))$-pseudoconcave functions retains all optimality properties of the scalar pseudoconcave functions, that is to say that they verify the global optimality of local optima, the global optimality of critical points and the sufficiency of the Kuhn-Tucker like optimality conditions. These are not new results and their proofs are given for the sake of completeness.

From now on we will use the following concepts of vector efficiency.

Definition 4. A point $x_0 \in A$ is said to be a *global C^*-efficient point* for f if:

$$\nexists x \in A, x \neq x_0, \text{ such that } f(x) \in f(x_0) + C^*;$$

A point $x_0 \in A$ is said to be a *local C^*-efficient point* for f if there exists a neighbourhood I_{x_0} of x_0 such that:

$$\nexists x \in A \cap I_{x_0}, x \neq x_0, \text{ with } f(x) \in f(x_0) + C^*$$

The first result which is extended to the vector case is the global efficiency of a local efficient point.

Theorem 5. *If $f : A \to \Re^m$, A convex set, is a locally (C^*, C^*)-quasiconcave function, then every local C^*-efficient point is also a global one.*

Proof. Suppose by contradiction that $x_0 \in A$ is a local C^*-efficient point for f but it is not also global, that is to say that $\exists x \in A$, $x \neq x_0$, such that $f(x) \in f(x_0) + C^*$; by means of the local (C^*, C^*)-quasiconcavity of f we then have that $\exists \lambda^* \in (0, 1]$ such that $f(x_0 + \lambda(x - x_0)) \in f(x_0) + C^*$ $\forall \lambda \in (0, \lambda^*)$ and this contradicts the local C^*-efficiency of x_0. □

Since any $(C^*, \text{Int}(C))$-pseudoconcave function is also locally (C^*, C^*)-quasiconcave, we have the following corollary.

Corollary 1. *If a differentiable function $f : A \to \Re^m$, A open convex set, is $(C^*, \text{Int}(C))$-pseudoconcave then every local C^*-efficient point is also a global one.*

Another important property which is extended to the vector case is the optimality of all critical points. Denoting with $C^+ = \{d \in \Re^m : d^T c \geq 0 \; \forall c \in C\} \subset \Re^m$ the positive polar cone of C, it is known that if $x_0 \in A$ is a local C^*-efficient point for f then (see for example [3,4]):

$$\exists \alpha \in C^+, \alpha \neq 0, \text{ such that } \alpha^T J_f(x_0) = 0$$

This suggests the following definition.

Definition 5. A point $x_0 \in A$ is said to be a *critical point* for f if:

$$\exists \alpha \in C^+, \alpha \neq 0, \text{ such that } \alpha^T J_f(x_0) = 0$$

By means of this definition we can prove the following result.

Theorem 6. *If $f : A \to \Re^m$, A open convex set, is a $(C^*, \text{Int}(C))$-pseudoconcave function then every critical point is also a global C^*-efficient one.*

Proof. Suppose by contradiction that $x_0 \in A$ is a critical point for f but not also a global C^*-efficient one, that is to say that $\exists x \in A$, $x \neq x_0$, such that $f(x) \in f(x_0) + C^*$; by means of the $(C^*, \text{Int}(C))$-pseudoconcavity of f we then have that $J_f(x_0)(x - x_0) \in \text{Int}(C)$. Since x_0 is a critical point there exists an $\alpha \in C^+$, $\alpha \neq 0$, such that $\alpha^T J_f(x_0) = 0$; by a known property of polar cones condition $J_f(x_0)(x - x_0) \in \text{Int}(C)$ implies that $\alpha^T J_f(x_0)(x - x_0) > 0$ which contradicts $\alpha^T J_f(x_0) = 0$. □

Let us now see how the defined classes of vector valued pseudoconcave and quasiconcave functions give us the chance to extend the sufficiency of the Kuhn-Tucker conditions.

With this aim let us consider the following vector optimization problem:

$$P : \begin{cases} C^*\text{-max } f(x) \\ g(x) \in V \end{cases}$$

where A is an open convex set, $g : A \to \Re^p$ is a vector valued differentiable function, $V \subset \Re^p$ is a convex cone and $x_0 \in A$ is such that $g(x_0) = 0$. With respect to the function g we will use the concept of weakly (C^*, C)-quasiconcavity studied in [6,8] [2]

[2] Let $C \subset \Re^m$ be a closed convex pointed cone with nonempty interior and let $C^* \subset \Re^m$ be any cone such that $\text{Int}(C) \subseteq C^* \subseteq C$. A function $f : A \to \Re^m$

Theorem 7. *Let us consider problem P; if $f : A \to \Re^m$ is a $(C^*, \mathrm{Int}(C))$-pseudoconcave function, $g : A \to \Re^p$ is a weakly (V,V)-quasiconcave function and $x_0 \in A$ verifies the following condition:*

$$\exists \alpha \in C^+, \alpha \neq 0, \exists \beta \in V^+ \quad \text{such that } \alpha^T J_f(x_0) + \beta^T J_g(x_0) = 0 \tag{5}$$

then x_0 is a global C^*-efficient point.

Proof. Suppose by contradiction that x_0 is not a global C^*-efficient point, that is to say that $\exists x \in A$, $x \neq x_0$, such that:

$$f(x) \in f(x_0) + C^* \text{ and } g(x) \in V;$$

then by means of the $(C^*, \mathrm{Int}(C))$-pseudoconcavity of f and the weak (V,V)-quasiconcavity of g we have:

$$J_f(x_0)(x - x_0) \in \mathrm{Int}(C) \text{ and } J_g(x_0)(x - x_0) \in V.$$

Using the multiplier vectors α and β of condition (5) we then have:

$$\alpha^T J_f(x_0)(x - x_0) > 0 \text{ and } \beta^T J_g(x_0)(x - x_0) \geq 0$$

so that $[\alpha^T J_f(x_0) + \beta^T J_g(x_0)](x - x_0) > 0$ which is a contradiction. □

5 First order characterizations

The well known result by Thompson and Parke [20] states that a scalar function f, defined on a convex set A, is [strictly] pseudoconcave if and only if the following implication holds $\forall x, x_0 \in A$, $x \neq x_0$:

$$f(x) > f(x_0)\ [\geq] \implies \begin{array}{c} \exists \xi_{x,x_0} > 0 \text{ such that } \forall \lambda \in (0,1) \\ f(x_0 + \lambda(x - x_0)) \geq f(x_0) + \lambda(1-\lambda)\xi_{x,x_0} \end{array}$$

In the vector case the following new result, which extends the one by Thompson and Parke, can be proved.

Theorem 8. *A differentiable function f, defined on an open convex set A, is a $(C^*, \mathrm{Int}(C))$-pseudoconcave function if and only if the following condition holds $\forall x, x_0 \in A$, $x \neq x_0$:*

$$f(x) \in f(x_0) + C^* \implies \begin{array}{c} \exists \xi_{x,x_0} \in \mathrm{Int}(C), \exists \lambda^* \in (0,1] \\ \textit{such that } \forall \lambda \in (0,\lambda^*) \\ f(x_0 + \lambda(x - x_0)) \in f(x_0) + \lambda(1-\lambda)\xi_{x,x_0} + C \end{array} \tag{6}$$

---

is said to be a *weakly (C^*, C)-quasiconcave function* if the following implication holds $\forall x, x_0 \in \Re^n$, $x \neq x_0$:

$$f(x) \in f(x_0) + C^* \implies J_f(x_0)(x - x_0) \in C$$

Note that in the scalar case the weakly (C^*, C)-quasiconcave functions coincide with the quasiconcave scalar functions.

Proof. The thesis follows directly by means of Theorem 1. □

Note that in [6,8] a function f over a convex set A has been defined to be a *strictly* $(C^*, \mathrm{Int}(C))$*-pseudoconcave function* if the following implication holds $\forall x, x_0 \in A$, $x \neq x_0$:

$$f(x) \in f(x_0) + C^* \quad \Longrightarrow \quad \begin{array}{c} \exists \xi_{x,x_0} \in \mathrm{Int}(C) \text{ such that } \forall \lambda \in (0,1) \\ f(x_0 + \lambda(x - x_0)) \in f(x_0) + \lambda(1-\lambda)\xi_{x,x_0} + C \end{array}$$

This class of functions comes out, by means of Theorem 8, to be strictly included in the one of $(C^*, \mathrm{Int}(C))$-pseudoconcave functions, as it is focused on by Example 1 (see [6,8]). Example 1 points out also the importance for $(C^*, \mathrm{Int}(C))$-pseudoconcave functions of the existence of a line segment property in the beginning ([3]) of the interval $(0,1)$, that is in $(0, \lambda^*)$, and points out also that (6) is a better definition for nonsmooth vector valued pseudoconcave function than the one given in [6,8].

In [13], see also [1], a continuously differentiable scalar function f, defined on an open convex set A, has been proved to be [strictly] pseudoconcave if and only if the following condition holds:

> $\forall x_0 \in A$ and $\forall d \in \Re^n$, $d \neq 0$, such that $\nabla f(x_0)^T d = 0$ the function $\phi(\lambda) = f(x_0 + \lambda d)$ attains a [strict] local maximum at $\lambda = 0$.

In the vector case it is possible to prove the following results.

Theorem 9. *Let f be a differentiable function defined over an open convex set A. If f is a $(C^*, \mathrm{Int}(C))$-pseudoconcave function then:*

> *$\forall x_0 \in A$ and $\forall d \in \Re^n$, $d \neq 0$, such that $J_f(x_0)d \in \mathrm{Fr}(C)$ the function $\phi(\lambda) = f(x_0 + \lambda d)$, $\lambda \geq 0$, attains a local C^*-efficient point at $\lambda = 0$.*

Proof. Suppose by contradiction that $\exists x_0 \in A$, $\exists d \in \Re^n$, $d \neq 0$, such that $J_f(x_0)d \in \mathrm{Fr}(C)$ and $\phi(\lambda)$ does not attain a local C^*-efficient point at $\lambda = 0$, that is to say that $\exists \overline{\lambda} > 0$ such that $f(x_0 + \overline{\lambda} d) \in f(x_0) + C^*$; by means of the $(C^*, \mathrm{Int}(C))$-pseudoconcavity of f it follows that $\overline{\lambda} J_f(x_0)d \in \mathrm{Int}(C)$ so that $J_f(x_0)d \in \mathrm{Int}(C)$ and this is a contradiction since $J_f(x_0)d \in \mathrm{Fr}(C)$. □

Theorem 10. *Let f be a differentiable function defined over an open convex set A. If the two following conditions hold:*

i) f is a locally (C^*, C^*)-quasiconcave function,

---

[3] Function f of Example 1 is $(C^*, \mathrm{Int}(C))$-pseudoconcave but it is not strictly $(C^*, \mathrm{Int}(C))$-pseudoconcave (see [6,8]) since the line segment property holds in $(0,1)$ but not in $(0,3)$, even if $f(3) \in f(0) + \mathrm{Int}(C)$.

ii) $\forall x_0 \in A$ *and* $\forall d \in \Re^n$, $d \neq 0$, *such that* $J_f(x_0)d \in \mathrm{Fr}(C)$ *the function* $\phi(\lambda) = f(x_0 + \lambda d)$, $\lambda \geq 0$, *attains a local* C^**-efficient point at* $\lambda = 0$.

then f *is a* $(C^*, \mathrm{Int}(C))$*-pseudoconcave function.*

Proof. Suppose by contradiction that there exist $x, x_0 \in \Re^n$, $x \neq x_0$, such that $f(x) \in f(x_0) + C^*$ and $J_f(x_0)(x - x_0) \notin \mathrm{Int}(C)$. From hypothesis *i)* it follows that $f(x_0 + \lambda(x - x_0)) \in f(x_0) + C^*$ $\forall \lambda \in (0, \lambda^*)$; this implies that $\lambda = 0$ is not a local C^*-efficient point for $\phi(\lambda)$ where $d = x - x_0$ and, by means of Theorem 2, that $J_f(x_0)(x - x_0) \in C$; we then have that $J_f(x_0)(x - x_0) \in \mathrm{Fr}(C)$ and this contradicts hypothesis *ii)*. □

The following example points out the importance of condition i) in Theorem 10.

Example 3. Consider the cone $C = \Re^2_+$ and the following twice differentiable function $f : [0, 2] \to \Re^2$:

$$f(x) = \begin{cases} [3, -1]^T - (x - 1)^4 [3, -1]^T & \text{if } x \in [0, 1] \\ [3, -1]^T + (x - 1)^4 [-2, 2]^T & \text{if } x \in (1, 2] \end{cases}$$

It results $J_f(x_0)d \in \mathrm{Fr}(C)$ only for $x = 1$ which is a global C^*-efficient point for f, so that condition ii) holds in Theorem 10; we also have that f is neither locally (C^*, C^*)-quasiconcave nor weakly (C^*, C)-quasiconcave nor $(C^*, \mathrm{Int}(C))$-pseudoconcave since $f(2) \in f(0) + \mathrm{Int}(C)$, $f(x) \notin f(0) + C$ $\forall x \in (0, 1)$, and $J_f(0)(x - 0) = 4(x - 0)[3, -1]^T \notin C$ $\forall x > 0$.

We finally provide the following result which characterizes a $(C^*, \mathrm{Int}(C))$-pseudoconcave function with no quasiconcavity requirement.

Theorem 11. *Let* f *be a differentiable function defined over an open convex set* A. *Function* f *is a* $(C^*, \mathrm{Int}(C))$*-pseudoconcave function if and only if the following condition holds:*

> $\forall x_0 \in A$ *and* $\forall d \in \Re^n$, $d \neq 0$, *such that* $J_f(x_0)d \notin \mathrm{Int}(C)$ *the function* $\phi(\lambda) = f(x_0 + \lambda d)$, $\lambda \geq 0$, *attains a global* C^**-efficient point at* $\lambda = 0$.

Proof. $\Rightarrow$) Suppose by contradiction that $\exists x_0 \in A$, $\exists d \in \Re^n$, $d \neq 0$, $\exists \overline{\lambda} > 0$ such that $J_f(x_0)d \notin \mathrm{Int}(C)$ and $f(x_0 + \overline{\lambda} d) \in f(x_0) + C^*$; by means of the $(C^*, \mathrm{Int}(C))$-pseudoconcavity of f it results $\overline{\lambda} J_f(x_0)d \in \mathrm{Int}(C)$ so that $J_f(x_0)d \in \mathrm{Int}(C)$ which is a contradiction.
$\Leftarrow$) Suppose by contradiction that there exist $x, x_0 \in \Re^n$, $x \neq x_0$, such that $f(x) \in f(x_0) + C^*$ and $J_f(x_0)(x - x_0) \notin \mathrm{Int}(C)$. Setting $d = x - x_0$, we get $J_f(x_0)d \notin \mathrm{Int}(C)$ and $\phi(1) \in \phi(0) + C^*$ which is a contradiction. □

6 Second order characterizations

In [13], see also [1], a twice continuously differentiable scalar function f, defined on an open convex set A, has been proved to be [strictly] pseudoconcave if and only if the following condition holds:

> $\forall x_0 \in A$ and $\forall d \in \Re^n$, $d \neq 0$, such that $\nabla f(x_0)^T d = 0$ either $d^T H_f(x_0) d < 0$ or $d^T H_f(x_0) d = 0$ and the function $\phi(\lambda) = f(x_0 + \lambda d)$ attains a [strict] local maximum at $\lambda = 0$.

In order to extend to the vector case this result, we first prove the following second order necessary condition which holds for the class of $(C^*, \text{Int}(C))$-pseudoconcave functions, for the class of (C^*, C^0)-pseudoconcave functions (class studied in [6,8] ([4]) which contains the $(C^*, \text{Int}(C))$-pseudoconcave functions), and for the class of locally $(\text{Int}(C), C)$-quasiconcave functions.

Theorem 12. *If function f verifies at least one of the following properties:*
i) f is a $(C^*, \text{Int}(C))$-pseudoconcave function,
ii) f is a (C^*, C^0)-pseudoconcave function,
iii) f is a locally $(\text{Int}(C), C)$-quasiconcave function,
then the following condition holds:

$$\forall x_0 \in A,\ \forall d \in \Re^n,\ d \neq 0,\ J_f(x_0)d = 0 \text{ implies } d^T H_f(x_0) d \notin \text{Int}(C)$$

Proof. Suppose by contradiction that there exists $w \in \Re^n$, $w \neq 0$, such that $J_f(x_0)w = 0$ and $w^T H_f(x_0) w \in \text{Int}(C)$. By means of the following second order Taylor expansion of f at x_0 it follows:

$$\frac{f(x_0 + tw) - f(x_0)}{t^2} = \frac{1}{t} J_f(x_0)w + \frac{1}{2} w^T H_f(x_0) w + \|w\|^2 \sigma(t, 0)$$

where $\lim_{t \to 0} \sigma(t, 0) = 0$. Since $J_f(x_0)w = 0$ it follows:

$$\begin{aligned} \lim_{t \to 0} \frac{f(x_0 + tw) - f(x_0)}{t^2} &= \lim_{t \to 0} \frac{1}{2} w^T H_f(x_0) w + \|w\|^2 \sigma(t, 0) \\ &= \frac{1}{2} w^T H_f(x_0) w \in \text{Int}(C) \end{aligned}$$

Then $\exists \epsilon > 0$ such that $f(x_0 + tw) \in f(x_0) + \text{Int}(C)$ $\forall t \in (-\epsilon, \epsilon)$. Let $t_1 \in (0, \epsilon)$ and let $x_1 = x_0 + t_1 w$ so that $f(x_1) \in f(x_0) + \text{Int}(C)$. If f is a (C^*, C^0)-pseudoconcave function [$(C^*, \text{Int}(C))$-pseudoconcave] it results $J_f(x_0)(x_1 - x_0) = t_1 J_f(x_0) w \in C^0$ [$\in \text{Int}(C)$] and this is a contradiction since $J_f(x_0)w = 0$;

[4] A function $f : A \to \Re^m$ is said to be a (C^*, C^0)-*pseudoconcave function* if the following implication holds $\forall x, x_0 \in \Re^n$, $x \neq x_0$:

$$f(x) \in f(x_0) + C^* \quad \Longrightarrow \quad J_f(x_0)(x - x_0) \in C^0$$

function f must then be, by means of the hypothesis, a locally $(\mathrm{Int}(C), C)$-quasiconcave function.
By means of the continuity of f $\exists \bar{\epsilon}$, $0 < \bar{\epsilon} \leq \epsilon$, such that $\forall t \in (-\bar{\epsilon}, 0)$ it is $f(x_0 + tw) \in f(x_0) + \mathrm{Int}(C)$ and $f(x_0 + tw) \in f(x_1) - \mathrm{Int}(C)$. By means of a well known result on the existence of the maximal points (see [2]) then $\exists t_2 \in (-\bar{\epsilon}, 0)$ such that, defined $x_2 = x_0 + t_2 w$, it is $f(x_0 + tw) \notin f(x_2) + C$ $\forall t \in (t_2, 0)$; note also that it results $f(x_2) \in f(x_0) + \mathrm{Int}(C)$ and $f(x_2) \in f(x_1) - \mathrm{Int}(C)$.
Since f is locally $(\mathrm{Int}(C), C)$-quasiconcave and $f(x_1) \in f(x_2) + \mathrm{Int}(C)$ then $\exists \lambda^* \in (0,1)$ such that $f(x_2 + \lambda(x_1 - x_2)) \in f(x_2) + C$ $\forall \lambda \in (0, \lambda^*)$, that is to say that $f(x_0 + tw) \in f(x_2) + C$ $\forall t \in (t_2, t_2 + \lambda^*(t_1 - t_2))$.
This is a contradiction since $f(x_0 + tw) \notin f(x_2) + C$ $\forall t \in (t_2, 0)$ and the proof is then complete. □

Theorem 13. *Function f is a $(C^*, \mathrm{Int}(C))$-pseudoconcave function if and only if the following conditions hold:*

i) f is a locally (C^*, C^*)-quasiconcave function,

ii) $\forall x_0 \in A$ and $\forall d \in \Re^n$, $d \neq 0$, such that $J_f(x_0)d = 0$ either $d^T H_f(x_0) d \notin C$ or $d^T H_f(x_0) d \in \mathrm{Fr}(C)$ and the function $\phi(\lambda) = f(x_0 + \lambda d)$, $\lambda \geq 0$, attains a local C^*-efficient point at $\lambda = 0$,

iii) $\forall x_0 \in A$ and $\forall d \in \Re^n$, $d \neq 0$, such that $J_f(x_0)d \in \mathrm{Fr}(C)$, $J_f(x_0)d \neq 0$, the function $\phi(\lambda) = f(x_0 + \lambda d)$, $\lambda \geq 0$, attains a local C^*-efficient point at $\lambda = 0$.

Proof. $\Rightarrow$) The result follows directly from Theorem 3, Theorem 12 and the definition of $(C^*, \mathrm{Int}(C))$-pseudoconcavity.
$\Leftarrow$) Suppose by contradiction that there exist $x, x_0 \in \Re^n$, $x \neq x_0$, such that $f(x) \in f(x_0) + C^*$ and $J_f(x_0)(x - x_0) \notin \mathrm{Int}(C)$. By means of hypothesis i) it results $f(x_0 + \lambda(x - x_0)) \in f(x_0) + C^*$ $\forall \lambda \in (0, \lambda^*)$; this implies that $\lambda = 0$ is not a local C^*-efficient point for $\phi(\lambda)$ and, by means of Theorem 2, that $J_f(x_0)(x - x_0) \in C$, which implies that $J_f(x_0)(x - x_0) \in \mathrm{Fr}(C)$.
If $J_f(x_0)(x - x_0) = 0$ then by means of hypothesis i) and the previous Theorem 2 it is $(x - x_0)^T H_f(x_0)(x - x_0) \in C$; this implies for hypothesis ii) that $(x - x_0)^T H_f(x_0)(x - x_0) \in \mathrm{Fr}(C)$ and that $\phi(\lambda)$ attains a local C^*-efficient point at $\lambda = 0$, which is a contradiction.
Suppose now $J_f(x_0)(x - x_0) \in \mathrm{Fr}(C)$, $J_f(x_0)(x - x_0) \neq 0$; hypothesis iii) then implies again that $\phi(\lambda)$ attains a local C^*-efficient point at $\lambda = 0$, which is a contradiction. The proof is then complete. □

Let us note that the previous Example 3 points out that none among the three conditions of Theorem 13 is redundant.

7 Further results

In this section we will show that the pseudoconcavity of a scalar or a vector valued function can be characterized by means of the optimality of any point x_0 with respect of a constrained scalar/vector optimization problem P_{x_0}. In the scalar case it is possible to prove the following result.

Theorem 14. *Let us consider a differentiable scalar function f, defined on an open convex set A, and the following problem, depending on $x_0 \in A$:*

$$P_{x_0} = \begin{cases} Max\ f(x) \\ \nabla f(x_0)^T(x - x_0) \leq 0 \end{cases}$$

Then the three following conditions are equivalent:

i) f is [strictly] pseudoconcave,
ii) x_0 is a [strict] global maximum for P_{x_0}, $\forall x_0 \in A$,
iii) x_0 is a [strict] local maximum for P_{x_0}, $\forall x_0 \in A$.

Proof. i) $\Rightarrow$ ii) Suppose by contradiction that $\exists x_0 \in A$ such that x_0 is not a [strict] global maximum for P_{x_0}, so that $\exists y \in A$, such that $\nabla f(x_0)^T(y - x_0) \leq 0$ and $f(y) > f(x_0)$ $[f(y) \geq f(x_0)]$; then by means of i) it results $\nabla f(x_0)^T(y - x_0) > 0$ which is a contradiction.
ii) $\Rightarrow$ iii) Trivial.
iii) $\Rightarrow$ i) Suppose by contradiction that f is not [strictly] pseudoconcave, that is to say that $\exists y, x_0 \in A$, $y \neq x_0$, with $f(y) > f(x_0)$ $[f(y) \geq f(x_0)]$ and $\nabla f(x_0)^T(y - x_0) \leq 0$; note that the whole line segment $[x_0, y]$ comes out to be feasible for P_{x_0}. Being the function $\phi(\lambda) = f(x_0 + \lambda(y - x_0))$ continuous over $[0,1]$, there exists the minimum value m_v for it, so that we can define the following $\overline{\lambda} \in [0,1]$:

$$\overline{\lambda} = sup\{\lambda \in [0,1] \text{ such that } f(x_0 + \lambda(y - x_0)) = m_v\};$$

given that $f(y) > f(x_0)$ $[f(y) \geq f(x_0)]$, that x_0 is a [strict] local maximum for P_{x_0} and that the segment $[x_0, y]$ is feasible for P_{x_0} it follows that $0 < \overline{\lambda} < 1$; by means of the continuity of f we can easily prove also that:

$$\begin{gathered} f(x_0 + \overline{\lambda}(y - x_0)) = m_v \text{ and that} \\ f(x_0 + \overline{\lambda}(y - x_0)) < f(x_0 + \lambda(y - x_0)) \quad \forall \lambda \in (\overline{\lambda}, 1]; \end{gathered} \tag{7}$$

it then results $\frac{d\phi}{d\lambda}(\overline{\lambda}) = 0 = \nabla f(\overline{x})^T(y - x_0)$ where $\overline{x} = x_0 + \overline{\lambda}(y - x_0)$. Let us consider now the problem $P_{\overline{x}}$; being:

$$\nabla f(\overline{x})^T(y - \overline{x}) = (1 - \overline{\lambda})\nabla f(\overline{x})^T(y - x_0) = 0$$

it comes out that the line segment $[\overline{x}, y] \subset [x_0, y]$ is feasible also for $P_{\overline{x}}$ so that, by means of the hypothesis, $\overline{x}$ is a [strict] local maximum for f over $[\overline{x}, y]$ and this contradicts (7). □

In the vector case the following analogous result holds.

Theorem 15. *Let us consider function f and the following problem, depending on the point x_0:*

$$P_{x_0} = \begin{cases} C^*\text{-}max\ f(x) \\ J_f(x_0)(x - x_0) \notin \text{Int}(C) \end{cases}$$

Then the three following conditions are equivalent:

i) f is $(C^*, \text{Int}(C))$-pseudoconcave,
ii) x_0 is a global C^*-efficient point for P_{x_0}, $\forall x_0 \in A$,
iii) f is locally (C^*, C^*)-quasiconcave and x_0 is a local C^*-efficient point for P_{x_0}, $\forall x_0 \in A$.

Proof. *i) ⇒ ii)* Suppose by contradiction that $\exists x_0 \in A$ such that x_0 is not a global C^*-efficient point for P_{x_0}, so that $\exists y \in A$, such that $J_f(x_0)(y - x_0) \notin \text{Int}(C)$ and $f(y) \in f(x_0) + C^*$; then by means of i) it results $J_f(x_0)(y - x_0) \in \text{Int}(C)$ which is a contradiction.
ii) ⇒ i) Suppose by contradiction that $\exists x_0, y \in A$, $y \neq x_0$, such that $f(y) \in f(x_0) + C^*$ and $J_f(x_0)(y - x_0) \notin \text{Int}(C)$; y comes out to be feasible for P_{x_0} and x_0 results not to be a global C^*-efficient point for P_{x_0}, which is a contradiction.
ii) ⇒ iii) The thesis follows since a global C^*-efficient point is also local and since ii) implies i) and any $(C^*, \text{Int}(C))$-pseudoconcave function is also locally (C^*, C^*)-quasiconcave for Theorem 3.
iii) ⇒ ii) Suppose by contradiction that $\exists x_0 \in A$ such that x_0 is not a global C^*-efficient point for P_{x_0}, so that $\exists y \in A$, such that $J_f(x_0)(y - x_0) \notin \text{Int}(C)$ and $f(y) \in f(x_0) + C^*$; by means of the local (C^*, C^*)-quasiconcavity of f it is $f(x_0 + \lambda(y - x_0)) \in f(x_0) + C^*$ $\forall \lambda \in (0, \lambda^*)$, $\lambda^* \in (0, 1)$, and this contradicts the local C^*-efficiency of x_0 for P_{x_0}. □

Acknowledgements. Careful reviews by the anonymous referees are gratefully acknowledged.

References

1. Avriel M., Diewert W.E., Schaible S. and I. Zang, "Generalized Concavity", Mathematical Concepts and Methods in Science and Engineering, vol.36, Plenum Press, New York, 1988.
2. Cambini A. and L. Martein, "On the existence of efficient points", *Optimization*, vol.28, pp.283-290, 1994.
3. Cambini A., Martein L., and R. Cambini, "Some optimality conditions in multiobjective programming", in *Multicriteria Analysis*, edited by J. Climaco, Springer-Verlag, Berlin, pp.168-178, 1997.

4. Cambini A., Martein L., and R. Cambini, "A new approach to second order optimality conditions in vector optimization", in *Advances in Multiple Objective and Goal Programming*, edited by R. Caballero, F. Ruiz and R. Steuer, Lecture Notes in Economics and Mathematical Systems, vol.455, Springer-Verlag, Berlin, pp.219-227, 1997.
5. Cambini A. and L. Martein, "Generalized concavity in multiobjective programming", in *Generalized Convexity, Generalized Monotonicity: Recent Results*, edited by J.-P. Crouzeix, J.-E. Martinez-Legaz and M. Volle, Nonconvex Optimization and Its Applications, vol.27, Kluwer Academic Publishers, Dordrecht, pp.453-468, 1998.
6. Cambini R., "Some new classes of generalized concave vector-valued functions", *Optimization*, vol.36, n.1, pp.11-24, 1996.
7. Cambini R., "Generalized Concavity and Optimality Conditions in Vector Optimization", in *Operations Research and its Applications*, edited by D.Z. Du, X.S. Zhang and K. Cheng, Lecture Notes in Operations Research, vol.2, World Publishing Corporation, Beijing, pp.172-180, 1996.
8. Cambini R., "Composition theorems for generalized concave vector valued functions", *Journal of Information and Opt. Sciences*, vol.19, n.1, pp.133-150, 1998.
9. Cambini R., "Generalized Concavity for Bicriteria Functions", in *Generalized Convexity, Generalized Monotonicity: Recent Results*, edited by J.-P. Crouzeix, J.-E. Martinez-Legaz and M. Volle, Nonconvex Optimization and Its Applications, vol.27, Kluwer Academic Publishers, Dordrecht, pp.439-451, 1998.
10. Cambini R. and S. Komlósi, "On the Scalarization of Pseudoconcavity and Pseudomonotonicity Concepts for Vector Valued Functions", in *Generalized Convexity, Generalized Monotonicity: Recent Results*, edited by J.-P. Crouzeix, J.-E. Martinez-Legaz and M. Volle, Nonconvex Optimization and Its Applications, vol.27, Kluwer Academic Publishers, Dordrecht, pp.277-290, 1998.
11. Cambini R. and S. Komlósi, "On Polar Generalized Monotonicity in Vector Optimization", *Optimization*, vol.47, pp.111-121, 2000.
12. Crouzeix J.P., "Characterizations of generalized convexity and generalized monotonicity, a survey", in *Generalized Convexity, Generalized Monotonicity: Recent Results*, edited by J.-P. Crouzeix, J.-E. Martinez-Legaz and M. Volle, Nonconvex Optimization and Its Applications, vol.27, Kluwer Academic Publishers, Dordrecht, pp.237-256, 1998.
13. Diewert W.E., Avriel M. and I. Zang, "Nine kinds of quasiconcavity and concavity", *J. Econ. Theory*, vol.25, pp.397-420, 1981.
14. Jahn J., "Introduction to the theory of vector optimization", Springer-Verlag, Berlin, 1994.
15. Komlósi S., "On pseudoconvex functions", *Acta Sci. Math. (Szeged)*, vol.57, pp.569-586, 1993.
16. Luc D.T., "Theory of Vector Optimization", Lecture Notes in Economics and Mathematical Systems, vol.319, Springer-Verlag, Berlin, 1988.
17. Luc D.T, "Generalized convexity and some applications to vector optimization", Proceedings of the XXI A.M.A.S.E.S. Conference, Appendix Volume, pp.61-76, 1997.
18. Mangasarian O.L., "Nonlinear Programming", McGraw-Hill, New York, 1969.
19. Martos B., "Nonlinear Programming, Theory and Methods", North-Holland, Amsterdam, 1975.
20. Thompson W.A. and D.W. Parke, "Some properties of generalized concave functions", *Oper. Res.*, vol.21, pp.305-313, 1973.

New Invexity-Type Conditions in Constrained Optimization

Giuseppe Caristi[1], Massimiliano Ferrara[1], and Anton Stefanescu[2]

[1] Faculty of Economics,
University of Messina
[2] Faculty of Mathematics,
University of Bucharest

Abstract. In the present paper we define weaker invexity-type properties and examine the relationships between the new concepts and other similar conditions. One obtains in this way necessary and sufficient conditions for Kuhn-Tucker sufficiency. Moreover one proves that the same conditions are sufficient for Wolfe duality.

2000 Mathematics Subject Classification. Primary 90C46; Secondary 52A41.

Keywords and phrases. invexity properties, constraint optimization, weak duality.

1 Introduction

As it is known, invexity has been introduced in optimization theory (Hanson (1981)), as a substitute for convexity in constrained optimization. Craven and Glover (1985) showed that any differentiable scalar function is invex if and only if every stationary point is a global minimizer. Thus, for unconstrained optimization, the concept of invexity is basically a restatement of the above property. However, in constrained optimization one requires that the objective and the constraint functions are invex with respect to the same vector function. As it was shown first by D.H. Martin (1985), in this context invexity is an unnecessarily strong condition for Kuhn-Tucker sufficiency. Several authors considered possible relaxations of invexity in order to obtain necessary and sufficient conditions for optimality criteria or for weak duality.

Our results deal with the same problem. The new concepts introduced below, extends invexity and are used to prove both the sufficiency of the Kuhn-Tucker conditions and the direct duality theorem.

Let X be a nonempty set in $\mathbf{R}^n$, $\varphi : X \to \mathbf{R}$ differentiable, and $\eta : X \times X \to \mathbf{R}^n$.

We recall that φ is said to be $\eta - invex$ at $y \in X$ if

$$\varphi(x) - \varphi(y) \geq \langle \eta(x, y), \nabla\varphi(y) \rangle, \text{ for all } x \in X.$$

φ is said to be $\eta - pseudoinvex$ at $y \in X$ if

$$x \in X, \langle \eta(x, y), \nabla\varphi(y) \rangle \geq 0 \Rightarrow \varphi(x) - \varphi(y) \geq 0$$

and φ is said to be $\eta - quasiinvex$ at $y \in X$ if

$$x \in X, \varphi(x) - \varphi(y) \leq 0 \Rightarrow \langle \eta(x,y), \nabla\varphi(y)\rangle \leq 0.$$

For the next definitions, X_0 will be a fixed subset of X.

Definition 1. • φ is $\eta - \inf - invex$ at $y \in X$, with respect to X_0, if

$$\inf_{x \in X_0} [\varphi(x) - \varphi(y)] \geq \inf_{x \in X_0} \langle \eta(x,y), \nabla\varphi(y)\rangle$$

• φ is $\eta - \sup - invex$ at $y \in X$, with respect to X_0, if

$$\sup_{x \in X_0} [\varphi(x) - \varphi(y)] \geq \sup_{x \in X_0} \langle \eta(x,y), \nabla\varphi(y)\rangle$$

• φ is $\eta - \inf - pseudoinvex$ at $y \in X$ with respect to X_0, if

$$\inf_{x \in X_0} \langle \eta(x,y), \nabla\varphi(y)\rangle \geq 0 \Rightarrow \inf_{x \in X_o} [\varphi(x) - \varphi(y)] \geq 0 \qquad (1)$$

• φ is $\eta - \sup - quasiinvex$ at $y \in X$ with respect to X_0, if

$$\sup_{x \in X_0} [\varphi(x) - \varphi(y)] \leq 0 \Rightarrow \sup_{x \in X_0} \langle \eta(x,y), \nabla\varphi(y)\rangle \leq 0. \qquad (2)$$

Definition 2. φ is $\eta - \inf - invex$ ($\eta - \sup - invex, \eta - \inf - pseudoinvex, \eta - \sup - quasiinvex$) on X, with respect to X_0, if it is $\eta - \inf - invex$ ($\eta - \sup - invex, \eta - \inf - pseudoinvex, \eta - \sup - quasiinvex$) at any point $y \in X$.

Remark 1. Obviously, $\eta - \inf - invexity$ implies $\eta - \inf - pseudoinvexity$ and $\eta - \sup - invexity$ implies $\eta - \sup - quasiinvexity$.

The above properties will be used in the classical framework of the scalar optimization problem:

$$(P) \inf_{x \in X_0} f(x), \text{where} X_0 = \{x \in X | g_j(x) \leq 0, j = 1, 2, ..., m\}$$

where f and g_j are differentiable.

Definition 3. The problem (P) is $\eta - inf - sup$-invex at $y \in X$, if f is $\eta - \inf - invex$ at y, with respect to X_0 and g_j, $j = 1, 2, ..., m$ are $\eta - \sup - invex$ at y, with respect to X_0. (P) is $\eta - inf - sup$-invex on X if it is $\eta - inf - sup$-invex at every point $y \in X$.

Let (D) the Wolfe dual in X of (P):

$$(D) \sup\nolimits_{(x,v) \in V} (f(x) + \sum_{j=1}^{m} v_j g_j(x))$$

where,

$$V = \{(x, v) \in X \times \mathbf{R}^m_+ \mid \nabla f(x) + \sum_{j=1}^{m} v_j \nabla g_j(x) = 0\}$$

In the next section we will show that if (P) is $\eta - inf - sup$-invex on X for some η, then any Kuhn-Tucker point is an optimum solution of (P) and the weak duality property holds.

2 Sufficiency of inf-sup invexity

Assuming f and g_j differentiable on X, *a Kuhn-Tucker point* is a pair $(x_0, v) \in X_0 \times \mathbf{R}^m_+$,satisfying the following two conditions:

$$\nabla f(x_0) + \sum_{j=1}^{m} v_j \nabla g_j(x_0) = 0 \tag{3}$$

$$\sum_{j=1}^{m} v_j g_j(x_0) = 0 \tag{4}$$

For $y \in X_0$, let us denote by $J_0(y)$ the set of active constraints at y; $J_0(y) = \{j \mid g_j(y) = 0\}$.

Theorem 1. *Let (x_0, v) be any Kuhn-Tucker point of the problem (P). If there exists some η, such that f is $\eta - inf$-pseudoinvex at $x_0$with respect to X_0, and for every $j \in J_0(x_0)$, g_j is $\eta - \sup -$quasiinvex at $y \in X$ with respect to X_0,then x_0 is an optimum solution of (P).*

Proof. By (3) and (4) one has:

$$\inf_{x \in X_0} \langle \eta(x, x_0), \nabla f(x_0) \rangle = \inf_{x \in X_0} \sum_{j \in J_0(x_0)} -v_j \langle \eta(x, x_0), \nabla g_j(x_0) \rangle$$

Obviously,

$$\inf_{x \in X_0} \sum_{j \in J_0(x_0)} -v_j \langle \eta(x, x_0), \nabla g_j(x_0) \rangle \geq \sum_{j \in J_0(x_0)} (-v_j) \sup_{x \in X_0} \langle \eta(x, x_0), \nabla g_j(x_0) \rangle$$

Since $\sup_{x \in X_0} [g_j(x) - g_j(x_0)] = 0$, whenever $j \in J_0(x_0)$, it follows finally from the invexity assumptions that:

$$\sum_{j \in J_0(x_0)} (-v_j) \sup_{x \in X_0} \langle \eta(x, x_0), \nabla g_j(x_0) \rangle \geq 0$$

and hence,

$$\inf_{x \in X_0} [f(x) - f(x_0)] \geq 0$$

□

Corollary 1. *Let (x_0, v) be any Kuhn-Tucker point of the problem (P). If there exists some η, such that f is $\eta - inf$-invex at x_0 with respect to X_0, and for every $j \in J_0(x_0)$, g_j is $\eta - \sup -$invex at $y \in X$ with respect to X_0, then x_0 is an optimum solution of (P).*

Corollary 2. *Assume that there exists some η, such that (P) is $\eta - inf - sup$-invex on X. Then for every Kuhn-Tucker point (x_0, v), x_0 is an optimum solution.*

Remark 2. An obvious relaxation of the above theorem can be obtained weakening the requirements for the constraint functions. In fact, it is sufficient to assume that

$$j \in J(x_0) \Rightarrow \sup_{x \in X_0} \langle \eta(x, x_0), \nabla g_j(x_0) \rangle \leq 0 \tag{5}$$

Consequently, one obtains:

Corollary 3. *Let (x_0, v) be any Kuhn-Tucker point of the problem (P). If there exists some η, such that the constraint functions satisfy (5) and the objective function satisfies:*

$$\inf_{x \in X_0} \langle \eta(x, x_0), \nabla f(x_0) \rangle \geq 0 \Rightarrow \inf_{x \in X_o} [f(x) - f(x_0)] \geq 0 \tag{6}$$

then x_0 is an optimum solution.

The next result concerns the weak duality property.

Theorem 2. *Assume that there exists some η, such that f is $\eta - inf -$pseudo invex on X and all g_j are $\eta - \sup$-invex on X. Then, for any feasible solution x of (P) $(x \in X_0)$, and for any feasible solution (y, v) of (D) $((y, v) \in V)$, one has:*

$$f(x) \geq f(y) + \sum_{j=1}^{m} v_j g_j(y)$$

Proof. One has:

$$\begin{aligned}
\inf\nolimits_{x \in X_0} & \langle \eta(x, y), \nabla f(y) \rangle - \textstyle\sum_{j=1}^{m} v_j g_j(y) = \\
& = \inf\nolimits_{x \in X_0} \langle \eta(x, y), \textstyle\sum_{j=1}^{m} (-v_j) \nabla g_j(y) \rangle - \textstyle\sum_{j=1}^{m} v_j g_j(y) \\
& \geq - \sup\nolimits_{x \in X_0} \textstyle\sum_{j=1}^{m} v_j \langle \eta(x, y), \nabla g_j(y) \rangle - \textstyle\sum_{j=1}^{m} v_j g_j(y) \\
& \geq - \textstyle\sum_{j=1}^{m} v_j \sup\nolimits_{x \in X_0} \langle \eta(x, y), \nabla g_j(y) \rangle - \textstyle\sum_{j=1}^{m} v_j g_j(y) \\
& \geq \textstyle\sum_{j=1}^{m} (-v_j) \sup\nolimits_{x \in X_0} [g_j(x) - g_j(y)] - \textstyle\sum_{j=1}^{m} v_j g_j(y) \\
& = \textstyle\sum_{j=1}^{m} (-v_j) \sup\nolimits_{x \in X_0} g_j(x) \geq 0
\end{aligned}$$

Hence, by the assumptions on f, it results:

$$\inf_{x \in X_0} [f(x) - f(y) - \sum_{j=1}^{m} v_j g_j(y)] \geq 0$$

□

Corollary 4. *If the problem (P) is $\eta - inf - $ sup-invex on X for some η, then it has the weak duality property.*

3 On the necessity of inf-sup-invexity

The invexity may appear as a general condition for a scalar function. However, when the invexity with respect to the same vector function is required for several functions, it may become a rather restrictive condition. Martin (1965) showed that if a mathematical programming problem with linear constraints which delimit a bounded feasible set is invex, then the objective function must be convex. This remark leads him to obtain weaker invexity-type conditions which are necessary and sufficient for the sufficiency of the Kuhn-Tucker conditions.

He called the problem (P) *Kuhn-Tucker invex on X,* if there exists a vector function $\eta : X \times X \to \mathbf{R}^n$, such that:

$$x, y \in X_0 \Rightarrow \begin{cases} f(x) - f(y) \geq \langle \eta(x,y), \nabla f(y) \rangle \\ \langle \eta(x,y), \nabla g_j(y) \rangle \leq 0, \text{whenever} g_j(y) = 0, \text{for} j = 1, 2, ..., m \end{cases}$$

Theorem 3. *(Martin (1985), Theorem 2.1) Every Kuhn-Tucker point of problem (P) is a global minimizer if and only if (P) is Kuhn-Tucker invex.*

Now, similar results follows easily using the concepts introduced in this paper.

Theorem 4. *A Kuhn-Tucker point (x_0, v) is a global minimizer of the problem (P), if and only if there exists a vector function $\eta : X \times X \to \mathbf{R}^n$, such that (5) and (6) hold.*

Proof. The sufficiency follows from Corollary 3 of Theorem 1. For the necessity one observes that Kuhn-Tucker invexity implies the conditions of the theorem. □

It is also obvious that,

Theorem 5. *Every Kuhn-Tucker point of problem (P) is a global minimizer if and only if there exists a vector function $\eta : X \times X \to \mathbf{R}^n$, such that f is $\eta-\inf-$pseudoinvex ($\eta - \inf -$invex) on X_0 with respect to X_0, and, for every $y \in X_0$, g_j is $\eta - \sup -$quasiinvex ($\eta - \sup -$invex) at y, whenever $j \in J_0(y)$.*

Finally, let us examine an example to show that our conditions actually don't imply convexity.

Example 1. Consider the problem:
$\inf_{x \in X_0}(-x_1 x_2)$, where, $X_0 = \{x \in \mathbf{R}^2 \mid x \geq 0, x_1 + x_2 \leq 1\}$.

Obviously, $y = (1/2, 1/2)$ is a Kuhn-Tucker point. One can easily verify that the problem is $\eta - \inf - \sup -invex$ at y, with respect to X_0,where, $\eta(x,y) = (x_1^2 - y_1, x_2^2 - y_2)$.

4 Comments

For the sake of the simplicity, in the following we will consider $X_0 = X$. Also, in order to simplify our terminology, any reference to X_0 will be omitted.

The relationship between different forms of invexity can be discussed either with reference to a specified scale function η, or referring to them as intrinsic properties.

Let η be a vector-valued function on $X \times X$.

The next proposition is obvious.

Proposition 1. *The following implications hold:*

$$[\varphi \mathrm{is} \eta - invex] \Rightarrow [\varphi \mathrm{is} \eta - \inf -invex] \Rightarrow [\varphi \mathrm{is} \eta - \inf -pseudoinvex].$$

and

$$[\varphi \mathrm{is} \eta - invex] \Rightarrow [\varphi \mathrm{is} \eta - pseudoinvex] \Rightarrow [\varphi \mathrm{is} \eta - \inf -pseudoinvex].$$

Let us denote by $E_i(\varphi, y)$ the set of all scale functions for which φ is invex at y; $E_i(\varphi, y) = \{\eta \mid \varphi \text{ is } \eta - invex \text{ at } y\}$. Similarly, $E_{ii}(\varphi, y)$, $E_p(\varphi, y)$ and $E_{ip}(\varphi, y)$ stand for the sets of all scale functions for which φ is $\eta - \inf -invex$, $\eta - pseudoinvex$, respectively, $\eta - \inf -pseudoinvex$.

The previous proposition says nothing else than that:

$$E_i(\varphi, y) \subseteq E_{ii}(\varphi, y) \subseteq E_{ip}(\varphi, y) \tag{7}$$

and

$$E_i(\varphi, y) \subseteq E_p(\varphi, y) \subseteq E_{ip}(\varphi, y) \tag{8}$$

The first inclusion in (8) may be strict, as it was shown in Kaul and Kaur (1985).

The following examples show that the strict inclusion may also hold in (7), or in the second part of (8). Moreover, the set $E_{ii}(\varphi, y) \setminus E_p(\varphi, y)$ may be non empty.

Example 2. Take $X = [-1, \infty)$, $\varphi(x) = x^3 + x$, $\eta(x,y) = -x$.

Obviously, $\eta \in E_{ii}(\varphi, 0)$, hence $\eta \in E_{ip}(\varphi, 0)$, but $\eta \notin E_p(\varphi, 0)$ (therefore, $\eta \notin E_i(\varphi, 0)$).

Example 3. $X = \mathbf{R}^2$, $\varphi(x) = -x_1 x_2$, $\eta(x, y) = (x_1^2 - y_1, x_2^2 - y_2)$.

Since $\inf_{x \in X} \langle \eta(x, (1/4, 1/4)), \nabla\varphi(1/4, 1/4)\rangle = -\infty$, it follows that $\eta \in E_{ii}(\varphi, (1/4, 1/4))$.

But, $\eta \notin E_p(\varphi, (1/4, 1/4))$, because

$$\langle \eta((1/2, 1/2), (1/4, 1/4)), \nabla\varphi(1/4, 1/4)\rangle = 0$$

and $\varphi(1/2, 1/2) - \varphi(1/4, 1/4) < 0$.

A second characterization of the relationships between the invexity-type properties does not refer to a specified scale function. We will say that φ is invex at y if there exists $\eta : X \times X \to \mathbf{R}$ such that φ is $\eta-$*invex* at y. All other forms of invexity are defined in the same way.

It is obvious that if φ is inf $-$*pseudoinvex* on X, then any stationary point is a global minimum point. Or, as it was shown in Craven and Glover (1985) (see also Ben-Israel and Mond (1986), Theorem 1), then φ is invex. Thus, by Proposition 1, one has:

Proposition 2. *φ is invex on $X \Leftrightarrow \varphi$ is pseudoinvex on $X \Leftrightarrow \varphi$ is inf-invex on $X \Leftrightarrow \varphi$ is* inf $-$*pseudoinvex on X.*

Now let us emphasize the relationships between invexity, quasi-invexity, sup-invexity and sup-quasiinvexity.

Denote by $E_{si}(\varphi, y)$, $E_q(\varphi, y)$, and $E_{sq}(\varphi, y)$ the sets of all scale functions for which φ is $\eta-\sup-$*invex*, $\eta-$*quasiinvex*, respectively, $\eta-\sup-$*quasiinvex* at y.

Proposition 3. *The following inclusions hold:*

$$E_i(\varphi, y) \subseteq E_{si}(\varphi, y) \subseteq E_{sq}(\varphi, y) \tag{9}$$

and

$$E_i(\varphi, y) \subseteq E_q(\varphi, y) \subseteq E_{sq}(\varphi, y) \tag{10}$$

The strict inclusions may hold in the above, as it follows from Kaul and Kaur (1985), and the next example.

Example 4. Let X, φ and η as in Example 3.

Since $\sup_{x \in X}[\varphi(x) - \varphi(-1, 0)] = +\infty$, it follows that φ is $\eta-\sup-$*quasiinvex* (and, hence $\eta-\sup-$*invex*) at $(-1, 0)$. On the other hand, $\varphi(1, 1) - \varphi(-1, 0) < 0$, while $\langle \eta((1, 1), (-1, 0)), \nabla\varphi(-1, 0)\rangle > 0$, so that φ is not $\eta-$*quasiinvex* (nor $\eta-$*invex*) at $(-1, 0)$.

A stronger result is the following.

Proposition 4. *Any invex function is sup-invex, but the converse is not always true.*

Indeed, the next example shows that a function may be sup-invex but not invex.

Example 5. Take $X = \mathbf{R}^2_+$, and φ and η as in the previous example.

Obvious, φ it $\eta-\sup-invex$ at the origin, but it is not invex at this point (because 0 is a stationary point which is not a global minimum.)

A question arises from the above discussions; what relevance have the results established in this section? In the framework of constraint optimization, invexity actually becomes a restrictive condition because it must be defined by a vector function with respect to one scale function. Thus, a larger set of associated scale functions for each component increases the chance of fitness for a common scale function.

To the end of this discussion, let us comment on the results of Section 3. From theorems 3 and 4 it follows that our conditions (5) and (6), on one hand, and the Kuhn-Tucker invexity, on the other hand, are equivalent conditions. However, it is easy to see that the set of all scale functions for which (5) and (6) are satisfied is larger than the set of the scale functions for which the problem (P) is Kuhn-Tucker invex. In the following example one shows that the inclusion may be strict.

Example 6. Let the problem (P) as in Example 1. As it was shown (5) and (6) are satisfied at the point $(1/2, 1/2)$ for $\eta(x, y) = (x_1^2 - y_1, x_2^2 - y_2)$. However, (P) is not Kuhn-Tucker invex with respect to this η. For $x = (0,0)$ and $y = (1/2, 1/2)$, $\varphi(x) - \varphi(y) = 1/4 < 1/2 = \langle \eta(x, y), \nabla\varphi(y) \rangle$.

References

1. A.Ben-Israel and B.Mond, What is invexity?, J. Austral. Math. Soc. Ser.B 28 (1986), 1-9.
2. B.D.Craven, Invex Functions and Constrained Local Minima, Bull. Austral. Math. Soc., 24 (1981), 357-366.
3. B.D.Craven and B.M.Glover, Invex functions and duality, J. Austral. Math. Soc. Ser.A 39 (1985), 97-99.
4. G.Giorgi, A Note on the Relationships Between Convexity and Invexity, J. Austral. Math. Soc. Ser.B 32 (1990), 97-99.
5. M.A.Hanson, On Sufficiency of the Kuhn-Tucker Conditions, Journal of Mathematical Analysis and Applications, 80 (1981), 545-550.
6. R.N.Kaul and S.Kaur, Optimality Criteria in Nonlinear Programming Involving Nonconvex Functions, Journal of Mathematical Analysis and Applications, 105 (1985), 104-112.
7. D.H.Martin, The Essence of Invexity, JOTA, 47 (1985), 65-76.

Stochastic s-(increasing) Convexity

Michel Denuit[1] and Claude Lefèvre[2]

[1] Institut de Statistique
Université Catholique de Louvain
Voie du Roman Pays, 20
B-1348 Louvain-la-Neuve, Belgium
denuit@stat.ucl.ac.be

[2] Institut de Statistique et de Recherche Opérationnelle
Université Libre de Bruxelles
CP 210, Boulevard du Triomphe
B-1050 Bruxelles, Belgium
clefevre@ulb.ac.be

Abstract. For many applications, especially in actuarial sciences, a problem of great practical interest is the stochastic comparison of mixed distributions and of compound sums. This question is investigated here with respect to the class of stochastic s-(increasing) convex orderings introduced recently. The analysis is based on a central property, called stochastic s-(increasing) convexity, for families of parametric distributions.

2000 Mathematics Subject Classification. Primary 62H20; Secondary 62P05.

Keywords and phrases. Stochastic comparisons, mixed distributions, compound sums, actuarial models, s-(increasing) convex functions/orderings/convexity.

1 Introduction

Let $\{X_\theta,\ \theta \in \Theta\}$ be a family of random variables valued in a subset $\mathcal{S} \subseteq I\!R$, and indexed by a single parameter real $\theta \in \Theta \subseteq I\!R$. The associated family of distributions is denoted by $\mathcal{P}(\Theta) = \{P_\theta,\ \theta \in \Theta\}$.

Now, given any function $\phi : \mathcal{S} \to I\!R$, let us define the function ϕ^* as

$$\phi^* : \Theta \to I\!R \,;\, \theta \mapsto E[\phi(X_\theta)]. \tag{1}$$

In the present paper, our main purpose is to investigate whether some specific property of the function ϕ can be transmitted to the function ϕ^*. More precisely, the property under study is the notion of s-(increasing) convex functions, which was introduced by Popoviciu (1933) in the theory of approximation and developed later in many research works. We also refer, e.g., to the books by Roberts and Varberg (1973) and Pečarić, Proschan and Tong (1992). Note that the 1-convex functions correspond to the increasing functions, and the 2-(increasing) convex functions to the (increasing) convex functions in the usual sense.

Recently, these classes of s-increasing convex functions have been used in probability and statistics in order to compare two random variables, or their distributions, in the so-called stochastic s-(increasing) convex orderings. We mention, e.g., Lefèvre and Utev (1996), Denuit and Lefèvre (1997), Denuit, Lefèvre and Shaked (1998), Denuit, Lefèvre and Utev (1999a,b) and Denuit, Lefèvre and Mesfioui (1999). In fact, these stochastic orderings will play, here again, an important role (see Section 4). For that reason, a succinct presentation of the theory is given in Section 2.

Coming back to (1), if the s-(increasing) convexity is transmitted from ϕ to ϕ^*, the family of distributions $\mathcal{P}(\Theta)$ is said to be stochastically s-(increasing) convex. This property of stochastic s-(increasing) convexity is analyzed in some details in Section 3. We will obtain some characterizations and sufficient conditions, and we will point out that a number of classical parametric distributions enjoy the property. Note that some results given in the actuarial literature will be so generalized. Moreover, we observe that the special cases where $s = 1$ and 2 have been examined by Shaked and Shanthikumar (1988), (1994, Chapter VI).

In Section 4, we apply some of the previous results for comparing mixed distributions and compound sums. This problem has many applications in theory and practice, especially in actuarial sciences. For illustration, we will derive lower and upper bounds, in the stochastic 2 and 3-convex orderings, for the distribution of the aggregate claim amounts over some period of time, when the claim frequency is mostly unknown, the only information available being its first two moments, with its largest value, and possibly the property that its density is decreasing.

To close, we indicate that in Denuit, Lefèvre and Utev (1999b), all this analysis based on the s-(increasing) convex functions has been developed in the more general framework of the generalized (increasing) convex functions with respect to an arbitrary Tchebycheff system of functions (in the sense of, e.g., Karlin and Studden (1966)). As underlined in that paper, however, such an extension can have the disadvantage to conceal certain properties that are closely related to the stochastic s-(increasing) convexity.

2 Stochastic s-(increasing) convex orderings

2.1 Integral stochastic orderings

Most stochastic orderings that allow to compare two random variables X and Y valued in $\mathcal{S} \subseteq \mathbb{R}$, are defined (or, at least, can be defined) by reference to a class $\mathcal{F}$ of measurable functions $\phi : \mathcal{S} \to \mathbb{R}$, namely

$$X \preceq^{\mathcal{S}}_{\mathcal{F}} Y \Leftrightarrow E\phi(X) \leq E\phi(Y) \text{ for all the functions } \phi \in \mathcal{F} \tag{2}$$

for which the expectations exist. Such orderings are usually referred to as integral stochastic orderings and obviously rely on the classical expected utility theory (viewing $\mathcal{F}$ as a class of utility functions). A rather general theory of these orderings can be found in Marshall (1991) and Müller (1997).

As particular cases of (2), the stochastic dominance is obtained when $\mathcal{F}$ represents the class of the increasing functions ϕ, and the stochastic (increasing) convex ordering when $\mathcal{F}$ is the class of the (increasing) convex functions ϕ (see, e.g., Shaked and Shanthikumar (1994)). We notice that the stop-loss order discussed in actuarial sciences corresponds to the stochastic increasing convex ordering (see, e.g., Kaas, Van Heerwaarden and Goovaerts (1994)).

2.2 Cones of s-(increasing) convex functions

Following Popoviciu (1933), given any positive integer s, the class $\overline{\mathcal{U}}^{\mathcal{S}}_{s-cx}$ of the s-convex functions on $\mathcal{S}$ is defined by

$$\overline{\mathcal{U}}^{\mathcal{S}}_{s-cx} = \{\phi : \mathcal{S} \to \mathbb{R} \,|\, [x_0, x_1, \ldots, x_s]\phi \geq 0 \;\forall\; x_0 < x_1 < \ldots < x_s \in \mathcal{S}\},$$

where $[\ldots]$ denotes the classical divided difference operator, which is defined recursively for $\phi : \mathcal{S} \to \mathbb{R}$ and $x_0 \neq x_1 \neq \ldots \neq x_s$ in $\mathcal{S}$ by

$$[x_0, x_1, \ldots, x_k]\phi = \frac{[x_1, x_2, \ldots, x_k]\phi - [x_0, x_1, \ldots, x_{k-1}]\phi}{x_k - x_0}, \quad k = 1, 2, \ldots, s-1,$$

starting from $[x_i]\phi = \phi(x_i)$ for $i = 0, 1, \ldots, s$. Note that the s-convexity is defined for any subset $\mathcal{S}$ of the real line $\mathbb{R}$ and does not require that $\mathcal{S}$ is an interval. The name "s-convex" comes from the fact that the functions in $\overline{\mathcal{U}}^{\mathcal{S}}_{2-cx}$, that is, the 2-convex functions, are convex in the usual sense when $\mathcal{S}$ is an interval.

The class $\overline{\mathcal{U}}^{\mathcal{S}}_{s-icx}$ of the s-increasing convex functions with domain $\mathcal{S}$ is then defined as the class of all the functions being simultaneously k-convex for $k = 1, 2, \ldots, s$, i.e.

$$\overline{\mathcal{U}}^{\mathcal{S}}_{s-icx} = \bigcap_{k=1}^{s} \overline{\mathcal{U}}^{\mathcal{S}}_{k-cx}.$$

2.3 Induced stochastic s-(increasing) convex orderings

The s-convex ordering and the s-increasing convex ordering are the integral stochastic orderings generated by the cone of the s-convex functions and by the cone of the s-increasing convex functions, respectively. To be more specific, given two random variables X and Y valued in $\mathcal{S}$, X is said to be smaller than Y in the s-convex ordering, denoted henceforth by $X \preceq^{\mathcal{S}}_{s-cx} Y$, when (2) holds with $\mathcal{F} = \overline{\mathcal{U}}^{\mathcal{S}}_{s-cx}$. Similarly, X is said to be smaller than Y in the s-increasing convex ordering, denoted by $X \preceq^{\mathcal{S}}_{s-icx} Y$, when (2) holds with $\mathcal{F} = \overline{\mathcal{U}}^{\mathcal{S}}_{s-icx}$.

As particular cases, we have that $\preceq^{\mathcal{S}}_{1-(i)cx}$ is just the stochastic dominance and $\preceq^{\mathcal{S}}_{2-(i)cx}$ is the usual (increasing) convex ordering. We mention that the stop-loss order of degree $s-1$ in actuarial sciences corresponds here to $\preceq_{s-icx}^{\mathbb{R}^+}$ (see, e.g., Kaas and Hesselager (1995)).

Now, the subset $\mathcal{S}$ in (2) can be taken arbitrarily. For particular choices of $\mathcal{S}$, however, the s-convex orderings can be characterized by replacing $\overline{\mathcal{U}}_{s-cx}^{\mathcal{S}}$ by some dense subclass in it. First, let us consider $\mathcal{S} = [a,b]$, $a < b \in \mathbb{R}$, b being possibly infinite. In this case, it is shown in Denuit, Lefèvre and Shaked (1998) that $\preceq_{s-cx}^{[a,b]}$ can be defined through (2) with for $\mathcal{F}$ the class

$$\mathcal{M}_{s-cx}^{[a,b]} = \{\pm(x-a)^k,\ \ k = 1,2,\ldots,s-1;\ \ (x-t)_+^{s-1},\ \ t \in [a,b]\}, \tag{3}$$

where $x_+ = x$ if $x > 0$ and 0 otherwise; for $s = 1$, by convention $(x-t)_+^0$ is the indicator function of $[t, +\infty[$. Alternatively, $\preceq_{s-cx}^{[a,b]}$ can be generated by the cone

$$\mathcal{U}_{s-cx}^{[a,b]} = \left\{\phi : [a,b] \to \mathbb{R} \,\middle|\, \phi^{(s)} \geq 0 \text{ on } [a,b]\right\}, \tag{4}$$

where $\phi^{(s)}$ denotes the s-th derivative of the function ϕ. Denoting by $\mathcal{C}^s([a,b])$ the class of the functions $\phi : [a,b] \to \mathbb{R}$ which are s times continuously differentiable, it can be proved that

$$\mathcal{U}_{s-cx}^{[a,b]} = \overline{\mathcal{U}}_{s-cx}^{[a,b]} \cap \mathcal{C}^s([a,b]).$$

From an analytical point of view $\mathcal{M}_{s-cx}^{[a,b]}$ consists of the extreme points of the class $\overline{\mathcal{U}}_{s-cx}^{[a,b]}$ while $\mathcal{U}_{s-cx}^{[a,b]}$ contains the "regular" s-convex functions.

On the other hand, when $\mathcal{S} = \mathcal{D}_n \equiv \{0,1,\ldots,n\}$, where n is some positive integer (by convention, $\mathcal{D}_\infty$ stands for the set $\mathbb{N}$), it can be shown (see Denuit and Lefèvre (1997) and Denuit, Lefèvre and Utev (1999a)) that

$$\overline{\mathcal{U}}_{s-cx}^{\mathcal{D}_n} = \{\phi : \mathcal{D}_n \to \mathbb{R} | \Delta^s\phi(i) \geq 0 \text{ on } \mathcal{D}_{n-s}\}, \tag{5}$$

where Δ^s denotes the usual s-th degree forward difference operator, which is defined recursively for a function $\phi : \mathcal{D}_n \to \mathbb{R}$ and $k \in \mathcal{D}_n$ by $\Delta^k\phi(i) = \Delta^{k-1}\phi(i+1) - \Delta^{k-1}\phi(i)$ for all $i \in \mathcal{D}_n$ such that $i + k \in \mathcal{D}_n$, i.e. for $i = 0,1,\ldots,n-k$ (by convention, $\Delta^1\phi \equiv \Delta\phi$ and $\Delta^0\phi \equiv \phi$). Moreover, $\preceq_{s-cx}^{\mathcal{D}_n}$ can also be defined through (2) with for $\mathcal{F}$ the class

$$\begin{aligned}\mathcal{M}_{s-cx}^{\mathcal{D}_n} = \Bigg\{ &\pm\binom{x}{i},\ \ i = 1,2,\ldots,s-1; \\ &\binom{x-i}{s-1},\ \ i = 1,2,\ldots,n-s+1\Bigg\},\end{aligned} \tag{6}$$

where $\binom{x}{i} = 0$ when $x < i$. Note that (5) and (6) can be seen as discrete analogs of (4) and (3), respectively.

A similar study can be carried out for the s-increasing convex order. More precisely, the analogs for $\preceq^{S}_{s-icx}$ of (3), (4), (5) and (6) above are respectively given by

$$\mathcal{M}^{[a,b]}_{s-icx} = \{(x-a)^k,\ \ k=1,2,\ldots,s-1;\ \ (x-t)_+^{s-1},\ \ t\in[a,b]\}, \tag{7}$$

$$\mathcal{U}^{[a,b]}_{s-icx} = \left\{\phi:[a,b]\to I\!R \,\middle|\, \phi^{(k)}\geq 0 \text{ on } [a,b] \text{ for } k=1,2,\ldots,s\right\}, \tag{8}$$

$$\overline{\mathcal{U}}^{\mathcal{D}_n}_{s-icx} = \{\phi:\mathcal{D}_n\to I\!R|\Delta^k\phi(i)\geq 0 \text{ on } \mathcal{D}_{n-k} \text{ for } k=1,2,\ldots,s\}, \tag{9}$$

$$\mathcal{M}^{\mathcal{D}_n}_{s-icx} = \left\{\binom{x}{i},\ \ i=1,2,\ldots,s-1;\right.$$
$$\left.\binom{x-i}{k-1},\ \ i=1,2,\ldots,n-k+1 \text{ for } k=1,2,\ldots,s\right\}. \tag{10}$$

3 Stochastic s-(increasing) convexity

3.1 Definition

Let us turn to our central question, namely the possible transmission of the properties of s-(increasing) convexity from ϕ to ϕ^* defined in (1).

Definition 1. If the s-convexity (resp. the s-increasing convexity) is transmitted from ϕ to ϕ^*, i.e.

$$\text{if } \phi^*\in\overline{\mathcal{U}}^{\Theta}_{s-cx} \text{ (resp. } \overline{\mathcal{U}}^{\Theta}_{s-icx}) \text{ whenever } \phi\in\overline{\mathcal{U}}^{S}_{s-cx} \text{ (resp. } \overline{\mathcal{U}}^{S}_{s-icx}),$$

provided that the expectation defining ϕ^* exists, then the family $\mathcal{P}(\Theta)$ (or $\{X_\theta,\ \theta\in\Theta\}$) is said to be stochastically s-convex (resp. stochastically s-increasing convex).

We observe that the stochastic 1-(increasing) convexity coincides with the stochastic increasingness, while the stochastic 2-(increasing) convexity is the usual stochastic (increasing) convexity (see, e.g., Shaked and Shanthikumar (1988, 1994)).

3.2 Characterizations

We are going to show that when $\mathcal{S}$ is an interval or an arithmetic grid, it suffices to check the transmission of the s-convexity, or of the s-increasing convexity, for the subclasses $\mathcal{U}^{S}_{s-cx}$ and $\mathcal{M}^{S}_{s-cx}$ of $\overline{\mathcal{U}}^{S}_{s-cx}$, or for the subclasses $\mathcal{U}^{S}_{s-icx}$ and $\mathcal{M}^{S}_{s-icx}$ of $\overline{\mathcal{U}}^{S}_{s-icx}$.

Proposition 1. *For $\mathcal{S} = [a,b]$, $a < b \in I\!R$, b possibly infinite, $\mathcal{P}(\Theta)$ is stochastically s-convex if, and only if,*

$$\phi^* \in \overline{\mathcal{U}}^{\Theta}_{s-cx} \text{ for all } \phi \in \mathcal{U}^{[a,b]}_{s-cx}, \tag{11}$$

or, equivalently, if, and only if,

$$\phi^* \in \overline{\mathcal{U}}^{\Theta}_{s-cx} \text{ for all } \phi \in \mathcal{M}^{[a,b]}_{s-cx}. \tag{12}$$

$\mathcal{P}(\Theta)$ is stochastically s-increasing convex if, and only if,

$$\phi^* \in \overline{\mathcal{U}}^{\Theta}_{s-icx} \text{ for all } \phi \in \mathcal{U}^{[a,b]}_{s-icx}, \tag{13}$$

or, equivalently, if, and only if,

$$\phi^* \in \overline{\mathcal{U}}^{\Theta}_{s-icx} \text{ for all } \phi \in \mathcal{M}^{[a,b]}_{s-icx}. \tag{14}$$

Proof. Let us begin with (11) and (13). It is well-known that the class $\overline{\mathcal{U}}^{[a,b]}_{s-cx}$ of the s-convex functions consists in all the pointwise limits of the sequences of functions with a non-negative derivative of degree s, that is those functions in $\mathcal{U}^{[a,b]}_{s-cx}$. Similarly, the elements of $\overline{\mathcal{U}}^{[a,b]}_{s-icx}$ are the pointwise limits of sequences in $\mathcal{U}^{[a,b]}_{s-icx}$. This yields (11) and (13) since $\overline{\mathcal{U}}^{\Theta}_{s-cx}$ and $\overline{\mathcal{U}}^{\Theta}_{s-icx}$ are closed in the topology of the pointwise convergence. Let us pass to (12) and (14). As shown in Popoviciu (1933), any $\phi \in \overline{\mathcal{U}}^{[a,b]}_{s-cx}$ can be obtained as the uniform limit of a sequence of functions $\{\varphi_n,\ n \geq s\}$ which are of the form

$$\varphi_n(x) = \phi(a) + \sum_{j=0}^{s-1} \alpha_j (x-a)^j + \sum_{j=0}^{n-s} \beta_j (x-t_j)_+^{s-1},$$

where $\alpha_0, \alpha_1, \ldots, \alpha_{s-1}$ are real constants, $\beta_0, \beta_1, \ldots, \beta_{n-s}$ are non-negative constants (so that $\varphi_n \in \overline{\mathcal{U}}^{[a,b]}_{s-cx}$), and $t_0 < t_1 < \ldots < t_{n-s} \in [a,b]$. Similarly, any $\phi \in \overline{\mathcal{U}}^{[a,b]}_{s-icx}$ is the uniform limit of a sequence of functions $\{\tilde{\varphi}_n,\ n \geq s\}$ which are of the form

$$\tilde{\varphi}_n(x) = \phi(a) + \sum_{j=0}^{s-1} \tilde{\alpha}_j (x-a)^j + \sum_{j=0}^{n-s} \tilde{\beta}_j (x-t_j)_+^{s-1},$$

where $\tilde{\alpha}_0, \tilde{\alpha}_1, \ldots, \tilde{\alpha}_{s-1}$ and $\tilde{\beta}_0, \tilde{\beta}_1, \ldots, \tilde{\beta}_{n-s}$ are non-negative constants (so that $\tilde{\varphi}_n \in \overline{\mathcal{U}}^{[a,b]}_{s-icx}$), and $t_0 < t_1 < \cdots < t_{n-s} \in [a,b]$. Thus (12) and (14) follow directly. □

Proposition 2. *For $\mathcal{S} = \mathcal{D}_n$, n possibly infinite, $\mathcal{P}(\Theta)$ is stochastically s-convex if, and only if,*

$$\phi^* \in \overline{\mathcal{U}}^{\Theta}_{s-cx} \text{ for all } \phi \in \mathcal{M}^{\mathcal{D}_n}_{s-cx},$$

while $\mathcal{P}(\Theta)$ is stochastically s-increasing convex if, and only if,

$$\phi^* \in \overline{\mathcal{U}}^{\Theta}_{s-icx} \textit{ for all } \phi \in \mathcal{M}^{\mathcal{D}_n}_{s-icx}.$$

Proof. We recall (see, e.g., Denuit and Lefèvre (1997)) that any function $\phi : \mathcal{D}_n \to I\!R$ can be expanded as

$$\phi(j) = \phi(0) + \binom{j}{1} \Delta\phi(0) + \binom{j}{2} \Delta^2\phi(0) + \ldots + \binom{j}{s-1} \Delta^{s-1}\phi(0)$$
$$+ \sum_{i=1}^{n-s+1} [\Delta^s \phi(i-1)] \binom{j-i}{s-1}, \quad j \in \mathcal{D}_n.$$

The results announced are then straightforward. □

3.3 A sufficient condition

Denuit, Lefèvre and Utev (1999b) have pointed the special role taken by the families of distributions that are totally positive. Specifically, let us assume that there exists a sigma-finite dominating measure μ for the family of probability distributions $\mathcal{P}(\Theta)$, and let f_θ denote the associated density function for P_θ. The family $\mathcal{P}(\Theta)$ is said to be totally positive of degree s (TP_s, in short) if the kernel obtained by considering the density $f_\theta(x)$ as a function of two variables (the parameter θ and the variable x) is TP_s, that is when the determinant of the matrix $\{f_{\theta_i}(x_j)\}_{i,j=1,2,\ldots,k}$ is non-negative for all $\theta_1 < \theta_2 < \ldots < \theta_k \in \Theta$ and $x_1 < x_2 < \ldots < x_k \in \mathcal{S}$, with $k = 1, 2, \ldots, s$; see, e.g., Karlin (1968).

Property 1. Let $\mathcal{P}(\Theta)$ be a TP_s family. If, in addition, for $k = 1, 2, \ldots, s-1$,

$$EQ_k(X_\theta) \equiv \int_{x\in\mathcal{S}} Q_k(x) f_\theta(x) d\mu(x) = \widetilde{Q}_k(\theta), \quad \theta \in \Theta,$$

where Q_k is an arbitrary real polynomial of exact degree k with positive highest coefficient, and $\widetilde{Q}_k$ is an associated real polynomial of the same type, then $\mathcal{P}(\Theta)$ is stochastically s-convex and stochastically s-increasing convex.

3.4 Families with a scale or location parameter

We start with the situation where the parameter θ involved in X_θ is a scale parameter.

Property 2. Let Y be a non-negative random variable, and define the random variables X_θ as

$$X_\theta = \theta Y \text{ with } \theta \in I\!R.$$

Then $\{X_\theta,\ \theta \in I\!R\}$ is stochastically s-convex and stochastically s-increasing convex, for any $s \geq 1$.

Proof. Let $\phi \in \mathcal{U}^{I\!R}_{s-cx}$. We then get

$$\frac{d^s}{d\theta^s}\phi^*(\theta) = \frac{d^s}{d\theta^s}E\phi(\theta Y) = E\left[Y^s\phi^{(s)}(\theta Y)\right] \geq 0,$$

hence the result. □

That property is no longer valid, in general, when Y is a real-valued random variable. An exception is given below.

Property 3. Let Y be a real-valued random variable with a symmetric density about 0, and define $X_\theta = \theta Y$ with $\theta \in I\!R$. Then $\{X_\theta,\ \theta \in I\!R\}$ is stochastically $2s$-convex, for any $s \geq 1$, while $\{X_\theta,\ \theta \in I\!R^+\}$ is stochastically $2s$-convex and stochastically $2s$-increasing convex, for any $s \geq 1$.

Proof. For $\phi \in \mathcal{U}^{I\!R}_{2s-cx}$, we have

$$\frac{d^{2s}}{d\theta^{2s}}\phi^*(\theta) = E\left[Y^{2s}\phi^{(2s)}(\theta Y)\right] \geq 0,$$

so that $\{X_\theta,\ \theta \in I\!R\}$ is stochastically $2s$-convex for all s. For $\phi \in \mathcal{U}^{I\!R}_{2s-icx}$, we obtain that for $k = 1, 2, \ldots, s$,

$$\frac{d^{2k}}{d\theta^{2k}}\phi^*(\theta) = E\left[Y^{2k}\phi^{(2k)}(\theta Y)\right] \geq 0,$$

and as $EY^{2k-1} = 0$ by hypothesis,

$$\frac{d^{2k-1}}{d\theta^{2k-1}}\phi^*(\theta) = E\left\{Y^{2k-1}[\phi^{(2k-1)}(\theta Y) - \phi^{(2k-1)}(0)]\right\},$$

where $[\phi^{(2k-1)}(\theta x) - \phi^{(2k-1)}(0)]x^{2k-1} \geq 0$ for all real x and all $\theta \geq 0$ since $\phi^{(2k)} \geq 0$. □

Let us now examine the case where the parameter θ involved in X_θ is a location parameter. The proof of the following result is immediate.

Property 4. Let Y be a real-valued random variable, and define the random variables X_θ as

$$X_\theta = Y + \theta \text{ with } \theta \in I\!R.$$

Then $\{X_\theta,\ \theta \in I\!R\}$ is stochastically s-convex and stochastically s-increasing convex, for any $s \geq 1$.

3.5 Partial and compound sums

We consider here that X_θ corresponds to some partial or compound sum of random variables. The following properties have been established in Denuit, Lefèvre and Utev (1999b).

Property 5. Let $\{Y_n, \ n \geq 1\}$ be a sequence of non-negative independent and identically distributed random variables and define the random variables X_θ as

$$X_\theta = \sum_{n=1}^{\theta} Y_n \text{ with } \theta \in I\!N.$$

Then $\{X_\theta, \ \theta \in I\!N\}$ is stochastically s-convex and stochastically s-increasing convex, for any $s \geq 1$.

Property 6. Let $\{Y_n, \ n \geq 1\}$ be a sequence of non-negative independent and identically distributed random variables. Let $\{N_\theta, \ \theta \in \Theta\}$ be a family of non-negative integer-valued random variables, independent of the Y_n's, and define the random variables X_θ as

$$X_\theta = \sum_{n=1}^{N_\theta} Y_n \text{ with } \theta \in \Theta.$$

If $\{N_\theta, \ \theta \in \Theta\}$ is stochastically s-increasing convex, then $\{X_\theta, \ \theta \in \Theta\}$ also is.

We mention that the result of Property 6 only holds for the stochastic s-increasing convexity and not for the stochastic s-convexity. The reason is that the proof uses the fact that the composition of two s-increasing convex functions is itself an s-increasing convex function, and that this does not hold for the s-convexity. The property below is closely related.

Property 7. Let $\{X_\theta, \ \theta \in I\!R\}$ be a stochastically s-increasing convex family of random variables. Then, for any s-increasing convex measurable function φ, the family of random variables $\{\varphi(X_\theta), \ \theta \in I\!R\}$ is also stochastically s-increasing.

3.6 Standard parametric families

Table 1 points out the possible stochastic s-convexity and s-increasing convexity for standard parametric families. This easily follows from the properties derived in this Section, or by a direct argument.

Examples of parametric distributions that are not stochastically s-convex are rather frequent. It can be checked, for instance, that if X_θ is an exponential random variable with mean $\theta \in \Theta$, then $\{\sqrt{X_\theta}, \ \theta \in \Theta\}$ is not stochastically s-convex (although being TP_s).

4 Stochastic comparisons of mixtures and compound sums

4.1 s-(increasing) convex ordering between mixtures

The property of stochastic s-(increasing) convexity provides a natural way for comparing mixed distributions through their mixing kernel, in the stochastic s-(increasing) convex ordering.

Probability Distribution	Stochastic s-(increasing) convexity
Poisson with mean λ	s-cx and s-icx in $\lambda \in \mathbb{R}^+$
Binomial(n,p) with mean np	s-cx and s-icx in $n \in \mathbb{N}$ s-cx and s-icx in $p \in [0,1]$
Negative Binomial(n,p) with mean np	s-cx and s-icx in $n \in \mathbb{N}$ s-cx and s-icx in $p \in [0,1]$
Discrete Uniform on $\mathcal{D}_n$	s-cx and s-icx in $n \in \mathbb{N}$
Continuous Uniform on $[0,\theta]$	s-cx and s-icx in $\theta \in \mathbb{R}^+$
Negative Exponential with mean λ	s-cx and s-icx in $\lambda \in \mathbb{R}^+$
Erlang(n,λ)	s-cx and s-icx in $n \in \mathbb{N}$ s-cx and s-icx in $\lambda \in \mathbb{R}^+$
Normal$(0,\sigma)$	$2s$-cx and $2s$-icx in $\sigma \in \mathbb{R}^+$
Normal(μ,σ_0)	s-cx and s-icx in $\mu \in \mathbb{R}$
LogNormal$(0,\sigma)$	$2s$-icx in $\sigma \in \mathbb{R}^+$
LogNormal(μ,σ_0)	s-icx in $\mu \in \mathbb{R}$
Chi-square(n)	s-cx and s-icx in $n \in \mathbb{N}$
Compound Poisson Sum(λ) with a.s.≥ 0 summands	s-icx in $\lambda \in \mathbb{R}^+$
Compound Binomial Sum(n,p) with a.s.≥ 0 summands	s-icx in $n \in \mathbb{N}$ s-icx in $p \in [0,1]$
Compound Neg. Binomial Sum(n,p) with a.s.≥ 0 summands	s-icx in $n \in \mathbb{N}$ s-icx in $p \in [0,1]$
Compound Discrete Uniform$(\mathcal{D}_n)$ with a.s.≥ 0 summands	s-icx in $n \in \mathbb{N}$

Table 1. Stochastic s-convexity (s-cx) and s-increasing convexity (s-icx) of some standard parametric distributions, for any $s \geq 1$.

Given the random variables $\{X_\theta,\ \theta \in \Theta\}$, let X_Λ denote a random variable distributed as a mixture of these X_θ's with mixing kernel Λ, i.e.

$$P[X_\Lambda \leq x] = \int_{\theta\in\Theta} P[X_\theta \leq x] dP[\Lambda \leq \theta], \quad x \in \mathbb{R}.$$

Property 8. If the family of distributions $\mathcal{P}(\Theta)$ is stochastically s-(increasing) convex, then

$$\Lambda_1 \preceq^{\Theta}_{s-(i)cx} \Lambda_2 \Rightarrow X_{\Lambda_1} \preceq^{\mathcal{S}}_{s-(i)cx} X_{\Lambda_2}. \tag{15}$$

Proof. For any function $\phi : \mathcal{S} \to \mathbb{R}$, we can write, by (1),

$$E[\phi(X_\Lambda)] = E[\phi^*(\Lambda)].$$

Now, let $\phi \in \overline{\mathcal{U}}^{\mathcal{S}}_{s-(i)cx}$. Since $\mathcal{P}(\Theta)$ is stochastically s-(increasing) convex, we know that $\phi^* \in \overline{\mathcal{U}}^{\Theta}_{s-(i)cx}$. Therefore, $\Lambda_1 \preceq^{\Theta}_{s-(i)cx} \Lambda_2$ implies that

$$E[\phi^*(\Lambda_1)] \leq E[\phi^*(\Lambda_2)],$$

or equivalently,

$$E[\phi(X_{\Lambda_1})] \leq E[\phi(X_{\Lambda_2})],$$

as required. □

Combining Table 1 and Property 8 directly yields the stochastic comparisons between mixtures given in Tables 2-3.

Probability distributions	Orderings among random parameters	Induced orderings between mixed distributions
Poisson(λ)	$\Lambda_1 \preceq_{s-(i)cx}^{\mathbb{R}^+} \Lambda_2$	Poisson(Λ_1) $\preceq_{s-(i)cx}^{\mathbb{N}}$ Poisson(Λ_2)
Binomial(n,p)	$N_1 \preceq_{s-(i)cx}^{\mathbb{N}_0} N_2$	Binomial(N_1,p) $\preceq_{s-(i)cx}^{\mathbb{N}}$ Binomial(N_2,p)
	$\pi_1 \preceq_{s-(i)cx}^{[0,1]} \pi_2$	Binomial(n,π_1) $\preceq_{s-(i)cx}^{\mathcal{D}_n}$ Binomial(n,π_2)
Negative Binomial(n,p)	$N_1 \preceq_{s-(i)cx}^{\mathbb{N}_0} N_2$	Neg. Bin.(N_1,p) $\preceq_{s-(i)cx}^{\mathbb{N}}$ Neg. Bin.(N_2,p)
	$\pi_1 \preceq_{s-(i)cx}^{[0,1]} \pi_2$	Neg. Bin.(n,π_1) $\preceq_{s-(i)cx}^{\mathbb{N}}$ Neg. Bin.(n,π_2)
Discrete Uniform(n)	$N_1 \preceq_{s-(i)cx}^{\mathbb{N}_0} N_2$	Disc. Unif.(N_1) $\preceq_{s-(i)cx}^{\mathbb{N}}$ Disc. Unif.(N_2)
Continuous Uniform $[0,\theta]$	$\Theta_1 \preceq_{s-(i)cx}^{\mathbb{R}^+} \Theta_2$	Cont. Unif.$[0,\Theta_1]$ $\preceq_{s-(i)cx}^{\mathbb{R}^+}$ Cont. Unif.$[0,\Theta_2]$
Negative Exponential(λ)	$\Lambda_1 \preceq_{s-(i)cx}^{\mathbb{R}_0^+} \Lambda_2$	Neg. Exp.(Λ_1) $\preceq_{s-(i)cx}^{\mathbb{R}^+}$ Neg. Exp.(Λ_2)
Erlang(n,λ)	$N_1 \preceq_{s-(i)cx}^{\mathbb{N}} N_2$	Erlang(N_1,λ) $\preceq_{s-(i)cx}^{\mathbb{R}^+}$ Erlang(N_2,λ)
	$\Lambda_1 \preceq_{s-(i)cx}^{\mathbb{R}_0^+} \Lambda_2$	Erlang(n,Λ_1) $\preceq_{s-(i)cx}^{\mathbb{R}^+}$ Erlang(n,Λ_2)

Table 2. Stochastic s-(increasing) convex orderings between mixed distributions, for any $s \geq 1$.

We note that various comparison properties derived in the actuarial literature (see, e.g., Goovaerts and De Vylder (1984), Hesselager (1998) and Hesselager, Wang and Willmot (1998)) correspond to particular cases of the above results.

Probability distributions	Orderings among random parameters	Induced orderings between mixed distributions
Normal$(0,\sigma)$	$S_1 \preceq^{\mathbb{R}^+}_{2s-(i)cx} S_2$	Normal$(0,S_1)$ $\preceq^{\mathbb{R}}_{2s-(i)cx}$Normal$(0,S_2)$
Normal(μ,σ_0)	$M_1 \preceq^{\mathbb{R}}_{s-(i)cx} M_2$	Normal(M_1,σ_0) $\preceq^{\mathbb{R}}_{s-(i)cx}$Normal$(M_2,\sigma_0)$
LogNormal$(0,\sigma)$	$S_1 \preceq^{\mathbb{R}^+}_{2s-icx} S_2$	LogNormal$(0,S_1)$ $\preceq^{\mathbb{R}^+}_{2s-icx}$LogNormal$(0,S_2)$
LogNormal(μ,σ_0)	$M_1 \preceq^{\mathbb{R}}_{s-icx} M_2$	LogNormal(M_1,σ_0) $\preceq^{\mathbb{R}^+}_{s-icx}$LogNormal$(M_2,\sigma_0)$
Chi-square(n)	$N_1 \preceq^{\mathbb{N}}_{s-(i)cx} N_2$	Chi-square(N_1) $\preceq^{\mathbb{R}^+}_{s-(i)cx}$Chi-square$(N_2)$
Comp. Poisson Sum(λ) with a.s.≥ 0 summands	$\Lambda_1 \preceq^{\mathbb{R}^+_0}_{s-icx} \Lambda_2$	C.P.S.(Λ_1) $\preceq^{\mathbb{R}^+}_{s-icx}$C.P.S.$(\Lambda_2)$
Comp. Bin. Sum(n,p) with a.s.≥ 0 summands	$N_1 \preceq^{\mathbb{N}_0}_{s-icx} N_2$	C.B.S.(N_1,p) $\preceq^{\mathbb{R}^+}_{s-icx}$C.B.S.$(N_2,p)$
	$\pi_1 \preceq^{[0,1]}_{s-icx} \pi_2$	C.B.S.(n,π_1) $\preceq^{\mathbb{R}^+}_{s-icx}$C.B.S.$(n,\pi_2)$
Comp. Neg. Bin. Sum(n,p) with a.s.≥ 0 summands	$N_1 \preceq^{\mathbb{N}_0}_{s-icx} N_2$	C.N.B.S.(N_1,p) $\preceq^{\mathbb{R}^+}_{s-icx}$C.N.B.S.$(N_2,p)$
	$\pi_1 \preceq^{[0,1]}_{s-icx} \pi_2$	C.N.B.S.(n,π_1) $\preceq^{\mathbb{R}^+}_{s-icx}$C.N.B.S.$(n,\pi_2)$

Table 3. Stochastic s-(increasing) convex orderings between mixed distributions, for any $s \geq 1$ (continued from Table 2).

4.2 s-(increasing) convex ordering between compound sums

Using Properties 5 and 8, we are able to compare compound sums, through their number of random variables, in the stochastic s-(increasing) convex ordering.

Property 9. Let $\{Y_n,\ n \geq 1\}$ be a sequence of non-negative independent and identically distributed random variables, and let N_1 and N_2 be two integer-valued random variables, independent from that sequence. Then,

$$N_1 \preceq^{\mathbb{N}}_{s-(i)cx} N_2 \Rightarrow \sum_{n=1}^{N_1} Y_n \preceq^{\mathbb{R}^+}_{s-(i)cx} \sum_{n=1}^{N_2} Y_n.$$

Moreover, let $\{Z_n,\ n \geq 1\}$ be another sequence of non-negative independent and identically distributed random variables such that $Z_n \preceq^{\mathbb{R}^+}_{s-(i)cx} Y_n$ for all

n. Then,

$$N_1 \preceq^{\mathbb{N}}_{s-(i)cx} N_2 \Rightarrow \sum_{n=1}^{N_1} Z_n \preceq^{\mathbb{R}^+}_{s-(i)cx} \sum_{n=1}^{N_2} Y_n.$$

4.3 An application in actuarial sciences

For illustration, let us consider the actuarial context in which $S = \sum_{n=1}^{N} Y_n$ represents the aggregate claim amounts over some period of time. Here N denotes the number of claims and the Y_n's are their severities. As usual, it is assumed that the Y_n's and N satisfy the conditions stipulated in Property 9.

Now, quite often in practice, the insurance has thousands of realizations for the Y_n's, so that a statistical model for the claim severities can be built adequately. On the contrary, the company has only one realization of N for each period, so that data on N are generally scarce. This leads us to examine the situations where the distribution of the Y_n's can be viewed as known, while the distribution of N is mostly unknown. Specifically, we will assume that the only information available on N are its first two moments, with its largest value.

On basis of this partial information, our goal is then to find stochastic bounds, lower and upper, for the aggregate claims S under interest. By Property 9, s-convex bounds on N will allow to construct s-convex bounds on S. Thus, the problem reduces to obtaining such lower and upper bounds on N. This has been done in Denuit and Lefèvre (1997) and Denuit, Lefèvre and Mesfioui (1999), yielding the results below.

Property 10. Suppose that the number of claims N is valued in $\mathcal{D}_n$ and has mean μ_1. Define $\xi_0 \in \mathcal{D}_{n-1}$ such that $\xi_0 < \mu_1 \leq \xi_0 + 1$. Then,

$$S^{(2)}_{\min} \preceq^{\mathbb{R}^+}_{2-cx} S \preceq^{\mathbb{R}^+}_{2-cx} S^{(2)}_{\max}, \tag{16}$$

where

$$S^{(2)}_{\min} = \begin{cases} \sum_{k=1}^{\xi_0} Y_k, & \text{with the probability } \xi_0 + 1 - \mu_1, \\ \sum_{k=1}^{\xi_0+1} Y_k, & \text{with the probability } \mu_1 - \xi_0, \end{cases}$$

and

$$S^{(2)}_{\max} = \begin{cases} 0, & \text{with the probability } 1 - \frac{\mu_1}{n}, \\ \sum_{k=1}^{n} Y_k, & \text{with the probability } \frac{\mu_1}{n}. \end{cases}$$

Suppose that, in addition, N has second moment μ_2. Define ξ_1 and $\xi_2 \in \mathcal{D}_{n-1}$ such that

$$\xi_1 < \frac{\mu_2}{\mu_1} \leq \xi_1 + 1 \text{ and } \xi_2 < \frac{n\mu_1 - \mu_2}{n - \mu_1} \leq \xi_2 + 1.$$

Then,

$$S^{(3)}_{\min} \preceq^{\mathbb{R}^+}_{3-cx} S \preceq^{\mathbb{R}^+}_{3-cx} S^{(3)}_{\max}, \tag{17}$$

where

$$S_{\min}^{(3)} = \begin{cases} 0, & \text{with the probability } p_1 = 1 - p_2 - p_3, \\ \sum_{k=1}^{\xi_1} Y_k, & \text{with the probability } p_2 = \frac{(\xi_1+1)\mu_1-\mu_2}{\xi_1}, \\ \sum_{k=1}^{\xi_1+1} Y_k, & \text{with the probability } p_3 = \frac{\mu_2-\xi_1\mu_1}{1+\xi_1}, \end{cases}$$

and

$$S_{\max}^{(3)} = \begin{cases} \sum_{k=1}^{\xi_2} Y_k, & \text{with the probability } q_1 = \frac{(1+\xi_2)(n-\mu_1)+\mu_2-n\mu_1}{n-\xi_2}, \\ \sum_{k=1}^{\xi_2+1} Y_k, & \text{with the probability } q_2 = \frac{(n+\xi_2)\mu_1-\mu_2-n\xi_2}{n-1-\xi_2}, \\ \sum_{k=1}^{n} Y_k, & \text{with the probability } q_3 = 1 - q_1 - q_2. \end{cases}$$

Note that $S_{\min}^{(2)}$ and $S_{\min}^{(3)}$ are defined independently of n, while $S_{\max}^{(2)}$ and $S_{\max}^{(3)}$ require the value of n. ¿From (16), we deduce, for instance, bounds on the stop-loss premium for S, namely

$$E\left(S_{\min}^{(2)} - d\right)_+ \leq E(S-d)_+ \leq E\left(S_{\max}^{(2)} - d\right)_+,$$

for any deductible d.

When, moreover, the density of N is known to be decreasing, the bounds (16) and (17) can be improved as follows.

Property 11. Suppose that the number of claims N is valued in $\mathcal{D}_n$, has a decreasing density and has mean μ_1. Define $\rho_0 \in \mathcal{D}_n$ such that $\rho_0 < 2\mu_1 \leq \rho_0 + 1$. Then,

$$S_{\min}^{(2)*} \preceq_{2-cx}^{\mathbb{R}^+} S \preceq_{2-cx}^{\mathbb{R}^+} S_{\max}^{(2)*}, \tag{18}$$

where

$$S_{\min}^{(2)*} = \sum_{k=1}^{N_{\min}^{(2)*}} Y_k \text{ with } f_{N_{\min}^{(2)*}}(k) = \begin{cases} \frac{2(\rho_0+1-\mu_1)}{(\rho_0+1)(\rho_0+2)}, & \text{for } k = 0, 1, \ldots, \rho_0, \\ \frac{2\mu_1-\rho_0}{\rho_0+2}, & \text{for } k = \rho_0 + 1, \\ 0, & \text{for } k \geq \rho_0 + 2, \end{cases}$$

and

$$S_{\max}^{(2)*} = \sum_{k=1}^{N_{\max}^{(2)*}} Y_k \text{ with } f_{N_{\max}^{(2)*}}(k) = \begin{cases} 1 - \frac{2\mu_1}{n+1}, & \text{for } k = 0, \\ \frac{2\mu_1}{n(n+1)}, & \text{for } k = 1, \ldots, n. \end{cases}$$

Suppose that, in addition, N has second moment μ_2. Define ρ_1 and $\rho_2 \in \mathcal{D}_{n-1}$ such that

$$\rho_1 < \frac{3\mu_2 - \mu_1}{2\mu_1} \leq \rho_1 + 1 \text{ and } \rho_2 < \frac{2n\mu_1 - 3\mu_2 + \mu_1}{n - 2\mu_1} \leq \rho_2 + 1.$$

Then,

$$S_{\min}^{(3)*} \preceq_{3-cx}^{\mathbb{R}^+} S \preceq_{3-cx}^{\mathbb{R}^+} S_{\max}^{(3)*}, \tag{19}$$

where

$$S_{\min}^{(3)*} = \sum_{k=1}^{N_{\min}^{(3)*}} Y_k \text{ with } f_{N_{\min}^{(3)*}}(k) = \begin{cases} \tilde{p}_1 + \frac{\tilde{p}_2}{\rho_1+1} + \frac{\tilde{p}_3}{\rho_1+2}, & \text{for } k = 0, \\ \frac{\tilde{p}_2}{\rho_1+1} + \frac{\tilde{p}_3}{\rho_1+2}, & \text{for } k = 1, \ldots, \rho_1, \\ \frac{\tilde{p}_3}{\rho_1+2}, & \text{for } k = \rho_1 + 1, \\ 0, & \text{for } k \geq \rho_1 + 2, \end{cases}$$

where $\tilde{p}_1 = 1 - \tilde{p}_2 - \tilde{p}_3$,

$$\tilde{p}_2 = \frac{2(\rho_1 + 1)\mu_1 - 3\mu_2 + \mu_1}{\rho_1} \text{ and } \tilde{p}_3 = \frac{3\mu_2 - \mu_1 - 2\rho_1\mu_1}{\rho_1 + 1},$$

and

$$S_{\max}^{(3)*} = \sum_{k=1}^{N_{\max}^{(3)*}} Y_k \text{ with } f_{N_{\max}^{(3)*}}(k) = \begin{cases} \frac{\tilde{q}_1}{\rho_2+1} + \frac{\tilde{q}_2}{\rho_2+2} + \frac{\tilde{q}_3}{n+1}, & \text{for } k = 0, 1, \ldots, \rho_2, \\ \frac{\tilde{q}_2}{\rho_2+2} + \frac{\tilde{q}_3}{n+1}, & \text{for } k = \rho_2 + 1, \\ \frac{\tilde{q}_3}{n+1}, & \text{for } k = \rho_2 + 2, \ldots, n, \end{cases}$$

where $\tilde{q}_3 = 1 - \tilde{q}_1 - \tilde{q}_2$,

$$\tilde{q}_1 = \frac{(1 + \rho_2)(n - 2\mu_1) + 3\mu_2 - \mu_1 - 2n\mu_1}{n - \rho_2},$$

and

$$\tilde{q}_2 = \frac{2(n + \rho_2)\mu_1 - 3\mu_2 + \mu_1 - n\rho_2}{n - 1 - \rho_2}.$$

References

1. Denuit, M., and Lefèvre, Cl. (1997) Some new classes of stochastic order relations among arithmetic random variables, with applications in actuarial sciences. Insurance: Mathematics and Economics **20**, 197–213.
2. Denuit, M., Lefèvre, Cl., and Mesfioui, M. (1999) On s-convex stochastic extrema for arithmetic risks. Insurance: Mathematics and Economics **25**, 143-155.
3. Denuit, M., Lefèvre, Cl., and Shaked, M. (1998) The s-convex orders among real random variables, with applications. Mathematical Inequalities and Their Applications **1**, 585–613.
4. Denuit, M., Lefèvre, Cl., and Utev, S. (1999a) Stochastic orderings of convex/concave type on an arbitrary grid. Mathematics of Operations Research **24**, 835-846.
5. Denuit, M., Lefèvre, Cl., and Utev, S. (1999b) Generalized stochastic convexity and stochastic orderings of mixtures. Probability in the Engineering and Informational Sciences **13**, 275–291.
6. Goovaerts, M.J., and De Vylder, F.E. (1984) Stop-loss ordering for scale and power mixture distributions. Scandinavian Actuarial Journal, 95–101.
7. Hesselager, O. (1998) Closure properties of some partial orderings under mixing. Insurance: Mathematics and Economics **22**, 163–170.

8. Hesselager, O., Wang, S., and Willmot, G. (1998) Exponential and scale mixtures and equilibrium distributions. Scandinavian Actuarial Journal, 125–142.
9. Kaas, R., and Hesselager, O. (1995) Ordering claim size distributions and mixed Poisson probabilities. Insurance: Mathematics and Economics **17**, 193–201.
10. Kaas, R., Van Heerwaarden, A.E. and Goovaerts, M.J. (1994) Ordering of Actuarial Risks. CAIRE. Brussels.
11. Karlin, S. (1968) Total Positivity. Stanford University Press. Stanford. California.
12. Karlin, S., and Studden, W.J. (1966) Tchebycheff Systems: with Applications in Analysis and Statistics. Wiley. New York.
13. Lefèvre, Cl., and Utev, S. (1996) Comparing sums of exchangeable Bernoulli random variables. Journal of Applied Probability **33**, 285–310.
14. Marshall, A.W. (1991) Multivariate stochastic orderings and generating cones of functions. In Stochastic Orders and Decision Under Risk, K. Mosler and M. Scarsini Editors, pp. 231–247. IMS Lecture Notes-Monograph Series, **19**.
15. Müller, A. (1997) Stochastic orderings generated by integrals: a unified study. Advances in Applied Probability **29**, 414–428.
16. Pečarić, J.E., Proschan, F., and Tong, Y.L. (1992) Convex Functions, Partial Orderings and Statistical Applications. Academic Press. New York.
17. Popoviciu, T. (1933) Sur quelques propriétés des fonctions d'une ou de deux variables réelles. Mathematica **8**, 1–85.
18. Roberts, A.W., and Varberg, D.E. (1973) Convex Functions. Academic Press. New York.
19. Shaked, M., and Shanthikumar, J.G. (1988) Stochastic convexity and its applications. Advances in Applied Probability **20**, 427–446.
20. Shaked, M., and Shanthikumar, J.G. (1994) Stochastic Orders and their Applications. Academic Press. New York.

Fixed Point Theorems, Coincidence Theorems and Variational Inequalities

Behzad Djafari-Rouhani[12], Enayet Tarafdar[3], and Peter Jonathan Watson[3]

[1] School of Mathematical Sciences,
Shahid Beheshti University,
Evin Tehran 19834, Iran.
b-rohani@cc.sbu.ac.ir

[2] Institute for Studies in Theoretical Physics and Mathematics, P.O. box 19395-5746, Tehran, Iran.

[3] Department of Mathematics,
The University of Queensland,
Queensland 4072, Australia.
eut@maths.uq.edu.au
pjw@maths.uq.edu.au

Abstract. We give a simple proof to an extension of a fixed point theorem of E. Tarafdar [4] for multivalued mappings and show its equivalence to a KKM type result. A noncompact coincidence point theorem is also established. Applications of these results to establish the existence of solutions to variational inequalities in not necessarily reflexive Banach spaces are also considered.

2000 Mathematics Subject Classification. Primary 47H04, 47H10; Secondary 47H05, 49J40.

Keywords and phrases. Fixed point theorem, KKM map, Coincidence theorem, pseudomonotone and u-hemicontinuous operators, variational inequality.

1 Introduction

The fixed point theorem in [4], which extends the Fan-Browder fixed point theorem, has many applications in mathematical analysis. The first result of this paper concerns a generalisation of this theorem which also simplifies the proof in [4]. We show this generalisation is equivalent to a KKM type result in the same way that Tarafdar's theorem is equivalent to the Fan-KKM theorem.

Using a coincidence result of Komiya [2], a noncompact coincidence point theorem is also proved. This result extends the fixed point theorem to multi-functions operating between different spaces. It also improves the coincidence theorem in [5].

The final section concerns an application to the variational inequality problem in Banach spaces without the assumption of reflexivity.

Let X and Y be topological vector spaces. A set-valued mapping $f : X \to 2^Y$ is said to be *upper semicontinuous* (usc for short) if for any $x \in X$ and any open set V containing $f(x)$ there exists an open set U containing x such that $f(z) \subset V$ for all $z \in U$.

By $\overline{\text{co}}A$ we mean the closed convex hull of the set $A \subset X$. A multifunction $F : X \to 2^X$ satisfying $\overline{\text{co}}(A) \subset \cup_{a\in A}F(a)$ for any finite subset A of X is said to be KKM. A topological space is said to be *contractible* if the identity mapping is homotopic to a constant function.

2 Main results

Theorem 1. *Let $X \neq \emptyset$ be a convex subset of a Hausdorff topological vector space E. Let $f : X \to 2^X$ be a multivalued mapping such that*

1. *for each $x \in X$, $f(x)$ is a nonempty convex subset of X;*
2. *for each $y \in X$, $f^{-1}(y) = \{x \in X : y \in f(x)\}$ contains an open set O_y which may be empty;*
3. $\cup_{y\in X} O_y = X$
4. *there exists a compact convex subset X_1 of X and points $\{\hat{x}_1, \ldots, \hat{x}_n\}$ in X such that*
$$D = \cap_{x\in X_1} O_x^c \subset \cup_{i=1}^n O_{\hat{x}_i}$$
(where O_x^c is the complement of O_x in X).

Then there exists a point $x_0 \in X$ such that $x_0 \in f(x_0)$.

Proof. Suppose first that $D = \emptyset$. Then it follows that for each $x \in X_1$, $f(x) \cap X_1 \neq \emptyset$. Indeed if this is not the case, that is, $f(x_0) \cap X_1 = \emptyset$ for some $x_0 \in X$, then $x \notin f(x_0)$ for all $x \in X_1$. This means $x_0 \notin f^{-1}(x)$ for all $x \in X_1$ so $x_0 \in O_x^c$ for all $x \in X_1$ which contradicts the assumption that $D = \emptyset$. So $f(x) \cap X_1 \neq \emptyset$ for all $x \in X_1$. Define $g : X_1 \to 2^{X_1}$ by $g(x) = f(x) \cap X_1$. Then g has convex values and $g^{-1}(y) = f^{-1}(y) \cap X_1 \supset O_y \cap X_1 = \hat{O}_y$ which is relatively open in X_1. As $D = \cap_{x\in X_1} O_x^c = \emptyset$ then clearly $\cup_{y\in X_1} \hat{O}_y = X_1$. Applying Lemma 1 of [4] gives the result in this case. Next suppose $D = \cap_{x\in X_1} O_x^c \neq \emptyset$. Define

$$X_2 = \overline{\text{co}}\{X_1 \cup \{\hat{x}_1, \ldots, \hat{x}_n\}\}$$

which is compact and convex. Then

$$\cap_{x\in X_2} O_x^c \subset (\cap_{x\in X_1} O_x^c) \cap (\cap_{i=1}^n O_{\hat{x}_i}^c) = \emptyset$$

by the hypothesis of the theorem. We may now repeat the arguments in the first part of the proof with X_2 in place of X_1 to get a fixed point of f. □

Theorem 1 is equivalent to the following KKM result:

Theorem 2. *Let X be a nonempty subset of a convex set Y in a Hausdorff topological vector space. Suppose $F : X \to 2^Y$ has closed values and such that $\overline{\text{co}}\{x_1, \ldots, x_m\} \subset \cup_{i=1}^m F(x_i)$ for any finite subset $\{x_1, \ldots, x_m\}$ of X. Further assume there exists a compact convex $X_1 \subset X$ and points $\{\hat{x}_1, \ldots, \hat{x}_n\} \subset X$ such that*

$$\cap_{x \in X_1} F(x) \subset \cup_{i=1}^n F(\hat{x}_i)^c.$$

Then

$$\cap_{x \in X} F(x) \neq \emptyset.$$

Proposition 1. *Theorem 1 and Theorem 2 are equivalent.*

Proof. To see Theorem 1 implies Theorem 2, assume the conditions of Theorem 2 but deny the conclusion; that is assume $\cap_{x \in X} F(x) = \emptyset$. Then the multifunction $f : Y \to 2^Y$ defined by $f(y) = \{x \in Y : y \notin F(x)\}$ has nonempty values. Also

$$f^{-1}(x) = \{y \in Y : x \in f(y)\} = \begin{cases} F(x)^c & \forall x \in X \\ Y & \forall x \in Y - X \end{cases}$$

where $F(x)^c = Y - F(x)$ and we define $F(x) = \emptyset, \forall x \in Y - X$. Therefore $f^{-1}(x)$ denoted by O_x is open in Y for each $x \in Y$, and $\cup_{x \in Y} O_x = Y$. Define $g : Y \to 2^Y$ by $g(y) = \text{co} f(y)$ so $g^{-1}(x) \supset f^{-1}(x) = O_x, \forall x \in Y$. By assumption there exists a compact convex X_1 and points $\{\hat{x}_1, \ldots, \hat{x}_n\}$ such that $\cap_{x \in X_1} O_x^c \subset \cup_{i=1}^n O_{\hat{x}_i}$. Applying Theorem 1 we have a fixed point of g which means there exists $x_0 \in Y$ with $x_0 = \sum_{j=1}^m \lambda_j x_j$ and $x_j \in f(x_0)$. By definition of f then $x_0 \notin F(x_i)$ so $\sum \lambda_j x_j \notin \cup_{j=1}^m F(x_j)$ which contradicts the assumptions of Theorem 2. To prove Theorem 2 implies Theorem 1, we assume the conditions of Theorem 1. Let $F(x) = O_x^c$ which is closed in X for all $x \in X$. Suppose $\cap_{x \in X_1} O_x^c = \emptyset$ which implies $\cap_{x \in X} O_x^c = \emptyset$. Then F cannot be KKM so there exists $\{x_1, \ldots, x_m\}$ such that $\overline{\text{co}}\{x_1, \ldots, x_m\}$ is not a subset of the corresponding union $\cup_{i=1}^m F(x_i)$. Thus there exists $x_0 = \sum_{i=1}^m \lambda_i x_i \notin \cup_{i=1}^m F(x_i)$. Then $x_0 \in \cap_{i=1}^m O_{x_i}$ and as $O_{x_i} \subset f^{-1}(x_i)$ we have $x_i \in f(x_0)$ for all $i = 1, \ldots, m$. As f has convex values, $x_0 \in f(x_0)$. Now suppose $\cap_{x \in X_1} O_x^c \subset \cup_{i=1}^n O_{\hat{x}_i}$ and is nonempty. Setting $X_2 = \overline{\text{co}}\{X_1 \cup \{\hat{x}_1, \ldots, \hat{x}_n\}\}$ we have

$$\cap_{x \in X_2} O_x^c \subset (\cap_{x \in X_1} O_x^c) \cap (\cap_{i=1}^n O_{\hat{x}_i}^c) = \emptyset.$$

Therefore $\cap_{X_2} O_x^c = \emptyset$ and the arguments above can be repeated to give the result. □

Remark 1. Theorem 1 is an improvement of Theorem 2 of [4] in the sense that condition 4 of Theorem 1 contains condition (iv) of Theorem 2 in [4] as a special case.

A further generalisation of Theorem 1 is as follows where we extend it to a coincidence theorem. For this, it is necessary to assume X is now a complete convex subset of a locally convex Hausdorff topological vector space.

Theorem 3. *Let X be a complete convex subset of a locally convex Hausdorff topological vector space E, and Y be convex set in Hausdorff topological vector space H. Let $g : Y \to 2^X$ be upper semicontinuous with nonempty compact contractible values and $f : X \to 2^Y$ have nonempty convex values such that*

(i) for each $y \in Y$, $f^{-1}(y)$ contains an open set $O_y \subset X$ (which may be empty for some $y \in Y$);
(ii) $\cup_{y\in Y} O_y = X$;
(iii) there exists a compact convex set $Y_1 \subset Y$ and points $\hat{y}_1, \ldots, \hat{y}_n$ in Y such that

$$D = \cap_{y\in Y_1} O_y^c \subset \cup_{i=1}^n O_{\hat{y}_i}.$$

Then there exists an $x_0 \in X$ such that $g^{-1}(x_0) \cap f(x_0) \neq \emptyset$.

Proof. First suppose $D = \emptyset$. In this case, $f(x) \cap Y_1 \neq \emptyset$ for all $x \in \overline{\text{co}}g(Y_1)$. To see this, suppose there is an $x_0 \in \overline{\text{co}}g(Y_1)$ such that $f(x_0) \cap Y_1 = \emptyset$. Then for each $y \in Y_1$, $y \notin f(x_0)$, i.e. $x_0 \notin f^{-1}(y)$ and as $O_y \subset f^{-1}(y)$, $x_0 \in O_y^c$ for all $y \in Y_1$. So $x_0 \in \cap_{y\in Y_1} O_y^c = D$ which contradicts our initial assumption that $D = \emptyset$. Thus we may define a nonempty set-valued mapping $h : \overline{\text{co}}g(Y_1) \to 2^{Y_1}$ by $h(x) = f(x) \cap Y_1$. Then $h(x)$ is convex and $h^{-1}(y) = \{x \in \overline{\text{co}}g(Y_1) : y \in h(x)\} = \{x \in \overline{\text{co}}g(Y_1) : y \in f(x) \cap Y_1\} = f^{-1}(y) \cap \overline{\text{co}}g(Y_1) \supset O_y \cap \overline{\text{co}}g(Y_1) = O_y^1$ which is relatively open in $\overline{\text{co}}g(Y_1)$. Moreover $\cup_{y\in Y_1} O_y^1 = \overline{\text{co}}g(Y_1)$ since $\cap_{y\in Y_1} O_y^c = \emptyset$. Thus by Lemma 1 of [5], there exists an $x_0 \in \overline{\text{co}}g(Y_1)$ such that $h(x_0) \cap g^{-1}(x_0) \neq \emptyset$, that is, $f(x_0) \cap g^{-1}(x_0) \neq \emptyset$. Next suppose $D \neq \emptyset$. Define $Y_2 = \overline{\text{co}}\{Y_1 \cup \{\hat{y}_1, \ldots, \hat{y}_n\}\}$ and note that

$$\cap_{y\in Y_2} O_y^c \subset (\cap_{y\in Y_1} O_y^c) \cap (\cap_{i=1}^n O_{\hat{y}_i}^c) = \emptyset.$$

Therefore the arguments above with Y_2 in place of Y_1 are sufficient to deduce the existence of a coincidence point for f and g. □

3 Applications to variational inequalities

We apply Theorem 1 to establish the existence of solutions to variational inequalities in not necessarily reflexive Banach spaces. That is, for X a nonreflexive Banach space, we consider a multivalued operator $A : K \subset X^{**} \to 2^{X^*}$ and we solve the problem of finding $x_0 \in K$ such that

$$\forall y \in K, \exists v \in A(x_0) \text{ with } (v, y - x_0) \geq 0.$$

The results of this section extend the results in [1] and [6] to multifunctions as well as extending compactness and continuity conditions.

Definition 1. An operator $A : X^{**} \to 2^{X^*}$ is said to be

1. monotone if for any $x, y \in X^{**}$ and any $v \in A(x)$ and $w \in A(y)$, it follows that

$$(v - w, x - y) \geq 0; \tag{1}$$

2. pseudomonotone if for any $x, y \in X^{**}$, and any $v \in A(x), w \in A(y)$,

$$(w, x-y) \geq 0 \text{ implies } (v, x-y) \geq 0; \tag{2}$$

3. weakly pseudomonotone if for any $x, y \in X^{**}$, and any $w \in A(y)$,

$$(w, x-y) \geq 0 \text{ implies } (v, x-y) \geq 0 \text{ for some } v \in A(x). \tag{3}$$

It is clear that monotone implies pseudomonotone which in turn implies weakly pseudomonotone.

Definition 2. Let $A : K \subset X^{**} \to 2^{X^*}$ be an operator. A is said to be u-hemicontinuous on K if for any $x, y \in K$ the mapping $\alpha \mapsto (A(x+\alpha(y-x)), y-x)$from $[0,1]$ to $\mathbb{R}$ is upper semicontinuous at 0^+.

The proof of the next Lemma can be found in [3].

Lemma 1 (Linearization Lemma). *Let $A : K \subset X^{**} \to 2^{X^*}$ where K is convex, nonempty. Consider the following problems:*

(I) $\exists x \in K$ *such that* $\forall y \in K, \exists v \in A(x)$ *with* $(v, y-x) \geq 0$;
(II) $\exists x \in K$ *such that* $\forall y \in K, \exists w \in A(y)$ *with* $(w, y-x) \geq 0$;
(III) $\exists x \in K$ *such that* $\forall y \in K, \forall w \in A(y), (w, y-x) \geq 0$.

Then

1. *problem (III) implies (II);*
2. *problem (II) implies (I) if A is u-hemicontinuous;*
3. *problem (I) implies (III) if A is pseudomonotone, and it implies (II) if A is weakly pseudomonotone.*

Theorem 4. *Let $A : K \subset X^{**} \to 2^{X^*}$ be pseudomonotone, u-hemicontinuous with nonempty values, where K is convex. Assume there exists points $\hat{x}_1, \ldots, \hat{x}_n$ in K, $\hat{v}_i \in A(\hat{x}_i)$ for all $i = 1, \ldots, n$, and a nonempty weak-star compact convex $X_1 \subset K$ such that*

$$D := \{x \in K : (v, y-x) \geq 0 \ \forall y \in X_1, \ \forall v \in A(y)\} \tag{4}$$

$$\subset \bigcup_{i=1}^{n} \{x \in K : (\hat{v}_i : \hat{x}_i - x) < 0\}. \tag{5}$$

Then there exists a solution to the Variational Inequality:

$$\exists x_0 \in K \text{ such that } \forall y \in K, \exists v \in A(x_0) \text{ with } (v, y-x_0) \geq 0. \tag{6}$$

Proof. Consider the condition

$$\forall x \in K \ \exists y \in K \text{ such that } (v, y-x) < 0 \text{ for some } v \in A(y). \quad (\star)$$

If this is not true, parts 1 and 2 of the Linearization Lemma imply we have a solution. So we may assume ($\star$) is true. Suppose there is no solution to (6). Then the multivalued function $F : K \to 2^K$ defined by

$$F(x) = \{y \in K : (v, y - x) < 0 \ \forall v \in A(x)\} \tag{7}$$

has nonempty convex values. We show the assumptions of Theorem 1 are satisfied, so F has a fixed point, which is clearly absurd. For $y \in K$ arbitrary,

$$\begin{aligned} F^{-1}(y) &= \{x \in K : y \in F(x)\} \\ &= \{x \in K : (v, y - x) < 0 \ \forall v \in A(x)\} \\ &\supset \{x \in K : (v, y - x) < 0 \ \exists v \in A(y)\} = O_y \end{aligned}$$

by the pseudomonotonicity of A. O_y is clearly weak-star open. Furthermore ($\star$) implies $\cup_{y \in K} O_y = K$. Finally,

$$D = \bigcap_{y \in X_1} O_y^c \subset \bigcup_{i=1}^{n} \{x \in K : (\hat{v}_i, \hat{x}_i - x) < 0\} \subset \bigcup_{i=1}^{n} O_{\hat{x}_i}$$

follows from (4)–(5). All the conditions of Theorem 1are satisfied and so there exists a fixed point of F. This contradiction proves the theorem. □

Acknowledgment. The first author would like to thank the Department of Mathematics, University of Queensland for the kind hospitality shown during his stay and acknowledges Research support in part by a grant from IPM.

References

1. S. Chang, B. S. Lee and Y. Chen, Variational Inequalities for Monotone operators in Nonreflexive Banach Spaces, *Appl. Math. Lett.*, **8**,(6), 29–34, (1995).
2. H. Komiya, Coincidence Theorem and Saddle Point Theorem, *Proc. Amer. Math. Soc.* **96**, 599-602,(1986).
3. I. V. Konnov and J. C. Yao, On the Generalized Vector Variational Inequality Problem, *J. Math. Anal. Appl.*, **206**, 42–58, (1997).
4. E. Tarafdar, A Fixed Point Theorem Equivalent to the Fan-Knaster-Kuratowski-Mazurkiewicz Theorem, *J. Math. Anal. Appl.* **128**,(2),475–479,(1987).
5. E. U. Tarafdar and P. J. Watson, A Coincidence Point Theorem and Related Results, *Appl. Math. Lett.* **11**,(1),37–40, (1998).
6. P. J. Watson, Variational Inequalities in Nonreflexive Banach Spaces, *Appl. Math. Lett.* **10**,(2), 45–48, (1997).

Representation of a Polynomial Function as a Difference of Convex Polynomials, with an Application

Albert Ferrer Biosca

Universitat Politècnica de Catalunya
Departament de Matemàtica Aplicada I, EUPB, Campus SUD
Barcelona, Spain
e-mail:albert@ma1.upc.es

Abstract. One of the main problems in global optimization is the multiextremal global optimization in which an objective function has several local minima and at least one global minimum. To deal with this problem, it is useful to have a convex difference representation of all the functions involved, as this allows for employing so-called d.c. optimization techniques.

Procedures that permit the calculation of the polynomial convex difference representation of any polynomial are presented and analyzed here, and an application is made to a real problem whose functions are polynomials of degrees up to four.

2000 Mathematics Subject Classification. Primary 90C26; Secondary 90C30.

Keywords and phrases. d.c. representation, d.c. programming problem, canonical d.c. program, polynomial homogenization, polynomial dehomogenization.

1 Motivation and introduction

Problems with multiple extremal points are obviously nonconvex. Several authors have put forward algorithms for solving the particular type of multiextremal point problem whose nonconvexity is only due to having reverse convex constraints. However, in practice, direct problem formulations are not convenient, and it is thus necessary to transform them into alternative ones that are more suitable for algorithmic purposes. In this report, the incentives for studying d.c. functions come from applications.

Definition 1. Let f be defined on a convex set $A \subset \boldsymbol{R}^n$. The function f will be said to be d.c. on A, if it is expressible as a difference of convex functions on A. On the other hand, a function f will be said to be d.c. at a point $a \in A$ if there exists a convex neighborhood B of a such that f is d.c. on $B \cap A$. If f is d.c. at every point of A, it will be said to be locally d.c. on A.

Definition 2. If f is a d.c. function on A convex set in $\boldsymbol{R}^n$, then $f = f_1 - f_2$, with f_1 and f_2 convex functions on A, is called a d.c. representation of f on A.

The class of d.c. functions on A, denoted by DC(A), is the vector space generated by the cone of convex functions on A. DC(A) has some interesting properties with respect to operations frequently encountered in optimization; for exemple, the maximum or minimum of a finite number of d.c. functions is a d.c. function, and taking either product or quotient of d.c. functions we obtain d.c. functions (Hiriart-Urruty [7]).

The class of d.c. functions is very rich. Some authors provide interesting theoretical d.c. representation results but no practical means to get them. Hartman [5] states that every locally d.c. function on $A \subset \boldsymbol{R}^n$ open or closed convex set is d.c. on A. As a consequence of this result, one can prove that every function $f \in C^2(A)$ (the class of functions whose second partial derivatives are continuous everywhere) on A open convex set is d.c. on A. Bougeard [2] proves that if $f \in C^2(A)$ then there exists a d.c. representation $f = f_1 - f_2$ where $f_1 \in C^2(A)$ and $f_2 \in C^\infty(A)$. Penot and Bougeard [12] establish a similar result with more global assumptions (A open convex set on a finite dimensional normed vector space and f a lower$-C^2$ function on A then $f = f_1 - f_2$ where f_1 and f_2 are convex and $f_2 \in C^\infty(A)$).

Since polynomials are functions of the class $C^2(A)$ one has that polynomials are d.c. functions on A. It turns out that any problem of minimizing a continuous real function over a compact convex set A can be approximated as closely as desired by a problem of minimizing a d.c. function over A. This property is a consequence of the Stone-Weierstrass Theorem, which states that continuous functions on compact sets are uniform limits of sequences of polynomials. This result provides an additional motivation for the d.c. representation of polynomials.

Nonconvex optimization problems often have the form of d.c. programming problems

$$\begin{aligned} \text{minimize } & f(x) \\ s.t.\ & g_k(x) \le 0,\ k = 1 \ldots m \\ & h_s(x) \le 0,\ s = 1 \ldots r \end{aligned} \tag{1}$$

where all functions f, g_k, $k = 1 \ldots m$ are d.c. functions on $\boldsymbol{R}^n$ and h_s, $s = 1 \ldots r$, are convex functions on $\boldsymbol{R}^n$. If we know a d.c. representation of each d.c. function, then (1) can be transformed into an equivalent canonical d.c. program

$$\begin{aligned} \text{minimize } & cx \\ s.t.\ & H(x) \le 0 \\ & G(x) \ge 0 \end{aligned} \tag{2}$$

with $c \in \boldsymbol{R}^n$ and both H and G are convex functions on $\boldsymbol{R}^n$ (Horst and Tuy [9]).

Several authors (Horst and Tuy [9], Konno, Thach and Tuy [10], Tuy [14], Strekalovsky and Tsevendorj [13]) have put forward algorithms for solving canonical d.c. programming problems whose nonconvexity is only due to having reverse convex constraints. The main problem we tackle is: how can we get a d.c. representation of a d.c. function?

While it is not too difficult to prove theoretically that a given function is d.c., it is often very problematical to obtain an effective d.c. representation of a d.c. function. We are interested in d.c. representations of polynomials. The aim of this paper is to give a new procedure to obtain a d.c. representation of polynomials, to compare it with the procedure described by Konno, Thach and Tuy [10], emphasizing its advantages and to make an application to the functions of a real problem, which are polynomials of degree up to four.

Let $\boldsymbol{R}_m[x_1, ..., x_n]$ be the vector space of polynomials of degree less than or equal to m and let $H_i[x_1, ..., x_n]$ be the vector space of homogeneous polynomials of degree i (generated by monomials $x_1^{t_1} x_2^{t_2} \dots x_n^{t_n}$, with $t_1 + t_2 + \dots + t_n = i$, which form a basis).

Lemma 1. *[11] The following properties are known:*

(i) If $i \leq m$ then $H_i[x_1, ..., x_n]$ is a vector subspace of $\boldsymbol{R}_m[x_1, ..., x_n]$

(ii) $s_i = dim\ H_i[x_1, ..., x_n] = C^i_{n+i-1} = \binom{n+i-1}{i}$

(iii) $\boldsymbol{R}_m[x_1, ..., x_n] = \bigoplus_{i=0}^m H_i[x_1, ..., x_n]$

Let $p(x_1, \dots, x_n) \in \boldsymbol{R}^m[x_1, ..., x_n]$. Thus one can write

$$p(x_1, \dots, x_n) = \sum_{i=0}^{m} q_i(x_1, \dots, x_n),$$

where $q_i(x_1, \dots, x_n) \in H_i[x_1, ..., x_n]$ $i = 0, 1, \dots, m$.

If $\{m_i^j(x_1, \dots, x_n);\ j = 1, 2, \dots, s_i\}$ is the usual monomial basis of the vector space of homogeneous polynomials of degree i then

$$q_i(x_1, \dots, x_n) = \sum_{j=1}^{s_i} \lambda_j m_i^j(x_1, \dots, x_n).$$

Using the procedure described by Konno, Thach and Tuy [10], which takes into account that $x_i^{t_i}$, $i = 1, \dots, n$ are convex functions on $\boldsymbol{R}^n_+$ and applying successively that given a couple of positive convex functions, ϕ_1 and ϕ_2 on $\boldsymbol{R}^n_+$, a d.c. representation of their product is

$$\phi_1(x)\phi_2(x) = \frac{1}{2}(\phi_1(x) + \phi_2(x))^2 - \frac{1}{2}(\phi_1(x)^2 + \phi_2(x)^2), \tag{3}$$

one obtains a d.c. representation for the monomials

$$x_1^{t_1} x_2^{t_2} \dots x_n^{t_n}, \text{ with } t_1 + t_2 + \dots + t_n = i,$$

on $\boldsymbol{R}^n_+$ (positive orthant). Hence one can get a d.c. representation for polynomial functions on the positive orthant.

We present an alternative procedure which uses the results in Corollaries 1 and 2 below. From Corollary 1 one only needs to make a suitable change to a basis of ith powers of homogeneous polynomials of degree 1 within $H_i[x_1,...,x_n]$ to get

$$q_i(x_1,\dots,x_n)=\sum_{j=1}^{s_i}\mu_j(a_1^jx_1+\cdots+a_n^jx_n)^i.$$

Thus, using Corollary 2 one can obtain a d.c. representation

$$q_i(x_1,...,x_n)=\sum_{j\in I^+}\mu_j\left(\sum_{k=1}^{n}a_k^jx_k\right)^i-\sum_{j\in I^-}(-\mu_j)\left(\sum_{k=1}^{n}a_k^jx_k\right)^i, \tag{4}$$

where $I^+=\{j;\mu_j>0\}$ and $I^-=\{j;\mu_j<0\}$ and both components are linear combinations of ith powers of homogeneous polynomials of degree 1. Hence, we can obtain an alternative d.c. representation of polynomials on suitable domains (see Lemma 3).

The important points to note here are that d.c. representations using powers of homogeneous polynomials of degree 1,

i) use basis change, within $H_i[x_1,...,x_n]$, to get a d.c. representation and not individual d.c. representation for monomials. Thus, each component of d.c. representations using powers of homogeneous polynomials of degree 1 generally has fewer summands than components of the d.c. representations by Konno, Thach and Tuy [10],
ii) allow the description of procedures to search for bases and to get d.c. representations of homogeneous components of a polynomial (see Appendix, where there is described a procedure using the MAPLE Symbolic Calculator). Alternative bases to get different d.c. representations of polynomials can be used,
iii) allow applications to real problems (see Sections 3-5 , where is described an application whose functions are polynomials of degrees up to four),
iv) enjoy a considerable stability, in the sense that every polynomial $p(x)$ can be expressed as a difference of convex polynomials, where both components are convex polynomials of the same degree as $p(x)$. This fact is an essential difference with d.c. representations by Konno, Thach and Tuy [10], where this property is not true. Thus, interesting questions such as:
   - what is the best d.c. representation of a d.c. function?,
   - how can we define that a representation is better that another one? and
   - is there any representation that is optimal in some sense?,

   can be formulated.

   According to Hiriart-Hurruty [7], optimal d.c. representations have to be sought in restricted classes of representations. Thus, the question: *"is it possible to get an optimal (in some sense) d.c. representation of a polynomial function of degree n?"* can be formulated within the context of the vector space of polynomial functions of degree n, which is a restricted class to get d.c. representations.

2 D.c. representation of a polynomial function

In this section we give an explicit decomposition of a polynomial as a difference of convex polynomials (Corollary 2), based on the fact that the set of mth powers of homogeneous polynomials of degree 1 is a generating set for the vector space of homogeneous polynomials of degree m (Proposition 1). The latter fact is well known, but we will include a proof as it will give us an alternative procedure to get d.c. representations of polynomials (we will give an exemple later on). Before proving Proposition 1 we will enunciate some lemmas. We will give the references where a proof can be found. Lemmas without specific references will be proved.

Lemma 2. *[1] Let $f : D \subset \boldsymbol{R}^n \longrightarrow \boldsymbol{R}$ be a convex function in convex set D and let $\Psi : B \subset \boldsymbol{R} \longrightarrow \boldsymbol{R}$ be a nondecreasing proper convex function in convex set B. If $f(D) \subset B$ then $\Psi \circ f : D \subset \boldsymbol{R}^n \longrightarrow \boldsymbol{R}$ is convex.*

Lemma 3. *Let $x \longmapsto g(x) := \left(\sum_{i=1}^n a_i \cdot x_i\right)^m$ with $m \in N$ be a function which is the mth power of a linear function; then, we define the closed half space:*

$$H_m^+ = \left\{(x_1, ..., x_n) \in \boldsymbol{R}^n; \quad \sum_{i=1}^n a_i \cdot x_i \geq 0\right\} \tag{5a}$$

$$H_m^- = \left\{(x_1, ..., x_n) \in \boldsymbol{R}^n; \quad \sum_{i=1}^n a_i \cdot x_i \leq 0\right\} \tag{5b}$$

The following results can be proved:

i) If m is even then g is a convex function on $\boldsymbol{R}^n$.
ii) If m is odd then g is convex on H_m^+ and concave on H_m^- ' .
iii) Every expression of the form

$$h(x) = \sum_{j=1}^s \lambda_j \left(\sum_{i=1}^n a_i^j x_i\right)^{m_j}, \tag{6}$$

with $\lambda_j > 0$ and $m_j \in I\!N$, is a convex function on the convex set

$$D = \left\{(x_1, ..., x_n) \in \boldsymbol{R}^n; \quad \sum_{i=1}^n a_i^j \cdot x_i \geq 0, j \in J\right\}, \tag{7}$$

where $J = \{j; m_j \text{ is odd}\}$.

Proof. Let $\Psi(t) = t^m$ a nondecreasing proper convex function in $\boldsymbol{R}_+$ and by applying Lemma 2 to the linear functions $\sum_{i=1}^n a_i \cdot x_i$ and $-(\sum_{i=1}^n a_i \cdot x_i)$ on H_m^+ and H_m^- respectively we can prove *i)* and *ii)*. About *iii)*, we define $D_j = H_{m_j}^+$ when m_j is odd and $D_j = \boldsymbol{R}^n$ when m_j is even, $j = 1 \ldots s$. Since

$D = \cap_{j=1}^{s} D_j$, it is easy to see that D is a convex subset of each convex set D_j, $j = 1 \ldots s$, hence each term $\left(\sum_{i=1}^{n} a_i^j x_i\right)^{m_j}$ is a convex function on D and as $\lambda_j > 0$, $j = 1 \ldots s$ one has that h is a convex function on D.

Lemma 4. *[4] Let $q(x_1, ..., x_n) \in \boldsymbol{R}_m[x_1, ..., x_n]$ be a polynomial of total degree m.*

i) Let $q = \sum_{i=0}^{m} q_i$ be the expansion of q as the sum of its homogeneous components where q_i has total degree i. Then

$$\overline{q}(x_0, x_1, ..., x_n) = \sum_{i=0}^{m} q_i(x_1, \ldots, x_n) x_0^{m-i} \tag{8}$$

is an homogeneous polynomial of total degree m in $H_m[x_0, x_1, ..., x_n]$. We will call $\overline{q}$ the homogenization of q.

ii) The homogenization of q can be computed using the formula

$$\overline{q}(x_0, x_1, \ldots, x_n) = x_0^m q\left(\frac{x_1}{x_0}, \ldots, \frac{x_n}{x_0}\right). \tag{9}$$

iii) Taking $x_0 = 1$ we thus get $q(x_1, \ldots, x_n) = \overline{q}(1, x_1, \ldots, x_n)$. For this reason we will call q the dehomogenization of the polynomial $\overline{q}$.

iv) Let $F(x_0, \ldots, x_n)$ be an homogeneous polynomial and let x_0^s be the highest power of x_0 dividing F. If $f(x_1, \ldots, x_n) = F(1, x_1, \ldots, x_n)$ is a dehomogenization of F, then

$$F = x_0^s \overline{f}. \tag{10}$$

(v) Let Φ be the function

$$\begin{array}{rcl} \Phi : \boldsymbol{R}_m[x_1, ..., x_n] & \longmapsto & H_m[x_0, x_1, ..., x_n] \\ q & \longmapsto & \Phi(q) = \overline{q} \end{array} \tag{11}$$

then Φ is a real vector space isomorphism.

The following Lemma is well known; we include a proof for the sake of completeness.

Lemma 5. *Let $\alpha_0, \alpha_1, ..., \alpha_m$ be $m + 1$ different elements of $\boldsymbol{R}$; then, the family of $m + 1$ polynomials*

$$B = \{(x + \alpha_0)^m, (x + \alpha_1)^m, ..., (x + \alpha_m)^m\}$$

is a vector space basis of the vector space $\boldsymbol{R}_m[x]$.

Proof. We know that the set

$$B_0 = \{1, x, x^2, ..., x^m\} \tag{12}$$

is a vector space basis of $\boldsymbol{R}_m[x]$. Then, we can write

$$(x+\alpha_i)^m = x^m + C_m^1 x^{m-1}\alpha_i + C_m^2 x^{m-2}\alpha_i^2 + \ldots + C_m^k x^{m-k}\alpha_i^k + \ldots + \alpha_i^m \quad (13)$$

and calculate the determinant

$$\Delta = \begin{vmatrix} 1 & \ldots & 1 & \ldots & 1 \\ C_m^1\alpha_0 & \ldots & C_m^1\alpha_i & \ldots & C_m^1\alpha_m \\ C_m^2\alpha_0^2 & \ldots & C_m^2\alpha_i^2 & \ldots & C_m^2\alpha_m^2 \\ \ldots & \ldots & \ldots & \ldots & \ldots \\ C_m^k\alpha_0^k & \ldots & C_m^k\alpha_i^k & \ldots & C_m^k\alpha_m^k \\ \ldots & \ldots & \ldots & \ldots & \ldots \\ C_m^m\alpha_0^m & \ldots & C_m^m\alpha_i^m & \ldots & C_m^m\alpha_m^m \end{vmatrix} = C_m^1 \ldots C_m^m \begin{vmatrix} 1 & \ldots & 1 & \ldots & 1 \\ \alpha_0 & \ldots & \alpha_i & \ldots & \alpha_m \\ \alpha_0^2 & \ldots & \alpha_i^2 & \ldots & \alpha_m^2 \\ \ldots & \ldots & \ldots & \ldots & \ldots \\ \alpha_0^m & \ldots & \alpha_i^m & \ldots & \alpha_m^m \end{vmatrix} \quad (14)$$

where the expression

$$\begin{vmatrix} 1 & \ldots & 1 & \ldots & 1 \\ \alpha_0 & \ldots & \alpha_i & \ldots & \alpha_m \\ \alpha_0^2 & \ldots & \alpha_i^2 & \ldots & \alpha_m^2 \\ \ldots & \ldots & \ldots & \ldots & \ldots \\ \alpha_0^m & \ldots & \alpha_i^m & \ldots & \alpha_m^m \end{vmatrix} = \prod_{i=0}^{m}\prod_{j=i+1}^{m}(\alpha_i - \alpha_j) \neq 0 \quad (15)$$

is the nonzero Vandermonde determinant which completes the proof.

Proposition 1. *([3] Ex.111, p.498) Let $H_m[x_1, x_2, \ldots, x_n]$ be the vector space of the homogeneous polynomials of total degree m. Then the set of mth powers of homogeneous polynomials of degree 1 $\{p^m; p \in H_1[x_1, x_2, \ldots, x_n]\}$ is a generating set for $H_m[x_0, x_1, \ldots, x_n]$. We have*

$$H_m[x_1, x_2, \ldots, x_n] = \langle p^m; p \in H_1[x_1, x_2, \ldots, x_n]\rangle \quad (16)$$

where the notation $\langle \ldots \rangle$ means the span of the set it contains.

Proof. The proof is by induction on n. It is trivial for the case $n = 1$ because $H_m[x_1] = \langle x_1^m \rangle$. For the case $n = 2$, let $h \in H_m[x_1, x_2]$. Since Φ of Lemma 4 is an isomorphism, one has $h = \Phi(q)$ for some $q \in \boldsymbol{R}_m[x_1, x_2]$. Let $\alpha_0, \alpha_1, \ldots, \alpha_m \in \boldsymbol{R}$ be a collection of different numbers by Lemma 5 there exist $\lambda_0, \lambda_1, \ldots, \lambda_m \in \boldsymbol{R}$ such that $q(x_2) = \sum_{i=0}^m \lambda_i (x_2 + \alpha_i)^m$, whence

$$h(x_1, x_2) = x_2^m q\left(\frac{x_1}{x_2}\right) = x_2^m \sum_{i=0}^m \lambda_i \left(\frac{x_1}{x_2} + \alpha_i\right)^m = \sum_{i=0}^m \lambda_i (x_1 + \alpha_i x_2)^m.$$

This proves that the polynomials $(x_1 + \alpha_i x_2)^m \quad i = 0, 1, \ldots, m$ generate $H_m[x_1, x_2]$.

Let us assume that our assertion holds for the case $n-1$; we will prove it for the case n.

Let $h(x_1, x_2, \ldots, x_{n-1}, 1) \in \boldsymbol{R}_m[x_1, \ldots, x_{n-1}]$ be the dehomogenization of the polynomial $h(x_1, x_2, ..., x_n) \in H_m[x_1, x_2, ..., x_n]$ with respect to the variable x_n. Then, we can write

$$h(x_1,\ x_2, ..., x_{n-1}, 1) = \sum_{s=0}^{m} h_s, \text{ with } h_s \in H_s[x_1, x_2, ..., x_{n-1}], \tag{17}$$

and by induction

$$h_s(x_1,\ x_2, ..., x_{n-1}) = \sum_i \mu_{s,i}(P_{s,i})^s \tag{18}$$

with

$$P_{s,i} = a_1^{s,i} x_1 + ... + a_{n-1}^{s,i} x_{n-1}. \tag{19}$$

Then, we have

$$h(x_1,\ x_2, \ldots, x_n) = \sum_{s=0}^{m} x_m^{m-s} h_s(x_1,\ x_2, \ldots, x_{n-1}) \tag{20}$$

and using the expression (18) within (20) we can write

$$h(x_1,\ x_2, ..., x_n) = \sum_{s=0}^{m} x_n^{m-s} \left(\sum_i \mu_{s,i}(P_{s,i})^s \right) = \sum_{i,\ 0 \le s \le m} \mu_{s,i} x_n^{m-s}(P_{s,i})^s. \tag{21}$$

Every homogeneous polynomial $x_n^{m-s}(P_{s,i})^s$ in the variables x_n and $P_{s,i}$ can be expressed as a linear combination of mth powers $(\alpha_{s,i,j}P_{s,i} + \beta_{s,i,j}x_n)^m$

$$x_n^{m-s}(P_{s,i})^s = \sum_j \rho_{s,i,j} (\alpha_{s,i,j}P_{s,i} + \beta_{s,i,j}x_n)^m. \tag{22}$$

Combining (19) with (22) we obtain

$$x_n^{m-s}(P_{s,i})^s = \sum_j \rho_{s,i,j} \left(\alpha_{s,i,j} a_1^{s,i} x_1 + ... + \alpha_{s,i,j} a_{n-1}^{s,i} x_{n-1} + \beta_{s,i,j} x_n \right)^m. \tag{23}$$

Substituting the expression (23) in (21) we obtain $h(x_1, x_2, ..., x_n)$ as a linear combination of mth powers of homogeneous polynomials of total degree 1, which proves the proposition.

Corollary 1. *Let $q_i(x_1, x_2, ..., x_n)$ be an homogeneous polynomial of total degree i; there exist basis of ith powers of homogeneous polynomials of total degree 1, $B^{(i} = \left\{ (a_1^j x_1 + \cdots + a_n^j x_n)^i,\ j = 1 \ldots s_i \right\}$ so that*

$$q_i(x_1, x_2, ..., x_n) = \sum_{j=1}^{s_i} \mu_j (a_1^j x_1 + \cdots + a_n^j x_n)^i, \tag{24}$$

with $\mu_j \in \boldsymbol{R}$, $j = 1, \ldots, n$. (Remember that s_i=dim $H_i[x_1, ..., x_n]$, see Lemma 1.)

Corollary 2. *Let $I^+ = \{j; \mu_j > 0\}$ and $I^- = \{j; \mu_j < 0\}$. Then, in the half space*

$$D_i = \left\{ (x_1, ..., x_n) \in \boldsymbol{R}^n; \quad \sum_{k=1}^{n} a_k^j \cdot x_k \geq 0, \ j = 1 \dots s_i \ with \ \mu_j \neq 0 \right\}$$

if i is even or $D_i = \boldsymbol{R}^n$ if i is odd, we have

$$q_i\left(x_1, x_2, ..., x_n\right) = \sum_{j \in I^+} \mu_j \left(\sum_{k=1}^{n} a_k^j x_k \right)^i - \sum_{j \in I^-} (-\mu_j) \left(\sum_{k=1}^{n} a_k^j x_k \right)^i, \qquad (25)$$

where both components are convex. Hence, every polynomial $p \in \boldsymbol{R}_m[x_1, ..., x_n]$, $p = \sum_{i=0}^{m} q_i$ can be expressed, on the convex set $D = \cap_{i=0}^{m} D_i$, as a convex polynomial function difference.

Proof. This result follows from Corollary 1 and Lemma 3.

Corollaries 1 and 2 provide a procedure to get d.c. representations of polynomials expressed in terms of suitable basis of ith powers (see Appendix). It should be noted that the proof of Proposition 1 gives us an additional procedure, different of searching for basis, to get d.c. representations of polynomials in $\boldsymbol{R}_m[x_1, \dots, x_n]$. One only needs to know d.c. representations of monomials in $H_i[x_1, x_2]$, $i = 2, \dots, m$.

For exemple, if we know that

$$xy = \frac{1}{4}(x+y)^2 - \frac{1}{4}(x-y)^2 \qquad (26)$$

and

$$x^2y^2 = \frac{1}{12}(x+y)^4 + \frac{1}{12}(x-y)^4 - (\frac{1}{6}x^4 + \frac{1}{6}y^4) \qquad (27)$$

then, we can obtain a d.c. representation of $xyz^2 \in \boldsymbol{R}_4[x, y, z]$.

Using (26) one can write

$$xyz^2 = (xy)z^2 = (\frac{1}{4}(x+y)^2 - \frac{1}{4}(x-y)^2)z^2 = \frac{1}{4}(x+y)^2z^2 - \frac{1}{4}(x-y)^2z^2.$$

Since (27) one has

$$(x+y)^2z^2 = \frac{1}{12}(x+y+z)^4 + \frac{1}{12}(x+y-z)^4 - (\frac{1}{6}(x+y)^4 + \frac{1}{6}z^4)$$

and

$$(x-y)^2z^2 = \frac{1}{12}(x-y+z)^4 + \frac{1}{12}(x-y-z)^4 - (\frac{1}{6}(x-y)^4 + \frac{1}{6}z^4),$$

whence

$$xyz^2 = \frac{1}{48}(x+y+z)^4 + \frac{1}{48}(x+y-z)^4 + \frac{1}{24}(x-y)^4 -$$

$$-(\frac{1}{24}(x+y)^4+\frac{1}{48}(x-y+z)^4+\frac{1}{48}(x-y-z)^4),$$

which is a d.c. representation of xyz^2 on $\boldsymbol{R}_4[x,y,z]$.

To use this procedure we need to know d.c. representations of monomials in $H_i[x_1,x_2]$, $i=2,\ldots,m$, which is not always possible. On the other hand, we must take into account that using this procedure the ith powers of the linear combination could be dependent, whence the importance of using basis of ith powers to get d.c. representations.

3 An application

An example of the application of these results is the Short-Term Hydrothermal Coordination of electricity generation Problem (STHCP). Its importance stems from the economic and technical implications that the solution to this problem has for electric utilities with a mixed (hydro and thermal) generation system (Heredia and Nabona[6]).

Given a short-term time period (from 24h to one week long) one wishes to find values for each hour (or shorter intervals) in the period so that a number of constraints are satisfied and the generation cost of thermal units is minimized. The model contains the replicated hydronetwork through which the temporary evolution of the reservoir system is represented. Figure 1 shows the network with only two reservoirs and where the time period has been subdivided into four intervals. We use j to indicate the jth reservoir, $j=1\ldots N_e$ and i to indicate the ith time interval, $i=1\ldots N_t$. It should be observed that, in each time interval i, the water discharge from reservoir R_1 to reservoir R_2 establishes a link between the reservoirs. On the other hand, the volume stored at the end of the time interval i and the volume stored at the beginning of the time interval $i+1$ are the same on each reservoir R_j, which establishes a link between each reservoir from the time interval i to $i+1$. The variables are the water discharges d_j^i from reservoir j over the ith interval and the volume stored v_j^i in reservoir j at the end of the ith time interval. The volumes stored at the beginning and at the end of the time period are known (they are not variables). Acceptable forecasts of electricity consumption l^i and of natural water inflow w_j^i into the reservoirs of the hydrogeneration system at each interval must be available.

The main feature in this formulation is that the power hydrogeneration function h_j^i at the reservoir j over the ith interval can be approximated by a polynomial function of degree 4 in the variables v_j^{i-1}, v_j^i and d_j^i (Heredia and Nabona [6]),

$$\begin{aligned}h_j^i = k^i * d_j^i * [&s_{vd} + \tfrac{s_{vl}}{2}(v_j^{i-1}+v_j^i) + \tfrac{s_{vq}}{3}(v_j^i - v_j^{i-1})^2 + \\ &+ s_{vq}v_j^{i-1} * v_j^i + \tfrac{s_{vc}}{4}((v_j^{i-1})^2 + (v_j^i)^2)(v_j^{i-1}+v_j^i) - \\ &- s_{dl} * d_j^i - s_{dq} * (d_j^i)^2]\end{aligned} \tag{28}$$

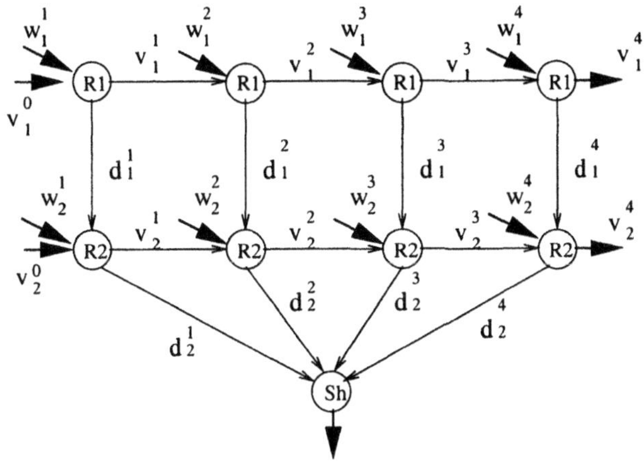

Fig. 1. Four intervals and two reservoirs replicated hydronetwork

where k^i, s_{vd}, s_{vl}, s_{vq}, s_{vc}, s_{dl} and s_{dq} are technological coefficients which depend of each reservoir.

3.1 Objective function, constraints and bounds

The objective function, which will be minimized, is the generation cost of thermal units,

$$f(\dots, v_j^{i-1}, v_j^i, d_j^i, \dots) = \sum_{i=1}^{N_t} c_i \left(l^i - \sum_{j=1}^{N_e} h_j^i(v_j^{i-1}, v_j^i, d_j^i) \right) . \tag{29}$$

We can see that it depends on power hydrogeneration functions $h_j^i(v_j^{i-1}, v_j^i, d_j^i)$ and is therefore not a linear function. Note that power hydrogeneration has no additional cost. The linear constraints are the flow balance equations at all nodes of the network,

$$v_j^i - v_j^{i-1} - d_{j-1}^i + d_j^i = w_j^i \quad j = 1, \dots, N_e,\ i = 1, \dots, N_t. \tag{30}$$

The nonlinear constraints are the thermal production with generation bounds,

$$\underline{g} \le l^i - \sum_{j=1}^{N_e} h_j^i \le \overline{g} \quad i = 1, \dots, N_t. \tag{31}$$

There are positive bounds on all variables,

$$\underline{d_j} \le d_j^i \le \overline{d_j} \qquad j = 1, \dots, N_e,\ i = 1, \dots, N_t \tag{32a}$$

$$\underline{v_j} \le v_j^i \le \overline{v_j} \quad j = 1, \dots, N_e,\ i = 1, \dots, N_t - 1. \tag{32b}$$

4 Example of d.c. representation using the power hydrogeneration polynomial

Ordering, in ascendent degrees, the power hydrogeneration polynomial (28) and substituting the coeficients $a = k^i$, $b = s_{vd}$, $c = 0.5s_{vl}$, $d = s_{vq}/3$, $e = 0.25s_{vc}$, $f = s_{dl}$ and $g = s_{dq}$ and the variables $x = d_j^i$, $y = v_j^{i-1}$ and $z = v_j^i$ leads to

$$h(x,y,z) = abx + acx(y+z) - afx^2 + adx(z-y)^2 -$$

$$-agx^3 + 3adxyz + aex(y^2+z^2)(y+z). \tag{33}$$

If we obtain a d.c. representation of its nonlinear homogeneous components then we can decompose it. Let h_2, h_3 and h_4 be power hydrogeneration homogeneous components of degree 2, 3 and 4 respectively. Hence, we have

$$h_2(x,y,z) = acx(y+z) - afx^2, \tag{34}$$

the homogeneous polynomial component of degree 2,

$$h_3(x,y,z) = adx(z-y)^2 - 3agx^3 + adxyz, \tag{35}$$

the homogeneous polynomial component of degree 3, and

$$h_4(x,y,z) = aex(y^2+z^2)(y+z), \tag{36}$$

the homogeneous polynomial component of degree 4.

We get suitable vector space basis, using the MAPLE Symbolic Calculator (see Appendix). Hence we can obtain a d.c. representation for each homogeneous component of h.

4.1 D.c. representation of the homogeneous polynomial component of degree 2

Let $B_0^{(2}$ be the initial basis in $H_2[x,y,z]$ with

$$B_0^{(2} = \{x^2, y^2, z^2, xy, xz, yz\}. \tag{37}$$

It is easy to prove that the set

$$B^{(2} = \left\{x^2, y^2, z^2, (x+y)^2, (x+z)^2, (y+z)^2\right\} \tag{38}$$

is a new $H_2[x,y,z]$ basis.

Every polynomial of $B^{(2}$ can be represented using polynomials of $B_0^{(2}$, as we can see in Table 1.

Table 1. Coordinates

$B^{(2}$	representation in $B_0^{(2}$	$B_0^{(2}$ Coordinates
x^2	x^2	$(1,0,0,0,0,0)$
y^2	y^2	$(0,1,0,0,0,0)$
z^2	z^2	$(0,0,1,0,0,0)$
$(x+y)^2$	x^2+y^2+2xy	$(1,1,0,2,0,0)$
$(x+z)^2$	x^2+z^2+2xz	$(1,0,1,0,2,0)$
$(y+z)^2$	y^2+z^2+2yz	$(0,1,1,0,0,2)$

Let C be the matrix which represents the change of coordinates from $B^{(2}$ (coordinates X) to $B_0^{(2}$ (coordinates Y).

$$C = \begin{pmatrix} 1&0&0&1&1&0 \\ 0&1&0&1&0&1 \\ 0&0&1&0&1&1 \\ 0&0&0&2&0&0 \\ 0&0&0&0&2&0 \\ 0&0&0&0&0&2 \end{pmatrix}. \tag{39}$$

Then, using the expression $X = C^{-1}Y$ we obtain the d.c. representation of $h_2(x,y,z)$.

$$h_2(x,y,z) = \frac{ac}{2}\left[(x+y)^2 + (x+z)^2\right] - \left[(ac+af)x^2 + \frac{ac}{2}\left(y^2+z^2\right)\right] \tag{40}$$

Note that, in this case, another possible d.c. representation is via the diagonalization of the symmetric matrix A such that $h_2(x) = X^tAX$. To simplify the expressions we suppose $a=1$, $c=2$ and $f=1$ in (34). Hence

$$A = \begin{pmatrix} -1&1&1 \\ 1&0&0 \\ 1&0&0 \end{pmatrix}. \tag{41}$$

The nonzero eigenvalues of A are 1 and -2. Moreover, the eigenvectors orthonormal matrix C of the A eigenvalues is

$$C = \begin{pmatrix} \frac{1}{\sqrt{3}} & \frac{-2}{\sqrt{6}} & 0 \\ \frac{1}{\sqrt{3}} & \frac{1}{\sqrt{6}} & \frac{-1}{\sqrt{2}} \\ \frac{1}{\sqrt{3}} & \frac{1}{\sqrt{6}} & \frac{1}{\sqrt{2}} \end{pmatrix}. \tag{42}$$

So that, using $(X^tC)D(C^tX)$, we obtain a shorter d.c. representation

$$h_2(x,y,z) = 1\left(\frac{x+y+z}{\sqrt{3}}\right)^2 - 2\left(\frac{-2x+y+z}{\sqrt{6}}\right)^2. \tag{43}$$

4.2 D.c. representation of the homogeneous polynomial component of degree 3

In this case, the initial basis is the set

$$B_0^{(3} = \{x^3, y^3, z^3, x^2y, x^2z, xy^2, xz^2, y^2z, yz^2, xyz\} \tag{44}$$

and we can choose

$$\begin{aligned} B^{(3} = \Big\{ & x^3, y^3, z^3, (x+2y)^3, (x+2z)^3, (y+2z)^3, \\ & (2x+y)^3, (2x+z)^3, (2y+z)^3, (x+y+z)^3 \Big\} \end{aligned} \tag{45}$$

as the new $H_3[x,y,z]$ basis. Then, the representation of the polynomial $h_3(x,y,z)$, in the basis $B^{(3}$, as a difference of convex polynomials is

$$\begin{aligned} h_3 = {} & \tfrac{ad}{18}\left(3(x+2z)^3 + (y+2z)^3 + 6(2x+y)^3 + (2y+z)^3\right) - \\ & - \left(a\left(g + \tfrac{5}{2}d\right)x^3 + \tfrac{ad}{18}\left(y^3 + 3z^3\right)\right). \end{aligned} \tag{46}$$

It should be noted that in practice, one is interested in d.c. representations of a polynomial only on the orthant $x \geq 0$. Taking into account that the odd cases lead to restrictions on the domain where the polynomial components are convex (see Lemma 3), one can use linear expressions, into m^{th} powers, with positive coefficients to obtain d.c. representations, which hold on the orthant $x \geq 0$.

4.3 D.c. representation of the homogeneous polynomial component of degree 4

The initial basis is

$$\begin{aligned} B_0^{(4} = \{ & x^4, y^4, z^4, x^3y, x^3z, xy^3, xz^3, y^3z, yz^3, \\ & x^2y^2, x^2z^2, y^2z^2, x^2yz, xy^2z, xyz^2\} \end{aligned} \tag{47}$$

and let

$$\begin{aligned} B^{(4} = \Big\{ & x^4, y^4, z^4, (x+y)^4, (x+z)^4, (y+z)^4, (y-x)^4, (z-x)^4 \\ & (y-z)^4, (x+y+z)^4, (x+y-z)^4, (x-y+z)^4, \\ & (x+2y+z)^4, (x+y+2z)^4, (x+2y+3z)^4 \Big\} \end{aligned} \tag{48}$$

be a vector space basis for $H_4[x,y,z]$. The representation of the polynomial $h_4(x,y,z)$ as a difference of convex polynomials is

$$\begin{aligned} h^{(4} = {} & \tfrac{ae}{24}\left(4x^4 + (y-z)^4 + (x+2y+z)^4 + (x+y+2z)^4\right) - \\ & - \tfrac{ae}{24}\left(8y^4 + 8z^4 + 5(y+z)^4 + (y-x)^4 + (z-x)^4 + 4(x+y+z)^4\right). \end{aligned} \tag{49}$$

5 D.c. representation of the short-term hydrothermal coordination model functions

As we know how to construct a d.c. representation of the power hydrogeneration functions h_j^i we can obtain a d.c. representation of the functions within (STHCP).

- **Objective function.** There is no loss of generality in assuming $c_l = 1$ in (29).

$$f = \sum_{i=1}^{N_t} \left(l^i - \sum_{j=1}^{N_e} h_j^i \right) \tag{50}$$

We denote $G_i = l^i - \sum_{j=1}^{N_e} h_j^i$. As we know how to calculate a d.c. representation of $h_j^i = f_i^j - g_i^j$ we can write

$$G_i = \left(l^i + \sum_{j=1}^{N_e} g_j^i \right) - \sum_{j=1}^{N_e} f_j^i \tag{51}$$

which is a d.c. representation of G_i. Hence, we can obtain a d.c. representation of the objective function f:

$$f = \sum_{i=1}^{N_t} G_i = \sum_{i=1}^{N_t} \left(l^i + \sum_{j=1}^{N_e} g_j^i \right) - \sum_{i=1}^{N_t} \left(\sum_{j=1}^{N_e} f_j^i \right) . \tag{52}$$

- Nonlinear constraints. We can use the same procedure to obtain a d.c. representation of the nonlinear constraints (31)

$$l^i - \sum_{j=1}^{N_e} h_j^i = \left(l^i + \sum_{j=1}^{N_e} g_j^i \right) - \left(\sum_{j=1}^{N_e} f_j^i \right) . \tag{53}$$

As we know a d.c. representation of functions within (STHCP), by means of successive transformations we can reduce it to an equivalent convex program with an additional reverse convex constraint and linear objective function (canonical d.c. programming problem (3)). Then one can attempt to use algorithms as describe in Horst, Phong and Thoai [8], Horst and Tuy [9], Konno, Thach and Tuy [10], Strekalovsky and Tsevendorj [13] or Tuy [14] to solve this kind of programming problems.

6 Appendix

In this section is described a procedure using the MAPLE Symbolic Calculator to search for basis and to get d.c. representations of each homogeneous components of a polynomial. The inputs are: number of variables (n), degree of homogeneous polynomial (m), basis of mth powers and the homogeneous polynomial which one want to get a d.c. representation. The procedure obtains several outputs to verify that the final solution is true. Finally, the outputs f_1 and f_2 are the d.c. components of the homogeneous polynomial. If a set of mth powers is not a basis then the procedure stops and it shows a message of error; thus one must search for alternative basis of mth powers.

As example of application we will get a d.c. representation of the homogeneous polynomial

$$x_1(x_3 - x_2)^2 - x_1^3 + 5x_1x_2x_3,$$

using the basis (45)

```
> restart:
with(linalg):
```

Input variables (n), degree of homogeneous polynomial (m):

```
n:=3:
m:=3:
```

```
d:=binomial(n+m-1, m):
u:=[seq(0,k=1..d)]:
xx:=[seq(x[k],k=1..n)]:
aa:=[seq(1,k=1..n)]:
xxx:=(evalm(aa &* xx))^m:
H:=[op(expand(xxx))]:
M:=[seq(H[k]/coeffs(H[k]),k=1..d)]:
```

$$M := [x_1{}^3,\ x_1{}^2\,x_2,\ x_1{}^2\,x_3,\ x_1\,x_2{}^2,\ x_1\,x_2\,x_3,\ x_1\,x_3{}^2,\ x_2{}^3,\ x_2{}^2\,x_3,\ x_2\,x_3{}^2,\ x_3{}^3]$$

```
a := array(1..d):
```

Input basis of mth powers:

```
a[1]:=xx[1]^3:
a[2]:=xx[2]^3:
a[3]:=xx[3]^3:
a[4]:=(xx[1]+2*xx[2])^3:
a[5]:=(xx[1]+2*xx[3])^3:
a[6]:=(xx[2]+2*xx[3])^3:
a[7]:=(2*xx[1]+xx[2])^3:
a[8]:=(2*xx[1]+xx[3])^3:
```

```
a[9]:=(2*xx[2]+xx[3])^3:
a[10]:=(xx[1]+xx[2]+xx[3])^3:
```

Input polynomial which one want to get d.c. representation

```
yyy:=xx[1]*(xx[3]-xx[2])^2-xx[1]^3+5*xx[1]*xx[2]*xx[3];
```

```
expand(yyy):
HH:=[op(expand(yyy))]:
nn:=nops(expand(yyy)):
```

$$yyy := x_1\,(x_3 - x_2)^2 - {x_1}^3 + 5\,x_1\,x_2\,x_3$$

$$x_1\,{x_3}^2 + 3\,x_1\,x_2\,x_3 + x_1\,{x_2}^2 - {x_1}^3$$

```
for i from 1 by 1 to nn
do
for k from 1 by 1 to d
do
if (member(M[k], {HH[i]/coeffs(HH[i])})) then
u[k]:=coeffs(HH[i]):
fi:
od:
od:
print(u);
```

$$[-1, 0, 0, 1, 3, 1, 0, 0, 0, 0]$$

```
v:= array(1..d,1..d):for j from 1 by 1 to d
do
for i from 1 by 1 to d
do
v[j,i]:=0
od;
od;
for j from 1 by 1 to d
do
if (j<n+1) then
L:=[expand(a[j])]:
nn:=1
else
L:=[op(expand(a[j]))]:
nn:=nops(expand(a[j])):
fi:
```

```
for i from 1 by 1 to nn
do
for k from 1 by 1 to d
do
if (member(M[k], {L[i]/coeffs(L[i])})) then
v[k,j]:=coeffs(L[i]):
fi:
od:
od:
od:
C:=evalm(v^(-1)):
ff:=evalm(C &* u):
ff1 := array(1..d):
ff2 := array(1..d):
for i from 1 by 1 to d
do
if (ff[i]>0) then
ff1[i]:=ff[i]:
ff2[i]:=0:
else
ff2[i]:=-ff[i]:
ff1[i]:= 0:
fi;
od:
f1:=evalm(a &* ff1);
f2:=evalm(a &* ff2);
f:=expand(f1-f2);
```

$$f1 := \frac{2}{3}{x_1}^3 + \frac{1}{6}{x_2}^3 + \frac{1}{6}{x_3}^3 + \frac{1}{36}(x_1+2x_2)^3 + \frac{1}{36}(x_1+2x_3)^3 + \frac{1}{2}(x_1+x_2+x_3)^3$$

$$f2 := \frac{1}{12}(x_2+2x_3)^3 + \frac{5}{36}(2x_1+x_2)^3 + \frac{5}{36}(2x_1+x_3)^3 + \frac{1}{12}(2x_2+x_3)^3$$

$$f := 3x_1x_2x_3 + x_1{x_2}^2 + x_1{x_3}^2 - {x_1}^3$$

Acknowledgment. This work was partially supported by CICYT grant TAP96-1044.

References

1. Avriel, M. (1976) Nonlinear programming Analysis and methods. Englewood Cliffs, NS Prentice-Hall.

2. Bougeard, M. (1978) Contribution à la théorie de Morse en dimension finie. Thèse de 3ème cycle de l'Université de Paris IX.
3. Chambadal, L., Ovaert, J.L. (1968) Algèbre linéaire et algèbre tensorielle. Dunod Université. Dunod, Paris.
4. Cox, D., Little, J., O'Shea, D. (1992) Ideals, varieties and algorithms. An introduction to computational algebraic geometry and commutative algebra. Undergraduate Texts in Mathematics. Springer-Verlag, New York.
5. Hartman, P. (1959) On functions representable as a difference of convex functions. Pacific Journal of Mathematics, **9**, 707-713.
6. Heredia, F.J., and Nabona, N. (1995) Optimum short-term hydrothermal scheduling with spinning reverse through network flows. IEEE Trans. on Power Systems, v. **10**, No. **3**, pp. 1642-1651.
7. Hiriart-Urruty, J.B.(1985) Generalized differentiability, duality and optimization for problems dealing with differences of convex functions. Lectures Notes in Economics and Mathematical Systems, **256**, 37-69, Springer-Verlag, Berlin.
8. Horst, R., Phong, T.Q., Thoai, Ng.V.(1990) On solving a d.c. programming problem by a sequence of linear programs. Annals of Operation Research, **25**, pp. 1-18.
9. Horst, R., and Tuy, H. (1990) Global optimization. Deterministic approaches. Springer-Verlag, Berlin.
10. Konno, H., Thach P.T., Tuy H. (1997) Optimization on low rank nonconvex structures. Kluwer Academic Publishers.
11. Lang, S. (1965) Algebra. Addison-Wesley Publishing Company.
12. Penot, J.P., Bougeard, M.L. (1988) Approximation and decomposition properties of some classes of locally D.C. functions. Mathematical Programming **41**, pp. 195-227.
13. Strekalovsky, A.S., Tsevendorj, I. (1998) Testing the $\mathcal{R}$-strategy for a reverse convex problem. Journal of Global Optimization **13**, pp. 61-74.
14. Tuy, H. (1998) Convex analysis and global optimization. Kluwer Academic Publishers.

Proper Efficiency and Generalized Convexity in Nonsmooth Vector Optimization Problems

Giorgio Giorgi[1] and Angelo Guerraggio[2]

[1] University of Pavia, Pavia, Italy
[2] Bocconi University of Milan, Milan, Italy

Abstract. In vector optimization problems several notions of proper efficiency have been proposed, in order to rule out some undesirable situations (tolerated by the definition of efficiency). The relationships between the above different definitions have been studied by many authors. In this paper we again show these relationships, under the assumption that the involved functions are not differentiable but only Lipschitzian.
The properties of Clarke's generalized derivative and subdifferential are here used, in order to get a picture similar to the one obtained for the differentiable case. Then, by means of suitable generalized convexity assumptions, we obtain some equivalences of the different notions of proper efficiency considered, with respect to the notion of efficiency. The results of this paper may be viewed as a generalization of the situation established for the smooth case.

2000 Mathematics Subject Classification. Primary 90C29; Secondary 90C25.

Keywords and phrases. Generalized convexity, Nonsmooth vector optimization.

1 Introduction

Several definitions of proper efficiency have been proposed by various authors (see the survey paper of Guerraggio, Molho, Zaffaroni (1994)) in order to rule out some situations tolerated in the notion of efficiency but scarcely meaningful. In the present work we shall be concerned with various definitions of proper efficiency for a vector optimization problem, where the functions involved are assumed to be Lipschitzian, in order to exploit the properties of the Clarke generalized derivative and subdifferential. We shall be concerned with the following vector optimization problem:

$$G(x^0) \subset T(S, x^0), \tag{P}$$

where $S = \{x \in X : g_i(x) \geqq 0, i = 1 \ldots m\}$, $f : X \to \mathbf{R}^p$, $g : X \to \mathbf{R}^m$, $X \subset \mathbf{R}^n$, X is open. If $x^0 \in S$, the set (supposed for simplicity not empty):

$$I(x^0) = \{i : g_i(x^0) = 0\}$$

denotes the index set of the *active constraints* at x^0; $g_I(x_0)$ is the vector whose components are $g_i(x_0)$, $i \in I(x_0)$. Given a set $A \subset \mathbf{R}^n$, the *closure* of A is denoted by $cl(A)$; $int(A)$ is the set of the *interior points* of A; $conv(A)$ is the *convex hull* of A. The *contingent cone* or *Bouligand tangent cone* to the set A at the point $x^0 \in cl(A)$ is the set:

$$T(A,x^0) = \{x \in \mathbf{R}^n : \exists \{\lambda_n\}, \lambda_n \in \mathbf{R}_+, \exists \{x^n\} \subset A,$$
$$x^n \to x^0 : \lim_{n \to +\infty} \lambda_n (x^n - x^0) = x\}.$$

We adopt the following convention to compare vectors of $\mathbf{R}^p$: we shall write:

i) $f(x) \geqq f(y)$, when $f_i(x) \geqq f_i(y)$, $\forall i = 1 \dots p$, i.e. $f(x) - f(y) \in \mathbf{R}^p_+$;
ii) $f(x) \geq f(y)$, when $f_i(x) \geqq f_i(y)$, $\forall i = 1 \dots p$, and there exists an index j such that $f_j(x) > f_j(y)$, i.e. when $f(x) - f(y) \in \mathbf{R}^p_+ \backslash \{0\}$;
iii) $f(x) > f(y)$, when $f_i(x) > f_i(y)$ $\forall i = 1 \dots p$, i.e. when $f(x) - f(y) \in int(\mathbf{R}^p_+)$.

The notions $\leqq$, $\leq$, $<$ are defined in a similar way.

Definition 1. The function $h : X \to \mathbf{R}$ is *Lipschitzian* on the open set $X \subset \mathbf{R}^n$ if there exists a positive constant k such that:

$$|h(x) - h(y)| \leq k||x - y||, \forall x, y \in X.$$

Definition 2. If $h : X \to \mathbf{R}$ is Lipschitzian on X, the *Clarke generalized derivative* of h at $x^0 \in X$, in the direction d, denoted by $h^0(x^0, d)$, is defined as:

$$h^0(x^0, d) = \limsup_{x \to x^0, t \to 0^+} \frac{h(x + td) - h(x)}{t}.$$

The *Clarke generalized subdifferential* of h at $x^0 \in X$ is defined as follows:

$$\partial_C h(x^0) = \{\psi \in \mathbf{R}^n : h^0(x^0, d) \geq \psi d, \forall d \in \mathbf{R}^n\}.$$

It is well known that, as a function of d, $h^0(x^0, d)$ is Lipschitzian on X, subadditive and positively homogeneous; $\partial_C h(x^0)$ is a nonempty, convex and compact subset of $\mathbf{R}^n$. Moreover we have, for every $d \in \mathbf{R}^n$:

$$h^0(x^0, d) = \max_{\psi \in \partial_C h(x^0)} \{\psi d\}$$

and, for every scalar α, one has:

$$\partial_C (\alpha h)(x) = \alpha \partial_C h(x).$$

Now let us consider a vector-valued function $f : X \to \mathbf{R}^p$; we assume that each f_i is Lipschitzian on X.

Definition 3. The *Clarke generalized Jacobian* of f at $x^0 \in X$, denoted $J_C f(x_0)$, is the convex hull of all (p, n) matrices Z obtained as the limit of a sequence of the form $\nabla f(x^i)$, where $x^i \to x^0$ and $x^i \in \Omega(f)$, $\Omega(f)$ being the set of points in X where f is differentiable (thanks to the Rademacher theorem, f is differentiable almost everywhere on X). Formally:

$$J_C f(x^0) = conv\{\lim_{i \to +\infty} \nabla f(x^i) : x^i \to x^0, x^i \in \Omega(f)\}.$$

If f is continuously differentiable at x^0, then $J_C f(x)$ reduces to the usual Jacobian matrix $\{\nabla f(x)\}$ and, when we have $p = 1$, $J_C f(x^0)$ reduces to the Clarke generalized subdifferential $\partial_C f(x^0)$. Moreover we have (the inclusion may be strict):

$$J_C f(x^0) \subset \partial_C f_1(x^0) \times \partial_C f_2(x^0) \times \ldots \times \partial_C f_p(x^0).$$

We shall denote by $\partial_C f(x^0)$ the cartesian product in the right-hand side of the above relation and shall call it the *Clarke generalized subgradient* of f at x^0. Obviously every element of $\partial_C f(x^0)$ can be interpreted as a matrix of order (p, n). We now extend to a nonsmooth setting the notions of convexity, pseudoconvexity and quasiconvexity.

Definition 4. Let $h : X \to \mathbf{R}$ be Lipschitzian on the open set $X \subset \mathbf{R}^n$.

a) h is **Clarke convex** on X or C-convex on X if:

$$h(x) - h(u) \geqq p(x - u), \forall p \in \partial_C h(u), \forall x, u \in X.$$

b) h is *C-pseudoconvex* on X if:

$$h(x) - h(u) < 0 \Rightarrow p(x - u) < 0, \forall p \in \partial_C h(u), \forall x, u \in X.$$

c) h is *C-quasiconvex* on X if:

$$h(x) - h(u) \leqq 0 \Rightarrow p(x - u) \leqq 0, \forall p \in \partial_C h(u), \forall x, u \in X.$$

The function $h : X \to \mathbf{R}$ is C-concave, C-pseudoconcave, C-quasiconcave if $-h$ is, respectively, C-convex, C-pseudoconvex, C-quasiconvex.
We can note that h is C-convex if and only if it is convex; also, h is C-pseudoconvex if and only if it is pseudoconvex according to the definition of pseudoconvexity used by various authors (see, e.g., Penot and Quang (1997)). Moreover, if h is C-convex on X, then h is both C-pseudoconvex and C-quasiconvex on X. As in the differentiable case, every C-pseudoconvex function $h : X \to \mathbf{R}$ satisfies the fundamental properties:

i) every local minimizer of h is a global minimizer;
ii) $0 \in \partial_C h(x^0) \Rightarrow h$ has a local minimizer at x^0.

Glover (1984) has proved the following assertion of Hiriart-Urruty (1979): if h is C-pseudoconvex, then it is semistricly quasiconvex and, being Lipschitzian and therefore continuous, also quasiconvex in the usual sense. Daniilidis and Hadjisavvas (1999) proved that a locally Lipschitzian function h is semistrictly quasiconvex if and only if

$$x, u \in X, z \in [u, x), h(x) - h(z) \leqq 0 \Rightarrow p(x - u) \leqq 0, \forall p \in \partial_C h(u).$$

Thus, semistrict quasiconvexity implies C-quasiconvexity. These notions are distinct since the function $h : \mathbf{R} \to \mathbf{R}$ defined by

$$h(x) = \begin{cases} -1, x \leqq -\frac{\pi}{2} \\ \sin x, -\frac{\pi}{2} < x < \frac{\pi}{2} \\ 1, \frac{\pi}{2} \leqq x \end{cases}$$

is C-quasiconvex but it is not semistrictly quasiconvex. Moreover Ellaia and Hassouni (1991) have proved that, if $h : X \to \mathbf{R}$ is Lipschitzian (or also locally Lipschitzian), then it is quasiconvex if and only if:

$$x, u \in X, h(x) - h(u) < 0 \Rightarrow p(x-u) \leqq 0, \forall p \in \partial_C h(u).$$

The same authors have shown that the strict inequality in the previous relation cannot be replaced by a weaker inequality, so our definition of C-quasiconvexity characterizes a proper subset of the set of (Lipschitzian) quasiconvex functions.

We give now, with reference to problem (P), the usual definition of efficient point.

Definition 5. A point $x^0 \in S$ is an *efficient point* for (P) if:

$$\nexists x \in S : f(x) - f(x^0) \leq 0,$$

i.e.:

$$f(S) - f(x^0) \cap \mathbf{R}^p_- = \{0\}$$

or, equivalently:

$$f(S) \cap (f(x^0) + \mathbf{R}^p_-) = \{f(x^0)\}.$$

2 Some definitions of proper efficiency for (P)

The concept of proper efficiency was introduced by H.W.Kuhn and A.W.Tucker in their classical paper of 1951 (see Kuhn and Tucker (1951)). We extend their definition for problem (P) where all functions -are assumed Lipschitzian on X.

Definition 6. A point $x^0 \in S \subset \mathbf{R}^n$ is said to be *properly efficient* in the sense of Kuhn-Tucker ($x^0 \in PE(S)_{KT}$) if it is efficient and for every matrices M and N, $M \in \partial_C f(x^0)$, $N \in \partial_C g_I(x^0)$, there exists no $y \in \mathbf{R}^n$ satisfying the following system:

$$\begin{cases} My \leq 0 \\ Ny \geqq 0. \end{cases}$$

Similarly the original definition of proper efficiency in the sense of Klinger (1967) requires the differentiability of the objective function. We now extend the said definition to a nonsmooth setting.

Definition 7. A point $x^0 \in S$ is said to be *properly efficient* in the sense of Klinger ($x^0 \in PE(S)_{KL}$) when it is efficient and, for every matrix $M \in \partial_C f(x^0)$ and for every sequence $\{t_k\}$ with $t_k \in \mathbf{R}_+$, $t_k \neq 0$, $t_k \to 0$, there exists no $y \in \mathbf{R}^n$ such that:

$$\begin{cases} My \leq 0 \\ g(x^0 + t_k y) \geqq 0 \end{cases}$$

for all $k \in \mathbf{N}$ large enough.

Definition 8. A point $x^0 \in S$ is said to be *locally properly efficient* in the sense of Borwein ($x^0 \in LPE(S)_{BO}$), when it is efficient and one has:

$$MT(S, x^0) \cap \mathbf{R}^p_- = \{0\}, \forall M \in \partial_C f(x^0).$$

The previous definition requires some comments, since the original Borwein's definition (see Borwein (1977)) is slightly different. It is well known that one has, for a differentiable objective function f:

$$\nabla f(x^0) T(S, x^0) \subset T(f(S), f(x^0)).$$

This relation that, under differentiability assumptions, is useful to obtain some inclusion results between various definitions of proper efficiency does not hold in the framework of Clarke generalized differentiability (if one substitutes $\nabla f(x^0)$ with a matrix $M \in \partial_C f(x^0)$).
Counterexample: $f(x) = |x|$, $x \in \mathbf{R}$, $x^0 = 0$, $S = \mathbf{R}$, $T(S, x^0) = \mathbf{R}$, $f(S) = \mathbf{R}_+$, so $T(f(S), f(x^0)) = \mathbf{R}_+$. Therefore there exists no $M \in \partial_C f(x^0) = [-1, 1]$ such that $MT(S, x^0) \subset T(f(S), f(x^0))$.
We have modified Borwein's original definition ($x^0 \in LPE(S)_{BO}$ if x^0 is efficient and one has $T(f(S), f(x^0)) \cap \mathbf{R}^p_- = \{0\}$) by substituting the set $T(f(S), f(x^0))$ with the smaller set $\nabla f(x^0) T(S, x^0)$ and then this one with $MT(S, x^0)$, $M \in \partial_C f(x^0)$, as we deal with Lipschitzian functions.
Obviously, the Kuhn-Tucker, Klinger and (locally) Borwein definitions do not complete the picture of the various definitions of proper efficiency proposed in the literature. For a more general survey of the various notions one can see Guerraggio, Molho and Zaffaroni (1994). In the present paper we work only with those definitions originally given under some differentiability assumptions (i.e. the Kuhn-Tucker and Klinger definitions) and with the (modified) Borwein definition, considering this one in order to link our implication results with the ones obtained for other definitions of proper efficiency, not involving differentiability (see again Guerraggio, Molho and Zaffaroni (1994)).

3 Relationships between the notions of proper efficiency introduced

It is well known that, if in problem (P) f and g are differentiable, then, denoting by $E(S)$ the set of efficient points, the following inclusion relationships hold (here the sets of properly efficient points are obviously defined in the original sense):

$$E(S) \supset PE(S)_{KL} \supset \begin{cases} LPE(S)_{BO} \\ PE(S)_{KT} \end{cases}$$

Moreover, a suitable constraint qualification gives $LPE(S)_{BO}$ as a subset of $PE(S)_{KT}$.
Now we want to extend the previous inclusions to the Lipschitzian case and to prove, in particular, that if f and g are Lipschitzian on $X \subset \mathbf{R}^n$, one has:

I) $PE(S)_{KT} \subset PE(S)_{KL}$.
II) $LPE(S)_{BO} \subset PE(S)_{KL}$.
III) $LPE(S)_{BO} \subset PE(S)_{KT}$.

(The last inclusion holds under a suitable constraints qualification).
We recall that, by definition, also for the above sets of proper efficient points the inclusion in $E(S)$ still holds.

Theorem 1. *Let f and g be Lipschitzian on $X \subset \mathbf{R}^n$. Then $PE(S)_{KT} \subset PE(S)_{KL}$.*

Proof. Let $x^0 \in PE(S)_{KT}$ and suppose on the contrary that $x^0 \notin PE(S)_{KL}$. Then there exist a matrix $M \in \partial_C f(x^0)$ and a sequence $\{t_k\}$, $t_k > 0$, $t_k \to 0^+$, such that:

$$\begin{cases} My \leq 0 \\ g(x^0 + t_k y) \geqq 0 \end{cases}$$

admits a solution $y \in \mathbf{R}^n$. In particular one will have:

$$g_i(x^0 + t_k y) \geqq 0, i \in I(x^0)$$

i.e.:

$$\frac{g_i(x^0 + t_k y) - g_i(x^0)}{t_k} \geqq 0, i \in I(x^0)$$

from which it follows:

$$\limsup_{t_k \to 0^+} \frac{g_i(x^0 + t_k y) - g_i(x^0)}{t_k} \geqq 0, i \in I(x^0).$$

Hence we have:

$$g_i^0(x^0, y) = \limsup_{x \to x^0, t_k \to 0^+} \frac{g_i(x + t_k y) - g_i(x)}{t_k} \geqq 0, i \in I(x^0).$$

That is, there exists a matrix $N \in \partial_C g_I(x^0)$ such that $Ny \geqq 0$, but $M \in \partial_C f(x^0)$ being such that $My \leq 0$, this just means that x^0 is not properly efficient in the sense of Kuhn-Tucker. □

Theorem 2. *Let f and g be Lipschitzian on $X \subset \mathbf{R}^n$. Then $LPE(S)_{BO} \subset PE(S)_{KL}$.*

Proof. Let $x^0 \in LPE(S)_{BO}$ and suppose on the contrary that $x^0 \notin PE(S)_{KL}$. This means that there exist a vector y, a matrix $M \in \partial_C f(x^0)$ and a positive sequence $\{t_k\} \to 0^+$ such that the system:

$$\begin{cases} My \leq 0 \\ g(x^0 + t_k y) \geqq 0 \end{cases}$$

admits a solution. Denote $x^0 + t_k y$ by y^k; therefore $y^k \in S$ and one has $y = \lambda_k(y^k - x^0)$, where $\lambda_k = 1/t_k$. So $\lim_{k \to +\infty} \lambda_k(y^k - x^0) = y$, i.e. $y \in T(S, x^0)$. The absurd conclusion follows, as $My \leq 0$ is equivalent to the existence of a matrix $M \in \partial_C f(x^0)$ such that:

$$MT(S, x^0) \cap \mathbf{R}^p_- \neq \{0\}.$$

□

The next result holds under an assumption that may be viewed as *Abadie constraint qualification* for the Lipschitzian case. Indeed, it can be proved (see Guerraggio, Molho and Zaffaroni (1994)) that, in the differentiable case, a similar theorem holds, under the assumption $C(x^0) \subset T(S, x^0)$, where $C(x^0)$ is the *linearizing cone* to S at x^0:

$$C(x^0) = \{y \in \mathbf{R}^n : \nabla g_i(x^0) y \geqq 0, i \in I(x^0)\}.$$

This justifies the following Abadie-type condition:

$$G(x^0) \subset T(S, x^0), \tag{1}$$

where $G(x^0) = \{y \in \mathbf{R}^n : \exists N \in \partial_C g_I(x^0) : Ny \geqq 0\}$.

Theorem 3. *Let f and g be Lipschitzian on $X \subset \mathbf{R}^n$ and let condition (1) be verified. Then we have $LPE(S)_{BO} \subset PE(S)_{KT}$.*

Proof. Let $x^0 \in LPE(S)_{BO}$ and suppose on the contrary that $x^0 \notin PE(S)_{KT}$. This means that there exist a vector y and matrices $M \in \partial_C f(x^0)$, $N \in \partial_C g_I(x^0)$ such that:

$$\begin{cases} My \leq 0 \\ Ny \geqq 0 \end{cases}$$

is satisfied for all $k \in \mathbf{N}$. The relation $Ny \geqq 0$ means that $y \in G(x^0)$ and, thanks to condition (1), that $y \in T(S, x^0)$. So the absurd conclusion follows, in the same lines of the proof of theorem 2. □

4 Proper efficiency and generalized convexity

If in the problem (P) f and g are differentiable, we obtain that certain sets of properly efficient points, are equivalent, under suitable convexity or generalized convexity assumptions. Again we assume that in (P) f and g are Lipschitzian on $X \subset \mathbf{R}^n$. We have the following results.

Theorem 4. *Let g_i, $i \in I(x^0)$, be C-pseudoconvex; then:*

$$PE(S)_{KL} \subset PE(S)_{KT}.$$

Proof. Suppose on the contrary that there exist $x^0 \in PE(S)_{KL} \setminus PE(S)_{KT}$. Therefore there exist $y \in \mathbf{R}^n$, $M \in \partial_C f(x^0)$ and $N \in \partial_C g_I(x^0)$, such that the system:

$$\begin{cases} My \leq 0 \\ Ny \geqq 0 \end{cases}$$

is satisfied. The second inequality of the above system is equivalent to the existence of row vectors $N_i \in \partial_C g_i(x^0)$, $i \in I(x^0)$, such that $N_i y \geqq 0$.
Let us consider any sequence $\{t_k\}$, with $t_k \to 0^+$; it follows $g_i(x^0 + t_k y) \geqq 0$ (with k large enough), $\forall i \in I(x^0)$. Indeed, if $g_i(x^0 + t_k y) < 0$, thanks to the C-pseudoconvexity of g_i, $i \in I(x^0)$, we have:

$$N_i t_k y < 0, \forall N_i \in \partial_C g_i(x^0), \forall i \in I(x^0).$$

But this contradicts the second inequality of the system. So we have found a vector $y \in \mathbf{R}^n$, a matrix $M \in \partial_C f(x^0)$ and a positive sequence $\{t_k\}$, $t_k \to 0^+$, such that the following system holds:

$$\begin{cases} My \leq 0 \\ g_i(x^0 + t_k y) \geqq 0, i \in I(x^0). \end{cases}$$

As the second inequalities of this system can be extended also to inactive constraints (which are continuous, being Lipschitzian), the inclusion contradicts the assumption $x^0 \in PE(S)_{KL}$. □

Now we prove a result concerning the equivalence between the set of the efficient points and the set of the properly efficient points, in the sense of Kuhn-Tucker.

Theorem 5. *Let f_i, $i = 1 \dots p$, be C-quasiconcave and g_i, $i \in I(x^0)$, be C-pseudoconvex; then:*

$$E(S) \subset PE(S)_{KT}.$$

Proof. Suppose that on the contrary there exists $x^0 \in E(S) \backslash PE(S)_{KT}$. Therefore there exists $y \in \mathbf{R}^n$, $M \in \partial_C f(x^0)$, $N \in \partial_C g_I(x^0)$ satisfying the system:

$$\begin{cases} My \leq 0 \\ \\ Ny \geqq 0. \end{cases}$$

The second inequality of the system, together with the continuity of the non active constraints, imply (see the proof of theorem 4) that $g(x^0 + t_k y) \geqq 0$, $t_k > 0$, $t_k \to 0^+$. Now we show that $f(x^0 + t_k y) \leq f(x^0)$; indeed, if there exists some index i such that $f_i(x^0 + t_k y) > f_i(x^0)$, then, thanks to the C-quasiconcavity of f_i, $i = 1 \dots p$, $M_i t_k y \geqq 0$, $\forall M_i \in \partial_C f_i(x^0)$. If $f_i(x^0 + t_k y) = f_i(x^0)$, $\forall i$, then thanks to the C-quasiconcavity we will again have $M_i t_k y \geqq 0$, $\forall M_i \in \partial f_i(x^0)$. In both cases we have a contradiction with the first inequality of the system. So we have found feasible points $x^0 + t_k y$, with $f(x^0 + t_k y) \leq f(x^0)$, which is absurd. □

Theorem 6. *Let g_i, $i \in I(x^0)$, be C-pseudoconvex. Then one has $PE(S)_{KL} \subset LPE(S)_{BO}$.*

Proof. Let first note that one has, under the assumption of the theorem:

$$PE(S)_{KL} = PE(S)_{KT}.$$

Now, suppose on the contrary that $x^0 \in PE(S)_{KT} \backslash LPE(S)_{BO}$. Since $x^0 \notin PE(S)_{BO}$, there exists $M \in \partial_C f(x^0)$ such that $\{0\} \subset MT(S,x) \cap \mathbf{R}^p_-$. Therefore there exists $y \in T(S, x^0)$ such that $My \in \mathbf{R}^p_- \backslash \{0\}$, i.e. $My \leq 0$.
Consider now the points $x^0 + t_k y$, where $\{t_k\}$ is any sequence converging to 0^+. We will prove that $g_i(x^0 + t_k y) \geqq 0$, $\forall i \in I(x^0)$. Suppose on the contrary that there exist an index $i \in I(x^0)$ and an index k such that $g_i(x^0 + t_k y) < 0$. Since $y \in T(S, x^0)$, there exist sequences $\{\lambda_n\}$ and $\{x^n\}$ with $\lambda_n > 0$, $x^n \to x^0$, $\lambda_n(x^n - x^0) \to y$ and $y^n = x^0 + t_k \lambda_n (x^n - x^0) \to x^0 + t_k y$. By continuity, $g_i(y_n) < 0$ for n sufficiently large. Also we have that $x_n = (t_k \lambda_n)^{-1} y^n + (1 - (t_k \lambda_n)^{-1}) x^0$ is a convex combination of y^n and x^0. Therefore the hypothesis on g_i gives $g_i(x^n) < g_i(x^0) = 0$, thus contradicting $x^n \in S$.
As $My \leq 0$ and $g(x^0 + t_k y) \geqq 0$, for $y \in T(S, x^0)$ it follows that $x^0 \notin PE(S)_{KL}$ and this is absurd. □

Acknowledgment. The authors wish to thank Prof. Nicolas Hadjisavvas and two anonymous referees for their useful suggestions.

References

1. Borwein, J. M. (1977) Proper efficient points for maximization with respect to cones. SIAM Journal on Control and Optimization **15**, 57–63
2. Borwein, J. M. (1980) The geometry of Pareto efficiency over cones. Mathematische Operationsforschung und Statistik, Ser. Optimization **11**, 235–248
3. Clarke, F. H. (1983) Nonsmooth analysis and optimization. John Wiley, New York
4. Daniilidis A., Hadjisavvas N. (1999) Characterization of nonsmooth semistrictly quasiconvex and strictly convex functions. J.O.T.A. **102**, 525–536
5. Ellaia R., Hassouni A. (1991) Characterization of nonsmooth functions through their generalized gradients. Optimization **22**, 401–416
6. Glover, B. M. (1984) Generalized convexity in nondifferentiable programming. Bull. Austral. Math. Soc. **30**, 193–218
7. Guerraggio A., Molho E., Zaffaroni A. (1994) On the notion of proper efficiency in vector optimization. J.O.T.A. **82**, 1–21
8. Hiriart-Urruty, J. B. (1979) New concepts in nondifferentiable programing. Bull. Soc. Math. de France **60**, 57–85
9. Klinger, A. (1967) Improper solutions of the vector maximum problem. Operations Research **15**, 570–572
10. Kuhn H. W., Tucker A. W. (1951) Nonlinear programming. Proceedings of the second Berkley symposium on Mathematical Statistics and Probability, Berkeley, Cal, 481–492
11. Luc D. T. (1996) Generalized Monotone Set Valued Maps and Support Bifunctions, Acta Math. Vietnamica **21**, 213–252
12. Penot J. P., Quang P. H. (1997) Generalized convexity of functions and generalized monotonicity of set-valued maps. J.O.T.A. **92**, 343–356

Duality for Fractional Min-max Problems Involving Arcwise Connected and Generalized Arcwise Connected Functions

Pankaj Gupta[1] and Davinder Bhatia[2]

[1] Department of Mathematics, Deen Dayal Upadhyaya College, Shivaji Marg, Karampura, New Delhi - 110 015, India.
[2] Department of Operational Research, Faculty of Mathematical Sciences, University of Delhi, Delhi - 110 007, India.

Abstract. In this paper, we introduce two duals for the fractional minmax programming problem of the form :

$$\text{Minimize}\,\text{Maximize}_{y\in Y}\ \frac{f(x,y)}{h(x,y)}$$
$$\text{subject to } g(x)\ \leqq 0,$$

and establish duality results under arcwise connectedness and generalized arcwise connectedness assumptions on the functions involved.

2000 Mathematics Subject Classification. Primary 90C47,49K35; Secondary 47N10.

Keywords and phrases. Minimax programming, arcwise connected functions, PCN functions, QCN functions,right derivative of functions with respect to an arc.

1 Introduction

The study of optimization problem has been of considerable interest because of their wide applications in different fields such as decision making, game theory, approximation theory, facility location problem, engineering designs and duality theory. Extensive treatment of minmax theory and its applications can be found in [8].

Necessary and sufficient conditions for static minmax programming problems were obtained by Schmitendorf [11]. Later, Tanimato [12] formulated two duals to the problem considered by Schmitendorf and established duality results under convexity conditions. Subsequently, optimality conditions and duality results were presented for various minmax problems in [3, 5, 6, 7] under different set-ups. In all the above mentioned references, the work was carried out under differentiability assumptions. Recently, Mehra and Bhatia [10] obtained necessary optimality conditions for the following static minmax programming problem

$$\textbf{(P)}\quad \text{Minimize}\,\text{Maximize}_{y\in Y}\ \phi(x,y)$$
$$\text{subject to } g(x)\ \leqq 0,$$
$$x\in X\ \subseteq \mathbb{R}^n,$$

in terms of right derivatives of the functions involved with respect to the same arc. They also formulated Mond-Weir type dual and developed duality results.

The purpose of this paper is to extend the duality results of [10] to the fractional case. We propose Bector type dual and Mond-Weir type dual to the fractional minmax problem and present duality results under arcwise connectedness and generalized arcwise connectedness assumptions on the functions involved.

2 Preliminaries

Let $\mathbb{R}^n$ denote the n-dimensional Euclidean space, and $\mathbb{R}^n_+$ its nonnegative orthant. For some nonempty subset Y, let $\mathbb{R}^Y = \pi_Y \mathbb{R}$ denote the product space in a product topology. The set $\mathbb{R}^Y_+ = \pi_Y \mathbb{R}_+$ denote the convex cone of all nonnegative functions on Y. Then the topological dual of $\mathbb{R}^Y_+$ is given by

$$(\mathbb{R}^Y_+) = \wedge = \{\ \lambda = (\lambda_y)_{y \in Y} : \exists \text{ a finite set } Y_0 \subseteq Y \text{such that} \\ \lambda_y = 0, \forall\, y \in Y \backslash Y_0 \text{ and } \lambda_y \geqq 0, \forall\, y \in Y_0\ \}$$

Definition 2.1. [1] A set $X \subseteq \mathbb{R}^n$ is said to be an arcwise connected (AC) set if for every pair of points $x^1, x^2 \in X$, there exists a continuous vector-valued function $H_{x^1,x^2} : [0,1] \to X$, called an arc, such that $H_{x^1,x^2}(0) = x^1, H_{x^1,x^2}(1) = x^2$.

Definition 2.2. [4] Let $\varphi : X \to \mathbb{R}$, where $X \subseteq \mathbb{R}^n$ is an AC set. Let $x^1, x^2 \in X$ and H_{x^1,x^2} be the arc connecting x^1 and x^2 in X. The function φ is said to possess a right derivative, denoted by $\varphi^+(H_{x^1,x^2}(0))$, with respect to an arc H_{x^1,x^2} at $\theta = 0$ if

$$\text{Lim}_{\theta \to 0+} \frac{\varphi(H_{x^1,x^2}(\theta)) - \varphi(x^1)}{\theta} \quad \text{exists.}$$

Definition 2.3. [4] Let $\phi : X \to \mathbb{R}$, where $X \subseteq \mathbb{R}^n$ is an AC set. Then
(a) ϕ is (CN) function if for every $x^1, x^2 \in X$, there exist an arc H_{x^1,x^2} in X and $\phi^+(H_{x^1,x^2}(0))$ such that $\phi(x^2) - \phi(x^1) \geqq \phi^+(H_{x^1,x^2}(0))$.
(b) ϕ is Q-connected (QCN) function if for every $x^1, x^2 \in X$, there exists an arc H_{x^1,x^2} in X and $\phi^+(H_{x^1,x^2}(0))$ such that

$$\phi(x^2) \leqq \phi(x^1) \Rightarrow \phi^+(H_{x^1,x^2}(0)) \leqq 0.$$

(c) ϕ is P-connected (PCN) function if for every $x^1, x^2 \in X$, there exist an arc H_{x^1,x^2} in X and $\phi^+(H_{x^1,x^2}(0))$ such that

$$\phi^+(H_{x^1,x^2}(0)) \geqq 0 \Rightarrow \phi(x^2) \geqq \phi(x^1).$$

(d) ϕ is a strictly P-connected (STPCN) function if for every $x^1, x^2 \in X$, there exist an arc H_{x^1,x^2} in X and $\phi^+(H_{x^1,x^2}(0))$ such that

$$\phi^+(H_{x^1,x^2}(0)) \geqq 0 \Rightarrow \phi(x^2) > \phi(x^1).$$

Definition 2.4. [9] Let $\varphi : X \to \mathbb{R}$ and $G : X \times Y \to \mathbb{R}$, where X and Y are arbitrary sets. The pair (φ, G) is called convexlike on X if for every $x^1, x^2 \in X$ there exist $x^3 \in X$ and $\theta \in (0,1)$ such that

$$\varphi(x^3) \leqq (1-\theta)\varphi(x^1) + \theta\varphi(x^2)$$

and

$$G(x^3, y) \leqq (1-\theta)G(x^1, y) + \theta\, G(x^2, y), \quad \forall\, y \in Y.$$

Mehra and Bhatia [10] obtained necessary optimality conditions for the problem (P) under the following assumptions.

1. X is an open AC subset of $\mathbb{R}^n$.
2. Y is a compact subset of $\mathbb{R}^m$.
3. $\phi(x, .)$ is continuous on $Y, \forall\, x \in X$ and $g_j(.), 1 \leqq j \leqq m$, is continuous on X.
4. Right derivatives of the functions $\phi(., y)$ and $g_j(.), 1 \leqq j \leqq m$, with respect to an arc $H_{x^1,x^2}(\theta)$ at $\theta = 0$ exist, $\forall\, x^1, x^2 \in X, \forall\, y \in Y$.
5. $\phi^+(H_{x^1,x^2}(0), .)$ is continuous on Y, $\forall\, x^1, x^2 \in X$.
   For $x^* \in X$, an optimal solution of (P), we set

$$I(x^*) = \{j : g_j(x^*) = 0\},$$

$$Y(x^*) = \left\{ y \in Y : \phi(x^*, y) = \sup_{z \in Y} \phi(x^*, z) \right\}.$$

For any $t > 0$ and $x \in X$, define

$$\varphi_j(x) = g_j^+(H_{x^*,x}(0)) + t, \quad j \in I(x^*)$$

$$G(x, y) = \phi^+(H_{x^*,x}(0), y) + t, \quad y \in Y(x^*)$$

$$\Omega(x^*, t) = \{\ (u, r) : r = (r_j)_{j \in I(x^*)}, u : Y \to \mathbb{R} \text{ and } \exists\, x \in X \text{ such that}$$

$$\varphi_j(x) \leqq r_j, j \in I(x^*), G(x, y) \leqq u(y),\ \forall\, y \in Y(x^*)\ \}$$

We shall use the following theorem giving the necessary optimality conditions for (P) and its corollary proved in [10].

Theorem 2.1. *Let x^* be a solution of problem (P). Further, let $(\varphi_j, G)_{j \in I(x^*)}$ be convexlike on X and let there exist a neighbourhood U of '0' in $\mathbb{R}^{Y(x^*)}$ and constants $\nu = (\nu_j)_{j \in I(x^*)}$ such that the set $\Omega(x^*, t) \cap \bar{U} \times \pi_{j \in I(x^*)}(-\infty, \nu_j]$ is nonempty closed set for every $t > 0$. Then there exist an integer $\alpha > 0$, scalars $\lambda_i \geqq 0, 1 \leqq i \leqq \alpha, \mu \geqq 0$ and vectors $y^i \in Y(x^*), 1 \leqq i \leqq \alpha$ such that*

$$\sum_{i=1}^{\alpha} \lambda_i \phi^+(H_{x^*,x}(0), y^i) + \sum_{j=1}^{m} \mu_j g_j^+(H_{x^*,x}(0)) \geqq 0, \qquad \forall\, x \in X$$

$$\mu_j g_j(x^*) = 0, \quad 1 \leqq j \leqq m$$

$$\sum_{i=1}^{\alpha} \lambda_i + \sum_{j=1}^{m} \mu_j \neq 0.$$

Corollary 2.1. *If there exists* $w^* \in X$ *such that* $g_j^+(H_{x^*,w^*}(0)) < 0$, $\forall j \in I(x^*)$, *then* $\sum_{i=1}^{\alpha} \lambda_i \neq 0$.

3 Problem formulation and duality model-I

The fractional minmax programming problem considered in this paper is as follows:

$$\text{(FP)} \qquad \text{Minimize Maximize}_{y \in Y} \frac{f(x,y)}{h(x,y)}$$
$$\text{subject to } g(x) \leqq 0,$$
$$x \in X.$$

Let $X^0 = \{x \in X : g_j(x) \leqq 0, 1 \leqq j \leqq m\}$ denote the set of feasible solutions of (FP), where $f, h : X \times Y \to \mathbb{R}, g : X \to \mathbb{R}^m$ with $f(x,y) \geqq 0$ and $h(x,y) > 0$ on $X^0 \times Y$. We also assume the following

A-1 : $X^0 \subseteq \mathbb{R}^n$ is an open AC set.

A-2 : $Y \subseteq \mathbb{R}^m$ is compact.

A-3 : $f(x,.)$ is continuous on $Y, \forall x \in X^0, h(.,.)$ is continuous on $X^0 \times Y$, and $g(.)$ is continuous on X^0.

A-4 : The right derivatives of the functions $f(.,y), h(.,y)$ and $g(.)$ with respect to same arc $H_{x^1,x^2}(\theta)$ at $\theta = 0$ exist, $\forall x^1, x^2 \in X^0, \forall y \in Y$.

For $x \in X^0$, let

$$I(x) = \{j : g_j(x) = 0\}, J(x) = \{1, 2, ..., m\} \backslash I(x),$$

$$Y(x) = \left\{ y \in Y : f(x,y)/h(x,y) = \sup_{z \in Y} f(x,z)/h(x,z) \right\}.$$

Let x^* be an optimal solution of the fractional minmax problem (FP). For any $t > 0$ and for $x \in X^0$, define

$$\psi_j(x) = g_j^+(H_{x^*,x}(0)) + t, \quad j \in I(x^*)$$

$$H(x,y) = (f/h)^+(H_{x^*,x}(0), y) + t, \quad y \in Y(x^*)$$

$$\Omega(x^*, t) = \{ (u,r) : r = (r_j)_{j \in I(x^*)}, u : Y \to \mathbb{R}, \exists x \in X \text{ such that}$$
$$\psi_j(x) \leqq r_j, j \in I(x^*), H(x,y) \leqq u(y), \forall y \in Y(x^*) \},$$

and set

$$\Gamma = \{ (\alpha, \lambda, y) : \alpha \text{isapositiveinteger}, \lambda \in \mathbb{R}_+^\alpha, \sum_{i=1}^{\alpha} \lambda_i = 1,$$
$$y = (y^1, ..., y^\alpha), y^i \in Y(x) \text{ for some } x \in X, 1 \leqq i \leqq \alpha \}.$$

Motivated by the dual model of Bector et al. [2] for the generalized fractional programming problems, we present the following Bector type dual for the fractional minmax programming problem (FP). Let $K(\alpha, \lambda, y)$ be the set of all tuples $(x, \mu) \in X \times \mathbb{R}_+^m$ satisfying the conditions

$$\left.\begin{array}{c}\left[\left(\sum_{i=1}^{\alpha} \lambda_i f(.,y^i) + \sum_{j=1}^{m} \mu_j g_j(.)\right) \Big/ \sum_{i=1}^{\alpha} \lambda_i h(.,y^i)\right]^+ \\ (H_{x,w}(0)) \geqq 0, \quad \forall w \in X^0 \\ \sum_{i=1}^{\alpha} \lambda_i f(x,y^i) + \sum_{j=1}^{m} \mu_j g_j(x) \geqq 0, \\ \{y^1, y^2, ..., y^\alpha\} \subset Y(x) \\ \mu \geqq 0, \lambda_i \geqq 0, 1 \leqq i \leqq \alpha, \sum_{i=1}^{\alpha} \lambda_i = 1. \end{array}\right] \tag{3.1}$$

The Bector type dual to (FP) is given as follows

$$\textbf{(BD)} \qquad \underset{(\alpha,\lambda,y)\in\Gamma\ (x,\mu)\in K(\alpha,\lambda,y)}{\text{Maximize Supremum}} \frac{\sum_{i=1}^{\alpha} \lambda_i f(x,y^i) + \sum_{j=1}^{m} \mu_j g_j(x)}{\sum_{i=1}^{\alpha} \lambda_i h(x,y^i)}$$

If for a triplet $(\alpha, \lambda, y) \in \Gamma$, the set $K(\alpha, \lambda, y)$ is empty, we define the supremum over it to be $-\infty$.

The following theorem establishes the duality relationship between problems (FP) and (BD).

Theorem 3.1. *Let x^* be an optimal solution of the problem (FP) and let there exists $w^* \in X^0$ such that $g_j^+(H_{x^*,w^*}(0)) < 0, j \in I(x^*)$. Further, let $(\psi_j, H)_{j\in I(x^*)}$ be convexlike on X, and let there exist a neighborhood U of '0' in $\mathbb{R}^{Y(x^*)}$ and constants $\nu = (\nu_j)_{j\in I(x^*)}$ such that the set $\Omega(x^*, t) \cap \bar{U} \times \pi_{j\in I(x^*)}(-\infty, \nu_j]$ is nonempty closed set for every $t > 0$. Then there exist $(\alpha^*, \lambda^*, y^*) \in \Gamma$, and $\mu^* \in \mathbb{R}_+^m$ with $(x^*, \mu^*) \in K(\alpha^*, \lambda^*, y^*)$. Further if $f(.,y), -h(.,y)$ and $g_j(.), 1 \leqq j \leqq m$ are CN with respect to the same arc, for every $y \in Y$, then (x^*, μ^*) and $(\alpha^*, \lambda^*, y^*)$ give an optimal solution to (BD). Also, the two problems (FP) and (BD) have the same extremal values.*

Proof. Since x^* is an optimal solution of (FP) and all the conditions of Theorem 2.1 and the corresponding Corollary 2.1 are satisfied, hence, there exist an integer $\alpha^* > 0$, scalars $\lambda^0 \in \mathbb{R}_+^{\alpha^*}, \mu^0 \in \mathbb{R}_+^m$ and vectors $y^{*^i} \in Y(x^*)$ with

$1 \leqq i \leqq \alpha^*$, such that

$$\sum_{i=1}^{\alpha^*} \lambda_i^0 (f/h)^+(H_{x^*,w}(0), y^{*^i}) + \sum_{j=1}^{m} \mu_j^0 g_j^+(H_{x^*,w}(0)) \geqq 0, \forall w \in X \quad (3.2)$$

$$\sum_{j=1}^{m} \mu_j^0 g_j(x^*) = 0, \quad 1 \leqq j \leqq m \quad (3.3)$$

$$\sum_{i=1}^{\alpha^*} \lambda_i^0 \neq 0. \quad (3.4)$$

As $h(.,.)$ is continuous on $X^0 \times Y$, hence (3.2) can be written as

$$\begin{aligned}\sum_{i=1}^{\alpha^*} (\lambda_i^0 / h(x^*, y^{*^i}))(f^+(H_{x^*,w}(0), y^{*^i}) \\ -(f(x^*, y^{*^i})/h(x^*, y^{*^i}))h^+(H_{x^*,w}(0), y^{*^i})) \\ + \sum_{j=1}^{m} \mu_j^0 g_j^+(H_{x^*,w}(0), y^{*^i})) \geqq 0, \quad \forall w \in X^0\end{aligned}$$

Taking

$$\widehat{\lambda}_i = \lambda_i^0 / h(x^*, y^{*^i}), \lambda_i^* = \widehat{\lambda}_i / \sum_{i=1}^{\alpha^*} \widehat{\lambda}_i, \mu_j^* = \mu_j^0 / \sum_{i=1}^{\alpha^*} \widehat{\lambda}_i,$$

we obtain

$$\sum_{i=1}^{\alpha^*} \lambda_i^* (f^+(H_{x^*w}(0), y^{*^i}) - (f(x^*, y^{*^i})/h(x^*, y^{*^i}))h^+(H_{x^*w}(0), y^{*^i}))$$

$$+ \sum_{j=1}^{m} \mu_j^* g_j^+(H_{x^*,w}(0)) \geqq 0, \quad \forall w \in X^0 \quad (3.5)$$

$$\sum_{j=1}^{m} \mu_j^* g_j(x^*) = 0 \quad (3.6)$$

$$\lambda_i^* \geqq 0, 1 \leqq i \leqq \alpha^*, \sum_{i=1}^{\alpha^*} \lambda_i^* = 1, \mu^* \geqq 0. \quad (3.7)$$

Form (3.5) and (3.6) it follows that

$$\sum_{i=1}^{\alpha^*} \lambda_i^* f^+(H_{x^*,w}(0), y^{*^i}) + \sum_{j=1}^{m} \mu_j^* g_j^+(H_{x^*,w}(0))$$

$$- \left(\sum_{i=1}^{\alpha^*} \lambda_i^* f(x^*, y^{*^i}) + \sum_{j=1}^{m} \mu_j^* g_j(x^*) \right) \left(\sum_{i=1}^{\alpha^*} \lambda_i^* h(x^*, y^{*^i}) \right)^{-1}$$

$$\sum_{i=1}^{\alpha^*} \lambda_i^* h^+(H_{x^*,w}(0), y^{*^i}) \geqq 0, \quad \forall w \in X^0$$

which on simplification yields

$$\left(\sum_{i=1}^{\alpha^*} \lambda_i^* f(., y^{*^i}) + \sum_{j=1}^{m} \mu_j^* g_j(.) \right) \Big/ \sum_{i=1}^{\alpha^*} \lambda_i^* h(., y^{*^i})^+ \qquad (3.8)$$

$$(H_{x^*,w}(0)) \geqq 0, \forall w \in X^0$$

Also, in view of (3.6) and the fact that $\lambda_i^* \geqq 0, f(x^*, y^{*^i}) \geqq 0, 1 \leqq i \leqq \alpha^*$, we have

$$\sum_{i=1}^{\alpha^*} \lambda_i^* f(x^*, y^{*^i}) + \sum_{j=1}^{m} \mu_j^* g_j(x^*) \geqq 0 \qquad (3.9)$$

Now, from (3.6)-(3.9), it follows that $(x^*, \mu^*) \in K(\alpha^*, \lambda^*, y^*)$.
Let (x, μ) be any element of $K(\alpha^*, \lambda^*, y^*)$, we have

$$\left(\left(\sum_{i=1}^{\alpha^*} \lambda_i^* f(., y^{*^i}) + \sum_{j=1}^{m} \mu_j g_j(.) \right) \Big/ \sum_{i=1}^{\alpha^*} \lambda_i^* h(., y^{*^i}) \right)^+ \qquad (3.10)$$

$$(H_{x^*,w}(0)) \geqq 0$$

$$\sum_{i=1}^{\alpha^*} \lambda_i^* f(x, y^{*^i}) + \sum_{j=1}^{m} \mu_j^* g_j(x) \geqq 0 \qquad (3.11)$$

$$\mu^* \geqq 0, \lambda_i^* \geqq 0, 1 \leqq i \leqq \alpha^*, \sum_{i=1}^{\alpha^*} \lambda_i^* = 1$$

Moreover, since $x^* \in X^0$, we have, $g_j(x^*) \leqq 0, 1 \leqq j \leqq m$. Hence,

$$\sum_{j=1}^{m} \mu_j g(x^*) \leqq 0 = \sum_{j=1}^{m} \mu_j^* g(x^*) \qquad (3.12)$$

Since $f(.,y^{*^i}), -h(.,y^{*^i}), 1 \leqq i \leqq \alpha^*$, and $g_j(.), 1 \leqq j \leqq m$ are CN with respect to the same arc, we have

$$f(x^*,y^{*^i}) \geqq f^+(H_{x,x^*}(0),y^{*^i}) + f(x,y^{*^i}) \tag{3.13}$$

$$-h(x^*,y^{*^i}) \geqq -h^+(H_{x,x^*}(0),y^{*^i}) - h(x,y^{*^i}) \tag{3.14}$$

$$g_j(x^*) \geqq g_j^+(H_{x,x^*}(0)) + g_j(x), \quad 1 \leqq j \leqq m, \tag{3.15}$$

As $\lambda_i^* \geqq 0, 1 \leqq i \leqq \alpha^*$ and $\mu_j \geqq 0, 1 \leqq j \leqq m$, it follows from (3.13) and (3.15) that

$$\sum_{i=1}^{\alpha^*} \lambda_i^* f(x^*,y^{*^i}) + \sum_{j=1}^{m} \mu_j g_j(x^*) \geqq \sum_{i=1}^{\alpha^*} \lambda_i^* f^+(H_{x,x^*}(0),y^{*^i})$$
$$+\sum_{j=1}^{m} \mu_j g_j^+((H_{x,x^*}(0)) + \sum_{i=1}^{\alpha^*} \lambda_i^* f(x,y^{*^i}) + \sum_{j=1}^{m} \mu_j g_j(x)$$

which alongwith the fact that $\sum_{i=1}^{\alpha^*} \lambda_i^* h(x^*,y^{*^i}) > 0$ implies

$$\frac{\sum_{i=1}^{\alpha^*} \lambda_i^* f(x^*,y^{*^i}) + \sum_{j=1}^{m} \mu_j g_j(x^*)}{\sum_{i=1}^{\alpha^*} \lambda_i^* h(x^*,y^{*^i})} \geqq \frac{\sum_{i=1}^{\alpha^*} \lambda_i^* f^+(H_{x,x^*}(0),y^{*^i}) + \sum_{j=1}^{m} \mu_j g_j^+(H_{x,x^*}(0))}{\sum_{i=1}^{\alpha^*} \lambda_i^* h(x^*,y^{*^i})}$$
$$+\frac{\sum_{i=1}^{\alpha^*} \lambda_i^* f(x,y^{*^i}) + \sum_{j=1}^{m} \mu_j g_j(x)}{\sum_{i=1}^{\alpha^*} \lambda_i^* h(x^*,y^{*^i})}$$

Thus, we get,

$$\frac{\sum_{i=1}^{\alpha^*} \lambda_i^* f(x^*, y^{*^i}) + \sum_{j=1}^{m} \mu_j g_j(x^*)}{\sum_{i=1}^{\alpha^*} \lambda_i^* h(x^*, y^{*^i})} - \frac{\sum_{i=1}^{\alpha^*} \lambda_i^* f(x, y^{*^i}) + \sum_{j=1}^{m} \mu_j g_j(x)}{\sum_{i=1}^{\alpha^*} \lambda_i^* h(x, y^{*^i})}$$

$$\geqq \frac{\sum_{i=1}^{\alpha^*} \lambda_i^* f^+(H_{x,x^*}(0), y^{*^i}) + \sum_{j=1}^{m} \mu_j g_j^+(H_{x,x^*}(0))}{\sum_{i=1}^{\alpha^*} \lambda_i^* h(x^*, y^{*^i})} +$$

$$\left(\sum_{i=1}^{\alpha^*} \lambda_i^* f(x, y^{*^i}) + \sum_{j=1}^{m} \mu_j g_j(x)\right) \left[\frac{1}{\sum_{i=1}^{\alpha^*} \lambda_i^* h(x^*, y^{*^i})} - \frac{1}{\sum_{i=1}^{\alpha^*} \lambda_i^* h(x, y^{*^i})}\right]$$

which on using (3.11), (3.14) and the fact that $\lambda_i^* \geqq 0$, $1 \leqq i \leqq \alpha^*$ yields

$$\frac{\sum_{i=1}^{\alpha^*} \lambda_i^* f(x^*, y^{*^i}) + \sum_{j=1}^{m} \mu_j g_j(x^*)}{\sum_{i=1}^{\alpha^*} \lambda_i^* h(x^*, y^{*^i})} - \frac{\sum_{i=1}^{\alpha^*} \lambda_i^* f(x, y^{*^i}) + \sum_{j=1}^{m} \mu_j g_j(x)}{\sum_{i=1}^{\alpha^*} \lambda_i^* h(x, y^{*^i})} \tag{3.16}$$

$$\geqq \frac{\sum_{i=1}^{\alpha^*} \lambda_i^* f^+(H_{x,x^*}(0), y^{*^i}) + \sum_{j=1}^{m} \mu_j g_j^+(H_{x,x^*}(0))}{\sum_{i=1}^{\alpha^*} \lambda_i^* h(x^*, y^{*^i})}$$

$$+ \left[\frac{\sum_{i=1}^{\alpha^*} \lambda_i^* f(x, y^{*^i}) + \sum_{j=1}^{m} \mu_j g_j(x)}{\sum_{i=1}^{\alpha^*} \lambda_i^* h(x^*, y^{*^i})}\right] \frac{\sum_{i=1}^{\alpha^*} \lambda_i^* h^+(H_{x,x^*}(0), y^{*^i})}{\sum_{i=1}^{\alpha^*} \lambda_i^* h(x, y^{*^i})}$$

On simplifying the right hand side of (3.16), we get

$$\frac{\sum_{i=1}^{\alpha^*} \lambda_i^* f(x^*, y^{*^i}) + \sum_{j=1}^{m} \mu_j g_j(x^*)}{\sum_{i=1}^{\alpha^*} \lambda_i^* h(x^*, y^{*^i})} - \frac{\sum_{i=1}^{\alpha^*} \lambda_i^* f(x, y^{*^i}) + \sum_{j=1}^{m} \mu_j g_j(x)}{\sum_{i=1}^{\alpha^*} \lambda_i^* h(x, y^{*^i})}$$

$$\geqq \frac{\sum_{i=1}^{\alpha^*} \lambda_i^* h(x, y^{*^i})}{\sum_{i=1}^{\alpha^*} \lambda_i^* h(x^*, y^{*^i})} \left[\frac{\sum_{i=1}^{\alpha^*} \lambda_i^* f(., y^{*^i}) + \sum_{j=1}^{m} \mu_j g_j(.)}{\sum_{i=1}^{\alpha^*} \lambda_i^* h(., y^{*^i})}\right]^+ (H_{x,x^*}(0)) \tag{3.17}$$

(3.17) alongwith (3.10) and (3.12) gives

$$\frac{\sum_{i=1}^{\alpha^*} \lambda_i^* f(x^*, y^{*^i}) + \sum_{j=1}^{m} \mu_j g_j(x^*)}{\sum_{i=1}^{\alpha^*} \lambda_i^* h(x^*, y^{*^i})} - \frac{\sum_{i=1}^{\alpha^*} \lambda_i^* f(x, y^{*^i}) + \sum_{j=1}^{m} \mu_j g_j(x)}{\sum_{i=1}^{\alpha^*} \lambda_i^* h(x, y^{*^i})} \geqq 0$$

Hence, (x^*, μ^*) is an optimal solution for the following problem

$$(\mathrm{BD})_{\eta^*} \qquad \underset{(x,\mu)\in K(\alpha^*,\lambda^*,y^*)}{\text{Maximize}} \left[\frac{\sum_{i=1}^{\alpha^*} \lambda_i^* f(x, y^{*^i}) + \sum_{j=1}^{m} \mu_j g_j(x)}{\sum_{i=1}^{\alpha^*} \lambda_i^* h(x, y^{*^i})} \right]$$

where $\eta^* = (\alpha^*, \lambda^*, y^*)$.
To complete the proof we must show that for any $(\alpha, \lambda, y) \in \Gamma$,

$$\underset{(x,\mu)\in K(\alpha,\lambda,y)}{\text{Sup}} \left[\frac{\sum_{i=1}^{\alpha^*} \lambda_i^* f(x, y^i) + \sum_{j=1}^{m} \mu_j g_j(x)}{\sum_{i=1}^{\alpha^*} \lambda_i^* h(x, y^i)} \right] \leqq \frac{\sum_{i=1}^{\alpha^*} \lambda_i^* f(x^*, y^{*^i}) + \sum_{j=1}^{m} \mu_j^* g_j(x^*)}{\sum_{i=1}^{\alpha^*} \lambda_i^* h(x^*, y^{*^i})} \tag{3.18}$$

Let (x, μ) be any element of $K(\alpha, \lambda, y)$, where $(\alpha, \lambda, y) \in \Gamma$.
Now, $y^{*^i} \in Y(x^*), 1 \leqq i \leqq \alpha^*$, and $y^k \in Y(x) \subset Y$, we have

$$\frac{f(x^*, y^{*^i})}{h(x^*, y^{*^i})} \geqq \frac{f(x^*, y^k)}{h(x^*, y^k)}$$

which alongwith $\lambda_i^* \geqq 0, 1 \leqq i \leqq \alpha^*, \lambda_k \geqq 0, 1 \leqq k \leqq \alpha$, yields

$$\frac{\sum_{i=1}^{\alpha^*} \lambda_i^* f(x^*, y^{*^i})}{\sum_{i=1}^{\alpha^*} \lambda_i^* h(x^*, y^{*^i})} \geqq \frac{\sum_{i=1}^{\alpha} \lambda_i f(x^*, y^i)}{\sum_{i=1}^{\alpha} \lambda_i h(x^*, y^i)} \tag{3.19}$$

(3.19) together with (3.6) implies

$$\frac{\sum_{i=1}^{\alpha^*} \lambda_i^* f(x^*, y^{*^i}) + \sum_{j=1}^{m} \mu_j^* g_j(x^*)}{\sum_{i=1}^{\alpha^*} \lambda_i^* h(x^*, y^{*^i})} \geqq \frac{\sum_{i=1}^{\alpha} \lambda_i f(x^*, y^i)}{\sum_{i=1}^{\alpha} \lambda_i h(x^*, y^i)}$$

Thus,

$$\frac{\sum_{i=1}^{\alpha^*}\lambda_i^* f(x^*,y^{*^i})+\sum_{j=1}^{m}\mu_j^* g_j(x^*)}{\sum_{i=1}^{\alpha^*}\lambda_i^* h(x^*,y^{*^i})} - \frac{\sum_{i=1}^{\alpha}\lambda_i f(x^*,y^i)+\sum_{j=1}^{m}\mu_j g_j(x)}{\sum_{i=1}^{\alpha}\lambda_i h(x^*,y^i)}$$

$$\geqq \frac{\sum_{i=1}^{\alpha}\lambda_i f(x^*,y^i)}{\sum_{i=1}^{\alpha}\lambda_i h(x^*,y^i)} - \frac{\sum_{i=1}^{\alpha}\lambda_i f(x^*,y^i)+\sum_{j=1}^{m}\mu_j g_j(x)}{\sum_{i=1}^{\alpha}\lambda_i h(x^*,y^i)}$$

$$\geqq \frac{\sum_{i=1}^{\alpha}\lambda_i f(x^*,y^i)+\sum_{j=1}^{m}\mu_j g_j(x^*)}{\sum_{i=1}^{\alpha}\lambda_i h(x^*,y^i)} - \frac{\sum_{i=1}^{\alpha}\lambda_i f(x^*,y^i)+\sum_{j=1}^{m}\mu_j g_j(x)}{\sum_{i=1}^{\alpha}\lambda_i h(x^*,y^i)}$$

$$\geqq \frac{\sum_{i=1}^{\alpha}\lambda_i h(x,y^i)}{\sum_{i=1}^{\alpha}\lambda_i h(x^*,y^i)}\left[\frac{\sum_{i=1}^{\alpha}\lambda_i f(.,y^i)+\sum_{j=1}^{m}\mu_j g_j(.)}{\sum_{i=1}^{\alpha}\lambda_i h(.,y^i)}\right](H_{x,x^*}(0))$$

$$\geqq 0$$

where the last inequality follows from the fact that $(x, \mu) \in K(\alpha, \lambda, y)$, $\lambda_i \geqq 0$, $h(x^*, y^i) \geqq 0$, $h(x, y^i) \geqq 0, 1 \leqq i \leqq \alpha$.
Hence,

$$\mathrm{Sup}_{z\in Y}\frac{f(x^*,z)}{h(x^*,z)} = \frac{f(x^*,y^i)}{h(x^*,y^i)}, \qquad 1 \leqq i \leqq \alpha^*$$

$$= \frac{\sum_{i=1}^{\alpha^*}\lambda_i^* f(x^*,y^{*^i})}{\sum_{i=1}^{\alpha^*}\lambda_i^* h(x^*,y^{*^i})} = \frac{\sum_{i=1}^{\alpha^*}\lambda_i^* f(x^*,y^{*^i})+\sum_{j=1}^{m}\mu_j^* g_j(x^*)}{\sum_{i=1}^{\alpha^*}\lambda_i^* h(x^*,y^{*^i})}$$

Thus, optimal values of primal problem (FP) and its dual problem (BD) are the same. This complete the proof. □

4 Duality model-II

We now introduce Mond-Weir type dual to the fractional minmax problem (FP).
Let $L(\alpha, \lambda, y)$ denote the set of all triplets $(x, \mu, v) \in X \times \mathbb{R}^m \times \mathbb{R}$ satisfying

$$\sum_{i=1}^{\alpha}\lambda_i f^+(H_{x,w}(0),y^i) - vh^+(H_{x,w}(0),y^i)$$
$$+\sum_{j=1}^{m}\mu_j g_j^+(H_{x,w}(0)) \geqq 0, \quad \forall w \in X^0 \tag{4.1}$$

$$\sum_{i=1}^{\alpha}\lambda_i(f(x,y^i) - vh(x,y^i)) \geqq 0, \tag{4.2}$$

$$\sum_{j=1}^{m} \mu_j g_j(x) \geqq 0, \tag{4.3}$$

$$\{y^1, y^2, ..., y^\alpha\} \in Y(x) \tag{4.4}$$

$$\lambda_i \geqq 0, 1 \leqq i \leqq \alpha, \sum_{i=1}^{\alpha} \lambda_i = 1, \mu \geqq 0, v \geqq 0 \tag{4.5}$$

In this model, we present the following dual to the problem (FP).

(MD) $\quad$ $\text{Maximize}_{(\alpha,\lambda,y)\in\Gamma}\ \text{Supremum}_{(x,\mu)\in L(\alpha,\lambda,y)}\ v$

If for a triplet $(\alpha, \lambda, y) \in \Gamma$, the set $L(\alpha, \lambda, y)$ is empty, then we define the supremum over it to be $-\infty$. The following theorem establishes the duality relationship between problems (FP) and (MD).

Theorem 4.1. *Let x^* be an optimal solution of the problem (FP) and let there exists $w^* \in X^0$ such that $g_j^+(H_{x^*,w^*}(0)) < 0, j \in I(x^*)$. Further, let $(\psi_j, H)_{j\in I(x^*)}$ be convexlike on X, and let there exist a neighbourhood U of '0' in $\mathbb{R}^{Y(x^*)}$ and constants $\nu = (\nu_j)_{j\in I(x^*)}$ such that the set $\Omega(x^*, t) \cap \bar{U} \times \pi_{j\in I(x^*)}(-\infty, \nu_j]$ is nonempty closed set for every $t > 0$. Then there exist $(\alpha^*, \lambda^*, y^*) \in \Gamma, v^* \in \mathbb{R}$ and $\mu^* \in \mathbb{R}_+^m$ with $(x^*, \mu^*, v^*) \in L(\alpha^*, \lambda^*, y^*)$. Further if for each fixed $x \in X^0$ and $(\hat{x}, \mu, v) \in L(\alpha, \lambda, v)$ any one of the following conditions hold*

(a) $f(., y^i)$ and $-h(., y^i), 1 \leqq i \leqq \alpha, g_j(.), j \in I(x^*)$ are CN with respect to the same arc.

(b) $\sum_{i=1}^{\alpha} \lambda_i(f(., y^i) - vh(., y^i))$ is PCN, $\sum_{j=1}^{m} \mu_j g_j(.)$ is QCN with respect to the same arc.

(c) $\sum_{i=1}^{\alpha} \lambda_i(f(., y^i) - vh(., y^i))$ is QCN, $\sum_{j=1}^{m} \mu_j g_j(.)$ is STPCN with respect to the same arc.

(d) $\sum_{i=1}^{\alpha} \lambda_i(f(., y^i) - vh(., y^i))$ is QCN, $g_j(.), 1 \leqq j \leqq m, j \neq s$, is QCN, $g_s(.)$ is STPCN with $\mu_s > 0$, with respect to the same arc.
Then (x^*, μ^*, v^*) and $(\alpha^*, \lambda^*, y^*)$ give an optimal solution to (MD). Also, the two problem (FP) and (MD) have the same extremal values.

Acknowledgment. The authors wish to express their deep gratitude to Professor R.N. Kaul, (Retd.) Professor, Department of Mathematics, University of Delhi, Delhi for his inspiration throughout the preparation of this paper. Thanks are also due to the unknown referees for their useful suggestions.

References

1. M.AVRIEL, AND I. ZANG, Generalized arcwise connected functions and characterization of local-global minimum properties. J. Optim. Theory Appl. 32(4) (1980), 407-425.
2. C.R. BECTOR, S. CHANDRA AND M.K. BECTOR, Generalized fractional programming duality : A parametric approach, J. Optim. Theory Appl. 60(1988), 243-260.
3. C.R. BECTOR, S. CHANDRA AND I.HUSAIN, Sufficient optimality conditions and duality for a continuous-time minmax programming problem, Asia-Pacific J. Oper. Res 9(1992), 55-76.
4. D.BHATIA, AND A. MEHRA, Optimality and duality involving arcwise connected and generalized arcwise connected functions. J. Optim. Theory Appl. 100(1) (1999), 181-194.
5. S. CHANDRA AND V. KUMAR, Duality in fractional minimax programming, J. Austral. Math. Soc (Ser.A) 58 (1995), 376-386.
6. J .M. DANSKIN, The Theory of Max-Min and Its Applications to Weapons Allocation Problems, Springer Verlag, Berlin, 1967.
7. N. DATTA AND D. BHATIA, Duality for a class of nondifferentiable mathematical programming problem in complex space, J. Math. Anal. Appl. 101(1984), 1-11.
8. V. F. DEMYANOV AND V. N. MALOZEMOV, Introduction to Minmax, Wiley, New York, 1974.
9. V. JEYAKUMAR AND J. GWINNER, Inequality system and optimization, J. Math. Anal. Appl. 159(1991), 51-71.
10. A. MEHRA AND D. BHATIA, Optimality and duality for minmax problems involving arcwise connected and generalized arcwise connected functions, J. Math. Anal. Appl. 231 (1999), 425-445.
11. W. E. SCHMITENDORF, Necessary conditions and sufficient conditions for static minmax problems, J. Math. Anal. Appl. 57 (1977), 683-693.
12. S. TANIMOTO, Duality for a class of nondifferentiable mathematical programming problems, J. Math. Anal. Appl. 79 (1981), 286-294.

Generalized Convexity for Unbounded Sets: The Enlarged Space

Guillermo L. Hansen[1] and Jean-Claude Dupin[2]

[1] Universidad de Buenos Aires
[2] Université de Valenciennes

Abstract. Following an idea suggested by Rockafellar in his classical book on Convex Analysis, the first author defined in his doctoral thesis the enlarged space, by adjoining to the ordinary n-dimensional space "improper points" defined by directions, and developed a theory of convexity in such a space. Later the second author introduced a new model for the enlarged space, suitable for applying tools from the theory of cones, and closely related to the model that Rockafellar & Wets called "ray model" of the "cosmic closure" of the Euclidean space in their recent book on Variational Analysis. In this paper we introduce that model and use it to get some new theorems and proofs.

2000 Mathematics Subject Classification. Primary 52A01, Secondary 52A20

Keywords and phrases. Generalized Convexity, Unbounded Convex Sets, Cosmic Closure, Enlarged Space.

1 Introduction

We shall present here what we call the *enlarged space*, which is similar to the *cosmic closure* introduced by Rockafellar and Wets in [12] and [13]. Note that our construction was developed independently of that of Rockafellar and Wets, while the first author was a graduate student in the eighties, and many of his results can be found in his doctoral dissertation [7]. Later, in 1994, the first author presented the basic facts of his theory at the *Congrès International sur la Convexité* at Marne-la-Vallée (France) [4].

We begin by recalling some basic definitions and facts. For details and proofs see [5] or the first author's thesis [7]. We shall use freely some results about cones, and as a general reference about this subject we refer to [8] or [2]. If $x \in \mathbb{R}^n$ and $\varepsilon > 0$, we shall denote by $U[x;\varepsilon]$ and $B[x;\varepsilon]$ the open and closed balls, respectively, with center x and radius ε, and $\mathbb{S}^{n-1}$ the unit sphere in $\mathbb{R}^n$. We define, in $\mathbb{R}^n \backslash \{0\}$, the equivalence relation $x \approx y$ if and only if $y = \lambda x$ for some $\lambda > 0$, and call $\mathbb{D}^n = (\mathbb{R}^n \backslash \{0\}) / \approx$. We call *enl* $\mathbb{R}^n = \mathbb{R}^n \cup \mathbb{D}^n$. The *points* of *enl* $\mathbb{R}^n$ are *proper* if they belong to $\mathbb{R}^n$, otherwise they are *improper*. If $x \in \mathbb{D}^n$ then $-x$ is the *antipodal* point of x, and if $A \subset \mathbb{D}^n$ then $-A = \{x \in \mathbb{D}^n : -x \in A\}$. We define the *unit mapping* $u : \mathbb{D}^n \longrightarrow \mathbb{S}^{n-1}$ by calling $u(d)$ the intersection of the ray from the origin in the direction of d with the unit sphere $\mathbb{S}^{n-1}$, and the *projection mapping*

$\pi : \mathbb{R}^n \setminus \{0\} \longrightarrow \mathbb{D}^n$ by calling $\pi(x)$ the equivalence class of x modulo $\approx$. If $x \neq 0$ and $d = \pi(x)$ we shall denote Δ_x or Δ_d the closed ray from the origin that passes through x. In general we shall use Δ to denote any closed ray from the origin. If $A \subset \mathbb{R}^n$ is non-empty the *conic hull* of A is the union $\mathbf{C}(A)$ of all rays from the origin that meet A.

The set *enl* $\mathbb{R}^n$ is called by Rockafellar-Wets the *cosmic closure* of $\mathbb{R}^n$, denoted by *csm* $\mathbb{R}^n$, and our set $\mathbb{D}^n$ of improper points is called *horizon* of $\mathbb{R}^n$, denoted by *hzn* $\mathbb{R}^n$. For the sake of compatibility we shall adhere in the sequel to these terminology and notation, and moreover we shall call *ordinary points* what we formerly called proper points, and *horizon points* our former improper points, that is the elements of *hzn* $\mathbb{R}^n$.

We shall now define a topology in *csm* $\mathbb{R}^n$. For every $d \in hzn\ \mathbb{R}^n$, $\varepsilon > 0$ and $\rho > 0$ we call $D[d;\varepsilon] = \pi\{U[u(d)];\varepsilon\}$ and $N[d;\varepsilon;\rho] = D[d;\varepsilon] \cup \bigcup_{\lambda \geq \rho} \lambda \cdot U[u(d);\varepsilon]$. The sets $U[x;\varepsilon]$ and $D[d;\varepsilon;\rho]$, for $x \in \mathbb{R}^n$, $d \in hzn\ \mathbb{R}^n$, $\varepsilon > 0$ and $\rho > 0$ constitute a neighborhood base for a Hausdorff topology in *csm* $\mathbb{R}^n$. In our former papers we called *enlarged space* the set *enl* $\mathbb{R}^n$ endowed with this topology and the usual affine structure in $\mathbb{R}^n$, but from now on we shall use the term *cosmic closure* for it, and also the notation *csm* $\mathbb{R}^n$. With the topology so defined it is a compact first countable space. If $A \subset csm\ \mathbb{R}^n$ is any set, we shall denote *int* A, *fr* A and *csm* A the interior, the frontier and the closure of A, respectively, the last notation being introduced by Rockafellar-Wets. These authors use sequences instead of neighborhoods to define the topology.

A main difference is that we introduce convexity by means of a generalized definition of segment, and subsequently we develop a rather complete theory of convex sets in the cosmic closure, which is our main goal. If a, $b \in csm\ \mathbb{R}^n$ the *closed segment* $[a;b]$ with endpoints a, b is: (i) the usual closed segment if a and b are both ordinary points, (ii) the closure of the ray from a in the direction of b if a is ordinary and b is a horizon point, and (iii) $\pi\{\lambda u(a) + (1-\lambda)u(b) : 0 \leq \lambda \leq 1\}$ if both a and b are horizon points. As usual the *open segment* $]a;b[$ is obtained from the above by removing its endpoints, and the meaning of $[a;b[$ and $]a;b]$ is obvious. A subset $C \subset csm\ \mathbb{R}^n$ is *convex* if $[a;b] \subset C$ whenever a, $b \in C$. Note that a pair of antipodal horizon points constitute a convex set that is not connected, but these are the only non-connected convex sets. The family of convex subsets of *csm* $\mathbb{R}^n$ has the usual intersection property, which allows us to define the *convex hull conv* A of a set A in the usual way. Convex sets are defined by Rockafellar-Wets in a different although equivalent way, but no attempt of development of a complete convexity theory is made.

Lines in *csm* $\mathbb{R}^n$ are defined in a similar way: if a, b are different points in *csm* $\mathbb{R}^n$, the *line through a and b* is: (i) the closure of the usual line through a and b if both points are ordinary, (ii) the closure of the set $\{a + \lambda u(b) : \lambda \in \mathbb{R}\}$ if a is ordinary and b is a horizon point, and (iii) the set $\pi\{\alpha u(a) + \beta u(b) : \alpha, \beta \in \mathbb{R}$ *not both* $0\}$ if a and b are both horizon points. A set $A \subset csm\ \mathbb{R}^n$

with at least two points is a *flat* if the line through any two different points of A is contained in A. The empty set and the ordinary singletons are flats by definition, and the family of all flats in $csm\ \mathbb{R}^n$ has the usual intersection property, which allows us to define the *affine hull aff* A of any subset $A \subset csm\ \mathbb{R}^n$ in the usual way. We say that a set $A \subset csm\ \mathbb{R}^n$ is *affinely dependent* if there exist a subset B of A, $B \neq A$, and an element $a \in A \backslash B$ such that $a \in aff\ B$. Otherwise we say that A is affinely independent. We introduced a sort of *generalized algebra* in the cosmic closure, by means of which we were able to prove (see [5] or [7]):

Theorem 1. *(Linear accessibility) If* $A \subset csm\ \mathbb{R}^n$ *is convex,* $x \in int\, A$ *and* $y \in csm\, A$*, with at least one of them an ordinary point, then* $]x; y[\subset int\, A$.

Theorem 2. *(Radon) If* $A \subset csm\ \mathbb{R}^n$ *then* A *is affinely dependent if and only if there exists a partition* $A = A_1 \cup A_2$ *such that one of the following conditions holds:*

(a) $conv\, A_1 \cap conv\, A_2 \neq \emptyset$

(b) $-A_1 \cap conv\, A_2 \neq \emptyset$

Moreover, in the second case, A_1 *can be chosen to be a horizon singleton.*

Theorem 3. *(Carathéodory) If* $A \subset csm\ \mathbb{R}^n$ *then*

$$conv\, A = \bigcup_{C \in \mathcal{F}} conv\, C$$

where $\mathcal{F}$ *is the family of all affinely independent subsets of* A.

Theorem 4. *(Helly) Let* $\mathcal{F}$ *be a family of convex subsets of* $csm\ \mathbb{R}^n$ *such that either* $\mathcal{F}$ *is finite or all its members are closed. If every subfamily of* $\mathcal{F}$ *of at most* $n+1$ *members has a non-empty intersection and one of the following conditions*

(a) The intersection of any subfamily of $n+1$ *members of* $\mathcal{F}$ *has ordinary points*

(b) No member of $\mathcal{F}$ *has a pair of antipodal horizon points*

holds, then the intersection of all members of $\mathcal{F}$ *is non-empty.*

2 Models for the cosmic closure

Several models for the cosmic closure can be given (see [3], [5], [7] and [13]). Three of them will be used here:

- A very simple but easily viewed one, called the *celestial model* in [13], consists in shrinking down the space $\mathbb{R}^n$ to the interior of the n-dimensional unit ball by means of the mapping $x \longmapsto x/(1+|x|)$, and identifying $hzn\ \mathbb{R}^n$ with the boundary of that ball by means of the unit mapping u.

Two more elaborated ones, closely related between them, were introduced by Rockafellar-Wets and the second author. Both of them view the cosmic closure $csm\ \mathbb{R}^n$ as a subset of $\mathbb{R}^{n+1}$. To introduce these models we shall use the following notation: in $\mathbb{R}^n \times \mathbb{R}$ we shall call $\mathbf{H}_0$ the hyperplane $\mathbb{R}^n \times \{0\}$, $\mathbf{H}_{-1}$ the parallel hyperplane $\mathbb{R}^n \times \{-1\}$ and $\mathbf{H}_0^-$ the halfspace of $\mathbb{R}^n \times \mathbb{R}$ bounded by $\mathbf{H}_0$ that includes $\mathbf{H}_{-1}$.

- The Dupin's *conic model* of *csm* $\mathbb{R}^n$ consists in identifying each point $x \in \mathbb{R}^n$ with the point $(x, -1)$ in $\mathbf{H}_{-1}$ and each $d \in hzn\ \mathbb{R}^n$ with the point $(u(d), 0)$ in $\mathbb{S}^n \cap \mathbf{H}_0 = \mathbb{S}^{n-1} \times \{0\} \subset \mathbf{H}_0$.
- In the Rockafellar and Wets' *ray space model*, differing slightly from that, points of *csm* $\mathbb{R}^n$ are identified with rays from the origin in $\mathbb{R}^n \times \mathbb{R}$ in the following way: each $x \in \mathbb{R}^n$ is identified with the ray in $\mathbb{R}^n \times \mathbb{R}$, issuing from the origin and passing through the point $(x, -1)$ in $\mathbf{H}_{-1}$, and each $d \in hzn\ \mathbb{R}^n$ with the ray from the origin in $\mathbb{R}^n \times \mathbb{R}$ passing through the point $(u(d), 0)$ in $\mathbb{S}^{n-1} \times \{0\} \subset \mathbf{H}_0$.

If A is any subset of *csm* $\mathbb{R}^n$, viewed in Dupin's model as the union of an *ordinary part* $A' \subset \mathbf{H}_{-1}$ and a *horizon part* $A^\infty \subset \mathbb{S}^{n-1} \cap \mathbf{H}_0$, the rays corresponding to the elements of A will constitute the cone $\mathbf{C}(A)$ in $\mathbb{R}^{n+1}$, and it is clear from the definitions that A is convex in *csm* $\mathbb{R}^n$ if and only if $\mathbf{C}(A)$ is convex in $\mathbb{R}^{n+1}$. This simple fact allows us to use results from the classical theory of convex cones to get results about convex sets in *csm* $\mathbb{R}^n$, and we shall do that quite freely.

In the ray space model, the empty subset of *csm* $\mathbb{R}^n$ corresponds by definition to the trivial apex-only cone, and the whole space *csm* $\mathbb{R}^n$ to the set of rays of the halfspace $\mathbf{H}_0^-$. It is important to note that, in that model, points of *csm* $\mathbb{R}^n$ are identified with rays, so that subsets of *csm* $\mathbb{R}^n$ are collections of rays, rather than subsets of $\mathbb{R}^{n+1}$. Thus, for example, disjoint subsets of *csm* $\mathbb{R}^n$, regarded as subsets of $\mathbb{R}^{n+1}$ have the origin in common, so that they are not disjoint subsets of $\mathbb{R}^{n+1}$. Nevertheless, the possibility of interpreting such collections of rays as cones in $\mathbb{R}^{n+1}$ will we fruitful, as we shall see. For example it is very easy to relate sets in both models: if A is a set in ray space model then $A \cap [\mathbf{H}_{-1} \cup (\mathbb{S}^{n-1} \cap \mathbf{H}_0)]$ is the corresponding representation of A in the Dupin's model, and if A is a set in that last model then $\mathbf{C}(A)$ is its representation in the ray space model.

A line in *csm* $\mathbb{R}^n$ corresponds in the ray space model to the intersection of a plane through the origin in $\mathbb{R}^{n+1}$ with $\mathbf{H}_0^-$, and more generally a flat in *csm* $\mathbb{R}^n$ to the intersection of a subspace of $\mathbb{R}^{n+1}$ with $\mathbf{H}_0^-$. If $A \subset$ *csm* $\mathbb{R}^n$ then *aff* A is the intersection with $\mathbf{H}_0^-$ of the subspace of $\mathbb{R}^{n+1}$ generated by A. A hyperplane $\mathbf{H}$ in *csm* $\mathbb{R}^n$ is the intersection with $\mathbf{H}_0^-$ of an n-dimensional subspace $\mathbf{S}$ of $\mathbb{R}^{n+1}$ other than $\mathbf{H}_0$. That subspace determines two *halfspaces* $\mathbf{S}^+$ and $\mathbf{S}^-$ of $\mathbb{R}^{n+1}$. We shall denote $\mathbf{H}^+ = \mathbf{S}^+ \cap \mathbf{H}_0^-$, $\mathbf{H}^- = \mathbf{S}^- \cap \mathbf{H}_0^-$, and shall say that $\mathbf{H}$ *separates in csm* $\mathbb{R}^n$ a set A from a set B if $A \subset \mathbf{H}^+$ and $B \subset \mathbf{H}^-$.

As for the topology of *csm* $\mathbb{R}^n$, if $\Delta = \Delta_x$ is a ray corresponding to a point $x \in$ *csm* $\mathbb{R}^n$, with $0 \neq x \in \mathbf{H}_0^-$, a neighborhood of Δ is any subset of *csm* $\mathbb{R}^n$ that includes a cone $\mathbf{C}(U[x; \varepsilon]) \cap \mathbf{H}_0^-$ for some $\varepsilon > 0$.

All these notions can easily be translated to the Dupin's model.

3 Separation

Lemma 1. *Let C, D be non-empty complementary convex subsets of $\mathbb{R}^n$. Then C and D are convex cones.*

Proof. Let $x_0 \in C$, $y_0 \in D$. If $]x_0; y_0[$ includes points of both C and D, by convexity we can find $x_1 \in [x_0; y_0]$ such that $[x_0; x_1[\subset C$ and $]x_1; y_0] \subset D$. Otherwise either $]x_0; y_0] \subset D$ or $[x_0; y_0[\subset C$, and we call $x_1 = x_0$ or $x_1 = y_0$, respectively. In any case, by complementary convexity, one of the open halflines with endpoint x_1 of the line Γ_1 through x_0 and y_0 is included in C and the other is included in D. We may assume that $x_1 \in C$. There must be points from both C and D outside Γ_1; let $y_1 \in D$ be one of such points. Then either the open halfline with endpoint x_1 through y_1 is included in D or there exists $x_2 \in [x_1; y_1]$ such that $[x_1; x_2[\subset C$ and $]x_2; y_1] \subset D$. If the open halfline with endpoint x_1 through y_1 is included in D we call $x_2 = x_1$. In any case, again by complementary convexity, one of the open halflines with endpoint x_2 of the line Γ_2 through x_1 and y_1 is included in C and the other is included in D. Let P be the plane determined by x_0, y_0 and y_1. It is easy to see that, by convexity, one of the open halfplanes of P determined by Γ is included in C and the other is included in D, so that $P \cap C$ and $P \cap D$ are cones with apex x_2. We may, by symmetry, assume that $x_2 \in C$. If $n > 2$ there must be points from both C and D outside P. We choose $y_2 \in D$ and repeat the construction. After n steps we get a point x_n such that C and D are cones with apex x_n. □

In what follows we shall work quite freely with the interpretation of subsets of *csm* $\mathbb{R}^n$ as cones in $\mathbf{H}_0^-$.

Theorem 5. *(Kakutani-Stone) Let A, B be non-empty disjoint convex subsets of csm $\mathbb{R}^n$. Then there exist complementary convex sets C, $D \subset$ csm $\mathbb{R}^n$ such that $A \subset C$ and $B \subset D$ if and only if at least one of the sets A, B has no antipodal horizon points.*

Proof. Assume that there exist complementary convex sets C, D such that $A \subset C$, $B \subset D$ and that both A, B have a pair of antipodal horizon points: let them be p, $q \in A$ and r, $s \in B$. Let $t \in]p; r[$. Since C, D are complementary, t must belong to one of them. Assume that $t \in C$. Then $r \in C$, since C is convex and $q \in A \subset C$. This is a contradiction because $r \in B \subset D$ and C and D are complementary. The case $t \in D$ is similar.

Conversely, assume that B has not antipodal horizon points. Then, in the ray space model A and B are convex cones having the origin of $\mathbb{R}^{n+1}$ as an apex. Since B has not antipodal horizon points its apex 0 is an extreme point of B, and therefore $B \setminus \{0\}$ and A are disjoint convex subsets of $\mathbb{R}^{n+1}$. Then there exist complementary convex subsets P, Q of $\mathbb{R}^{n+1}$ such that $A \subset P$ and $B \setminus \{0\} \subset Q$. The sets P, Q are convex cones by the preceding lemma, and so are $P \cap \mathbf{H}_0^-$ and $Q \cap \mathbf{H}_0^-$. We shall show that 0 is a common apex. Let $x \in Q$.

If there exists a $\lambda > 0$ such that $\lambda x \notin Q$ then $\lambda x \in P$. Since P is a cone and $0 \in A \subset P$ then $x \in P$, which contradicts the complementariness of P and Q. Then Q is a cone with apex 0. Let now $x \in P$. If there exists a $\lambda > 0$ such that $\lambda x \in Q$ then also $x \in Q$ because Q is a cone with apex 0, again a contradiction. Let $C = P$, $D = Q \cup \{0\}$. Then C and D are complementary convex subsets of $csm\ \mathbb{R}^n$ such that $A \subset C$ and $B \subset D$. □

As said in the preceding section, we shall call A' and A^∞, respectively, the ordinary and horizon parts of a set $A \subset csm\ \mathbb{R}^n$, that is $A' = A \cap \mathbb{R}^n$, $A^\infty = A \cap hzn\ \mathbb{R}^n$.

Theorem 6. *(Separation by hyperplanes) Let A, B be two convex subsets of $csm\ \mathbb{R}^n$, each of them with ordinary points. If A' and B' can be separated by a hyperplane in the classical sense as subsets of $\mathbb{R}^n$, then A and B can be separated in $csm\ \mathbb{R}^n$ by a hyperplane.*

Proof. Let $\mathbf{H}$ be a hyperplane in $\mathbb{R}^n$ that separates A' from B'. More precisely assume that $A' \subset \mathbf{H}^+$, $B' \subset \mathbf{H}^-$. If $d \in A^\infty$ then for every $x \in A'$ the segment $[x; d] \subset A$, so that $[x; d[\subset A' \subset \mathbf{H}^+$. Therefore $d \in [x; d] = cl\ [x; d[\subset cl\ \mathbf{H}^+$, whence $A = A' \cup A^\infty \subset cl\ \mathbf{H}^+$. Similarly $B \subset cl\ \mathbf{H}^-$. □

Corollary 1. *Let A, B be two convex subsets of $csm\ \mathbb{R}^n$, each of them with ordinary points. If $0 \notin int\ (A' - B')$ then A and B can be separated in $csm\ \mathbb{R}^n$ by a hyperplane.*

Remark 1. The condition that the two sets have ordinary points in the above theorem cannot be relaxed. Consider the following example in $csm\ \mathbb{R}^3$: let $A = \pi\{(\xi, \nu, \zeta) : \xi^2 + \nu^2 = 1, \zeta = 0\}$, $B = cl\ \{(\xi, \nu, \zeta) : \xi = \nu = 0\}$. Then A and B are disjoint closed convex sets that cannot be separated by a hyperplane. Recall that the set of all horizon points is not a hyperplane.

Strict separation is more interesting in the cosmic closure. The closures (in $csm\ \mathbb{R}^n$) of the usual counterexamples to strict separation of disjoint closed convex sets in $\mathbb{R}^n$, e.g. the closures of the sets

$$A = \{(\xi, \upsilon) \in \mathbb{R}^2 : \xi > 0, \upsilon \geq 1/\xi\} \tag{1}$$

$$B = \{(\xi, \upsilon) \in \mathbb{R}^2 : \upsilon \leq 0\} \tag{2}$$

have a common horizon point. This suggests, to assure that two closed convex sets in $csm\ \mathbb{R}^n$ be strictly separated, to assume that the sets are disjoint, but this assumption leaves aside such simple and interesting cases as:

$$A = \{(\xi, \upsilon) \in \mathbb{R}^2 : \xi \geq 1\} \tag{3}$$

$$B = \{(\xi, \upsilon) \in \mathbb{R}^2 : \xi \leq -1\} \tag{4}$$

which can be strictly separated in $\mathbb{R}^2$ although their closures in $csm\ \mathbb{R}^2$ have points in common. The right assumption is related to the way in which the common horizon points appear in both sets.

Before stating the strict separation theorem, we need a previous lemma:

Lemma 2. *Let $A \subset csm\ \mathbb{R}^n$ be convex. Then A is closed in $csm\ \mathbb{R}^n$ if and only if $\mathbf{C}(A)$ is closed in $\mathbb{R}^{n+1}$.*

Proof. It suffices to show that if A is closed in $csm\ \mathbb{R}^n$ then $\mathbf{C}(A)$ is closed in $\mathbb{R}^{n+1}$, because the other implication is immediate. Let then $x \in cl\ \mathbf{C}(A)$, $x \neq 0$. Then $x \in \mathbf{H}_0^-$ because $\mathbf{C}(A) \subset \mathbf{H}_0^-$ and $\mathbf{H}_0^-$ is closed, and therefore $x \in \mathbf{H}_0$ or $x \in \mathbf{H}_0^- \backslash \mathbf{H}_0$. If $x \in \mathbf{H}_0$ then there exists $u \in \mathbb{S}^{n-1} \cap \mathbf{H}_0$ such that $u \in \Delta_x$. Let V be any neighborhood of u in $csm\ \mathbb{R}^n$. Then $V \supset \mathbf{C}(U[u;\varepsilon]) \cap [(\mathbb{S}^{n-1} \cap \mathbf{H}_0) \cup \mathbf{H}_{-1}]$ for some $\varepsilon > 0$. Since $\mathbf{C}(U[u;\varepsilon])$ is a neighborhood of $x \in cl\ \mathbf{C}(A)$ there exists $y \in \mathbf{C}(A) \cap \mathbf{C}(U[u;\varepsilon])$ and therefore there exists

$$u_y \in \mathbf{C}(A) \cap \mathbf{C}(U[u;\varepsilon]) \cap (\mathbb{S}^{n-1} \cap \mathbf{H}_0) \subset A \cap V$$

or there exists

$$v_y \in \mathbf{C}(A) \cap \mathbf{C}(U[u;\varepsilon]) \cap \mathbf{H}_{-1} \subset A \cap V$$

whence $u \in A$ because A is closed and therefore $x \in \mathbf{C}(A)$.

If $x \in \mathbf{H}_0^- \backslash \mathbf{H}_0$ let $v \in \Delta_x \cap \mathbf{H}_{-1}$ and V be any neighborhood of v in $\mathbb{R}^{n+1}$. Then $\mathbf{C}(V)$ is a neighborhood of x and therefore there exists $y \in \mathbf{C}(A) \cap \mathbf{C}(V)$ because $x \in cl\ \mathbf{C}(A)$, whence $\Delta_y \cap \mathbf{H}_{-1} \in A \cap V$. Since A is closed this implies $v \in A$ and therefore $x \in \mathbf{C}(A)$. □

Theorem 7. *(Strict separation) Let A, B be two closed convex sets in $csm\mathbb{R}^n$, each of them with ordinary points. If $A \cap B = A^\infty \cap B^\infty = -A^\infty \cap -B^\infty$ then there exists a hyperplane $\mathbf{H}$ such that $(A^\infty \cap -A^\infty) \cup (B^\infty \cap -B^\infty) \subset \mathbf{H}$, $A \backslash (A^\infty \cap -A^\infty) \subset \mathbf{H}^+ \backslash \mathbf{H}$ and $B \backslash (B^\infty \cap -B^\infty) \subset \mathbf{H}^- \backslash \mathbf{H}$.*

Proof. It suffices to apply Theorem 1 of [1] (see also Theorem 2.5 of [9]) to the cones $\mathbf{C}(A)$ and $\mathbf{C}(B)$ in $\mathbb{R}^{n+1}$:

$$\mathbf{C}(A) \cap \mathbf{C}(B) = \mathbf{C}(A \cap B) = \mathbf{C}(A^\infty \cap B^\infty) = \mathbf{C}(-A^\infty \cap B^\infty)$$

is a subspace of $\mathbb{R}^{n+1}$. By Klee-Bair-Gwinner theorem there exists then a linear form f such that:

$$f < 0 \quad in\ \mathbf{C}(A) \backslash [\mathbf{C}(A) \cap -\mathbf{C}(A)] \tag{5}$$

$$f = 0 \quad in\ [\mathbf{C}(A) \cap -\mathbf{C}(A)] \cup [\mathbf{C}(B) \cap -\mathbf{C}(B)] \tag{6}$$

$$f \geq 0 \quad in\ \mathbf{C}(B) \backslash [\mathbf{C}(B) \cap -\mathbf{C}(B)] \tag{7}$$

Inverting the roles of A and B we get a linear form g such that

$$g < 0 \quad in\ \mathbf{C}(B) \backslash [\mathbf{C}(B) \cap -\mathbf{C}(B)] \tag{8}$$

$$g = 0 \quad in\ [\mathbf{C}(A) \cap -\mathbf{C}(A)] \cup [\mathbf{C}(B) \cap -\mathbf{C}(B)] \tag{9}$$

$$g \geq 0 \quad in\ \mathbf{C}(A) \backslash [\mathbf{C}(A) \cap -\mathbf{C}(A)] \tag{10}$$

Then the linear form $f - g$ verifies

$$f - g < 0 \quad in\ \mathbf{C}(A) \backslash [\mathbf{C}(A) \cap -\mathbf{C}(A)] \tag{11}$$

$$f - g = 0 \quad in\ [\mathbf{C}(A) \cap -\mathbf{C}(A)] \cup [\mathbf{C}(B) \cap -\mathbf{C}(B)] \tag{12}$$

$$f - g > 0 \quad in\ \mathbf{C}(B) \backslash [\mathbf{C}(B) \cap -\mathbf{C}(B)] \tag{13}$$

which is equivalent to the thesis. The conditon that both A and B have ordinary points is necessary to assure that the hyperplane $\ker(f - g)$ in $\mathbb{R}^{n+1}$ does not coincide with $\mathbf{H}_0$. □

Remark 2. Again the condition that the two sets have ordinary points is necessary. Consider the following example in $csm\,\mathbb{R}^2$: $A = cl\ \{(\xi, v) \in \mathbb{R}^2 : \xi = 0\}$, and $B = hzn\ \mathbb{R}^2$.

Corollary 2. *Let A, B be two disjoint closed convex sets in csm $\mathbb{R}^n$, each of them with ordinary points. Then there exists a hyperplane $\mathbf{H}$ such that $A \subset \mathbf{H}^+ \backslash \mathbf{H}$ and $B \subset \mathbf{H}^- \backslash \mathbf{H}$.*

4 Extremal subsets - Minkowski's theorem

Let $E \subset A \subset csm\ \mathbb{R}^n$ be convex sets. Then E is an *extremal subset* of A if whenever x, $y \in A$ and $]x; y[\cap E \neq \emptyset$ it follows that x, $y \in E$. One point extremal subsets will be called *extreme points*. The set of all extreme points of a convex set A will be denoted $ext\ A$. In the following we shall also use the concepts of extremal ray and extreme point in the classical sense, and to avoid confusion we shall denote $ext_0 A$ and $rext_0 A$ the sets of (classical) extreme points and extremal rays of a subset A of $\mathbb{R}^n$ (or $\mathbb{R}^{n+1}$). It is easy to see that if A is a closed convex subset of $csm\ \mathbb{R}^n$ without antipodal horizon points, and we consider $csm\ \mathbb{R}^n$ as in Dupin's model, a ray $\Delta \in rext_0 \mathbf{C}(A)$ if and only if $\Delta \cap [\mathbf{H}_{-1} \cup (\mathbb{S}^{n-1} \cap \mathbf{H}_0)] \in ext\ A$ (see [6]). Recall that a cone $\mathbf{C} \subset \mathbb{R}^n$ is said *pointed* if it includes its apex, and a *salient cone* if it is a convex pointed cone which does not include lines through its apex (see [2], [8]).

Theorem 8. *(Minkowski) Let $A \subset csm\ \mathbb{R}^n$ be a closed convex set without antipodal horizon points. Then $A = conv\ ext\ A$.*

Proof. Since $ext\ A \subset A$ and A is convex, then $conv\ ext\ A \subset A$. For the reverse inclusion we shall use Dupin's model. By lemma 10 $\mathbf{C}(A)$ is a closed convex salient cone with apex 0 in $\mathbb{R}^{n+1}$, and its apex 0 is its unique extreme point ([8], 4.7 Rem. III). From Klee's theorem ([10] th. 2.5) it follows that

$$\mathbf{C}(A) = conv[ext_0 \mathbf{C}(A) \cup rext_0 \mathbf{C}(A)] = \\ conv[\{0\} \cup rext_0 \mathbf{C}(A)] = conv\ rext_0 \mathbf{C}(A)$$

because the origin belongs to every extremal ray of $\mathbf{C}(A)$. Then if $x \in \mathbf{C}(A)$ there exist extremal rays $\Delta_1, \cdots, \Delta_p$ of $\mathbf{C}(A)$ such that $x \in conv(\Delta_1 \cup \cdots \cup \Delta_p) = \Delta_1 + \cdots + \Delta_p$, whence, if $x \in \mathbf{C}(A) \cap [\mathbf{H}_{-1} \cup (\mathbb{S}^{n-1} \cap \mathbf{H}_0)]$

$$x \in (\Delta_1 + \cdots + \Delta_p) \cap [\mathbf{H}_{-1} \cup (\mathbb{S}^{n-1} \cap \mathbf{H}_0)] = conv\ (a_1, \cdots, a_p)$$

where $a_i = \Delta_i \cap [\mathbf{H}_{-1} \cup (\mathbb{S}^{n-1} \cap \mathbf{H}_0)]$ is an extreme point of A for every $i = 1, \cdots, p$. □

Acknowledgment. The first author was partially supported by Argentine Research Project 01/TX38.

References

1. Bair, J. et Gwinner, J.: *Sur la séparation vraie des cônes convexes*, Arkiv för Mat. **16** (1978) 207-212
2. Bourbaki, N.: *Espaces Vectoriels Topologiques*, Hermann, 1966
3. Dupin, J.-C.: *A propos de "l 'espace élargi " de G. Hansen*, private communication, Journées Internationales de Convexité, Liège-Valenciennes, 1995
4. Hansen, G.L.: *The enlarged affine space: a tool for the unified study of bounded and unbounded convex sets*, presented at the International Conference on Convexity, Marne-La-Vallée, France, Sept. 5 - Sept. 9, 1994
5. Hansen, G.L.: *Unbounded convex sets: the enlarged affine space*, Revista de Matemática y Física Teórica, Serie A, **32** (1998) 217-236
6. Hansen, G.L.: *Nouvelle représentation de l 'espace élargi*, Convexa 97-Jornadas de Convexidad UBA (1997) 1-4
7. Hansen, G.L.: *Conjuntos convexos no-acotados* (in Spanish), Tesis Doctoral, Universidad de Buenos Aires, 1990
8. Jongmans, F.: *Etude des cônes associés à un ensemble*, séminaire stencilé, Liège, 1983-1984
9. Klee, V.L.: *Separation properties of convex cones*, Proc. Amer. Math. Soc. **6** (1955) 313-316
10. Klee, V.L.: *Extremal structure of convex sets*, Arch. Math. **VIII** (1957) 234-240
11. Rockafellar, R.T.: *Convex analysis*, Princeton University Press, 1970
12. Rockafellar, R.T. and Wets, R.J-B.: *Cosmic Convergence*, in Optimization and Non-linear Analysis, edited by A. Ioffe, M. Marcus and S. Reich, Pitman Research Notes in Mathematics, 244, pp. 249-272, Longman Scientific and Technical, Harlow, UK, 1992
13. Rockafellar, R.T. and Wets, R.J-B.: *Variational analysis*, Springer Verlag, 1998

A Note on Minty Variational Inequalities and Generalized Monotonicity

Reinhard John

Wirtschaftstheorie II,
University of Bonn,
Adenauerallee 24-42,
D-53113 Bonn, Germany
rjohn@wiwi.uni-bonn.de.

Abstract. Some notions of generalized monotonicity for multi-valued mappings are characterized in terms of properties of the associated Minty variational inequalities. In particular, it is shown that the Minty variational inequality problem derived from a map F defined on a convex domain is solvable on any nonempty, compact, and convex subdomain if and only if F is properly quasimonotone.

2000 Mathematics Subject Classification. 47J20, 49J40, 90C26

Keywords and phrases. Variational inequalities, quasimonotonicity, pseudomonotonicity.

1 Introduction

Let X be a nonempty subset of $\mathbb{R}^l$ and let F be a multi-valued mapping from X into $\mathbb{R}^l$, i.e. F is a set-valued mapping with domain X and nonempty values $F(x) \subseteq \mathbb{R}^l$. Furthermore, let Y be a nonempty subset of X.

The *Minty variational inequality problem* $MVIP(F,Y)$ associated with F and Y is to find $\overline{x} \in Y$ such that

$$x^* \cdot (\overline{x} - x) \leq 0$$

for all $x \in Y$ and all $x^* \in F(x)$. $\overline{x}$ is then called a *solution* to $MVIP(F,Y)$ or a *Minty solution* with respect to F and Y.

In some recent contributions [3, 9, 10] this problem has been termed a "dual" variational inequality problem in order to indicate its close relationship to the classical ("primal") *Stampacchia variational inequality problem* $SVIP(F,Y)$ associated with F and Y which consists in finding $\overline{x} \in Y$ such that there is $\overline{x}^* \in F(\overline{x})$ with

$$\overline{x}^* \cdot (x - \overline{x}) \geq 0$$

for all $x \in Y$.

The terminology utilized here is the same as in [7] and is due to Giannessi [5].

The relevance of the Minty problem to applications was pointed out in [6, 7]. There, it has been shown for the case of a single-valued continuous map F defined on an open convex domain that its solutions can be considered as the subset of stable equilibria within the set of all equilibria (represented by the Stampacchia solutions) of a dynamical system associated with F.

It is well known that, in contrast to the Stampacchia problem, compactness and convexity of Y and continuity assumptions on F do not guarantee the existence of a solution to $MVIP(F, Y)$ (see e.g. [7], Example 2.1.) and that some generalized monotonicity conditions on F are needed.

However, as shown by Konnov [9], even if F is single-valued, continuous, and quasimonotone and is defined on a compact and convex set, a Minty solution may not exist. On the other hand, in the same contribution he proved that semistrict quasimonotonicity is sufficient for existence.

This result has been improved by Daniilidis and Hadjisavvas [3] who have shown that semistrict quasimonotonicity can be replaced by the weaker notion of proper quasimonotonicity which they had introduced in [2].

There, a multi-valued map F from $X \subseteq \mathbb{R}^l$ into $\mathbb{R}^l$ has been called *properly quasimonotone* if for any $x = \sum_{i=1}^n \lambda_i x_i$ with $x_i \in X$, $\lambda_i > 0$, and $\sum_{i=1}^n \lambda_i = 1$, there exists $j \in \{1, \dots, n\}$ such that for all $x_j^* \in F(x_j)$

$$x_j^* \cdot (x - x_j) \leq 0.$$

It is straightforward to check that proper quasimonotonicity of F implies that F is *quasimonotone*, i.e. for all $x, y \in X$ and all $x^* \in F(x), y^* \in F(y)$

$$x^* \cdot (y - x) > 0 \text{ implies } y^* \cdot (y - x) \geq 0.$$

The main purpose of this note is to show that no further improvement of the result by Daniilidis and Hadjisavvas is possible. More precisely, we shall prove that a multi-valued mapping from a convex set $X \subseteq \mathbb{R}^l$ into $\mathbb{R}^l$ is properly quasimonotone if and only if for any nonempty, convex, and compact subset Y of X there is a Minty solution with respect to F and Y.

While the proof of the "only if"-part in [3] is a straightforward application of Fan's Lemma, we would like to point out that only a more elementary lemma by Berge ([1], p. 172) is needed.

The next section contains the main result which also yields immediately a characterization of quasimonotone maps in terms of the solvability of Minty variational inequality problems on compact line segments as domains. Section 3 concludes with the simple observation that the notion of pseudomonotonicity can be *characterized* by the property that Stampacchia solutions are always Minty solutions.

2 Characterization of proper quasimonotonicity

Throughout this section, F denotes a multi-valued mapping from a nonempty convex set $X \subseteq \mathbb{R}^l$ into $\mathbb{R}^l$.

F induces a multi-valued mapping G_F from X into X by the definition

$$G_F(x) = \{y \in X \mid \forall x^* \in F(x) : x^* \cdot (y - x) \leq 0\}.$$

Observe that $G_F(x)$ is a closed and convex subset of X for every $x \in X$. Furthermore, proper quasimonotonicity of F and solutions to $MVIP(F,Y)$ with $Y \subseteq X$ can be easily expressed in terms of G_F.

Indeed, it is straightforward to see that F is properly quasimonotone if and only if G_F is a *KKM map* [4], i.e. for any $x_1, \ldots, x_n \in X$ the convex hull $co\{x_1, \ldots, x_n\}$ of $\{x_1, \ldots, x_n\}$ is contained in $\bigcup_{i=1}^n G_F(x_i)$. It is also obvious by definition that $\bar{x}$ is a solution to $MVIP(F,Y)$ if and only if

$$\bar{x} \in Y \cap \bigcap_{x \in Y} G_F(x) = \bigcap_{x \in Y} (G_F(x) \cap Y).$$

The crucial step for proving the main result is to show that G_F being a KKM map is equivalent to a finite intersection property of G_F. For this we employ the following lemma (cf. Berge [1], p.172).

Lemma 1. *Let $X_1, \cdots, X_n$ be convex and compact subsets of $\mathbb{R}^l$ such that their union is convex. If $\bigcap_{i \neq j} X_i \neq \emptyset$ for $j = 1, \ldots, n$, then $\bigcap_{i=1}^n X_i \neq \emptyset$.*

We emphasize that, compared to the well-known KKM-Lemma, its proof is elementary in the sense that it uses only the separation property of convex sets. This is of course due to the assumption that the sets X_i are convex. For the same reason the next statement claims an equivalence that does not hold for maps which are just closed-valued.

Proposition 1. *Let G be a multi-valued mapping from $X \subseteq \mathbb{R}^l$ into X with closed and convex values $G(x)$ for all $x \in X$. Then G is a KKM map if and only if it satisfies the following condition:*

For any $n \geq 1$ and any $x_1, \ldots, x_n \in X$ there exists $x \in co\{x_1, \ldots, x_n\}$ such that $x \in \bigcap_{i=1}^n G(x_i)$. (FIP)

Proof. Assume first that G is a KKM map. We prove (FIP) by induction on n.

For $n = 1$ the KKM-Property means $x \in G(x)$ which is equivalent to the validity of (FIP) for $n = 1$.

Assume that (FIP) holds for any $n - 1$ elements of X and consider $x_1, \ldots, x_n \in X$. If we define $X_i = G(x_i) \cap co\{x_1, \ldots, x_n\}$ for $i = 1, \cdots, n$, then all sets X_i are convex and compact. Moreover,

$$\bigcup_{i=1}^n X_i = co\{x_1, \cdots, x_n\} \cap \bigcup_{i=1}^n G(x_i).$$

By the KKM-property of G, $co\{x_1, \ldots, x_n\} \subseteq \bigcup_{i=1}^n G(x_i)$, hence, $\bigcup_{i=1}^n X_i = co\{x_1, \ldots, x_n\}$ is convex. The induction hypothesis implies

$$\bigcap_{i \neq j} X_i \neq \emptyset \text{ for } j = 1, \ldots, n.$$

Thus, it follows from the Lemma that $\bigcap_{i=1}^n X_i \neq \emptyset$, i.e. (FIP) holds for n.

We assume now that G satisfies (FIP) and prove again by induction that G is a KKM map.

For $n = 1$, (FIP) means that $x \in G(x)$. As already mentioned before, this is equivalent to the KKM-property for $n = 1$.

Assume that the KKM-property is satisfied for any $n-1$ elements of X and consider $x_1, \ldots, x_n \in X$. We have to prove that $co\{x_1, \ldots, x_n\} \subseteq \bigcup_{i=1}^n G(x_i)$.

Let $x = \sum_{i=1}^n \lambda_i x_i \in co\{x_1, \ldots, x_n\}$. If some $\lambda_j = 0$, then x is an element of the convex hull of $n-1$ points in X and, by the induction hypothesis, $x \in G(x_i)$ for some i.

Assume now that $\lambda_i > 0$ for $i = 1, \cdots, n$. Since G satisfies (FIP) there exists $\lambda'_1, \ldots, \lambda'_n$ such that $\sum_{i=1}^n \lambda'_i = 1, \lambda'_i \geq 0$ for $i = 1, \ldots, n$, and $x' = \sum_{i=1}^n \lambda'_i x_i \in \bigcap_{i=1}^n G(x_i)$.

If $x' = x$ then, obviously, $x \in \bigcup_{i=1}^n G(x_i)$. Otherwise define

$$\mu := max\{\frac{\lambda'_i}{\lambda_i} | i = 1, \ldots, n\}.$$

Since $\lambda_i > 0$ for all i, μ is a well-defined real number. Moreover, $\lambda'_i \leq \lambda_i$ for all i is excluded (this would imply $\lambda'_i = \lambda_i$ for all i and, consequently, $x' = x$). Thus, it follows that $\mu > 1$.

Setting $\nu = \frac{1}{\mu}$ we obtain $0 < \nu < 1$ and, by definition, $\nu\lambda'_i \leq \lambda_i$ for all i with $\nu\lambda'_j = \lambda_j$ for some j.

If we define $\lambda''_i := \frac{\lambda_i - \nu\lambda'_i}{1-\nu}$ for $i = 1, \ldots, n$, then $\lambda''_i \geq 0$ for all i and $\lambda''_j = 0$. In addition,

$$\sum_{i=1}^n \lambda''_i = \frac{\sum_{i=1}^n \lambda_i - \nu \sum_{i=1}^n \lambda'_i}{1-\nu} = 1.$$

By the induction hypothesis, $x'' = \sum_{i=1}^n \lambda''_i x_i \in G(x_k)$ for some $k \neq j$. Since $x' \in \bigcap_{i=1}^n G(x_i) \subseteq G(x_k)$ and $G(x_k)$ is convex, $\nu x' + (1-\nu)x'' \in G(x_k)$. From

$$\begin{aligned} \nu x' + (1-\nu)x'' &= \textstyle\sum_{i=1}^n (\nu\lambda'_i + (1-\nu)\lambda''_i)x_i \\ &= \textstyle\sum_{i=1}^n (\nu\lambda'_i + (1-\nu)\frac{\lambda_i - \nu\lambda'_i}{1-\nu})x_i = \sum_{i=1}^n \lambda_i x_i = x \end{aligned}$$

it follows that $x \in G(x_k) \subseteq \bigcup_{i=1}^n G(x_i)$. □

As a straightforward consequence, we obtain

Theorem 1. *Let F be a multi-valued mapping from a nonempty convex set $X \subseteq \mathbb{R}^l$ into $\mathbb{R}^l$. Then the following conditions are equivalent:*

(i) F is properly quasimonotone.
(ii) For any $n \geq 1$ and any $x_1, \ldots, x_n \in X$ there exist $x \in co\{x_1, \ldots, x_n\}$ such that for $i = 1, \ldots, n$ and every $x_i^* \in F(x_i)$

$$x_i^* \cdot (x - x_i) \leq 0.$$

(iii) For any nonempty, convex, and compact subset Y of X there is a solution to MVIP(F,Y).

Proof. Observe first that (ii) is a restatement of (FIP) for G_F. Thus, the Proposition implies that (i) and (ii) are equivalent.

In order to prove that (ii) implies (iii), let Y be a nonempty, convex, and compact subset of X. It has to be shown that $\bigcap_{x \in Y}(G_F(x) \cap Y) \neq \emptyset$.

Consider finitely many points $x_1, \ldots, x_n \in Y$. Since G_F satisfies (FIP) there exists $x \in co\{x_1, \ldots, x_n\}$ such that $x \in \bigcap_{i=1}^n G_F(x_i)$. Convexity of Y implies that $x \in Y$. Hence, $\bigcap_{i=1}^n (G_F(x_i) \cap Y) \neq \emptyset$.

Since Y is compact and the sets $G_F(x) \cap Y$ are closed in Y, it follows that $\bigcap_{x \in Y}(G_F(x) \cap Y) \neq \emptyset$.

Assume now that (iii) holds and let $x_1, \ldots, x_n \in X$. Since $co\{x_1, \ldots, x_n\}$ is nonempty, convex, and compact, there is a solution to

$$MVIP(F, co\{x_1, \ldots, x_n\}),$$

i.e. there exists $\overline{x} \in co\{x_1, \ldots, x_n\}$ such that $\overline{x} \in G_F(x)$ for every $x \in co\{x_1, \ldots, x_n\}$. Consequently, $\overline{x} \in \bigcap_{i=1}^n G_F(x_i)$. Thus, we have shown that G_F satisfies (FIP). □

From Theorem 1 we derive immediately the following characterization of quasimonotonicity.

Corollary 1. *Let F be a multi-valued mapping from a nonempty convex set $X \subseteq \mathbb{R}^l$ into $\mathbb{R}^l$. Then F is quasimonotone if and only if for any closed interval $[x, y] = co\{x, y\}$ in X there is a solution to $MVIP(F, [x, y])$.*

Proof. It is trivial and well known that F is quasimonotone if the restriction of F to any line segment in X is quasimonotone. As can be easily shown, quasimonotonicity is equivalent to proper quasimonotonicity on such domains. Since nonempty, convex, and compact subsets of line segments in X are closed intervals $[x, y]$, the claim is proved by Theorem 1. □

3 Characterization of pseudomonotonicity

This section concludes with some simple observations about the characterization of pseudomonotonicity in terms of the relationship between Minty and Stampacchia variational inequalities.

Recall that a multivalued mapping F from $X \subseteq \mathbb{R}^l$ into $\mathbb{R}^l$ is *pseudomonotone*, if for all $x, y \in X$ and all $x^* \in F(x), y^* \in F(y)$

$$x^* \cdot (y - x) \geq 0 \text{ implies } y^* \cdot (y - x) \geq 0.$$

For any $Y \subseteq X$, let $M(F,Y)$ (resp. $S(F,Y)$) denote the set of solutions to $MVIP(F,Y)$ (resp. $SVIP(F,Y)$). For the case of a single-valued and continuous map F defined on a convex domain X it has been shown in [6] that pseudomonotonicity of F can be characterized by any of the following properties (a) and (b):

(a) $S(F,Y) \subseteq M(F,Y)$ for every convex $Y \subseteq X$.

(b) $S(F,Y)$ is convex for every convex $Y \subseteq X$.

Unfortunately, the characterization by (b) cannot be extended to the case of a multi-valued mapping. This is shown by the following

Example 1. Let $X = [0,1] \subseteq \mathbb{R}$ and let F be defined by $F(x) = [-1,1] \subseteq \mathbb{R}$. As a constant mapping, F is certainly continuous. Furthermore, $0 \in F(x)$ for each $x \in X$ implies that $S(F,Y) = Y$ for any $Y \subseteq X$, i.e. (b) holds. However, F is not pseudomonotone since for $x = 0, y = 1, x^* = 1, y^* = -1$ we obtain $x^* \cdot (y - x) > 0$ and $y^* \cdot (y - x) < 0$.

On the other hand, condition (a) also characterizes pseudomonotonicity in the case of a multi-valued map. This is shown by the almost trivial next result.

Theorem 2. *Let F be multi-valued mapping from a convex set $X \subseteq \mathbb{R}^l$ into $\mathbb{R}^l$. Then the following conditions are equivalent:*

(i) F is pseudomonotone.
(ii) $S(F,Y) \subseteq M(F,Y)$ for every $Y \subseteq X$.
(iii) $S(F,Y) \subseteq M(F,Y)$ for every convex $Y \subseteq X$.
(iv) $S(F,[x,y]) \subseteq M(F,[x,y])$ for every $x,y \in X$.
(v) $S(F,\{x,y\}) \subseteq M(F,\{x,y\})$ for every $x,y \in X$.

Proof. The implication (i) $\Rightarrow$ (ii) is well known and follows immediately from the definitions: If $\overline{x} \in S(F,Y)$ then there is $\overline{x}^* \in F(\overline{x})$ such that $\overline{x}^* \cdot (x - \overline{x}) \geq 0$ for all $x \in Y$. Pseudomonotonicity of F implies $x^* \cdot (x - \overline{x}) \geq 0$ or, equivalently, $x^* \cdot (\overline{x} - x) \leq 0$ for all $x \in Y$ and all $x^* \in F(x)$, i.e. $\overline{x} \in M(F,Y)$.

The implications (ii) $\Rightarrow$ (iii) and (iii) $\Rightarrow$ (iv) are trivial. Assume now that (iv) holds and that, without loss of generality, $x \in S(F,\{x,y\})$. By definition, there exists $x^* \in F(x)$ such that $x^* \cdot (y - x) \geq 0$. This implies for any $z = \lambda y + (1-\lambda)x \in [x,y]$ that $x^* \cdot (z - x) = \lambda x^* \cdot (y - x) \geq 0$, i.e. $x \in S(F,[x,y])$. By (iv), $x \in M(F,[x,y])$. Hence, by definition, $x \in M(F,\{x,y\})$. Thus, (v) is satisfied.

Finally, assume that (v) holds and let $x,y \in X, x^* \in F(x), y^* \in F(y)$. If $x^* \cdot (y - x) \geq 0$ then, since trivially $x^*(x - x) \geq 0, x \in S(F,\{x,y\})$. It follows from (v) that $x \in M(F,\{x,y\})$ which implies by definition that $y^* \cdot (x - y) \leq 0$, or $y^* \cdot (y - x) \geq 0$. Hence, F is pseudomonotone, i.e. (i) holds. $\square$

Observe that for the case of a nonconvex domain X the proof of the implications (i) $\Rightarrow$ (ii) and (v) $\Rightarrow$ (i) remain still valid, implying the equivalence of (i),(ii) and (v). It is also obvious that the proof can be easily adapted to characterize strict pseudomonotonicity. In that case, $M(F,Y)$ has to be replaced by the set of solutions $\overline{x}$ to the *strict* Minty variational inequality problem, i.e. by all $\overline{x} \in Y$ such that $x^* \cdot (\overline{x} - x) < 0$ for each $x \in Y$ with $x \neq y$ and each $x^* \in F(x)$.

Finally, it should be noticed that both Theorems 1 and 2 require a property of $MVIP(F,Y)$ for *all* Y in their statements (iii) (resp. also (ii) in Theorem 2). This is essential since it is not difficult to give examples where $S(F,X) = M(F,X) \neq \emptyset$ although F is not even quasimonotone (see e.g. Example 2.1. in [8]).

Acknowledgment. I would like to thank Nicolas Hadjisavvas, Igor Konnov, and an anonymous referee for their helpful comments.

References

1. Berge, C. (1966): *Espaces topologiques, fonctions multivoques* (2e édition), Dunod, Paris.
2. Daniilidis, A. and N. Hadjisavvas (1999): "On the Subdifferentials of Quasiconvex and Pseudoconvex Functions and Cyclic Monotonicity", *J. Math. Anal. and Appl. 237*, 30-42.
3. Daniilidis, A.and N. Hadjisavvas (1999): "Characterization of Nonsmooth Semistrictly Quasiconvex and Strictly Quasiconvex Functions", *J. Opt. Theory and Appl. 102*, 525-536.
4. Dugundji, J. and A. Granas (1978): "KKM Maps and Variational Inequalities", *Annali della Scuola Normale Superiore di Pisa, Serie IV, 5*, 679-682.
5. Giannessi, F.(1998): "On Minty Variational Principle", in F. Giannessi, S. Komlósi, and T. Rapcsák (eds.): *New Trends in Mathematical Programming*, Kluwer, Dordrecht.
6. John, R.(1998): "Variational Inequalities and Pseudomonotone Functions: Some Characterizations", in J.P. Crouzeix, J.E. Martinez-Legaz, and M. Volle (eds.): *Generalized Convexity, Generalized Monotonicity*, Kluwer, Dordrecht.
7. Komlósi, S. (1999): "On the Stampacchia and Minty Variational Inequalities", in G. Giorgi and F.A. Rossi (eds.): *Generalized Convexity and Optimization for Economic and Financial Decisions*, Pitagora Editrice, Bologna.
8. Konnov, I.(1997): "A Class of Combined Iterative Methods for Solving Variational Inequalities", *J. Opt. Theory and Appl. 94*, 677-693.
9. Konnov, I.(1998): "On Quasimonotone Variational Inequalities", *J. Opt. Theory and Appl. 99*, 165-181.
10. Marcotte, P. and D.L. Zhu (1999): "A Cutting Plane for Quasimonotone Variational Inequalities", Université de Montréal.

On Vector Equilibrium and Vector Variational Inequality Problems

Igor V. Konnov

Department of Applied Mathematics,
Kazan State University,
Kazan, Russia.

Abstract. The scalar equilibrium problem has numerous applications in Mathematical Physics, Economics and Game Theory and includes optimization and variational inequality problems. The vector equilibrium problem (VEP) is a generalization of the scalar one and it is also extensively investigated. In this work we consider relationships between VEP and other general vector problems. First we investigate ways of transforming VEP into a vector variational inequality problem under a K-space setting and obtain the relationships between monotonicity type properties of the corresponding functions. Next, we present a new gap function for VEP which allows one to reduce VEP to a vector optimization problem with single-valued function. We also consider applications of these results to vector saddle point and inverse vector optimization problems.

2000 Mathematics Subject Classification. 47J20, 47H07, 90C29, 65K10.

Keywords and phrases. Vector equilibrium problem, vector variational inequality problem, inverse vector optimization problem, generalized monotonicity.

1 Introduction

Let X be a real linear topological space and let Y be a real linear topological space with a partial order $\geq$ induced by a pointed closed convex and solid cone C. Therefore, $y' \geq y''$ is equivalent to $y' - y'' \in C$, $y' > y''$ is equivalent to $y' - y'' \in \text{int}C$, and $y' \not> y''$ is equivalent to $y' - y'' \notin \text{int}C$. Let K be a nonempty convex subset of X and let $\Phi : K \times K \to Y$ be a bifunction, such that $\Phi(x, x) \geq 0$ for every $x \in K$. Then we can consider the *vector equilibrium problem* (in short, VEP) that is to find an element $x^* \in K$ such that

$$\Phi(x^*, y) \not< 0 \qquad \forall y \in K. \tag{1}$$

We denote by K^e the solution set of this problem. VEP (1) is a natural extension of the scalar equilibrium problem, which corresponds to the case where $Y = R$. The existence of solutions for VEP (1) was investigated by many authors; e.g., see [1] – [4] and references therein.

Together with VEP (1) it is natural to consider its dual formulation, which is to find an element $y^* \in K$ such that

$$\Phi(x, y^*) \not> 0 \quad \forall x \in K. \tag{2}$$

We denote by K_d^e the solution set of this problem. It is well known that problems (1) and (2) are closely related to each other. More precisely, they are equivalent under certain continuity and (generalized) monotonicity properties on Φ. In the scalar case, the equilibrium problems generalize the optimization, fixed point, complementarity, and variational inequality problems.We recall the definition of the variational inequality problem in the vector case.

Let $L(X, Y)$ be the space of linear continuous operators from X into Y and let $T : K \to 2^{L(X,Y)}$ be a mapping. The *vector variational inequality problem* (in short, VVIP) is the problem of finding an element $x^* \in K$ such that

$$\forall y \in K, \exists t^* \in T(x^*) : t^*(y - x^*) \not< 0. \tag{3}$$

We denote by K^v the solution set of this problem. VVIP of the form (3) was also extensively investigated by many authors; e.g., see [5], [6], [2], [3] and references therein.

By analogy with the dual form of VEP, one can define the dual form of VVIP. Namely, the *dual vector variational inequality problem* is the problem of finding an element $x^* \in K$ such that

$$\forall y \in K, \forall t \in T(y) : t(y - x^*) \not< 0. \tag{4}$$

We denote by K_d^v the solution set of this problem. Note that in the case where T is single-valued, setting

$$\Phi(x, y) = T(x)(y - x), \tag{5}$$

we obtain $K^v = K^e$ and $K_d^v = K_d^e$. In the multivalued vector case, however, there is no such simple means to reduce VVIP (3) to VEP (1).

It is clear that the variational inequality format is simpler than the equilibrium one and it is often more suitable for investigation and solution. In fact, VVIP is always linear in y, moreover, both problems (3) and (4) have clear geometrical sense. In this paper, we intend to establish relationships among several vector problems. In the next section, we give some definitions and preliminary results. In Section 3, we consider ways of transforming VEP into VVIP in K-spaces and investigate monotonicity type properties of the cost mapping of the corresponding VVIP. In Section 4, we present a new gap function for VEP which allows one to reduce VEP to a vector optimization problem with single-valued function. In Section 5, we reduce an inverse vector optimization problem to VEP and VVIP under a finite-dimensional Euclidean space setting.

2 Preliminaries

In this section, we give some definitions and recall properties of solutions for VEP and VVIP. First we consider VEP (1) and give definitions for monotonicity and continuity type properties.

Definition 1. Let $F : K \to Y$ be a mapping. F is said to be

(i) *convex*, if, for each pair of points $x \in K, y \in K$ and for all $\alpha \in [0,1]$, we have

$$F(\alpha x + (1-\alpha)y) \leq \alpha F(x) + (1-\alpha)F(y);$$

(ii) *u-hemicontinuous*, if, for each pair of points $x \in K, y \in K$ and for all $\alpha \in [0,1]$, the mapping $\alpha \mapsto F(\alpha x + (1-\alpha)y)$ is continuous at 0^+.

Definition 2. Let $\Phi : K \times K \to Y$ and $\Psi : K \times K \to Y$ be bifunctions. Φ is said to be

(i) *monotone with respect to* Ψ, if for each pair of points $x \in K, y \in K$, we have

$$\Phi(x,y) + \Psi(y,x) \leq 0;$$

(ii) *pseudomonotone with respect to* Ψ, if, for each pair of points $x \in K, y \in K$, we have

$$\Phi(x,y) > 0 \Longrightarrow \Psi(y,x) < 0;$$

(iii) *quasimonotone with respect to* Ψ, if, for each pair of points $x \in K, y \in K$, we have

$$\Phi(x,y) > 0 \Longrightarrow \Psi(y,x) \leq 0.$$

Φ is said to be *monotone* (respectively, *pseudomonotone, quasimonotone*), if it is monotone (respectively, pseudomonotone, quasimonotone) with respect to itself.

From the above definitions it follows that each monotone bifunction is pseudomonotone, and that each pseudomonotone bifunction is quasimonotone, but the reverse assertions are not true in general. Recall the known relationships between the problems (1) and (2), e.g., see [1].

Proposition 1. *Let K be a nonempty convex subset of X. Suppose that $\Phi : K \times K \to Y$ is a bifunction such that $\Phi(x,x) \geq 0$ for every $x \in K$. Then:*

(i) if $\Phi(\cdot,y)$ is u-hemicontinuous for every $y \in K$ and $\Phi(x,\cdot)$ is convex for every $x \in K$, then $K_d^e \subseteq K^e$;

(ii) if Φ is pseudomonotone, then $K^e \subseteq K_d^e$.

Now we turn to VVIP (3). Taking into account (5) we can give analogues of the corresponding properties from Definitions 1 and 2.

Definition 3. Let $T : K \to 2^{L(X,Y)}$ be a mapping. T is said to be *u-hemicontinuous*, if for each pair of points $x \in K, y \in K$ and for all $\alpha \in [0,1]$, the mapping $\alpha \mapsto T(x+\alpha z)z$ with $z = y - x$ is upper semicontinuous at 0^+.

Definition 4. Let $T : K \to 2^{L(X,Y)}$ and $S : K \to 2^{L(X,Y)}$ be mappings. T is said to be

(i) *monotone with respect to* S, if, for each pair of points $x \in K, y \in K$ and for all $t' \in T(x), s'' \in S(y)$, we have

$$(s'' - t')(y - x) \geq 0;$$

(ii) *pseudomonotone with respect to* S, if, for each pair of points $x \in K, y \in K$ and for all $t' \in T(x), s'' \in S(y)$, we have

$$t'(y-x) > 0 \Longrightarrow s''(y-x) > 0;$$

(iii) *quasimonotone with respect to* S, if, for each pair of points $x \in K, y \in K$ and for all $t' \in T(x), s'' \in S(y)$, we have

$$t'(y-x) > 0 \Longrightarrow s''(y-x) \geq 0.$$

T is said to be *monotone* (respectively, *pseudomonotone, quasimonotone*), if it is monotone (respectively, pseudomonotone, quasimonotone) with respect to itself.

It is clear that relationships among the above concepts of (generalized) monotonicity are the same as those for bifunctions. Moreover, in the case where T, S are single-valued, $\Phi(x,y) = T(x)(y-x)$ and $\Psi(x,y) = S(x)(y-x)$, Definitions 2 and 4 coincide.

Note that most of the above concepts of (generalized) monotonicity for bifunctions and mappings have been considered by many authors; e.g., see [1] – [6]. The following result can be viewed as a modification of Proposition 1.

Proposition 2. *[6] Let K be a nonempty convex subset of X and let $T : K \to 2^{L(X,Y)}$ be a mapping. Then:*
(i) if T is u-hemicontinuous, then $K^v_d \subseteq K^v$;
(ii) if T is pseudomonotone, then $K^v \subseteq K^v_d$.

Let $S : K \to L(X,Y)$ and $F : K \to Y$ be mappings. Then we can define the following *mixed vector variational inequality problem* (in short, MVVIP) that is to find an element $x^* \in K$ such that

$$S(x^*)(y-x^*) + F(y) - F(x^*) \not< 0 \qquad \forall y \in K. \tag{6}$$

We denote by K^m the solution set of this problem. It is clear that (6) coincides with (1) in the case where

$$\Phi(x,y) = S(x)(y-x) + F(y) - F(x) \tag{7}$$

and that (6) can be viewed as an intermediate problem between VEP (1) and VVIP (3). Evidently, MVVIP is an extension of the corresponding scalar problem; see [7]. In the single-valued case, it was considered in [8]. More general mixed vector problems were considered in [9]. Taking into account (7) and (2), it is natural to consider the dual formulation of MVVIP, which is to find an element $x^* \in K$ such that

$$S(y)(y-x^*) + F(y) - F(x^*) \not< 0 \quad \forall y \in K. \tag{8}$$

We denote by K^m_d the solution set of this problem. Now, using Propositions 1 and 2, we establish the same relationship between the solution sets of problems (6) and (8).

Proposition 3. *Let K be a nonempty convex subset of X and let $S : K \to L(X,Y)$ and $F : K \to Y$ be mappings. Then:*
(i) if S and $-F$ are u-hemicontinuous and F is convex, $K_d^m \subseteq K^m$;
(ii) if S is monotone, $K^m \subseteq K_d^m$.

Due to (7), the proof follows from Proposition 1.

3 Vector problems in K-spaces

In this section, we recall properties of K-spaces and obtain the relationship between vector optimization and vector variational inequality problems. This relationship allows us to reduce VEP to the usual VVIP.

Let Y be a real linear topological space with a partial order $\geq$ induced by a pointed closed convex and solid cone C. Given a set $Z \subseteq Y$, we denote by $U(Z)$ the set of its upper bounds, i.e.,

$$U(Z) = \{u \in Y \mid u \geq z \quad \forall z \in Z\}.$$

Then we can determine the supremum of Z, denoted by $\sup Z$, as follows:

$$v = \sup Z \iff v \in U(Z) \text{ and } \forall u \in U(Z), u \geq v;$$

see [10, p.24]. If each subset Z of Y with the property $U(Z) \neq \emptyset$ has a supremum, then Y is called a Kantorovich space (in short, K-space); e.g., see [10, p.137]. Denote by $l(X,Y)$ the space of linear operators from X into Y. Throughout this section, we suppose that K *is a nonempty convex subset in a real linear topological space X and that Y is a real topological K-space.*

Definition 5. Let $P : X \to Y$ be a mapping. P is said to be *sublinear* if
(i) for each pair of points $x, y \in X$, we have

$$P(x + y) \leq P(x) + P(y);$$

(ii) for all $x \in X$ and $\lambda > 0$, we have

$$P(\lambda x) = \lambda P(x).$$

Together with the usual mappings with the values in Y, we shall consider also mappings with the values in $Y \bigcup \{+\infty\}$, so that if $F : X \to Y \bigcup \{+\infty\}$, then we in general have

$$\text{dom}(F) = \{x \in X \mid F(x) < +\infty\} \subset X.$$

Let U be a convex subset in X and let $F : U \to Y \bigcup \{+\infty\}$ be a convex mapping such that $K \subseteq \text{intdom}(F)$. Then, one can define the subdifferential of F at each point $x \in K$:

$$\partial F(x) = \{A \in l(X,Y) \mid A(y - x) \leq F(y) - F(x) \quad \forall y \in X\};$$

and, also, the directional derivative

$$F'(x,h) = \inf_{\alpha>0} \frac{1}{\alpha}(F(x+\alpha h) - F(x))$$

for each $h \in X$, which is sublinear in h. Therefore, using the Hahn-Banach-Kantorovich theorem, it is not too hard to obtain the following results; e.g., see [10, Chapter II].

Proposition 4. *Let $F : U \to Y \bigcup \{+\infty\}$ be a convex mapping such that $K \subseteq \operatorname{intdom}(F)$. Take an arbitrary point $x \in K$. Then:*
(i) $\partial F(x)$ is nonempty;
(ii) $\partial F(x) = \{A \in l(X,Y) \mid A(y-x) \leq F'(x,y-x) \quad \forall y \in X\}$;
(iii) for each $h \in X$, we have

$$F'(x,h) = \sup_{A \in \partial F(x)} Ah,$$

and, for each $h \in X$, there exists an $A_h \in \partial F(x)$ such that

$$F'(x,h) = A_h h.$$

Recall that the *Pareto vector optimization problem* (in short, VOP) is to find an element $x^* \in K$ such that there is no element $y \in K$ with $F(y) < F(x^*)$, or briefly,

$$\min \longrightarrow \{F(y) \mid y \in K\}. \tag{9}$$

We denote by K° the solution set of this problem. The necessary and sufficient condition for optimality can be formulated as follows.

Proposition 5. *Let $F : U \to Y \bigcup \{+\infty\}$ be a convex mapping such that $K \subseteq \operatorname{intdom}(F)$. Fix $x^* \in K$. Then the following assertions are equivalent:*
(i) $x^* \in K^\circ$;
(ii)

$$F'(x^*, y-x^*) \not< 0 \quad \forall y \in K; \tag{10}$$

(iii)

$$\forall y \in K, \exists t^* \in \partial F(x^*) : t^*(y-x^*) \not< 0. \tag{11}$$

Proof. The implication (iii) $\Longrightarrow$ (i) follows from the definition of the subdifferential. Next, suppose that (10) holds, but there exists $y \in K$ such that

$$t^*(y-x^*) < 0 \quad \forall t^* \in \partial F(x^*).$$

However, this is a contradiction with Proposition 4 (iii). Hence, (ii) $\Longrightarrow$ (iii). Next, suppose that $x^* \in K^\circ$, but there exists $y \in K$ such that $F'(x^*, y-x^*) < 0$. Set $h = y - x^*$. Then, by definition,

$$F(x^* + \alpha h) - F(x^*) < 0 \quad \text{and} \quad x^* + \alpha h \in K$$

for $\alpha > 0$ small enough, a contradiction. Thus, (i) $\Longrightarrow$ (ii) and the proof is complete. $\square$

It is clear that (11) coincides with (3) in the case where $T = \partial F$. Thus, Proposition 5 shows that each convex VOP is equivalent to VVIP (3). Applying the above results to equilibrium problems, we can obtain relationships between VEP and VVIP.

Theorem 1. *Let* $\Phi : U \times U \to Y \bigcup \{+\infty\}$ *be a bifunction such that* $\Phi(x,x) = 0$ *for every* $x \in U$, $K \times K \subseteq \operatorname{intdom}(\Phi)$. *Suppose that* $\Phi(x,\cdot)$ *is convex for all* $x \in U$. *Let the mapping* $T : U \to 2^{l(X,Y)}$ *be defined by*

$$T(x) = \partial_y \Phi(x,y)|_{y=x} \,. \tag{12}$$

Then:

(i) $K^e = K^v$;

(ii) $K^e_d \subseteq K^v_d$;

(iii) if $K^e \subseteq K^e_d$ *and the mapping* T, *defined by (12), is u-hemicontinuous,* $K^e_d = K^v_d = K^v = K^e$.

Proof. Since each $x^* \in K^e$ is a solution to VOP (9) with $F(x) = \Phi(x^*, x)$, assertion (i) follows from Proposition 5. Next, let $x^* \in K^e_d$. Take any $x \in K$. Then $\Phi(x,x^*) \not> 0$ and

$$\Phi(x,x^*) = \Phi(x,x^*) - \Phi(x,x) \geq t(x^* - x)$$

for all $t \in T(x)$. It follows that

$$t(x^* - x) \not> 0 \quad \forall t \in T(x),$$

hence $x^* \in K^v_d$, i.e., assertion (ii) holds. In case (iii), using (ii), (i), and Proposition 2 (i), we now have

$$K^e_d \subseteq K^v_d \subseteq K^v = K^e.$$

But, by assumption, $K^e \subseteq K^e_d$ and assertion (iii) holds too. □

Note that, on account of Proposition 1, $K^e_d \subseteq K^e$ if Φ is pseudomonotone.

Thus, VEP (1) can be reduced to VVIP (3) with T being defined in (12). We now derive monotonicity type properties for this mapping. These results can be viewed as extensions of those in [11, Proposition 2.2] and [12, p. 108], respectively, for the vector case.

Theorem 2. *Let* $\Phi : U \times U \to Y \bigcup \{+\infty\}$ *and* $\Psi : U \times U \to Y \bigcup \{+\infty\}$ *be bifunctions such that* $\Phi(x,x) = 0$ *and* $\Psi(x,x) = 0$ *for every* $x \in U$ *and let* $K \times K \subseteq \operatorname{intdom}(\Phi)$ *and* $K \times K \subseteq \operatorname{intdom}(\Psi)$. *Suppose that* $\Phi(x,\cdot)$ *and* $\Psi(x,\cdot)$ *are convex for every* $x \in K$ *and that* $T : U \to 2^{l(X,Y)}$ *is defined by (12) and* $S : U \to 2^{l(X,Y)}$ *is defined by*

$$S(x) = \partial_y \Psi(x,y)|_{y=x} \,.$$

If Φ *is monotone (respectively, pseudomonotone, quasimonotone) with respect to* Ψ *on* K, *then* T *is monotone (respectively, pseudomonotone, quasimonotone) with respect to* S *on* K.

Proof. Fix $x, y \in K$. Then, by definition, we have

$$\Phi(x,y) - \Phi(x,x) \geq t'(y-x) \quad \forall t' \in T(x) \tag{13}$$

and

$$\Psi(y,x) - \Psi(y,y) \geq s''(x-y) \quad \forall s'' \in S(y). \tag{14}$$

If Φ is monotone with respect to Ψ on K, adding (13) and (14) yields

$$0 \geq \Phi(x,y) + \Psi(y,x) \geq (t' - s'')(y-x),$$

i.e., T is monotone with respect to S on K. Next, let Φ be pseudomonotone with respect to Ψ on K and let $t'(y-x) > 0$ for some $t' \in T(x)$. Then (13) gives $\Phi(x,y) > 0$, hence, by pseudomonotonicity with respect to Ψ, $\Psi(y,x) < 0$. Now, from (14) we have $s''(x-y) < 0$ for all $s'' \in S(y)$, i.e., T is pseudomonotone with respect to S on K. Analogously, we obtain quasimonotonicity with respect to S for T. □

Setting $\Psi \equiv \Phi$ in Theorem 2, we obviously obtain the monotonicity type properties for T in (12) if the same hold for Φ.

The monotonicity of T can also be obtained under convexity type assumptions as the following theorem states.

Theorem 3. *Let $\Phi : U \times U \to Y \bigcup \{+\infty\}$ be a bifunction such that $\Phi(x,x) = 0$ for every $x \in U$, $K \times K \subseteq \mathrm{intdom}(\Phi)$. Suppose that $\Phi(x,\cdot)$ and $-\Phi(\cdot,y)$ are convex for all $x, y \in K$. If T is defined by (12), then it is monotone on K.*

Proof. Fix $x, y \in K$. Take any $\alpha \in (0,1)$ and set $x_\alpha = \alpha x + (1-\alpha)y$. It is clear that $x_\alpha \in K$ since K is convex. Due to convexity of $-\Phi(\cdot, x_\alpha)$, we have

$$0 = \Phi(x_\alpha, x_\alpha) \geq \alpha\Phi(x, x_\alpha) + (1-\alpha)\Phi(y, x_\alpha)$$

or equivalently,

$$\alpha\left[\Phi(x,x_\alpha) - \Phi(x,x)\right] \leq (1-\alpha)\left[\Phi(y,y) - \Phi(y,x_\alpha)\right]. \tag{15}$$

On the other hand, by definition,

$$\Phi(x,x_\alpha) - \Phi(x,x) \geq t'(x_\alpha - x)$$

and

$$\Phi(y,x_\alpha) - \Phi(y,y) \geq t''(x_\alpha - y)$$

for all $t' \in T(x)$ and $t'' \in T(y)$. Using these inequalities together with (15) yields

$$\begin{aligned}\alpha t'(x_\alpha - x) &\leq \alpha\left[\Phi(x,x_\alpha) - \Phi(x,x)\right] \\ &\leq (1-\alpha)\left[\Phi(y,y) - \Phi(y,x_\alpha)\right] \\ &\leq (1-\alpha)t''(y - x_\alpha).\end{aligned}$$

Since $(x_\alpha - x) = (1-\alpha)(y-x)$ and $y - x_\alpha = \alpha(y-x)$, it follows that

$$\alpha(1-\alpha)t'(y-x) \leq \alpha(1-\alpha)t''(y-x),$$

or equivalently,

$$(t''-t')(y-x) \geq 0.$$

Hence, T is monotone on K. □

Theorem 3 enables us to easily obtain the monotonicity of cost mappings for many VVIP's associated to vector problems. For example, let us consider VOP (9) with F being convex. Setting $\Phi(x,y) = F(y) - F(x)$, we now have $K^e = K^\circ$, moreover, all the assumptions of Theorem 3 hold with $T = \partial F$. Therefore, Theorem 3 in particular yields the monotonicity of the subdifferential of each convex mapping. Now we consider MVVIP (6).

Theorem 4. *Let $S : U \to l(X,Y)$ be a monotone mapping and let $F : U \to Y \bigcup \{+\infty\}$ be a convex mapping, $K \subseteq \mathrm{intdom}(F)$. Then MVVIP (6) is equivalent to VVIP (3) where $T = S + \partial F$ is monotone.*

Proof. It is clear that MVVIP (6) is equivalent to VEP (1) with Φ being defined in (7). Φ is then monotone and $\Phi(x,\cdot)$ is convex for every $x \in U$. On account of Theorem 4 (i), MVVIP (6) is now equivalent to VVIP (3) with T being defined in (12). Moreover, due to Theorem 5 with $\Psi \equiv \Phi$, we see that T is monotone. Applying now the extension of the Moreau-Rockafellar theorem (see [10, Section II.3.3]), we obtain

$$T(x) = \partial_y \Phi(x,y)\big|_{y=x} = S(x) + \partial F(x),$$

and the result follows. □

4 Gap function for vector equilibrium problems

In this section, we establish a vector optimization formulation for VEP's. A gap function for single-valued VVIP's was introduced by Chen et al in [13]. Further, several gap functions for VEP's were introduced and studied in [14], [15]. These gap functions enable one to convert single-valued VVIP (or VEP) into a vector optimization problem for a set-valued mapping. Modifying the approach of [14] and [15], we now give a single-valued vector optimization formulation for VEP (1).

Throughout this section we suppose that K *is a nonempty convex subset in a real linear topological space* X, Y *is a real locally convex space with a partial order* $\geq$ *induced by a pointed closed convex and solid cone* C, *and that* $\Phi : K \times K \to Y$ *is a bifunction such that* $\Phi(x,x) = 0$ *for every* $x \in K$.

Since Y is a locally convex space, we can define the dual cone C^* to C. Since C is solid and pointed, we have $C^* \neq \{0\}$, besides, C^* has a weak* compact base, i.e., there exists $B \subseteq C^*$, B is convex and weak* compact, such that $0 \notin B$ and $C^* = \bigcup_{t \geq 0} tB$. We fix such a base and set, for $u \in Y$,

$$\sigma(u) = \max_{t \in B}(t, u) \quad \text{and} \quad \sigma^*(u) = \min_{t \in B}(t, u)$$

(see [16]). Then, for all $u \in Y$, we have

$$u < 0 \Leftrightarrow \sigma(u) < 0, u \leq 0 \Leftrightarrow \sigma(u) \leq 0;$$
$$u \not< 0 \Leftrightarrow \sigma(u) \geq 0, u \not\leq 0 \Leftrightarrow \sigma(u) > 0;$$

and

$$u > 0 \Leftrightarrow \sigma^*(u) > 0, u \geq 0 \Leftrightarrow \sigma^*(u) \geq 0;$$
$$u \not> 0 \Leftrightarrow \sigma^*(u) \leq 0, u \not\geq 0 \Leftrightarrow \sigma^*(u) < 0.$$

Besides, if $u', u'' \in Y$ and $u' \leq u''$, then $\sigma(u') \leq \sigma(u'')$ and $\sigma^*(u') \leq \sigma^*(u'')$. For instance, we prove this property for σ. By definition, there exists $t' \in B$ such that $(t', u') = \sigma(u')$. It follows that $(t', u' - u'') \leq 0$ and $\sigma(u') \leq (t', u'') \leq \sigma(u'')$, as desired.

Let us consider a bifunction $H : K \times K \to Y$ satisfying the following conditions:

(a) $H(x, x) = 0$ *for every* $x \in K$;
(b) $H(x, y) > 0$ *for all distinct* $x, y \in K$;
(c) *for all* $x, y', y'' \in K$ *and for all* $\alpha \in [0, 1]$, *we have*

$$H(x, \alpha y' + (1-\alpha)y'') \leq \alpha H(x, y') + (1-\alpha)H(x, y'') - \alpha(1 - \kappa\alpha^{\tau-1})H(y'', y')$$

where $\tau > 1$ *and* $\kappa \in (0, 1]$.

Together with VEP (1), we shall consider the following VEP associated to $\Phi + H$: Find $x^* \in K$ such that

$$\Phi(x^*, y) + H(x^*, y) \not< 0 \quad \forall y \in K. \tag{16}$$

We denote by K_H^e the solution set of this problem. For brevity, set

$$\Psi(x, y) = \Phi(x, y) + H(x, y).$$

We first give an equivalence result for VEP's (1) and (16). A similar result was obtained in [15, Proposition 3.1], but we give the proof for the convenience of the reader.

Proposition 6. *Suppose* $\Phi(x, \cdot)$ *is convex for each* $x \in K$ *and* H *satisfies (a) – (c). Then* $K^e = K_H^e$.

Proof. Suppose x^* solves (VEP)(1). If $x^* \notin K_H^e$, then there exists $y \in K$ such that

$$\Phi(x^*, y) + H(x^*, y) < 0.$$

It follows that $x^* \neq y$ and

$$\Phi(x^*, y) < -H(x^*, y) < 0,$$

a contradiction.

Next, suppose x^* solves (VEP)(16). For contradiction, suppose that $x^* \notin K^e$. Then there exists $y \in K$ such that $\Phi(x^*, y) < 0$. Set $y_\alpha = \alpha y + (1-\alpha)x^*$. For each $\alpha \in (0,1)$, we have

$$\begin{aligned}\Phi(x^*, y_\alpha) + H(x^*, y_\alpha) &\leq \alpha\Phi(x^*, y) + (1-\alpha)\Phi(x^*, x^*) + \alpha H(x^*, y) \\ &\quad +(1-\alpha)H(x^*, x^*) - \alpha(1 - \kappa\alpha^{\tau-1})H(x^*, y) \\ &= \alpha\left[\Phi(x^*, y) + \kappa\alpha^{\tau-1}H(x^*, y)\right].\end{aligned}$$

Hence, there exists $\alpha' \in (0,1)$ such that

$$\Phi(x^*, y) + \kappa\alpha^{\tau-1}H(x^*, y) < 0,$$

when $\alpha \in (0, \alpha')$. It follows that

$$\Phi(x^*, y_\alpha) + H(x^*, y_\alpha) \leq \alpha\left[\Phi(x^*, y) + \kappa\alpha^{\tau-1}H(x^*, y)\right] < 0,$$

a contradiction, since $y_\alpha \in K$. □

Fix $x \in K$ and consider the scalar function

$$f_x(y) = \sigma[\Psi(x, y)].$$

Proposition 7. *Let all the assumptions of Proposition 6 hold. Then $f_x : K \to R$ is strictly convex.*

Proof. Fix $y', y'' \in K$ and $\alpha \in (0,1)$. Then, by assumption,

$$\Psi(x, \alpha y' + (1-\alpha)y'') \leq \alpha\Psi(x, y') + (1-\alpha)\Psi(x, y'') - \alpha(1 - \kappa\alpha^{\tau-1})H(y'', y').$$

It follows that

$$\begin{aligned}f_x(\alpha y' + (1-\alpha)y'') &\leq \sigma\big[\alpha\Psi(x, y') + (1-\alpha)\Psi(x, y'') - \alpha(1 - \kappa\alpha^{\tau-1}) \\ &\qquad\qquad H(y'', y')\big] \\ &\leq \alpha\sigma[\Psi(x, y')] + (1-\alpha)\sigma[\Psi(x, y'')] - \\ &\qquad\qquad \alpha(1 - \kappa\alpha^{\tau-1})\sigma^*[H(y'', y')] \\ &= \alpha f_x(y') + (1-\alpha)f_x(y'') - \alpha(1 - \kappa\alpha^{\tau-1})\sigma^*[H(y'', y')]\end{aligned}$$

If $y' \neq y''$, then $H(y'', y') > 0$ due to (b), hence $\sigma^*[H(y'', y')] > 0$. Thus, f_x is strictly convex. □

In order to formulate our variational principle, we consider the scalar optimization problem

$$\min_{y \in K} \longrightarrow f_x(y). \tag{17}$$

Taking into account Proposition 7, we can define the single-valued map $y : K \to K$ as follows: $y(x)$ is a unique solution of problem (17), i.e.,

$$y(x) = \{y' \in K \mid f_x(y') \leq f_x(y) \quad \forall y \in K\}.$$

For brevity, set

$$\psi(x) = \Psi(x, y(x)).$$

Lemma 1. *Let all the assumptions of Proposition 6 hold. Then, for each $x \in K$, we have*

$$\psi(x) \leq 0. \tag{18}$$

Proof. By definition,

$$\sigma[\psi(x)] = \sigma[\Psi(x, y(x))] \leq \sigma[\Psi(x, x)] = 0.$$

Therefore, $\psi(x) \leq 0$, as desired. □

We now define the following *ideal vector optimization problem*:

$$\max_{x \in K} \longrightarrow_I \psi(x), \tag{19}$$

that is, to find all $x_0 \in K$ for which $\psi(x_0) \geq \psi(x)$ for all $x \in K$. We denote by K_ψ° the solution set of problem (19).

Theorem 5. *Suppose $\Phi(x, \cdot)$ is convex for each $x \in K$ and H satisfies (a) – (c). Then:*
(i) $x^* \in K_H^e \Longleftrightarrow \psi(x^*) = 0 \Longleftrightarrow x^* = y(x^*)$.
(ii) If $K_H^e \neq \emptyset$, $K_H^e = K_\psi^\circ$.

Proof. Suppose x^* solves VEP (16). Then, for every $y \in K$, we have $\Psi(x^*, y) \not< 0$ and $f_{x^*}(y) \geq 0$. On the other hand, from (18) it follows that $f_{x^*}(y(x^*)) \leq 0$. Therefore,

$$f_{x^*}(y(x^*)) = \sigma[\Psi(x^*, x^*)] = \sigma[\Psi(x^*, y(x^*))] = 0$$

and we see that $x^* = y(x^*)$ and, by definition, $\psi(x^*) = 0$. Conversely, let $\psi(x^*) = 0$. Assume for contradiction that $x^* \notin K_H^e$, then there exists $y \in K$ such that $\Psi(x^*, y) < 0$, hence that $f_{x^*}(y) < 0$ and $f_{x^*}(y(x^*)) < 0$. It follows that $\psi(x^*) = \Psi(x^*, y(x^*)) < 0$, a contradiction. Therefore, we must have $x^* \in K_H^e$ and assertion (i) is true.

Next, let $K_H^e \neq \emptyset$. Take any $x^* \in K_H^e$, then $\psi(x^*) = 0$ due to (i). From (18) we have, for each $x \in K$, $\psi(x) \leq \psi(x^*) = 0$, i.e., $x^* \in K_\psi^\circ$. Conversely, take any $x^* \in K_\psi^\circ$. From (18) it follows that $\psi(x^*) \leq 0$. Assume for contradiction that $\psi(x^*) \neq 0$. Since $K_H^e \neq \emptyset$, there is $x' \in K_H^e$, but $\psi(x') = 0$ due to (i). It follows that $\psi(x^*) \leq \psi(x')$, a contradiction. Thus, assertion (ii) is also true. □

Combining Theorem 5 and Proposition 6, we obtain the following immediately.

Theorem 6. *Suppose $\Phi(x, \cdot)$ is convex for each $x \in K$ and H satisfies (a) – (c). Then:*
(i) $x^* \in K^e \Longleftrightarrow \psi(x^*) = 0 \Longleftrightarrow x^* = y(x^*)$.
(ii) If $K^e \neq \emptyset$, $K^e = K_\psi^\circ$.

Thus, VEP (1) can be reduced to the ideal vector optimization problem (19) whose solution exists under the usual assumptions.

5 Inverse vector optimization and vector saddle point problems

In this section, we consider an inverse vector optimization problem under a Euclidean space setting and its relationships with other vector problems.

Let U and V be nonempty, convex and closed sets in Euclidean spaces R^n and R^s, respectively, and let $L : U \times V \to R^m$ be a bifunction. It is clear that the space $Y = R^m$ is an ordered space with a partial order $\geq$ induced by the pointed closed convex and solid cone $C = R^m_+$, moreover, R^m is then a K-space.

Definition 6. Let $L : U \times V \to R^m$ be a bifunction.

(i) A point $w^* = (u^*, v^*) \in U \times V$ is called a *saddle point* of L if

$$L(u^*, v) \not> L(u^*, v^*) \not> L(u, v^*) \quad \forall u \in U, \forall v \in V. \tag{20}$$

(ii) A point $w^* = (u^*, v^*) \in U \times V$ is called a *regular saddle point* of L, if

$$L(u^*, v) \not> L(u, v^*) \quad \forall u \in U, \forall v \in V. \tag{21}$$

It is clear that each regular saddle point is a saddle point, but the reverse assertion is not true in general, as the following example illustrates.

Example 1. Set $a = (2,5)^T, b = (3,3)^T$ and $c = (1,4)^T$. Then $a \not> b$ and $b \not> c$, but $a > c$. In other words, the relation $\not>$ is not transitive.

Nevertheless, there exists a broad class of bifunctions such that (20) and (21) are equivalent. Let $f : U \to R^m$ and $h : U \to R^s$ be functions. Then one can define the following VOP:

$$\min \longrightarrow \{f(u) \mid h(u) \leq 0, u \in U\} \tag{22}$$

and the corresponding vector Lagrange function

$$L(u, v) = f(u) + \langle v, h(u) \rangle e \tag{23}$$

where $v \in R^s_+$ and $e = (1, \ldots, 1)^T \in R^m$. Moreover, one can also define the following scalar Lagrange function

$$M(u, v, z) = \langle z, f(u) \rangle + \langle v, h(u) \rangle \tag{24}$$

where $z \in S^m_+ = \{z \in R^m_+ \mid \sum_{i=1}^{m} z_i = 1\}$, e.g., see [17, Chapter 4].

Definition 7. A point $w^* = (u^*, v^*) \in U \times V$ is called a *scalar saddle point* of the function M defined by (24) if there is $z^* \in S^m_+$ such that

$$M(u^*, v, z^*) \leq M(u^*, v^*, z^*) \leq M(u, v^*, z^*) \quad \forall u \in U, \forall v \in V. \tag{25}$$

We now give an equivalence result for saddle points and optimality conditions for VOP (22).

Proposition 8. *Let U be a nonempty convex and closed set in R^n and let functions L and M be defined by (23) and (24), respectively. Set $V = R^s_+$ and suppose that f and h are convex. Then:*

(i) the following assertions are equivalent:
(a) (u^*, v^*) is a regular saddle point of L,
(b) (u^*, v^*) is a saddle point of L,
(c) (u^*, v^*) is a scalar saddle point of M,
(d) it holds that

$$\langle v - v^*, h(u^*)\rangle \leq 0 \quad \forall v \in V \quad \text{and} \quad L(u^*, v^*) \not> L(u, v^*) \quad \forall u \in U; \tag{26}$$

(ii) (a) each of assertions (a) – (d) of (i) implies that u^* solves VOP (22),
(b) if h is affine and U is a polyhedral set, for each solution u^* of VOP (22) there is a point $v^* \in V$ such that all the assertions (a) –(d) of (i) are true.

Proof. In case (i), (a) $\Longrightarrow$(b) and (b) $\Longrightarrow$ (d) by definition. Next, (26) implies that

$$L(u^*, v) \leq L(u^*, v^*) \quad \text{and} \quad L(u^*, v^*) \not> L(u, v^*)$$

for all $u \in U, v \in V$, hence that $L(u^*, v) \not> L(u, v^*)$. Therefore, (d) $\Longrightarrow$(a). Obviously, the first relations in (25) and (26) are equivalent. The equivalence of the second relations in (25) and (26) follows from the well-known scalar optimality condition for VOP's; e.g., see [17, Theorem 2.2.2]. Thus, (c) $\Longleftrightarrow$ (d) and (i) is true.

In case (ii), assertion (a) is well known; e.g., see [17, Theorem 4.2.1]. Next, under the assumptions of (b), if u^* is a solution to VOP (22), then there is $z^* \in S^m_+$ such that it solves the following scalar optimization problem:

$$\min \longrightarrow \{\langle z^*, f(u)\rangle \mid h(u) \leq 0, u \in U\};$$

e.g., see [17, Theorem 2.2.2]. By using the standard optimality conditions of scalar optimization (e.g., see [18, Chapter 4, Theorem 3.2]), we now conclude that there exists $v^* \in V$ such that (25) holds. Due to (i) we see that (ii) is now also true. □

It is clear that the *regular saddle point problem* (RSDP for short) is a particular case of VEP. In fact, setting

$$K = U \times V, x = (u, v), y = (u', v'), \Phi(x, y) = L(u', v) - L(u, v');$$

we see that RSDP (21) becomes equivalent to VEP (1) with Φ being monotone. Moreover, $\Phi(x, \cdot)$ is now convex under the assumptions of Proposition 8, hence RSDP (21) is then equivalent to VVIP (3) with monotone cost mapping due to Theorems 1 and 2.

In [19,20], an approach to converting an inverse scalar linear programming problem into a saddle point problem was proposed. An extension of this approach to vector linear programming problems was suggested in [21]. We now consider a further extension of this approach for a class of vector nonlinear problems.

Fix $x \in R^n$ and consider the following VOP:

$$\min \longrightarrow \{F(x)u + S(u) \mid Au \geq b(x), u \in R^s_+\}, \tag{27}$$

where $F : R^n \to R^{m\times s}$ and $b : R^n \to R^l$ are continuous mappings, $S : R^s \to R^m$ is a convex mapping, A is an $l \times s$ matrix, and $n = s + l$. Thus, we have $x = (u, v)$ where $u \in R^s$ and $v \in R^l$. We denote by $U(x)$ the solution set of VOP (27). Then we can define the *inverse vector optimization problem* (IVOP for short) which is to find $x^* = (u^*, v^*) \in R^n_+$ such that u^* is a solution to VOP (27) associated to x^*, i.e., $u^* \in U(x^*)$. We denote by X^* the solution set of this problem.

Let us consider the vector Lagrange function of VOP (27):

$$\tilde{L}(u, v, x) = F(x)u + S(u) + \langle v, b(x) - Au\rangle e, \tag{28}$$

where $e = (1, \dots, 1)^T \in R^m$. It is clear that (28) is an analogue of (23).

Proposition 9. *If $x^* = (u^*, v^*)$ solves RSDP (21) where $U = R^s_+, V = R^l_+, L(u, v) = \tilde{L}(u, v, x^*)$, then $x^* \in X^*$.*

Proof. The result follows immediately from Proposition 8 (ii) (a) with $f(u) = F(x^*)u + S(u)$ and $h(u) = b(x^*) - Au$. □

Set

$$\Phi(x, y) = \tilde{L}(u', v, x) - \tilde{L}(u, v', x), K = R^n_+, x = (u, v), y = (u', v'). \tag{29}$$

Combining Proposition 9 and Theorem 1, we now obtain the following sufficient conditions of optimality for IVOP.

Theorem 7. *(i) The following assertions are equivalent:*

(a) $x^* = (u^*, v^*)$ solves VEP (1), (29),

(b) $x^* = (u^*, v^*)$ solves VVIP (3), (29), (12).

(ii) If x^* solves VEP (1), (29), or VVIP (3), (29), (12), then $x^* \in X^*$.

Let us now consider monotonicity properties of Φ in (29). By definition, we have

$$\Phi(x, y) + \Phi(y, x) = (F(x) - F(y))(u' - u) + \langle b(y) - b(x), v' - v\rangle e,$$

hence Φ is monotone if and only if the mapping $G : R^n \to R^{m \times n}$, defined by

$$G(x) = (F(x), -e\ b(x)^T), \tag{30}$$

is monotone. Moreover, since $\Phi(x, \cdot)$ is convex for each $x \in K$, from Theorem 2 it follows that if G in (30) is monotone, then so is T in (12). Thus, the IVOP associated to VOP (27) can be reduced either to VEP or to VVIP with monotone cost mappings.

Acknowledgment. The author is grateful to referees and Prof. N. Hadjisavvas for their valuable comments.

References

1. M. Bianchi, N. Hadjisavvas, and S. Schaible, Vector equilibrium problems with generalized monotone bifunctions, *J. Optim. Theory Appl.*, **92** (1997), 527–542.
2. W. Oettli and D. Schläger, Generalized vectorial equilibria and generalized monotonicity, in: *"Functional Analysis with Current Applications in Science, Technology and Industry"* (M. Brokate and A.H. Siddiqi, eds), Pitman Research Notes in Mathematical Series, No.377, Addison Wesley Longman Ltd., Essex (1998), 145–154.
3. N. Hadjisavvas and S. Schaible, Quasimonotonicity and pseudomonotonicity in variational inequalities and equilibrium problems, in: *"Generalized Convexity, Generalized Monotonicity"* (J.-P. Crouzeix, J.E. Martinez-Legaz and M. Volle, eds.), Kluwer Academic Publ., Dordrecht - Boston - London (1998), 257–275.
4. I.V. Konnov and J.C. Yao, Existence of solutions for generalized vector equilibrium problems, *J. Math. Anal. and Appl.*, **233** (1999), 328–335.
5. G. Y. Chen and B. D. Craven, A vector variational inequality and optimization over an efficient set, *Zeitschrift für Oper. Res.*, **3** (1990), 1–12.
6. I.V. Konnov and J.C. Yao, On the generalized vector variational inequality problem, *J. Math. Anal. and Appl.*, **206** (1997), 42–58.
7. F. E. Browder, On the unification of the calculus of variations and the theory of monotone nonlinear operators in Banach spaces, *Proc. Nat. Acad. Sci. USA*, **56** (1966), 419–425.
8. X. Q. Yang, Vector variational inequality and its duality, *Nonlinear Anal., Theory, Methods and Appl.*, **21** (1993), 869–877.
9. X.P. Ding, Generalized vector quasi-variational-like inequalities with discontinuous nonmonotone mappings, *Adv. Nonlin. Variat. Ineq.*, **1** (1998), 1–18.
10. G.P. Akilov and S.S. Kutateladze, *Ordered Vector Spaces*, Nauka, Novosibirsk (1978) (in Russian).
11. I.V. Konnov, *Methods for Solving Finite-Dimensional Variational Inequalities*, DAS, Kazan (1998) (in Russian).
12. S.P. Uryas'yev, *Adaptive Algorithms of Stochastic Optimization and Game Theory*, Nauka, Moscow (1990) (in Russian).
13. G.Y. Chen, C.J. Goh, and X.Q. Yang, On gap functions for vector variational inequalities, in: " *Vector Variational Inequality and Vector Equilibria. Mathematical Theories*" (F. Giannessi, ed.), Kluwer Academic Publ., Dordrecht - Boston - London (1999), 55–72.

14. Q.H. Ansari, I.V. Konnov, and J.C. Yao, Variational principles for vector equilibrium problems (submitted to JOTA).
15. Q.H. Ansari, I.V. Konnov, and J.C. Yao, Existence of solutions and variational principles for vector equilibrium problems, Preprint No. 99-1, Institute of Problems in Informatics, Kazan (1999).
16. W. Oettli, A remark on vector-valued equilibria and generalized monotonicity, *Acta Mathem. Vietnamica*, **22** (1997), 213–221.
17. V.V. Podinovskii and V.D. Nogin, *Pareto-Optimal Solutions of Multicriteria Problems*, Nauka, Moscow (1982) (in Russian).
18. A.G. Sukharev, A.V. Timokhov, and V.V. Fedorov, *A Course of Optimization Methods*, Nauka, Moscow (1986) (in Russian).
19. V.A. Bulavskii, Quasi-linear programming and vector optimization, *Doklady Akademii Nauk SSSR*, **257** (1981), 788–791 (in Russian).
20. E.A. Gol'shtein and N.V. Tret'yakov, *Augmented Lagrange Functions*, Nauka, Moscow (1989) (in Russian).
21. I.V. Konnov, Vector variational inequalities and vector optimization, in: "*Multiple Criteria and Game Problems under Uncertainty*" (V.A. Uglov and S.V. Sidorov, eds.), Russian Correspondence Institute of Textile and Light Industry Publishing Office, Moscow (1996), 44.

Stochastic Orders Generated by Generalized Convex Functions

Alfred Müller

Universität Karlsruhe,
Institut für Wirtschaftstheorie und OR,
Kaiserstr. 12, Geb. 20.21
76128 Karlsruhe, Germany
email: `mueller@wior.uni-karlsruhe.de`

Abstract. We give a survey of the use of concepts of generalized convex functions in the theory of stochastic order relations. We describe the use of (a, b)-concavity, antistarshapedness, and t-quasi-unimodality in the theory of individual decisions under risk. In the case of multivariate distributions, we show how supermodularity and directional convexity arise naturally in the context of dependence orderings. Moreover, we state some new results and some open problems connected with increasing quasiconcave ordering and linear convex ordering.

2000 Mathematics Subject Classification. Primary 60E15; Secondary 26A51, 26B25.

Keywords and phrases. stochastic orders, dependence, duality, generalized convexity, directional convexity, supermodularity.

1 Introduction

Stochastic orders are an important tool in probability theory and statistics, with applications in such diverse areas as economics, operations research and biology. For a comprehensive treatment of this subject including a variety of applications we refer to the books of Shaked and Shanthikumar (1994) or Szekli (1995). A survey of the literature up to 1992 is given in the bibliography of Mosler and Scarsini (1994), which contains several thousands of references.

Many of the stochastic orders, which are in common use, are defined as follows. Let $(S, \mathcal{A})$ be some measure space, and let $\mathfrak{F}$ be some class of measurable functions $f : S \to \mathbb{R}$. Then we say that for two random variables with values in $(S, \mathcal{A})$ the relation

$$X \leq_{\mathfrak{F}} Y \quad \text{holds, if } Ef(X) \leq Ef(Y) \ \text{ for all } f \in \mathfrak{F}, \tag{1}$$

holds whenever the expectation is well defined. Notice that we did not specify $(S, \mathcal{A})$, so that X, Y can be discrete or continuous univariate random variables, random vectors or even stochastic processes.

There are two main fields, where such orderings are important. In applied probability one often encounters the problem that realistic modelling leads to intractable stochastic models. This is the case e.g. in complex queueing models or Markovian decision processes. Therefore approximations and bounds for their characteristics are of practical interest. Mostly these characteristics are an integral of a function f with respect to some probability measure P. Introducing additional (more or less unrealistic) assumptions in the model (like independence, stationarity or specific parametric distributions) often leads to an approximation, for which the characteristic can be evaluated. However, this is now the integral of f with respect to some perturbed probability measure $\tilde{P}$. Of course it is an important question, if it can be shown that this approximation leads to a bound for our characteristic. Here stochastic order relations of the above mentioned type play a major role. In fact, this problem has been the motivation for a lot of work on stochastic order relations, initiated mainly by Stoyan (1983). Whitt (1986) and Müller (1997a) used the notion *integral stochastic order* for the type of stochastic order relations introduced in 1.

The other field, where they play a dominant role, is the theory of decisions under risk. This branch of economics deals with the problem how an agent decides, if she has to choose between different risky alternatives. A decision leads to a random amount of some goods. This means that the agent has to compare random variables. The most studied approach to this problem is the so called subjective expected utility theory. It states that the decision maker has a utility function u, such that she prefers the alternative with an random yield of Y to one that yields X, if and only if $Eu(X) \leq Eu(Y)$. In practice, however, it is impossible to determine somebodies utility function exactly. Therefore the following important question naturally arises. Is it possible to predict the choice of the agent, if we have only partial knowledge of her utility function, say that we only know that it belongs to some class $\mathfrak{F}$ of utility functions? According to the definition in 1 it is obvious that she will prefer Y to X, if $X \leq_{\mathfrak{F}} Y$. This shows, why these stochastic order relations are an important tool in the investigation of decisions under risk. In this field of research they can be found under the key word *stochastic dominance.*

In many aspects of the theory and the applications of these stochastic orders concepts of generalized convexity play an important role. It is the aim of this paper to give a survey of these concepts, and how and where they are used. In section 2 we describe a general theory of integral stochastic orders, which is nothing else than duality theory of convex cones in a pair of linear spaces endowed with a bilinear mapping. In section 3 we describe some examples for univariate probability measures, where generalized convex functions have been used. Comparison methods for multivariate distributions are described in section 4. Here concepts of generalized convexity appear naturally, when one deals with ordering of dependence structures. We close with a section, where we present some results on quasiconcave stochastic order and linear convex order.

2 General theory of integral stochastic orders

In this section we describe some general results for integral stochastic orders, which are based on duality theory. For more details we refer to Müller (1997a), where all the necessary proofs can be found. Some similar, but less general results are also given in Border (1991) and Castagnoli and Maccheroni (1998).

We start with the necessary notations. Let $(S, \mathcal{A})$ be an arbitrary measure space and let $b : S \to [1, \infty)$ be a measurable function, called *weight function.* We consider the linear space $\mathfrak{B}_b$ of measurable functions $f : S \to \mathbb{R}$, for which

$$||f||_b := \sup_{s\in S} \frac{|f(s)|}{b(s)} < \infty.$$

For a signed measure μ on $\mathcal{A}$ we denote the positive and negative variation by μ^+ resp. μ^-. As usual $|\mu| := \mu^+ + \mu^-$ is the total variation. Integrals will often be written as a bilinear form $\langle f, \mu\rangle := \int f \, d\mu := \int f \, d\mu^+ - \int f \, d\mu^-$. Notice that $\langle f, \mu\rangle$ exists and is finite if and only if $\langle |f|, |\mu|\rangle = \langle |f|, \mu^+\rangle + \langle |f|, \mu^-\rangle < \infty$.

The linear space of all signed measures μ on $\mathcal{A}$ with $\langle b, |\mu|\rangle < \infty$ is denoted by $\mathbb{M}_b$. We write $\mathbb{P}$ for the subset of all probability measures (p.m.) on $\mathcal{A}$, and $\mathbb{P}_b := \mathbb{P} \cap \mathbb{M}_b$ is the restriction of $\mathbb{M}_b$ to $\mathbb{P}$. Thus $\mathbb{P}_b$ is simply the set of all probability measures, for which $\int b \, dP$ is finite. $\mathbb{P}_b$ is nonvoid as it contains all p.m.'s with finite support.

It is clear that integration is a bilinear mapping on the linear spaces $\mathbb{M}_b$ and $\mathfrak{B}_b$, and in fact as a first result we state that they form a strict duality in the sense of Choquet (1969).

Lemma 1. *$\mathbb{M}_b$ and $\mathfrak{B}_b$ are in strict duality under the bilinear mapping*

$$\begin{array}{rcl} \langle\cdot,\cdot\rangle : \mathbb{M}_b \times \mathfrak{B}_b & \to & \mathbb{R} \\ \langle\mu, f\rangle & := & \int f \, d\mu \end{array} . \tag{2}$$

Now we will turn back to the theory of stochastic order relations. Let $\mathfrak{F} \subset \mathfrak{B}$ be a class of real-valued functions. Then we define on $\mathbb{P}_b$ the order relation $\leq_{\mathfrak{F}}$ by

$$P \leq_{\mathfrak{F}} Q \quad \text{if} \quad \int f \, dP \leq \int f \, dQ \quad \forall\, f \in \mathfrak{F}. \tag{3}$$

We denote $\leq_{\mathfrak{F}}$ as an *integral stochastic order* and $\mathfrak{F}$ is called *generator.*

Using signed measures, the right hand side of (3) can be rewritten as $\int f \, d(Q - P) \geq 0 \quad \forall f \in \mathfrak{F}$, which means that the signed measure $Q - P$ lies in $\mathfrak{F}^+$, the *dual cone* to $\mathfrak{F}$, which as usual is defined as $\mathfrak{F}^+ := \{\mu \in \mathbb{M}_b : \langle\mu, f\rangle \geq 0 \; \forall f \in \mathfrak{F}\}$.

There may be different classes of functions, which generate the same stochastic order. For checking $P \leq_{\mathfrak{F}} Q$ it is desirable to have "small" generators, whereas "large" generators are interesting for applications. We will now define the *maximal generator* of an integral stochastic order relation.

Definition 1. Let b be a weight function and let $\mathfrak{F} \subset \mathfrak{B}_b$ be an arbitrary generator of an order $\leq_{\mathfrak{F}}$ on $\mathbb{P}_b$.
The set $\mathfrak{R}_{\mathfrak{F}} := \{f \in \mathfrak{B}_b : P, Q \in \mathbb{P}_b,\ P \leq_{\mathfrak{F}} Q \ \Rightarrow \ \int f\, dP \leq \int f\, dQ\}$ is called *maximal generator* of $\leq_{\mathfrak{F}}$ (in $\mathfrak{B}_b$).

Using the bipolar Theorem for convex cones, we can characterize the maximal generator as follows. If we endow $\mathfrak{B}_b$ with the weak topology, which is induced by the duality described in Lemma 1, then we get the following result.

Theorem 1. *$\mathfrak{R}_{\mathfrak{F}}$ is the closure of the convex cone, which is spanned by $\mathfrak{F}$ and the constant functions.*

Theorem 1 is rather of theoretical nature, but we easily get the following sufficient condition for $\mathfrak{F} = \mathfrak{R}_{\mathfrak{F}}$, which sometimes is very easy to check.

Corollary 1. *If $\mathfrak{F} \subset \mathfrak{V} \subset \mathfrak{R}_{\mathfrak{F}}$, and $\mathfrak{V}$ is a convex cone containing the constant functions and closed under pointwise convergence, then $\mathfrak{V} = \mathfrak{R}_{\mathfrak{F}}$.*

There are two main reasons, why the maximal generator is important. It not only shows which integrals can be compared by a given stochastic order, it also is important to show interesting properties of the order relations, which we will consider now. Most of following properties have been defined by Stoyan (1983). Some of them can also be found in Marshall (1991).

Definition 2. Let S be an ordered metric vector space and $b : S \to [1, \infty)$ a weight function. Let $\leq_{\mathfrak{F}}$ be some (pre-)order on $\mathbb{P}_b$. Then $\leq_{\mathfrak{F}}$ has
a) Property **(R)**, if $a \leq b$ implies $\delta_a \leq_{\mathfrak{F}} \delta_b$;
b) Property **(M)**, if $X \leq_{\mathfrak{F}} Y$ implies $aX \leq_{\mathfrak{F}} aY$ for all $a \geq 0$;
c) Property **(T)**, if $X \leq_{\mathfrak{F}} X + a$ holds for all r.v. X and all positive a;
d) Property **(C)**, if $P_1 \leq_{\mathfrak{F}} P_2$ implies $P_1 * Q \leq_{\mathfrak{F}} P_2 * Q$ for all p.m. Q;
e) Property **(MI)**, if $P_\theta \leq_{\mathfrak{F}} Q_\theta$, $\theta \in \Theta$, implies $P \leq_{\mathfrak{F}} Q$, where $P = \int P_\theta \mu(d\theta)$ and $Q = \int Q_\theta\ \mu(d\theta)$ for some measure μ on Θ, i.e. P, Q are the μ-mixtures of the families $P_\theta, Q_\theta,\ \theta \in \Theta$.
f) Property **(W)**, if $\leq_{\mathfrak{F}}$ is closed with respect to weak convergence, i.e. if $P_n \leq_{\mathfrak{F}} Q_n$ holds for all $n \in \mathbb{N}$ and the sequences $(P_n), (Q_n)$ converge weakly to P resp. Q, then $P \leq_{\mathfrak{F}} Q$.

It can be shown that all these properties can be traced back to properties of the generators $\mathfrak{F}$ respectively $\mathfrak{R}_{\mathfrak{F}}$. As usual, we denote the set of all bounded continuous functions by $\mathfrak{C}_b$.

Theorem 2.
a) Property (R) holds, if and only if all functions in $\mathfrak{F}$ are increasing.
b) Property (M) holds, if and only if $\mathfrak{R}_{\mathfrak{F}}$ is scale invariant, i.e. $f \in \mathfrak{R}_{\mathfrak{F}}$ and

$a > 0$ implies $f_a \in \mathfrak{R}_{\mathfrak{F}}$, where $f_a(x) := f(ax)$.
c) Property (T) holds, if and only if all functions in $\mathfrak{F}$ are increasing.
d) Property (C) holds, if and only if $\mathfrak{R}_{\mathfrak{F}}$ is invariant under translations.
e) Property (MI) holds always!
f) Property (W) holds, if and only if there is a generator $\mathfrak{B} \subset \mathfrak{C}_b$ for $\leq_{\mathfrak{F}}$.

3 Univariate examples

The best known univariate stochastic order relation is the *usual stochastic order* $\leq_{st}$ which is generated by the class of all bounded increasing functions $f : \mathbb{R} \to \mathbb{R}$. In economics this is usually called *first order stochastic dominance* (FSD). This order relation compares the size of the random variables. As a weight function we can simply use $b \equiv 1$. Then it follows immediately from Corollary 1 that the set of all bounded increasing functions is the maximal generator. It is easy to see that a small generator is given by the indicator functions $1_{(t,\infty)}$, $t \in \mathbb{R}$. This implies that for checking $\leq_{st}$ it is sufficient to check, if the distribution functions can be compared pointwise. It is also easy to see that $\leq_{st}$ has all the desirable properties described in Definition 2.

Another important example is *second order stochastic dominance* (SSD). Here the maximal generator consists of all increasing concave functions, and therefore this order relation is also called *increasing concave ordering*, written as $\leq_{icv}$. This is important for comparing the variability resp. riskiness of distributions. It is the main tool to describe risk aversion of decision makers. Here we need a non-trivial weight function, e.g. $b(x) = 1 + |x|$. A small generator is given by the functions $f_a(x) = \min\{x - a, 0\}$, $a \in \mathbb{R}$. Hence we have

$$X \leq_{icv} Y \iff E\min\{X - a, 0\} \leq E\min\{Y - a, 0\} \ \forall a \in \mathbb{R},$$

which is easy to check. We want to remark that this ordering is the one preferred by economists, since they usually use utility. On the other hand, people working in actuarial sciences and finance typically speak of *risks*, i.e. of disutilities. Therefore they prefer the use of *increasing convex ordering* (written $\leq_{icx}$), which is equivalent, since

$$X \leq_{icv} Y \quad \text{iff} \quad -X \leq_{icx} -Y.$$

In actuarial sciences $\leq_{icx}$ is also known as *stop-loss order.* Another useful concept is the *convex order* $\leq_{cx}$ generated by all convex function, which sometimes is called *variability ordering*, since it compares the variability of random variables with the same mean, see e.g. Ross (1983, p. 270ff.).

Sometimes these concepts are not sufficient, and then one is naturally lead to generalizations of convexity. One stream of research is to consider not only first and second order properties of functions, but also properties of higher order differences resp. derivatives. This leads to the notions of *third order stochastic dominance* and more general *n-th order stochastic dominance.*

These concepts have been introduced in the economic literature by Whitmore (1970) and Fishburn (1976). In the applied probability literature strongly related concepts have been used by Rolski (1976), and Denuit et al. (1998, 1999) have described several interesting applications of these concepts in actuarial sciences.

These higher order convex (concave) functions are a subset of the convex (concave) functions. On the other hand, in the theory of individual decisions under risk one is interested in supersets of the class of increasing concave functions, since it is well known that most decision makers are not always risk averse. In some situations they are (e.g. when buying insurance), but in other situations they are risk seeking, e.g. when playing lotteries. To describe this by utility functions, one needs functions that are not everywhere concave. Friedman and Savage (1948) suggested functions, which they called concave-convex-concave.

More recently, Landsberger and Meilijson (1990a, b) derived axiomatically a class of utility functions, called *star-shaped*, which is suited for this situation. Their work has been generalized in Müller (1998b), where the concepts of (a,b)-concave functions and of t-quasiunimodal functions have been introduced. For the rest of this section we want to describe these concepts, and how they naturally arise in the theory in decisions under risk.

In their seminal paper, Rothschild and Stiglitz (1970) introduced the phrase *mean preserving spread* for a very intuitive notion of comparative risk.

Definition 3. A distribution function G differs from a distribution function F by a *mean preserving spread* (MPS), if they have the same mean and if there is an interval (a,b), such that G assigns no greater probability than F to any open subinterval of (a,b) and G assigns as much probability as F to any open subinterval either to the left or to the right of (a,b). If we want to make the dependence on the interval (a,b) explicit, then we will speak of an (a,b)-spread.

Rothschild and Stiglitz (1970) and Müller (1998a) have shown that for distributions with the same mean $F \leq_{icv} G$ holds, if and only if G can be obtained from F by a sequence of mean preserving spreads.

If we assume now that a decision maker is only risk averse to the MPS for some intervals (a,b), but not to all of them, then we get stochastic dominance relations based on generalized concave functions. We can characterize the decision makers, which are consistent with (a,b)-spreads for a fixed interval (a,b) by introducing the notion of (a,b)-concavity.

Definition 4. A function $f : \mathbb{R} \to \mathbb{R}$ is said to be (a,b)-concave if $f(x) \geq \ell(x)$ for $a \leq x \leq b$ and $f(x) \leq \ell(x)$ otherwise, where ℓ is the line through $(a, f(a))$ and $(b, f(b))$.

Then we get the following result. For a proof of this and all subsequent results of this section we refer to Müller (1998b).

Theorem 3. *For all distribution functions F, G and all $a < b$ it holds: G differs from F by a mean preserving (a,b)-spread, if and only if $\int f\, dF \geq \int f\, dG$ for all (a,b)-concave functions $f : \mathbb{R} \to \mathbb{R}$.*

Now we will consider the case, where the distribution G can be obtained from the distribution F by a sequence of (a,b)-spreads, where the intervals (a,b) are restricted to an arbitrary subset $M \subset S := \{(a,b) : a < b\}$. We will denote this as a *mean preserving M-increase in risk.* Then we get the following characterization.

Theorem 4. *If a distribution function G differs from a distribution function F by a mean preserving M-increase in risk, then*

$$\int u\, dF \geq \int u\, dG \quad \textit{for all } u \in \mathfrak{U}_M,$$

where $\mathfrak{U}_M$ is the set of all functions that are (a,b)-concave for all $(a,b) \in M$, and $\mathfrak{U}_M$ is the largest class of functions with this property.

Landsberger and Meilijson (1990a, b) considered the special case $M = L_\nu := \{(a,b) \in S : a \leq \nu \leq b\}$ for some fixed ν, what they called mean preserving spread about ν. There main result is stated in the next Theorem, which is an easy consequence of Theorem 4.

Theorem 5. *If $M = L_\nu := \{(a,b) \in S : a \leq \nu \leq b\}$, then $\mathfrak{U}_M$ is the set of all functions that are antistarshaped at ν, i.e. $\mathfrak{U}_M$ is the set of all functions U with*

$$x \to \frac{U(x) - U(\nu)}{x - \nu}, \quad x \neq \nu \textit{ is non-increasing.} \tag{4}$$

These functions are called antistarshaped, since it is easy to see that f is antistarshaped if and only if the epigraph of $-f$ is a starshaped set with respect to the point $(\nu, -f(\nu))$.

There is another intuitive notion of a spread, for which in Müller (1998b) the notion *t-dispersal* has been introduced. It has first been consider by Diamond and Stiglitz (1974), and can also been found in the actuarial literature under the name dangerousness (see e.g. van Heerwaarden (1991)). It is defined as follows.

Definition 5. A distribution function G differs from a distribution function F by a *t-dispersal*, if $G(x) - F(x) \geq 0$ for $x < t$ and $G(x) - F(x) \leq 0$ for $x \geq t$. A t-dispersal is said to be *mean preserving* if F and G have the same mean.

The utility functions consistent with this notion of comparative risk exhibit some property of generalized quasi-concavity, or generalized unimodality, as it has been called in Müller (1998b).

Definition 6. a) A function $f : \mathbb{R} \to \mathbb{R}$ is called *unimodal* with mode $t \in \mathbb{R}$, if f is increasing on $(-\infty, t]$ and decreasing on $[t, \infty)$.
b) We say that a function $f : \mathbb{R} \to \mathbb{R}$ is *t-quasiunimodal* for $t \in \mathbb{R}$, if there is a linear function ℓ such that $f - \ell$ is unimodal with mode t. The set of all t-quasiunimodal functions will be denoted by Ω_t.

Remark: It is easy to see that a differentiable function is t-quasiunimodal if and only if $f'(x) \geq f'(y)$ for all $x < t < y$.

Theorem 6. *G is a mean preserving t-dispersal of F if and only if* $\int f\, dF \geq \int f\, dG$ *for all t-quasiunimodal functions.*

4 Multivariate examples

Now we consider the comparison of random vectors with values in some euclidian space $\mathbb{R}^d$. As in the univariate case we can define $\mathbf{X} \leq_{st} \mathbf{Y}$, if $Ef(\mathbf{X}) \leq Ef(\mathbf{Y})$ holds for all increasing functions, and $\mathbf{X} \leq_{cx} [\leq_{icv}, \leq_{icx}] \mathbf{Y}$, if $Ef(\mathbf{X}) \leq Ef(\mathbf{Y})$ for all convex [increasing concave, increasing convex] functions. It is easy to see that these generators are the maximal ones, and these orderings have a lot of nice properties, so they are quite useful for some purposes. On the other hand, they also have some drawbacks. One flaw is that they do not have useful small generators, so that there is no feasible algorithm to check them for empirical distributions. In the case of convex order it is even known that there is no hope to find a small generator, since Johansen (1972, 1974) has shown that the extreme rays in the cone of all convex functions are dense within that set!

Another drawback is that they are not very useful for dealing with random vectors which exhibit dependencies between the components. For this situations there are much more interesting stochastic orders which are generated by some classes of generalized convex functions. We will introduce now the so called *supermodular* and *directionally convex order.* They have been considered recently by Bäuerle and Müller (1998), Bäuerle and Rolski (1998), Meester and Shanthikumar (1993), Müller (1997b), Müller and Scarsini (1999, 2000), Shaked and Shanthikumar (1997) and Szekli (1995), among others. We will show now, how these orders naturally arise in the context of dependence orderings.

Supermodular and directionally convex functions exhibit some second order properties, which are different from convexity. They can be defined as follows.

Definition 7. a) For a function $f : \mathbb{R}^d \to \mathbb{R}$ define the difference operators $\Delta_i^\varepsilon f(\mathbf{x}) := f(\mathbf{x} + \varepsilon \mathbf{e}_i) - f(\mathbf{x})$, where $\mathbf{e}_i$ is the ith unit vector and $\varepsilon > 0$.
b) A function $f : \mathbb{R}^d \to \mathbb{R}$ is said to be *supermodular*, if

$$\Delta_i^\varepsilon \Delta_j^\delta f(\mathbf{x}) \geq 0$$

holds for all $\mathbf{x} \in \mathbb{R}^d$, $1 \le i < j \le d$ and all $\varepsilon, \delta \ge 0$.

c) A function $f : \mathbb{R}^d \to \mathbb{R}$ is said to be *directionally convex*, if

$$\Delta_i^\varepsilon \Delta_j^\delta f(\mathbf{x}) \ge 0$$

holds for all $\mathbf{x} \in \mathbb{R}^d$, $1 \le i, j \le d$ and all $\varepsilon, \delta > 0$.

We collect some important properties and characterizations in the next two theorems. First we collect some useful properties of supermodular functions.

Theorem 7. *a) A function $f : \mathbb{R}^d \to \mathbb{R}$ is supermodular, if and only if*

$$f(\mathbf{x} \wedge \mathbf{y}) + f(\mathbf{x} \vee \mathbf{y}) \;\ge\; f(\mathbf{x}) + f(\mathbf{y}) \quad \forall \mathbf{x}, \mathbf{y} \in \mathbb{R}^d,$$

where the lattice operators $\wedge$ and $\vee$ are defined as

$$\mathbf{x} \wedge \mathbf{y} := (\min\{x_1, y_1\}, ..., \min\{x_d, y_d\})$$

and

$$\mathbf{x} \vee \mathbf{y} := (\max\{x_1, y_1\}, ..., \max\{x_d, y_d\}).$$

b) If f is twice differentiable, then f is supermodular if and only if

$$\frac{\partial^2}{\partial x_i \partial x_j} f(\mathbf{x}) \;\ge\; 0 \text{ for all } \mathbf{x} \in \mathbb{R}^d,\ 1 \le i < j \le d.$$

c) If $g_1, ..., g_d : \mathbb{R} \to \mathbb{R}$ are increasing functions and f is supermodular, then $f(g_1(\cdot), ..., g_d(\cdot))$ is also supermodular.
d) If $f : \mathbb{R}^d \to \mathbb{R}$ is increasing and supermodular, and $\phi : \mathbb{R} \to \mathbb{R}$ is increasing and convex, then $\phi \circ f$ is supermodular.

Theorem 8. *a) The following conditions are equivalent.*

1. *f is directionally convex.*
2. *f is supermodular and, in addition it is convex in each coordinate, when the other coordinates are held fixed.*
3. *For all $\mathbf{x}_i \in \mathbb{R}^d$, $i = 1, 2, 3, 4$, such that $\mathbf{x}_1 \le \mathbf{x}_2 \le \mathbf{x}_4$, $\mathbf{x}_1 \le \mathbf{x}_3 \le \mathbf{x}_4$ and $\mathbf{x}_1 + \mathbf{x}_4 = \mathbf{x}_2 + \mathbf{x}_3$,*

$$f(\mathbf{x}_1) + f(\mathbf{x}_4) \le f(\mathbf{x}_2) + f(\mathbf{x}_3).$$

4. *For all $\mathbf{x}_1, \mathbf{x}_2 \in \mathbb{R}^d$ with $\mathbf{x}_1 \le \mathbf{x}_2$ and all $\mathbf{y} \ge \mathbf{0}$,*

$$f(\mathbf{x}_1 + \mathbf{y}) - f(\mathbf{x}_1) \le f(\mathbf{x}_2 + \mathbf{y}) - f(\mathbf{x}_2).$$

b) If f is twice differentiable, then f is directionally convex if and only if

$$\frac{\partial^2}{\partial x_i \partial x_j} f(\mathbf{x}) \;\ge\; 0 \text{ for all } \mathbf{x} \in \mathbb{R}^d,\ 1 \le i, j \le d.$$

Notice that the characterizations 3. and 4. are very natural extensions of the notion of a univariate convex function to higher dimensions. They are also the reason why this concept is called *directionally convex*, since it combines properties of convexity with lattice properties.

Next we will show, why these concepts are useful for comparing dependencies. Let us first introduce, how dependencies within random vectors can be described. The distribution function F of a random vector $\mathbf{X} = (X_1, ..., X_d)$ can be characterized by their marginal distributions $F_1, ..., F_d$ and their copulae function C.

$$F(x_1, ..., x_d) = C(F_1(x_1), ..., F_d(x_d)),$$

where the copula $C : [0,1]^d \to [0,1]$ is a distribution function of a random vector $\mathbf{U} = (U_1, ..., U_d)$ with uniform marginals. The copula function shows how the marginals are connected. The most important examples for copulae are

1. Independent copula: $U_1, ..., U_d$ independent. Hence

$$C(\mathbf{u}) = u_1 \cdot ... \cdot u_d.$$

2. Comonotone copula (also called upper Fréchet bound):
Here $\mathbf{U} = (U_1, U_1, ..., U_1)$. Hence

$$C(\mathbf{u}) = C^+(\mathbf{u}) := \min\{u_1 ...., u_d\}.$$

Now we define the *supermodular stochastic order*

$$\mathbf{X} \leq_{sm} \mathbf{Y} \quad \text{iff} \quad Ef(\mathbf{X}) \leq Ef(\mathbf{Y}) \text{ for all supermodular } f,$$

and the *directionally convex order*

$$\mathbf{X} \leq_{dcx} \mathbf{Y} \quad \text{iff} \quad Ef(\mathbf{X}) \leq Ef(\mathbf{Y}) \text{ for all directionally convex } f.$$

Supermodular order is the stochastic order relation best suited for the comparison of random vectors with fixed marginals and different copulae. It has the following nice properties:

1. If $\mathbf{X} \leq_{sm} \mathbf{Y}$, then $\mathbf{X}$ and $\mathbf{Y}$ have the same marginals.

2. If $\mathbf{X} \leq_{sm} \mathbf{Y}$, then $Cov(X_i, X_j) \leq Cov(Y_i, Y_j)$ for all $i, j = 1, ..., d$.

3. $\leq_{sm}$ is closed under convolution, mixtures, weak convergence, permutations, marginalization, and scale transforms of the marginals, i.e. it depends only on the copula. In fact, supermodular order fulfils all nine axioms of a dependence order as introduced by Joe (1997). This has been shown in Müller and Scarsini (2000).

One of the most interesting results, which distinguishes supermodular order from other dependence orderings like orthant orders is the following theorem, which is easy to prove.

Theorem 9.

$$\mathbf{X} \leq_{sm} \mathbf{Y} \quad \Rightarrow \quad \sum_{i=1}^{d} X_i \leq_{cx} \sum_{i=1}^{d} Y_i.$$

This result is important in the context of finance and actuarial sciences, where $\mathbf{X}$ and $\mathbf{Y}$ describe two portfolios of d risks, which differ in their dependence structure. The theorem shows that stronger dependence in the supermodular sense leads to higher risk. For more details about applications in this context we refer to Bäuerle and Müller (1998) and Müller (1997b). Some other applications can be found e.g. in Bäuerle and Rolski (1998) and Szekli (1995).

Now we want to derive a similar result for random vectors with the same copula, but with different marginals. One could expect that for two portfolios of risks with the same dependence structure, more risky marginals should lead to more risky portfolios. This, however, is not even true in the case of the normal distribution, see e.g. Müller and Scarsini (1999, Prop.4.4).

A useful result can only be shown under the condition that the copula exhibits some notion of positive dependence, which we will define now. The random variables $X_1, X_2, \ldots, X_d$ (or their joint distribution function) are said to be *conditionally increasing in sequence* (CIS) if $P[X_i > t | X_1 = x_1, X_2 = x_2, \ldots, X_{i-1} = x_{i-1}]$ is an increasing function of the variables $x_1, x_2, \ldots, x_{i-1}$ for all $t \in \mathbb{R}$, and a random vector $\mathbf{X} = (X_1, ..., X_d)$ is said to be *conditionally increasing* (CI), if $\mathbf{X}_\pi := (X_{\pi(1)}, ..., X_{\pi(d)})$ is CIS for all permutations π.

Under this condition the following result has been derived by Müller and Scarsini (1999).

Theorem 10. *Let $\mathbf{X}, \mathbf{Y}$ be random vectors with a common CI copula C and assume that $X_i \leq_{cx} Y_i$, $1 \leq i \leq d$. Then $\mathbf{X} \leq_{dcx} \mathbf{Y}$.*

As an immediate corollary we get

Corollary 2. *Let $\mathbf{X}, \mathbf{Y}$ fulfill the conditions of Theorem 10. Then, for all non-negative $\alpha_1, \ldots, \alpha_d$,*

$$\sum_{i=1}^{d} \alpha_i X_i \leq_{cx} \sum_{i=1}^{d} \alpha_i Y_i.$$

We want to mention that these results become wrong, if we replace CI by CIS! Moreover, the result is sharp in the following sense.

Theorem 11. *If $X_i \leq_{cx} Y_i$, $1 \leq i \leq d$ implies $Ef(\mathbf{X}) \leq Ef(\mathbf{Y})$ for all random vectors $\mathbf{X}, \mathbf{Y}$ with comonotone copula C^+, then f is directionally convex.*

This shows that the notion of directional convexity arises naturally in this context! For some other applications of directionally convex order we refer to Bäuerle and Rolski (1998) and to Meester and Shanthikumar (1993).

5 Other stochastic orders based on generalized convexity

In this last section we first want to mention some results about quasiconcave orderings.

Definition 8. Let $\mathbf{X}, \mathbf{Y}$ be random vectors. Then we say that
a) $\mathbf{X}$ precedes $\mathbf{Y}$ in quasi-concave order (written $\mathbf{X} \leq_{qcv} \mathbf{Y}$), if $Ef(\mathbf{X}) \leq Ef(\mathbf{Y})$ for all quasi-concave functions f.
b) $\mathbf{X}$ precedes $\mathbf{Y}$ in increasing quasi-concave order (written $\mathbf{X} \leq_{iqcv} \mathbf{Y}$), if $Ef(\mathbf{X}) \leq Ef(\mathbf{Y})$ for all increasing quasi-concave functions f.

The ordering $\leq_{qcv}$, however, is trivial, as we will see in the following Theorem, which seems to be new.

Theorem 12.
$\mathbf{X} \leq_{qcv} \mathbf{Y}$ holds, if and only if $\mathbf{X}$ and $\mathbf{Y}$ have the same distribution.

Proof. For all $\mathbf{a} \in \mathbb{R}^d$ the sets $\{\mathbf{x} \in \mathbb{R}^d : \mathbf{a}^T\mathbf{x} > t\}$, $t \in \mathbb{R}$, and $\{\mathbf{x} \in \mathbb{R}^d : \mathbf{a}^T\mathbf{x} \leq t\}$, $t \in \mathbb{R}$, are convex, and hence the corresponding indicator functions are quasi-concave. Thus $\mathbf{X} \leq_{qcv} \mathbf{Y}$ implies $P(\mathbf{a}^T\mathbf{X} > t) \leq P(\mathbf{a}^T\mathbf{Y} > t)$ and $P(\mathbf{a}^T\mathbf{X} \leq t) \leq P(\mathbf{a}^T\mathbf{Y} \leq t)$ for all $t \in \mathbb{R}$, i.e. $\mathbf{a}^T\mathbf{X}$ and $\mathbf{a}^T\mathbf{Y}$ have the same distribution. Therefore $Ee^{i\mathbf{a}^T\mathbf{X}} = Ee^{i\mathbf{a}^T\mathbf{Y}}$ for all $\mathbf{a} \in \mathbb{R}^d$, and since a distribution is uniquely determined by its characteristic function, $\mathbf{X}$ and $\mathbf{Y}$ must have the same distributions. □

According to Theorem 12 we can restrict our investigation from now on to the ordering $\leq_{iqcv}$. It is obvious that any increasing concave function is also increasing and quasi-concave, hence we have the following implication.

Theorem 13.
$\mathbf{X} \leq_{iqcv} \mathbf{Y} \not\rightarrow \mathbf{X} \leq_{icv} \mathbf{Y}$.

On the other hand, there is the following relation between increasing quasi-concave order and the usual stochastic order.

Theorem 14.
a) If $\mathbf{X} \leq_{st} \mathbf{Y}$ then $\mathbf{X} \leq_{iqcv} \mathbf{Y}$.
b) If $\mathbf{X} \leq_{iqcv} \mathbf{Y}$ then $X_i \leq_{st} Y_i$ for all $1 \leq i \leq d$.

Proof. a) is obvious. To see b), it is sufficient to observe that the level sets of an increasing function, which depends only on x_i, are halfspaces and therefore convex. Hence any increasing function, which depends only on one variable, is quasiconcave. □

A small generator of this order relation is given in the next theorem, which slightly generalizes a result of Levhari et al. (1975).

Theorem 15. *A small generator of increasing quasi-concave order is given by the indicator functions of closed increasing convex sets.*

The characterization of the maximal generator, however, is still an open problem. We can show the following result.

Theorem 16. *All functions $f : \mathbb{R}^d \to \mathbb{R}$ in the maximal generator of increasing quasi-concave order have the following property:*

$$f(\mathbf{x}) + f(\mathbf{y}) \leq f(\alpha\mathbf{x} + (1-\alpha)\mathbf{y}) + f(\mathbf{x} \vee \mathbf{y}) \tag{5}$$

for all $\alpha \in [0,1]$ and all $\mathbf{x}, \mathbf{y} \in \mathbb{R}^d$.

Proof. Let f be an increasing quasi-concave function. Then

$$\min\{f(\mathbf{x}), f(\mathbf{y})\} \ \leq \ f(\alpha\mathbf{x} + (1-\alpha)\mathbf{y})$$

for all $\alpha \in [0,1]$ and all $\mathbf{x}, \mathbf{y} \in \mathbb{R}^d$. This implies

$$\begin{aligned} f(\mathbf{x}) + f(\mathbf{y}) &= \min\{f(\mathbf{x}), f(\mathbf{y})\} + \max\{f(\mathbf{x}), f(\mathbf{y})\} \\ &\leq f(\alpha\mathbf{x} + (1-\alpha)\mathbf{y}) + f(\mathbf{x} \vee \mathbf{y}). \end{aligned}$$

Hence all increasing quasi-concave functions have property (5). On the other hand, the set of all functions with (5) is a convex cone, closed under pointwise convergence and containing the constant functions. Thus it follows from Corollary 1 that it is the maximal generator of some integral stochastic order. Hence the maximal generator of $\leq_{iqcv}$ must be a subset of it. □

Conjecture: We conjecture that the set of all functions with property (5) is in fact the maximal generator of $\leq_{iqcv}$. However, we are not yet able to prove this. But it is easy to see that an indicator function has property (5), if and only if it is an indicator of an increasing convex set. This observation supports the conjecture.

In the case of multivariate normal distribution the following characterization can be shown. We omit the straightforward proof.

Theorem 17. *Let $\mathbf{X} \sim \mathcal{N}(\boldsymbol{\mu}, \boldsymbol{\Sigma})$ and $\mathbf{X}' \sim \mathcal{N}(\boldsymbol{\mu}', \boldsymbol{\Sigma}')$ be d-dimensional normally distributed random vectors. Then $\mathbf{X} \leq_{iqcv} \mathbf{X}'$ if and only if $\mu_i \leq \mu_i'$ for all $1 \leq i \leq d$ and $\boldsymbol{\Sigma} = \boldsymbol{\Sigma}'$.*

There are still more stochastic orders based on generalized convex functions. As a final example we want to mention here the so called *linear convex order* $\leq_{lcx}$. It has been considered e.g. by Koshevoy and Mosler (1998) and by Scarsini (1998). Its natural generator is the set of all functions

$$\mathfrak{F} = \{g \circ h | g : \mathbb{R} \to \mathbb{R} \text{ convex and } h : \mathbb{R}^d \to \mathbb{R} \text{ linear}\}.$$

It is easy to see that a small generator is given by the functions

$$f(\mathbf{x}) = \max\{\mathbf{a}^T\mathbf{x} - b, 0\}, \quad \mathbf{a} \in \mathbb{R}^d,\ b \in \mathbb{R}.$$

However, here again it is an open problem to characterize the maximal generator. It is only known that the maximal generator does not contain all convex functions. This can be derived from an example given in Elton and Hill (1992).

References

1. BÄUERLE, N., A. MÜLLER (1998). Modeling and comparing dependencies in multivariate risk portfolios. *ASTIN Bulletin* **28**, 59 - 76.
2. BÄUERLE, N., T. ROLSKI (1998). A monotonicity result for the work-load in Markov-modulated queues. *Journal of Applied Probability* **35**, 741 - 747.
3. BORDER, K.C. (1991). Functional Analytic Tools for Expected Utility Theory. In Aliprantis, Border, Luxemburg (eds.) *Positive Operators, Riesz Spaces and Economics, Studies in Econ. Theory* **2**, 69 - 88.
4. CASTAGNOLI, E., F. MACCHERONI (1998). Generalized Stochastic Dominance and Unanimous Preferences In Giorgi G., F. Rossi (eds.) *Generalized convexity and optimization for economic and financial decisions*, 111 - 120.
5. CHOQUET G. (1969). *Lectures on Analysis I, II, III.* W. A. Benjamin.
6. DENUIT, M., DE VYLDER, E., AND C. LEFÈVRE (1999). Extremal generators and extremal distributions for the continuous s-convex stochastic orderings. *Insurance: Mathematics And Economics* **24**, 201 - 217.
7. DENUIT, M., C. LEFÈVRE, M. SHAKED (1998). The s-convex orders among real random variables, with applications. *Mathematical Inequalities and their Applications* **1**, 585 - 613.
8. DIAMOND, P., J. STIGLITZ. (1974). Increases in Risk and in Risk Aversion. *Journal of Economic Theory* **8**, 337 - 360.
9. ELTON, J., T.P. HILL (1992). Fusions of a probability distribution. *Annals of Probability* **20**, 421 - 454.
10. FISHBURN, P.C. (1976). Continua of stochastic dominance relations for bounded probability distributions. *Journal of Mathematical Economics* **3**, 295 - 311.
11. FRIEDMAN, M., L. SAVAGE (1948). The utility analysis of choices involving risk. *Journal of Political Economy* **56**, 279 - 304.
12. VAN HEERWARDEN A.E. (1991). *Ordering of risks: Theory and Actuarial Applications.* Tinbergen Institute Series no. 20.
13. JOE, H. (1997). *Multivariate Models and Dependence Concepts.* Chapman and Hall, London.
14. JOHANSEN, S. (1972). A representation theorem for a convex cone of quasi convex functions. *Mathematica Scandinavica* **30**, 297 - 312.
15. JOHANSEN, S. (1974). The extremal convex functions. *Mathematica Scandinavica* **34**, 61 - 68.
16. KOSHEVOY, G., K. MOSLER (1998). Lift zonoids, random convex hulls and the variability of random vectors. *Bernoulli* **4**, 377 - 399.

17. LANDSBERGER, M., I. MEILIJSON (1990a). A tale of two tales: An alternative characterization of comparative risk. *Journal of Risk and Uncertainty* **3**, 65 - 82.
18. LANDSBERGER, M., I. MEILIJSON (1990b). Lotteries, insurance, and star-shaped utility functions. *Journal of Economic Theory* **52**, 1 - 17.
19. LEVHARI, D., J. PAROUSH, B. PELEG (1975). Efficiency analysis for multivariate distributions. *Review of Economic Studies* **42**, 87 - 91.
20. MARSHALL A. W. (1991). Multivariate Stochastic orderings and generating cones of functions. In Mosler K. und Scarsini M. (eds.), *Stochastic Orders and Decision under Risk.* IMS Lecture Notes, Volume **19**, 231 - 247 .
21. MEESTER, L.E., J.G. SHANTHIKUMAR (1993). Regularity of stochastic processes: A theory of directional convexity. *Probability in the Engineering and Informational Sciences* **7**, 343 - 360.
22. MOSLER K., M. SCARSINI (1994). *Stochastic Orders and Applications: A classified Bibliography.* Lecture Notes in Econ. and Math. Syst., Volume **401**.
23. MÜLLER, A. (1997a). Stochastic orders generated by integrals: A unified study. *Advances in Applied Probability* **29**, 414 - 428.
24. MÜLLER, A. (1997b). Stop-loss order for portfolios of dependent risks. *Insurance: Mathematics and Economics* **21**, 219 - 223.
25. MÜLLER, A. (1998a). Comparing risks with unbounded distributions. *Journal of Mathematical Economics.* **30**, 229 - 239.
26. MÜLLER, A. (1998b). Another tale of two tails: On characterizations of comparative risk. *Journal of Risk and Uncertainty* **16**, 187 - 197.
27. MÜLLER, A., M. SCARSINI (1999). Stochastic comparison of random vectors with fixed copulae. WIOR-Report 554, University Karlsruhe.
28. MÜLLER, A., M. SCARSINI (2000). Some remarks on the supermodular order. *Journal of Multivariate Analysis* **73**, 107 - 119.
29. ROLSKI T. (1976). Order relations in the set of probability distribution functions and their applications in queueing theory. *Dissertationes Mathematicae* **132**.
30. ROSS, S.M. (1983). *Stochastic Processes.* Wiley.
31. ROTHSCHILD, M., J. STIGLITZ (1970). Increasing risk, I: A definition. *Journal of Economic Theory* **2**, 225 - 243.
32. SCARSINI, M. (1998). Multivariate convex orderings, dependence, and stochastic equality. *Journal of Applied Probability* **35**, 93 - 103.
33. SHAKED, M., J.G. SHANTHIKUMAR (1994). *Stochastic Orders and their Applications.* Academic Press, London.
34. SHAKED, M., J.G. SHANTHIKUMAR (1997). Supermodular stochastic orders and positive dependence of random vectors. *Journal of Multivariate Analysis* **61**, 86 - 101.
35. STOYAN D. (1983). *Comparison Methods for Queues and Other Stochastic Models.* Wiley.
36. SZEKLI R. (1995). *Stochastic Ordering and Dependence in Applied Probability.* Lecture Notes in Statistics, Volume **97**. Springer, New York.
37. WHITMORE, G.A. (1970). Third degree stochastic dominance. *American Economic Review* **60**, 457 - 459.
38. WHITT, W. (1986). Stochastic comparisons for non-Markov processes. *Mathematics of Operations Research* **11**, 608-618.

Separation Theorems for Convex Sets and Convex Functions with Invariance Properties

Zsolt Páles

Institute of Mathematics and Informatics,
University of Debrecen,
H-4010 Debrecen, Pf. 12, Hungary

Abstract. In this paper, a generalization of Stone's celebrated separation theorem is offered. It is shown that if the given disjoint convex sets are invariant with respect to a commuting family of affine transformations, then they can be separated by complementary convex sets enjoying the same invariance properties. The recession cone of the separating sets can also be nonsmaller than that of the data. As applications, we investigate the separability of affine invariant convex sets. It turns out that the separating affine function inherits invariance properties from the data. The results obtained generalize the Hahn–Banach and the Dubovitskii–Milyutin separation theorems. Sandwich theorems are also considered for convex-concave and for sublinear-superlinear pairs of functions admitting further invariance properties. In this way, the Hahn–Banach extension theorem can also be generalized.

2000 Mathematics Subject Classification. Primary 46A22; Secondary 52A05.

Keywords and phrases. Separation of convex sets, Stone's theorem, affine transformation, invariant convex set, separation by hyperplanes, Dubovitskii–Milyutin separation theorem, sandwich theorem.

1 Introduction

One of the most basic separation theorems due to M. H. Stone [6, Lemma 2B, p. 7] states the following: *If A and B are nonempty disjoint convex subsets of a linear space, then there exist convex complementary subsets A_0 and B_0 of X such that $A \subset A_0$ and $B \subset B_0$.*

(Two sets are called complementary in X if they form a partition of X, that is if they are disjoint and their union equals X.)

The proof of this result can be explained in a few lines. Using Zorn's lemma, one can find disjoint convex sets A_0 and B_0 such that $A \subset A_0$ and $B \subset B_0$ that are maximal with respect to inclusion. It remains to show that $A_0 \cup B_0 = X$. If, on the contrary, there exists $x \in X \setminus (A_0 \cup B_0)$, then denote by $\widehat{A}$ and $\widehat{B}$ the convex hull of $A_0 \cup \{x\}$ and $B_0 \cup \{x\}$, respectively. Now, due to the maximality of A_0 and B_0, we have that $\widehat{A} \cap B_0 \neq \emptyset$ and $\widehat{B} \cap A_0 \neq \emptyset$. Thus there exist points $a, a_0 \in A_0$ and $b, b_0 \in B_0$ such that

$$b_0 \in [a, x], \qquad a_0 \in [b, x].$$

(Here $[u, v]$ denotes the segment joining u and v.) Denoting by p the trivially existing intersection of the segments $[b, b_0]$ and $[a, a_0]$, we can see, by the convexity of A_0 and B_0, that p belongs to $A_0 \cap B_0$, which contradicts the disjointness of A_0 and B_0. Thus, indeed, $A_0 \cup B_0 = X$.

In this paper we investigate the following problem: If A and B admit some nice properties, can A_0 and B_0 be constructed such that they also enjoy the same properties? More precisely if A and B are invariant under a family of transformations, we are interested in finding A_0 and B_0 with the same invariance property.

If $T : X \to X$ is a selfmap of X, then a subset $A \subset X$ is said to be *T-invariant* whenever $T(A) \subset A$. If $\mathcal{T}$ is a family of selfmaps of X, then A is called *$\mathcal{T}$-invariant* if A is T-invariant for each $T \in \mathcal{T}$.

The most general transformations considered in our investigations are those that preserve the convexity structure of convex sets, i.e., *affine transformations.* A map $T : X \to X$ is called *affine* if

$$T(\lambda x + (1-\lambda)y) = \lambda T(x) + (1-\lambda)T(y) \qquad (x, y \in X, \lambda \in [0,1]).$$

An elementary argument shows that $T : X \to X$ is affine if and only if the map $\Lambda_T : X \to X$ defined by $\Lambda_T(x) := T(x) - T(0)$ is linear. The linear map so defined is called the *linear part of* T. It is useful to note that, for an affine map T, the following identity is also valid

$$T(\lambda x + (1-\lambda)y + z) = \lambda T(x) + (1-\lambda)T(y) + \Lambda_T(z) \; (x, y, z \in X, \lambda \in \mathbb{R}).$$

If $\mathcal{T}$ is a family of affine maps, then the family of the linear parts of the affine maps from $\mathcal{T}$ is denoted by $\Lambda_{\mathcal{T}}$.

Two maps $S, T : X \to X$ are said to *commute* if $S \circ T = T \circ S$, that is, if $S(T(x)) = T(S(x))$ for all $x \in X$. It is easy to verify that if S and T are affine maps, then they commute if and only if

$$\Lambda_S \circ \Lambda_T = \Lambda_T \circ \Lambda_S \qquad \text{and} \qquad \Lambda_S(T(0)) - T(0) = \Lambda_T(S(0)) - S(0).$$

A family of affine maps is called *commuting* if any two members of this family commute with each other.

Any vector $v \in X$ induces an affine transformation (called translation by v) defined by $\mathcal{T}_v(x) := x + v$. Given a convex set $A \subset X$, the set of those vectors such that A is invariant with respect to the induced translation is called the *recession cone of* A, that is,

$$\operatorname{rec}(A) := \{v \in X \mid \mathcal{T}_v(A) = v + A \subset A\}.$$

It is easy to check that $\operatorname{rec}(A)$ is always a convex cone containing the zero of X. The correspondence $A \mapsto \operatorname{rec}(A)$ is unfortunately not monotonic, that is, $A \subset B$ does not imply $\operatorname{rec}(A) \subset \operatorname{rec}(B)$ in general. Hence, the recession cone of A_0 in Stone's theorem can be smaller than that of A.

The outline of this paper is as follows. In the next section we present generalizations of Stone's theorem. It is shown that if the given disjoint convex sets are invariant with respect to a commuting family of affine transformations, then they can be separated by complementary convex sets enjoying the same invariance properties. The recession cone of the separating sets can also be nonsmaller than that of the data. In Section 3, we investigate the separability of affine invariant convex sets. It turns out that the separating affine function inherits invariance properties from the data. The results obtained generalize the Hahn–Banach and the Dubovitskii–Milyutin separation theorems. In the last section, sandwich theorems are considered for convex-concave and for sublinear-superlinear pairs of functions admitting further invariance properties. In this way, the Hahn–Banach extension theorem can also be generalized. For a recent survey on the Hahn–Banach theorem and for further references, see [1].

2 Main results

The main result of this paper is contained in the following generalization of Stone's theorem.

Theorem 1. *Let X be a linear space and let $\mathfrak{T}$ be a commuting family of affine maps of X into itself. Assume that A and B are nonempty disjoint convex $\mathfrak{T}$-invariant subsets of X, in addition, let C and D be nonempty $\Lambda_{\mathfrak{T}}$-invariant convex subcones of* $\operatorname{rec}(A)$ *and* $\operatorname{rec}(B)$, *respectively. Then there exist complementary $\mathfrak{T}$-invariant convex sets A_0 and B_0 such that*

$$A \subset A_0, \qquad B \subset B_0, \qquad C \subset \operatorname{rec}(A_0), \qquad D \subset \operatorname{rec}(B_0).$$

Proof. First, we show that $\mathfrak{T}$ can be replaced by a larger set $\mathfrak{T}^*$ of commuting affine maps so that all the conditions of theorem remain valid and $\mathfrak{T}^*$ is convex and closed under composition. We construct $\mathfrak{T}^*$ in two steps.

First let

$$\widehat{\mathfrak{T}} := \{T_1 \circ \cdots \circ T_k \mid k \in \mathbb{N},\ T_1, \ldots, T_k \in \mathfrak{T}\} \cup \{\mathrm{id}\}.$$

Then $\widehat{\mathfrak{T}}$ is again a family of affine maps which is closed under composition. It is also commuting since $\mathfrak{T}$ is commuting. Clearly, A and B are $\widehat{\mathfrak{T}}$-invariant and also C and D are $\Lambda_{\widehat{\mathfrak{T}}}$-invariant cones.

Now define $\mathfrak{T}^*$ as the convex hull of $\widehat{\mathfrak{T}}$, that is, let

$$\mathfrak{T}^* = \operatorname{conv} \widehat{\mathfrak{T}} = \Big\{ \sum_{i=1}^{k} \lambda_i T_i \mid k \in \mathbb{N}, \lambda_1, \ldots, \lambda_k \geq 0, \sum_{i=1}^{k} \lambda_i = 1, T_1, \ldots, T_k \in \widehat{\mathfrak{T}} \Big\}.$$

Then $\mathfrak{T}^*$ is a commuting family of affine operators. Using the convexity of A, B, C and D, we can see that A, B are $\mathfrak{T}^*$-invariant sets, and C, D are $\Lambda_{\mathfrak{T}^*}$-invariant. It is also obvious that $\mathfrak{T}^*$ is closed under composition.

In order to simplify the notation, we may assume that $\mathcal{T} = \mathcal{T}^*$. We can also suppose that $0 \in C$ and $0 \in D$.

Now consider the set $\mathcal{H}$ of those pairs $(\widehat{A}, \widehat{B})$ that satisfy the statements of the theorem except that $\widehat{A} \cup \widehat{B} = X$, that is, let

$$\mathcal{H} := \{(\widehat{A}, \widehat{B}) \mid \widehat{A}, \widehat{B} \subset X \text{ are convex, disjoint, } A \subset \widehat{A},\ B \subset \widehat{B} \\ \text{and for any } T \in \mathcal{T}^*,\ T(\widehat{A}) \subset \widehat{A},\ T(\widehat{B}) \subset \widehat{B},\ \widehat{A} + C \subset \widehat{A},\ \widehat{B} + D \subset \widehat{B}\}.$$

Then $\mathcal{H}$ is not empty, because $(A, B) \in \mathcal{H}$. On the other hand, if $\{(A_i, B_i)\}_{i \in I}$ is a chain of elements of $\mathcal{H}$ with respect to the componentwise inclusion, then

$$\Big(\bigcup_{i \in I} A_i, \bigcup_{i \in I} B_i\Big) \in \mathcal{H}.$$

Hence, by Zorn's lemma, $\mathcal{H}$ contains a maximal element, say (A_0, B_0). To complete the proof, it suffices to show that $A_0 \cup B_0 = X$.

For, assume on the contrary that $x \in X \setminus (A_0 \cup B_0)$. Define

$$\widehat{A} := \operatorname{join}(A_0, \mathcal{T}(x) + C) := \bigcup \{[a, T(x) + c] \mid a \in A_0,\ T \in \mathcal{T},\ c \in C\},$$
$$\widehat{B} := \operatorname{join}(B_0, \mathcal{T}(x) + D) := \bigcup \{[b, T(x) + d] \mid b \in B_0,\ T \in \mathcal{T},\ d \in D\}.$$

Then $x \in \widehat{A}$ since $0 \in C$, $\mathrm{id} \in \mathcal{T}$. Hence, the inclusion $A_0 \subset \widehat{A}$ is proper. Thus, due to the maximality of (A_0, B_0), the pair $(\widehat{A}, B_0)$ cannot be in $\mathcal{H}$.

On the other hand, due to the convexity of A_0, C and $\mathcal{T}$, the set $\widehat{A}$ is also convex. If $\widehat{a} = \lambda a + (1 - \lambda)(T(x) + c)$ is an arbitrary element of $\widehat{A}$ and $S \in \mathcal{T}$, then

$$S(\widehat{a}) = \lambda S(a) + (1 - \lambda)\big(S(T(x)) + \Lambda_S(c)\big) \in [S(a), S \circ T(x) + \lambda_S(c)] \subset \widehat{A}.$$

Hence $\widehat{A}$ is $\mathcal{T}$-invariant. If $c_0 \in C$ and $\widehat{a}$ is the same as above, then

$$\widehat{a} + c_0 = \lambda(a + c_0) + (1 - \lambda)(T(x) + c + c_0) \in [a + c_0, T(x) + c + c_0] \in \widehat{A}.$$

Thus $\widehat{A} + C \subset \widehat{A}$ is also valid. The noninclusion $(\widehat{A}, B_0) \notin \mathcal{H}$ therefore yields that $\widehat{A}$ intersects B_0. Hence there exist elements $\lambda \in [0, 1[$, $a \in A_0$, $c \in C$ and $T \in \mathcal{T}$ such that

$$\lambda a + (1 - \lambda)(T(x) + c) = b_0 \in B_0.$$

Repeating the above argument with $\widehat{B}$ instead of $\widehat{A}$, we similarly obtain the existence of elements $\mu \in [0, 1[$, $b \in B_0$, $d \in D$ and $S \in \mathcal{T}$ such that

$$\mu b + (1 - \mu)(S(x) + d) = a_0 \in A_0.$$

Now, to reach the desired contradiction, we show that $A_0 \cap B_0$ is nonempty.

The set A_0 being $\mathfrak{T}$-invariant, we have that $T(a_0), S(a) \in A_0$. By convexity, it follows that

$$\frac{1-\lambda}{1-\lambda\mu}T(a_0) + \frac{\lambda-\lambda\mu}{1-\lambda\mu}S(a) \in A_0.$$

The set C is a $\Lambda_{\mathfrak{T}}$-invariant cone and $A_0 + C \subset A_0$, hence

$$\widehat{a} := \frac{1-\lambda}{1-\lambda\mu}T(a_0) + \frac{\lambda-\lambda\mu}{1-\lambda\mu}S(a) + \frac{(1-\lambda)(1-\mu)}{1-\lambda\mu}\Lambda_S(c) \in A_0.$$

On the other hand, we have

$$\begin{aligned}
\widehat{a} &= \frac{1-\lambda}{1-\lambda\mu}T(\mu b + (1-\mu)(S(x)+d)) + \frac{\lambda-\lambda\mu}{1-\lambda\mu}S(a) + \frac{(1-\lambda)(1-\mu)}{1-\lambda\mu}\Lambda_S(c) \\
&= \frac{1-\mu}{1-\lambda\mu}S(\lambda a + (1-\lambda)(T(x)+c)) + \frac{\mu-\lambda\mu}{1-\lambda\mu}T(b) + \frac{(1-\lambda)(1-\mu)}{1-\lambda\mu}\Lambda_T(d) \\
&= \frac{1-\mu}{1-\lambda\mu}S(b_0) + \frac{\mu-\lambda\mu}{1-\lambda\mu}T(b) + \frac{(1-\lambda)(1-\mu)}{1-\lambda\mu}\Lambda_T(d) =: \widehat{b}.
\end{aligned}$$

Here, the second equality was obtained using the affine property of T and S and also $T \circ S = S \circ T$. Exactly as we have proved $\widehat{a} \in A_0$, one can see the inclusion $\widehat{b} \in B$.

Thus A_0 and B_0 are not disjoint. The contradiction obtained validates $A_0 \cup B_0 = X$, completing the proof of the theorem. □

Now we list some immediate consequences of the above result. The first corollary is the special case $C = D = \{0\}$ of Theorem 1.

Corollary 1. *Let X be a linear space and let $\mathfrak{T}$ be a commuting family of affine maps of X into itself. Assume that A and B are nonempty disjoint convex $\mathfrak{T}$-invariant subsets of X. Then there exist complementary $\mathfrak{T}$-invariant convex sets A_0 and B_0 such that $A \subset A_0$, $B \subset B_0$.*

The second corollary is the special case $\mathfrak{T} = \{\mathrm{id}\}$, $C = \mathrm{rec}(A)$, $D = \mathrm{rec}(B)$.

Corollary 2. *Let X be a linear space and assume that A, B are nonempty disjoint convex subsets of X. Then there exist complementary convex sets A_0 and B_0 such that*

$$A \subset A_0, \qquad B \subset B_0, \qquad \mathrm{rec}(A) \subset \mathrm{rec}(A_0), \qquad \mathrm{rec}(B) \subset \mathrm{rec}(B_0).$$

Clearly, the above corollaries also generalize Stone's theorem. The third consequence of Theorem 1 is a generalization for invariant convex cones.

Corollary 3. *Let X be a linear space and let $\mathfrak{T}$ be a commuting family of linear selfmaps of X. Assume that A and B are nonempty disjoint convex $\mathfrak{T}$-invariant cones of X, Then there exist complementary $\mathfrak{T}$-invariant convex cones A_0 and B_0 such that $A \subset A_0$, $B \subset B_0$.*

Proof. For each positive λ, the operator $\lambda \cdot \mathrm{id}$ commutes with the elements of $\mathcal{T}$. Hence

$$\mathcal{T}^* = \mathcal{T} \cup \{\lambda \cdot \mathrm{id} \mid \lambda > 0\}$$

is a commuting family of affine maps. The sets A, B being cones, are also $\mathcal{T}^*$-invariant. Thus, by Corollary 1, there exist complementary convex $\mathcal{T}^*$-invariant sets A_0, B_0 such that $A \subset A_0$, $B \subset B_0$. However, due to the $\lambda \cdot \mathrm{id}$ invariance for any $\lambda > 0$, the sets A_0 and B_0 are also cones. □

Remark 1. Stone-type theorems do not exist for midpoint convex sets, the separation of disjoint midpoint convex sets is possible if and only if their rational convex hulls are disjoint. For details, see [8]. There are many problems of interest that are analogous to the separation of convex sets. For instance, the separation of disjoint subsemigroups of an Abelian semigroup is studied in [10]. Results on the separation of sets that are closed with respect to commuting families of operations are obtained in [12].

3 Separation by hyperplanes

In this section, we present analytic forms of the separation of convex sets.

If $A \subset X$ is a convex set, then the *core* or in other terms, the *algebraic interior of* A is defined by

$$\mathrm{cor}(A) := \{a \in A \mid \forall x \in X\, \exists \lambda \in]0,1[\text{ such that } [a, (1-\lambda)a + \lambda x] \subset A\}.$$

(See Holmes [6, 2C, p. 7].) Intuitively, $a \in A$ is in the core of A if it is possible to move towards any point of X while staying in A. If X is a topological linear space, then the interior of A is always contained in $\mathrm{cor}(A)$.

To obtain the main results of this section, we need the following lemma.

Lemma 1. *Let X be a linear space, A be a nonempty subset of X and $f : X \to \mathbb{R}$ be a nonconstant affine function such that*

$$f|_A \le 0 \le f|_{X \setminus A}.$$

If $g : X \to \mathbb{R}$ is an affine function such that $\sup_A g = 0$, then there exists a constant $\alpha \ge 0$ such that $g = \alpha f$.

Proof. First we show that, for any $x \in X$, the equality $f(x) = 0$ implies $g(x) = 0$.

Let $x \in X$ be such that $f(x) = 0$. The function f being nonconstant, there exists an element $x_0 \in X$ such that $f(x_0) < 0$. Let $0 < t \le 1$. Then

$$f((1-t)x + tx_0) = (1-t)f(x) + tf(x_0) = tf(x_0) < 0.$$

Therefore, for all $t \in]0,1]$, the element $(1-t)x + tx_0$ cannot belong to $X \setminus A$. Thus, it belongs to A, whence

$$(1-t)g(x) + tg(x_0) = g((1-t)x + tx_0) \le 0 \qquad (t \in]0,1]).$$

Taking the limit $t \to 0$, we get $g(x) \le 0$.

Now we prove that $g(x) < 0$ is impossible. If $g(x) < 0$, then there exists an element a of A such that $g(x) < g(a) \le 0$. For some small positive t_0, with the notation $a_0 = (1 - t_0)a + t_0 x_0$, we have $g(x) < g(a_0)$ and also $f(a_0) < 0$. Thus, for all positive s,

$$f(sa_0 + (1-s)x) = sf(a_0) + (1-s)f(x) = sf(a_0) < 0.$$

Hence $sa_0 + (1-s)x \in A$ for all $s > 0$. This yields

$$s(g(a_0)-g(x))+g(x) = sg(a_0)+(1-s)g(x) = g(sa_0+(1-s)x) \le 0 \qquad (s > 0),$$

which is an obvious contradiction of $g(x) < g(a_0)$. The contradiction proves that $g(x) < 0$ is impossible.

To complete the proof, we show that $g = \alpha f$ with $\alpha := g(x_0)/f(x_0)$. Let $u \in X$ be arbitrary. Assume first that $f(u) \ne f(x_0)$. Then, for the vector

$$x = \frac{f(x_0)}{f(x_0) - f(u)} u - \frac{f(u)}{f(x_0) - f(u)} x_0,$$

we have $f(x) = 0$. Hence, $g(x) = 0$, which yields

$$0 = g(x) = \frac{f(x_0)}{f(x_0) - f(u)} g(u) - \frac{f(u)}{f(x_0) - f(u)} g(x_0),$$

i.e.,

$$g(u) = \frac{g(x_0)f(u)}{f(x_0)} = \alpha f(u).$$

If $f(u) = f(x_0)$, then choosing an element w such that $0 \ne f(w) \ne f(u) = f(x_0)$, and applying the above equality with w instead of x_0, we get

$$g(u) = \frac{g(w)f(u)}{f(w)} = \frac{\alpha f(w)f(u)}{f(w)} = \alpha f(u).$$

Since $f(x_0) < 0$, x_0 belongs to A. Hence $g(x_0) \le 0$, whence $\alpha \ge 0$ follows. Thus the proof is complete. □

Theorem 2. *Let X be a linear space and let $\mathcal{T}$ be a commuting family of affine maps of X into itself. Assume that A and B are nonempty disjoint convex $\mathcal{T}$-invariant subsets of X. Suppose that either X is finite dimensional or* $\operatorname{cor}(A) \cup \operatorname{cor}(B)$ *is nonempty. Then there exist a nonconstant affine function $f : X \to \mathbb{R}$ and a nonnegative function $\alpha : \mathcal{T} \to [0, \infty[$ such that*

$$f|_A \le 0, \qquad 0 \le f|_B, \tag{1}$$

and

$$f \circ T = \alpha(T) f \qquad (T \in \mathcal{T}). \tag{2}$$

Proof. Using Theorem 1, or Corollary 1, we can find complementary $\mathcal{T}$-invariant convex sets A_0 and B_0 such that $A \subset A_0$ and $B \subset B_0$. If X is not finite dimensional, then $\mathrm{cor}(A_0) \cup \mathrm{cor}(B_0) \neq \emptyset$. Applying [6, Theorem 4B] (which can be derived directly from the classical Stone theorem, without using Zorn's lemma), there exist a nonzero linear function $\varphi : X \to \mathbb{R}$ and a constant $c \in \mathbb{R}$ such that

$$\varphi|_{A_0} \le c \le \varphi|_{B_0}. \tag{3}$$

By (3), the function $f := \varphi - c$ is now an affine function satisfying (1). The sets A_0 and B_0 being $\mathcal{T}$-invariant, we have

$$f \circ T|_{A_0} \le 0 \le f \circ T|_{B_0} \qquad (T \in \mathcal{T}).$$

Observe that f and $g := f \circ T$ satisfies the conditions of Lemma 1. Indeed, the image of g is a convex subset of $\mathbb{R}$ containing 0. Hence, the inequality $\sup_{A_0} g < 0 \le \inf_{B_0} g$ is impossible. Therefore, there exists a nonnegative constant $\alpha(T)$ such that $f \circ T = \alpha(T) f$. Thus (2) is also proved and the proof of the theorem is complete. □

The separation theorem for convex cones now can easily be deduced.

Corollary 4. *Let X be a linear space and let $\mathcal{T}$ be a commuting family of linear selfmaps of X. Assume that A and B are nonempty disjoint convex $\mathcal{T}$-invariant cones of X. Suppose that either X is finite dimensional or* $\mathrm{cor}(A) \cup \mathrm{cor}(B)$ *is nonempty. Then there exist a non identically zero linear function $f : X \to \mathbb{R}$ and a nonnegative function $\alpha : \mathcal{T} \to [0, \infty[$ such that (1) and (2) hold.*

Proof. Denote

$$\mathcal{T}^* = \mathcal{T} \cup \{\lambda \cdot \mathrm{id} \mid \lambda > 0\}.$$

Then, due to the linearity of the elements of $\mathcal{T}$, this new family is a commuting family of affine maps. The sets A, B being cones, are also $\mathcal{T}^*$-invariant. Thus, by Theorem 2, there exist an affine function $f : X \to \mathbb{R}$ and a nonnegative function $\alpha : \mathcal{T}^* \to [0, \infty[$ such that (1), (2) are valid and, for all $\lambda > 0$,

$$f \circ (\lambda \cdot \mathrm{id}) = \alpha(\lambda \cdot \mathrm{id}) f.$$

If f is of the form $\varphi - c$, where $\varphi : X \to \mathbb{R}$ is a nonconstant linear function and c is a constant, then the above equation yields

$$\lambda\varphi - c = \alpha(\lambda \cdot \mathrm{id})(\varphi - c) \qquad (\lambda > 0).$$

Hence $\alpha(\lambda \cdot \mathrm{id}) = \lambda$ and $c = 0$. Thus f must be a linear function as asserted. □

Remark 2. If in the above result X is a topological vector space, and one of the sets A or B has interior points, then the continuity of f can also be stated, since then the linear part of f will be bounded (from above or below by (1)) on an open subset of X.

Remark 3. Analogous separation theorems for disjoint subsemigroups of commutative semigroups and their applications can be found in [9].

A generalization of the so-called Dubovitskii-Milyutin separation theorem can also be obtained.

Theorem 3. *Let X be a linear space and let $\mathcal{T}$ be a commuting family of affine maps of X into itself. Assume that $A_0, A_1, \ldots, A_n$ are nonempty convex $\mathcal{T}$-invariant subsets of X such that their intersection is empty. Suppose that either X is finite dimensional or* $\operatorname{cor}(A_1), \ldots, \operatorname{cor}(A_n)$ *are nonempty sets. Then there exist affine functions $f_0, \ldots, f_n : X \to \mathbb{R}$ (not all identically constant) and a nonnegative function $\alpha : \mathcal{T} \to [0, \infty[$ such that*

$$f_i|_{A_i} \le 0 \qquad (i = 0, \ldots, n), \tag{4}$$

$$f_i \circ T = \alpha(T) f_i \qquad (T \in \mathcal{T}, i = 0, \ldots, n), \tag{5}$$

and

$$f_0 + f_1 + \cdots + f_n = 0. \tag{6}$$

Proof. Using Theorem 1 or Corollary 1 repeatedly, we can assume that the complements of $A_0, A_1, \ldots, A_n$ are also $\mathcal{T}$-invariant. Indeed, $A = A_0$ and $B = A_1 \cap \cdots \cap A_n$ are disjoint $\mathcal{T}$-invariant convex subsets of X. By Corollary 1, there exists complementary $\mathcal{T}$-invariant convex sets $A_0 \subset A_0^*$ and $B \subset B_0$. Then we have that the sets $A_0^*, A_1, \ldots, A_n$ are nonempty $\mathcal{T}$-invariant convex sets whose intersection is empty; furthermore A_0 is a subset of A_0^* and $X \setminus A_0^*$ is also a $\mathcal{T}$-invariant convex set. Thus, for $i = 0$, the set A_i can be replaced by a larger $\mathcal{T}$-invariant convex set that satisfies all the conditions of the theorem and whose complement is also $\mathcal{T}$-invariant and convex. Arguing similarly, this can be done for $i = 1, \ldots, n$ as well. Now define

$$A := A_1 \times \cdots \times A_n, \qquad B := \{(x_1, \ldots, x_n) \mid x_1 = \cdots = x_n \in A_0\}.$$

Then, due to the emptiness of the intersection of $A_0, \ldots, A_n$, the sets A and B are disjoint subsets of X^n. They are also convex, and, if X is not finite dimensional, then the core of A is nonempty.

If, for $T \in \mathcal{T}$, the map $T^* : X^n \to X^n$ is defined as

$$T^*(x_1, \ldots, x_n) = (T(x_1), \ldots, T(x_n)) \qquad ((x_1, \ldots, x_n) \in X^n),$$

then A and B are $\mathcal{T}^*$-invariant with respect to the family of transformations $\mathcal{T}^* := \{T^* \mid T \in \mathcal{T}\}$ which is a commuting family of affine maps.

Thus, by Theorem 2, there exists a nonconstant affine map $f : X^n \to \mathbb{R}$ and a nonnegative function $\alpha^* : \mathcal{T}^* \to [0, \infty[$ such that $f|_A \le 0 \le f|_B$ holds and

$$f \circ T^* = \alpha^*(T^*) f \qquad (T^* \in \mathcal{T}^*). \tag{7}$$

Define the function α on $\mathcal{T}$ by $\alpha(T) := \alpha^*(T^*)$. The affine function f can be expressed in terms of some linear functions $\varphi_1, \dots, \varphi_n : X \to \mathbb{R}$ and a constant $c \in \mathbb{R}$ in the following form

$$f(x_1, \dots, x_n) = \varphi_1(x_1) + \cdots + \varphi_n(x_n) + c \qquad ((x_1, \dots, x_n) \in X^n). \tag{8}$$

It follows from $f|_A \leq 0$ that φ_i is bounded above on A_i for each fixed $i = 1, \dots, n$. Hence, with $c_i := \sup_{A_i} \varphi_i$, the function $f_i := \varphi_i - c_i$ satisfies (4) with equality for $i = 1, \dots, n$. Therefore, either f_i is identically zero or nonconstant. Now (8) can be rewritten as

$$f(x_1, \dots, x_n) = f_1(x_1) + \cdots + f_n(x_n) + c_0 \qquad ((x_1, \dots, x_n) \in X^n), \tag{9}$$

where $c_0 := c + c_1 + \cdots + c_n$. Since $\sup_A f \leq 0$, we have that $c_0 \leq 0$.

On the other hand, using that A_i and its complement are $\mathcal{T}$-invariant convex sets, and applying Theorem 2, we can state the existence of a nonconstant affine function $F_i : X \to \mathbb{R}$ and a nonnegative function $\alpha_i : \mathcal{T} \to [0, \infty[$ such that

$$F_i|_{A_i} \leq 0 \leq F_i|_{X \setminus A_i} \qquad (i = 1, \dots, n)$$

and

$$F_i \circ T = \alpha_i(T) F_i \qquad (T \in \mathcal{T}, i = 1, \dots, n). \tag{10}$$

Thus the conditions of Lemma 1 are satisfied for the functions F_i and f_i. Therefore f_i is a nonnegative multiple of F_i. It follows from (10) that

$$f_i \circ T = \alpha_i(T) f_i \qquad (T \in \mathcal{T}, i = 1, \dots, n). \tag{11}$$

Hence, by (7), (8), and (11), we get

$$\begin{aligned}
\alpha(T)(f_1(x_1) + & \cdots + f_n(x_n) + c_0) \\
&= \alpha^*(T^*) f(x_1, \dots, x_n) \\
&= f \circ T^*(x_1, \dots, x_n) \\
&= f(T(x_1), \dots, T(x_n)) \\
&= f_1 \circ T(x_1) + \cdots + f_n \circ T(x_n) + c_0 \\
&= \alpha_1(T) f_1(x_1) + \cdots + \alpha_n(T) f_n(x_n) + c_0
\end{aligned}$$

for $x_1, \dots, x_n \in X$ and $T \in \mathcal{T}$. If $f_i \equiv 0$, then we clearly have (5). In the other case, the range of $f_i(x_i)$ is $\mathbb{R}$ if x_i varies in X. Hence, the above identity yields (5) for the rest of the indices in $\{1, \dots, n\}$. Denote

$$f_0 := -(f_1 + \cdots + f_n)$$

Then we get (6), and thus (5) also follows for $i = 0$. To obtain (4) also for the case $i = 0$, observe that $f|_B \geq 0$ yields

$$-f_0(x) = f_1(x) + \cdots + f_n(x) = f(x, \dots, x) - c_0 \geq -c_0 \geq 0 \qquad (x \in A_0),$$

which is the desired inequality. The proof is thus complete. □

The Dubovitskii-Milyutin separation theorem for cones can be deduced from Theorem 3. The method of the proof is the same as that of Corollary 4.

Corollary 5. *Let X be a linear space and let $\mathcal{T}$ be a commuting family of linear selfmaps of X. Assume that $A_0, A_1, \ldots, A_n$ are nonempty convex $\mathcal{T}$-invariant cones of X such that their intersection is empty. Suppose that either X is finite dimensional or* $\operatorname{cor}(A_1), \ldots, \operatorname{cor}(A_n)$ *are nonempty sets. Then there exist linear functions $f_0, \ldots, f_n : X \to \mathbb{R}$ (not all identically zero) and a nonnegative function $\alpha : \mathcal{T} \to [0, \infty[$ such that (4), (5) and (6) hold.*

Remark 4. If X is a topological vector space, and the nonemptiness of the interior of $A_1, \ldots, A_n$ is assumed then the continuity of $f_0, f_1, \ldots, f_n$ (in Theorem 4 and Corollary 5) can also be stated, since then the linear part of $f_1, \ldots, f_n$ will be bounded from above by (4) on an open subset of X. The continuity of f_0 is then a consequence of (6).

Remark 5. If $\mathcal{T} = \{\mathrm{id}\}$, then the above results reduce to the well known separation theorem of Dubovitskii and Milyutin (see [2], [5]). For generalizations of that result in the semigroup setting see [11].

4 Sandwich theorems

In this part, we investigate the separation of convex and concave functions. We allow here extended real valued functions and define convexity, concavity in the usual way with the conventions

$$(-\infty) + x = (-\infty) + (-\infty) = c \cdot (-\infty) = -\infty, \qquad 0 \cdot (-\infty) = 0$$

for $x \in \mathbb{R}$, $c > 0$. An extended real valued function is called *affine* if it is convex and concave; it is said to be *linear* if it is sub- and superlinear and sublinearity means subadditivity and positive homogeneity.

Theorem 4. *Let X be a linear space and let $\mathcal{T}$ be a commuting family of affine maps of X into itself. Assume that $p, q : X \to [-\infty, \infty[$ are such that p is convex, q is concave, $q \le p$, and there exist $\alpha : \mathcal{T} \to [0, \infty[$ and $\beta : \mathcal{T} \to \mathbb{R}$ such that*

$$p \circ T \le \alpha(T)p + \beta(T), \qquad q \circ T \ge \alpha(T)q + \beta(T) \qquad (T \in \mathcal{T}). \tag{12}$$

Assume that the family of affine maps $\{\tau_T : \mathbb{R} \to \mathbb{R} \mid T \in \mathcal{T}\}$ defined by $\tau_T(t) := \alpha(T)t + \beta(T)$ is commuting. Then there exists an affine function $f : X \to [-\infty, \infty[$ such that

$$q \le f \le p \tag{13}$$

and

$$f \circ T = \alpha(T)f + \beta(T) \qquad (T \in \mathcal{T}). \tag{14}$$

Proof. If q is identically $-\infty$, then $f = -\infty$ satisfies the requirements of the theorem. Thus, we may assume that q is not identically $-\infty$. Then, due to its convexity, $p(x) > -\infty$ for all $x \in X$.

Define

$$A = \mathrm{epi}(p) := \{(x,t) \in X \times \mathbb{R} \mid p(x) \le t\}, \quad B = \{(x,t) \in X \times \mathbb{R} \mid t < q(x)\}.$$

Then, by the convexity of p and concavity of q, the sets A and B are convex. By $q \le p$, we have that A is disjoint from B. We show that

$$\mathrm{cor}\, A = \{(x,t) \in X \times \mathbb{R} \mid p(x) < t\},$$

and hence, $\mathrm{cor}(A) \neq \emptyset$.

For, let $(x_0, t_0) \in X \times \mathbb{R}$ be such that $p(x_0) < t_0$ and let $(x,t) \in X \times \mathbb{R}$ be arbitrary. Then

$$p((1-\lambda)x_0 + \lambda x) \le (1-\lambda)p(x_0) + \lambda p(x) \le (1-\lambda)t_0 + \lambda t$$

if λ is a sufficiently small nonnegative number, say if $\lambda \in [0, \varepsilon]$. Thus, for these values of λ,

$$(1-\lambda)(x_0,t_0) + \lambda(x,t) = \big((1-\lambda)x_0 + \lambda x, (1-\lambda)t_0 + \lambda t\big) \in \mathrm{epi}(p),$$

that is

$$[(x_0,t_0), (1-\varepsilon)(x_0,t_0) + \varepsilon(x,t)] \subset \mathrm{epi}(p).$$

Thus (x_0, t_0) belongs to the core of A.

Define $T^* : X \times \mathbb{R} \to X \times \mathbb{R}$ by

$$T^*(x,t) := (T(x), \tau_T(t)) = (T(x), \alpha(T)t + \beta(T)) \qquad (T \in \mathcal{T}).$$

Then $\mathcal{T}^* := \{T^* \mid T \in \mathcal{T}\}$ is a commuting family of affine maps. Due to assumption (12) and the nonnegativity of α, we have that A and B are $\mathcal{T}^*$-invariant sets.

Thus we are in the position to apply Theorem 2. Therefore, there exists a nonconstant affine function $F : X \times \mathbb{R} \to \mathbb{R}$ such that

$$F|_A \le 0 \le F|_B \tag{15}$$

and there exists a function $\alpha^* : \mathcal{T} \to [0, \infty[$ such that

$$F \circ T^* = \alpha^*(T^*)F \qquad (T^* \in \mathcal{T}^*). \tag{16}$$

The function F being affine, there exist a linear function $\varphi : X \to \mathbb{R}$ and constants $c, d \in R$ such that

$$F(x,t) = \varphi(x) + dt + c.$$

The first inequality in (15) yields

$$\varphi(x) + dt + c \le 0 \qquad \text{if } x \in X,\ p(x) \le t. \tag{17}$$

Taking the limit $t \to \infty$, we get that $d \le 0$. However, $d = 0$ is also impossible, otherwise (17) implies the boundedness of φ, whence it follows that F is constant. Thus $d < 0$. Substituting $t = p(x)$ in (17), we get that

$$f(x) := \frac{\varphi(x) + c}{-d} \le p(x) \qquad (x \in X). \tag{18}$$

The function f so defined satisfies the second inequality in (13).

By the second inequality in (15), we have

$$\varphi(x) + dt + c \ge 0 \qquad \text{if } x \in X,\ t < q(x). \tag{19}$$

If $q(x) = -\infty$ then $q(x) \le f(x)$ is trivial. If $q(x) > -\infty$ then taking the limit $t \to q(x)$ from below in (19), we get

$$\varphi(x) + dq(x) + c \ge 0 \qquad \text{if } x \in X, q(x) > -\infty.$$

This yields $q(x) \le f(x)$ also in the case $q(x) > -\infty$. Thus (13) is proved.

The invariance property (16) of F reduces to

$$\varphi \circ T(x) + d(\alpha(T)t + \beta(T)) + c = \alpha^*(T^*)(\varphi(x) + dt + c) \qquad (x \in X,\ t \in \mathbb{R}).$$

Thus, substituting $x = 0$ and $t = 0$, we get that $\alpha^*(T^*) = \alpha(T)$,

$$\varphi \circ T = \alpha(T)\varphi, \qquad d\beta(T) + c = \alpha(T)c.$$

Hence

$$\begin{aligned} f \circ T &= \frac{\varphi \circ T + c}{-d} = \frac{\alpha(T)\varphi + \alpha(T)c - d\beta(T)}{-d} \\ &= \alpha(T)\frac{\varphi + c}{-d} + \beta(T) = \alpha(T)f + \beta(T). \end{aligned}$$

Thus f satisfies (14) as well. □

Remark 6. One can describe the cases when the affine maps τ_T defined in the theorem form a commuting family. Given $S, T \in \mathcal{T}$, the maps τ_S and τ_T commute if and only if

$$\beta(S)(\alpha(T) - 1) = \beta(T)(\alpha(S) - 1),$$

that is, if the two dimensional vectors $(\alpha(T) - 1, \beta(T))$ and $(\alpha(S) - 1, \beta(S))$ are parallel to each other. Therefore, $\{\tau_T \mid T \in \mathcal{T}\}$ is commuting if and only if the set of two dimensional vectors

$$\{(\alpha(T) - 1, \beta(T)) \mid T \in (T)\}$$

has rank one, that is, if they are scalar multiples of some nonzero vector (u, v). In other words, there exists a function $\gamma : \mathcal{T} \to \mathbb{R}$ such that

$$\alpha(T) - 1 = \gamma(T)u, \qquad \beta(T) = \gamma(T)v \qquad (T \in \mathcal{T}).$$

There are essentially two different cases. If $u \neq 0$, then $\beta(T) = v(\alpha(T) - 1)/u = c(1 - \alpha(T))$ and α and $c \in \mathbb{R}$ are arbitrary. If $u = 0$, then $\alpha \equiv 1$ and β is arbitrary.

Using the sandwich theorem for convex functions, it is easy to deduce the sandwich theorem for sublinear, superlinear functionals following the ideas of the proof of Corollary 4.

Corollary 6. *Let X be a linear space and let $\mathcal{T}$ be a commuting family of linear selfmaps of X. Assume that $p, q : X \to [-\infty, \infty[$ are such that p is sublinear, q is superlinear, $q \leq p$, and there exists a function $\alpha : \mathcal{T} \to [0, \infty[$ such that*

$$p \circ T \leq \alpha(T)p, \qquad q \circ T \geq \alpha(T)q \qquad (T \in \mathcal{T}). \tag{20}$$

Then there exists a linear function $f : X \to [-\infty, \infty[$ such that (13) and

$$f \circ T = \alpha(T)f \qquad (T \in \mathcal{T}). \tag{21}$$

hold.

Remark 7. The result of Theorem 4 can also be derived using the generalization of Rodé's sandwich theorem due to Volkmann and Weigel. Rodé's theorem (c.f. Rodé [14]) is enough to obtain Corollary 6 only. The results in [14] are, however, much more general than ours from other points of view. They cover almost all of the known sandwich theorems, e.g. those obtained by Fuchsteiner [3], [4] for the separation of sub- and superadditive functions defined on Abelian semigroups. For a relatively simple proof of this essential result, see König [7].

Acknowledgment. This research has been supported by the Hungarian National Research Science Foundation (OTKA) Grant T-030082 and by the High Educational Research and Development Fund (FKFP) Grant 0310/1997.

References

1. G. Buskes, *The Hahn-Banach theorem surveyed*, Dissertationes Math. **327** (1993), 49 pp.
2. A. Ya. Dubovitskii and A. A. Milyutin, *Second variations in extremal problems with constraints*, Dokl. Akad. Nauk SSSR, **160**(1) (1965), 18-21.
3. B. Fuchssteiner, *Sandwich theorems and lattice semigroups*, J. Functional Analysis **16** (1974), 1–14.
4. B. Fuchssteiner and W. Lusky, *Convex cones*, North-Holland Math. Studies, Vol. **56**, Amsterdam – New York – Oxford, 1981.

5. I. V. Girsanov, *Lectures on Mathematical Theory of Extremum Problems*, Lecture Notes in Economics and Mathematical Systems 67, Springer Verlag, Berlin–Heidelberg–New York, 1972.
6. R. B. Holmes, *Geometric Functional Analysis and its Applications*, Graduate Texts in Mathematics 24, Springer Verlag, New York/Heidelberg/Berlin, 1975.
7. H. König, *On the abstract Hahn-Banach theorem due to Rodé*, Aequationes Math. **34** (1987), 89–95.
8. Zs. Páles, *On the separation of midpoint convex sets*, C. R. Math. Rep. Acad. Sci. Canada **8** (1986), 309-312.
9. Zs. Páles, *Hahn-Banach theorem for separation of semigroups and its applications*, Aequationes Math. **37** (1989), 141-161.
10. Zs. Páles, *A Stone-type theorem for Abelian semigroups*, Arch. Math. **52** (1989), 265-268.
11. Zs. Páles, *A generalization of the Dubovitskii-Milyutin separation theorem for Abelian semigroups*, Arch. Math. **52** (1989), 384-392.
12. Zs. Páles, *Geometric versions of Rodé's theorem*, Radovi Matematički **8** (1992/1998), 1-13.
13. A. W. Roberts and D. E. Varberg, *Convex Functions*, Academic Press, New York and London, 1973.
14. G. Rodé, *Eine abstrakte Version des Satzes von Hahn-Banach*, Arch. Math. **31** (1978), 474–481.
15. P. Volkmann and H. Weigel, *Systeme von Funktionalgleichungen*, Arch. Math. **37** (1981), 443-449.

Convexity and Generalized Convexity Methods for the Study of Hamilton-Jacobi Equations

Jean-Paul Penot[1] and Michel Volle[2]

[1] Université de Pau, Faculté des Sciences,
Laboratoire de Mathématiques Appliquées, UPRES A 5033
Av. de l'Université 64000 PAU, France
[2] Université d'Avignon, Faculté des Sciences,
Département de Mathématiques
33 rue Louis Pasteur 84000 AVIGNON, France

Abstract. We present a digest of recent results about the first-order Hamilton-Jacobi equation. We use explicit formulas of the Hopf and Lax-Oleinik types, stressing the role of quasiconvex duality: here the usual Fenchel conjugacy is replaced with quasiconvex conjugacies known from some years and the usual inf-convolution is replaced by the sublevel convolution. The role of the full theory of variational convergences (epi-convergence and sublevel convergence) is put in light for the verification of initial conditions. We observe that duality methods and variational convergences are not limited to the case of finite-valued functions as in the classical approaches to Hamilton-Jacobi equations. This extension allows to deal with problems arising from various cases, such as optimal control problems, attainability or obstacle problems.

2000 Mathematics Subject Classification. 35C05, 35CD05, 35E10, 35F30, 49L25, 49J52.

Keywords and phrases. conjugate, convexity, duality, Hamilton-Jacobi equation, subdifferential, viscosity solution.

1 Introduction

The Hamilton-Jacobi equation has been extensively studied during the last decades (see [1], [4]-[14], [19]-[22], [24]-[32], [34]-[39], [58]-[62]...) and several monographs ([5], [6], [27], [37], [59]) are devoted to it, partially or entirely. In particular, the notion of viscosity solution has given a new start to existence and uniqueness questions ([22], [21], [19], [6], [5], [27], [37]...).

Given a normed vector space (n.v.s.) X, with dual X^* and functions $g : X \to \overline{\mathbb{R}} := \mathbb{R} \cup \{-\infty, +\infty\}$, $H : X^* \times \mathbb{R} \to \overline{\mathbb{R}}$, the evolution Hamilton-Jacobi equation consists in finding solutions to the system

$$\frac{\partial u}{\partial t}(x,t) + H(Du(x,t), u(x,t)) = 0 \qquad (1)$$

$$u(x,0) = g(x) \qquad (2)$$

where $u : X \times \mathbb{R}_+ \to \overline{\mathbb{R}}$ is the unknown function, and Du (resp. $\frac{\partial u}{\partial t}$) denotes the derivative of u with respect to its first (resp. second) variable. When H depends on its first variable only, we identify it with the one variable function from which it derives. Note that we accept solutions, Hamiltonians and initial value functions taking the values $+\infty, -\infty$ or being discontinuous.

These equations have been successfully studied by many different means from partial differential equations techniques to control theory and game theory. In particular, when preparing [61], [62], [48], [53] we were not aware of a number of works presenting bridges between the study of partial differential equations and nonsmooth analysis ([5], [8], [13], [18], [24], [29], [59]...).

In the present paper we survey the use of tools from duality theory to find explicit solutions of (1)-(2). Duality techniques have been in use for solving Hamilton-Jacobi equations for a long time (see for instance [30], [38], [27]...), but not in a systematic way. In [10]-[12] the use of quasiconvex dualities has been introduced in this field; this line has been pursued in [1], [8], [61], [62]. A systematic use of duality methods has been suggested in [48] and developed in the convex case in [53] (for the state variable x), [31], [32] (for the variables (x, t)); thorough surveys with new results and proofs are given in [1], [8] and [54]. Here we give a short account of these results, referring to [1], [8] and the expanded version of the present paper [54] for the proofs and more complete developments. As in these references, the analogy between the convex case and the quasiconvex case is the guideline. However, for the sake of simplicity, we do not deal with all the quasiconvex dualities considered there. Also, we do not fully adopt here one of the new features of [54] which consists in replacing the conjugates of g and H by functions which are just related to these conjugates by inequalities. This technique allows to find more easily supersolutions and lower solutions and to use comparison results, but makes the presentation more complex.

As in [32] and [54] we choose to separate assumptions for (1) from assumptions which ensure (2); of course, the union of these assumptions gives conditions to solve the system (1)-(2). Our choice clarifies the roles of the different assumptions and enables to detect subsolutions and supersolutions in situations which would be excluded by the conjunction of the two sets of assumptions; along with comparison theorems such as [5] Th. 3.7 p. 56, [6] Th. 4.9 p. 100, one can get in this way estimates about solutions.

2 Preliminaries: generalized convex dualities

Various notions of conjugate function are available. Here, for a function f on $X \times \mathbb{R}$, conjugacy will be with respect to the first variable. The convex (or Legendre-Fenchel) conjugate is the natural tool for the convex case; it is given by

$$f^*(p,q) := \sup_{x \in X} (\langle p, x \rangle - f(x,q)) \qquad \text{for } (p,q) \in X^* \times \mathbb{R},$$

Here $\langle p, x\rangle$, also denoted by $p.x$, denotes the value of p on $x \in X$.

In the quasiconvex case, it is more difficult to recover a function $f : X \to \overline{\mathbb{R}}$ through biconjugacy. Thus, one usually defines the conjugate on $Y = X^* \times \mathbb{R}$ rather than on X^*. Moreover, the sublevel sets of f are bound to play a crucial role since the information we have is located in such sets. Among several possibilities, we select the correspondence $f \mapsto f^\sharp$ given by

$$f^\sharp(p,q) := \sup\{p.x : x \in [f \leq q]\} \quad \text{for} \quad y := (p,q) \in Y = X^* \times \mathbb{R},$$
$$[f \leq q] := f^{-1}([-\infty, q]).$$

which has been considered in [52] and used in connection with Hamilton-Jacobi equations in [8]-[10], [61], [62]. In fact, the "sharp" correspondence $D : f \mapsto f^\sharp$ is not a *conjugacy* (i.e. the relation $D(f + c) = D(f) - c$ is not always satisfied) and not even a duality, in the sense of [43]: the relation

$$D(\inf_{i \in I} f_i) = \sup_{i \in I} D(f_i)$$

defining *dualities* is not always satisfied. This relation is satisfied by the "flat" duality $f \mapsto f^\flat$ given by

$$f^\flat(p,q) := \sup\{p.x : x \in [f < q]\}, \quad \text{for} \quad y := (p,q) \in Y = X^* \times \mathbb{R},$$
$$[f < q] := f^{-1}([-\infty, q[)$$

which is closely related. Nonetheless, $f \mapsto f^\sharp$ is a *quasi-duality* in the sense of [48] that it is determined by a generating function $G_\sharp : X \times Y \times \overline{\mathbb{R}} \to \overline{\mathbb{R}}$, which is nonincreasing in its third variable, through the representation

$$f^\sharp(y) = \sup_{x \in X} G_\sharp(x, y, f(x)) \quad \text{for } y = (p,q) \in Y,$$

with $G_\sharp(x,p,q,r) = p.x - \iota_{[-\infty,q]}(r)$, where the *indicator function* ι_S of a subset S of a space T is the function given by $\iota_S(t) = 0$ for $t \in S$, $\iota_S(t) = +\infty$ for $t \in T \backslash S$. Thus, the extension presented in [48] of the process of [43] to quasi-dualities enables one to exhibit a reverse correspondence, also denoted by $k \mapsto k^\sharp$, from $\overline{\mathbb{R}}^Y$ to $\overline{\mathbb{R}}^X$, given by $k^\sharp(x) = \sup_{y \in Y} G'_\sharp(y, x, k(y))$, where $y := (p,q) \in Y$,

$$G'_\sharp(y,x,s) := \inf\{r \in \overline{\mathbb{R}} : G_\sharp(x,y,r) \leq s\}.$$

Since $G_\sharp$ is nonincreasing in its third variable, for any $k \in \overline{\mathbb{R}}^Y$ one has

$$\begin{aligned} k^\sharp(x) &= \sup\{\sup\{t \in \overline{\mathbb{R}} : G_\sharp(x,y,t) > k(y)\} : y \in Y\} \\ &= \sup\{t \in \overline{\mathbb{R}} : \exists y \in Y,\ G_\sharp(x,y,t) > k(y)\} \\ &= \sup\{t \in \overline{\mathbb{R}} : \exists (p,q) \in Y,\ q \geq t,\ p.x > k(p,q)\} \\ &= \sup\{q \in \overline{\mathbb{R}} :\ k_q^*(x) > 0\}. \end{aligned}$$

A similar computation shows that the flat conjugate $k^\flat$ of k is given by the same formula. Thus, one has the following equivalences which prove the second assertion of the next lemma (the first one being obvious):

$$k^\sharp(x) \le r \Longleftrightarrow k^\flat(x) \le r \Longleftrightarrow x \in \bigcap_{s>r} [k_s^* \le 0]. \tag{3}$$

Lemma 1. *For any function $f : X \to \overline{\mathbb{R}}$, the functions $f^\flat$ and $f^\sharp$ are l.s.c. sublinear functions of their first variables (or are identically equal to $-\infty$) and are nondecreasing functions of their second variables. Moreover, $f^\flat$ is jointly l.s.c.*

For any function $k : X^* \times \mathbb{R} \to \overline{\mathbb{R}}$, the function $k^\flat = k^\sharp$ is l.s.c. and quasiconvex.

The two musical correspondences are closely linked since for any $f : X \to \overline{\mathbb{R}}$ and any $q > \inf f$ one has

$$f^\flat(p,q) = \sup_{r<q} f^\sharp(p,r) = \sup_{r<q} f^\flat(p,r).$$

It is natural to introduce the class of functions f for which $f^\flat = f^\sharp$; we call it the class of *harmonious functions*. It contains the interesting class of quasiconvex functions f which can be written as $f = \varphi \circ g$ where g is a proper convex function such that $\inf g = -\infty$ (or $\inf g$ is not attained) and φ is an increasing function on $\mathbb{R}$ (this follows from [56] Prop. 5.3). More generally, we have the following characterization.

Proposition 1. *A function $f : X \to \overline{\mathbb{R}}$ is harmonious if and only if for each $q \in \mathbb{R}$ one has $\overline{co}([f < q]) = \overline{co}([f \le q])$. In particular, an harmonious function f has no minimizer with finite value and a l.s.c. quasiconvex function is harmonious if, and only if, it has no local minimizer with finite value.*

Proof. Since the support function of a subset A of X is equal to the support function of its closed convex hull $\overline{co}(A)$, the condition is sufficient. It is necessary because two closed convex subsets which have the same support functions are equal. Moreover, when $f^\flat = f^\sharp$, the infimum $q := \inf f(X)$ of f cannot be finite and attained since otherwise, for any $p \in X^*$ one would have $f^\flat(p,q) = -\infty$ and $f^\sharp(p,q) > -\infty$. The last assertion follows from the characterization of l.s.c functions f without local minimizers as functions such that for each $q \in \mathbb{R}$ the closure of $[f < q]$ is $[f \le q]$ (see [63], [64], [60]) □

Let us give some more attention to second conjugates (defined as conjugates of conjugates).

Lemma 2. *The flat and the sharp quasi-dualities defined above are hemi-dualities in the sense that for any function $f \in \overline{\mathbb{R}}^X$ one has*

$$f^{\flat\flat} \le f, \quad f^{\sharp\sharp} \le f.$$

However, for $k \in \overline{\mathbb{R}}^Y$, the inequality $k^{\sharp\sharp} \leq k$ is not always satisfied, so that some care is required.

The characterization of functions on X which coincide with their second conjugates is simple.

Proposition 2. *([1], [62], [48]) For each $f : X \to \overline{\mathbb{R}}$, the l.s.c. quasiconvex hull of f is $f^{\flat\flat} = f^{\sharp\sharp}$. In particular f is l.s.c. quasiconvex iff $f = f^{\flat\flat}$ iff $f = f^{\sharp\sharp}$.*

Proof. Let h be the l.s.c. quasiconvex hull of f. It is known (see [23]) that for any $r \in \mathbb{R}$ one has

$$[h \leq r] = \bigcap_{s>r} \overline{co}([f < s] = \bigcap_{s>r} \overline{co}([f \leq s]).$$

Now, by (3) with $k = f^\flat$ (resp. $f^\sharp$), one has

$$[f^{\flat\flat} \leq r] = \bigcap_{s>r} [k_s^* \leq 0] = \bigcap_{s>r} \overline{co}([f < s])$$

$$(\text{resp. } [f^{\sharp\sharp} \leq r] = \bigcap_{s>r} [k_s^* \leq 0] = \bigcap_{s>r} \overline{co}([f \leq s])),$$

so that $h = f^{\flat\flat} = f^{\sharp\sharp}$. □

Let us observe that since $k^\sharp = k^\flat$ for any $k \in \overline{\mathbb{R}}^Y$, one also has $h = f^{\sharp\flat} = f^{\flat\sharp}$.

A characterization of the functions k on $X^* \times \mathbb{R}$ which are equal to their flat biconjugates $k^{\flat\flat}$ is easy.

Proposition 3. *([61]) A function k on $Y := X \times \mathbb{R}$ is equal to $k^{\flat\flat}$ iff it is l.s.c. on Y, is a proper (or identically $-\infty$) sublinear function of its first variable and is nondecreasing in its second variable.*

It would be desirable to dispose of a similar characterization for the sharp duality. Such an aim is pursued in [49]. A sufficient condition appears in the following lemma.

Lemma 3. *([1]) When $k : X^* \times \mathbb{R} \to \overline{\mathbb{R}}$ is nondecreasing in its second variable, for each $x \in X$ one has $k^\flat(x) = k^\sharp(x) = \inf\{q \in \mathbb{R}\colon k_q^*(x) \leq 0\}$ and, for each $q \in \mathbb{R}$, denoting by k_{q+0} the function given by $k_{q+0}(p) := k(p, q+0) := \lim_{r \searrow q} k(p,r) := \inf_{r>q} k(p,r)$ one has*

$$[k^\flat \leq q] = [k^\sharp \leq q] = [k_{q+0}^* \leq 0] = \partial k_{q+0}(0). \tag{4}$$

In particular, when for each $p \in X^*$ the function $k(p,\cdot)$ is u.s.c. and nondecreasing, one has

$$k^\flat(x) = k^\sharp(x) = \min\{q \in \mathbb{R} : \ k_q^*(x) \leq 0\},$$

in the usual sense that the infimum is attained when it is finite and in that case $k^{\sharp\sharp} \leq k$.

If moreover $k_q := k(\cdot, q)$ *is sublinear and weak* l.s.c. for each* $q \in \mathbb{R}$, *then* $k^{\sharp\sharp} = k$.

Proof. The first assertion is obvious as $q \mapsto k_q^*(x)$ is nonincreasing. One has $k^\sharp(x) \leq q$ iff for each $r > q$ and each $p \in X^*$ one has $p.x \leq k(p, r)$ iff for each $p \in X^*$ one has $p.x \leq k(p, q+0)$ iff $k_{q+0}^*(x) \leq 0$. When for each $p \in X^*$ the function $k(p, \cdot)$ is u.s.c. and nondecreasing, one has $k_{q+0} = k_q$, hence, by (4), for each $(p, q) \in X^* \times \mathbb{R}$,

$$\begin{aligned} k^{\sharp\sharp}(p,q) &= \sup\{p.x : x \in [k_{q+0}^* \leq 0]\} \\ &= \sup\{p.x : x \in [k_q^* \leq 0]\} \leq k_q(p). \end{aligned}$$

When, furthermore, k_q is sublinear and weak* l.s.c., it is the supremum of the continuous linear forms x majorized by k_q i.e. in $[k_q^* \leq 0]$. Then $k^{\sharp\sharp}(\cdot, q) = k_q(\cdot)$. □

A necessary condition for a function k on $X^* \times \mathbb{R}$ to be equal to its biconjugate $k^{\sharp\sharp}$ (or, more generally, to be the conjugate $f^\sharp$ of some function f on X) is contained in the following statement.

Lemma 4. *([1]) If h denotes the l.s.c. quasiconvex hull of a function* $f : X \to \overline{\mathbb{R}}$, *the functions* $k = f^\flat$, $f^\sharp$ *are such that* $k_{q+0}^* = (h_q^\sharp)^*$, k_{q+0} *being* $\inf_{r>q} k_r$. *Moreover* $h_q^\sharp$ *is the l.s.c. hull of* k_{q+0}. *In particular, when f is l.s.c. and quasiconvex, the function* $f_q^\sharp$ *is the l.s.c. hull of* $f_{q+0}^\sharp$.

Another sufficient condition holds under reflexivity, quasiconvexity and coercivity assumptions (recall that g is said to be coercive if $g(x) \to +\infty$ as $\|x\| \to +\infty$).

Lemma 5. *If X is reflexive and if g is a l.s.c. coercive quasiconvex function, then, for each* $p \in X^*$, *the functions* $G(p, \cdot) := g^\sharp(p, \cdot)$ *are upper semicontinuous (u.s.c.) and nondecreasing. Moreover one has* $G^{\sharp\sharp} = G$.

Proof. The function $g^\sharp(p, \cdot)$ is clearly nondecreasing; suppose $g^\sharp(p, \cdot)$ is not u.s.c. at some $q \in \mathbb{R}$. Then there exist some real number $r > g^\sharp(p, q)$ and some sequence (q_n) with limit q such that $g^\sharp(p, q_n) > r$ for each n. Let $x_n \in [g \leq q_n]$ be such that $p.x_n \geq r$. Since (x_n) is contained in the weakly compact set $[g \leq \sup q_n]$, the sequence (x_n) has a weak cluster point x_∞. Since g is weakly l.s.c. one has $g(x_\infty) \leq r$ and $p.x_\infty \geq r$, a contradiction with $r > g^\sharp(p, q)$. The last assertion follows from Lemma 3. □

Note that one cannot drop the quasiconvexity and coercivity assumptions, as shown by the following example.

Example 1. Let $X = \mathbb{R}^2$ and let g be given by $g(x_1, x_2) = 0$ for $x_2 \leq 0$, $g(x_1, x_2) = x_1 - \log x_2$ for $x_2 > 0$. Then, for $p := (0, 1)$ the function $g^\sharp(p, \cdot)$ is not u.s.c. at 0 since $g^\sharp(p, 0) = 0$ and $g^\sharp(p, q) = +\infty$ for $q > 0$.

Two other quasi-dualities are considered in [1] and [51], given by

$$f^{\top}(p,q) := \sup \{p.x - f(x) : x \in [f < q]\},$$

$$f^{\%}(p,q) := \sup \{p.x - f(x) : x \in [f \leq q]\}.$$

These quasi-dualities are adapted to an important subclass of the class of l.s.c. quasiconvex functions we call the class of "truncavex" functions; a function being *truncavex* if it is a supremum of a family of truncated continuous affine functions, i.e. the supremum of a family of functions of the form $g \wedge c$, where g is a continuous affine function on X and c is a constant. This class is a proper subclass of the class of l.s.c. quasiconvex functions: the function $x \mapsto x^3$ on $\mathbb{R}$ does not belongs to it. This class of functions has interesting duality properties (see [41], [42], [52]) but will not be dealt with in this short survey.

The dualities described above give rise to subdifferentials, following a general process devised by Martinez-Legaz and Singer in a series of papers ([43]-[45]). This process is extended to quasi-dualities in [48]. We evoke some of them in the next section, without attempting to relate them to the quasi-dualities presented here.

3 Generalized solutions

Since what can be considered as a solution to equation (1) is not necessarily smooth, one has to define concepts of generalized solutions. Historically, the first notions used differentiability in a dense subset. A crucial step forward has been performed by Crandall and Lions ([22], [21], see also [19]) who introduced the concept of viscosity solution. This concept can be described by using the *viscosity subdifferential* of f at z which is the set of derivatives at z of functions φ of class C^1 such that $f - \varphi$ attains a local minimum at z. In spaces with Fréchet smooth Lipschitz bump functions, this subdifferential coincides with the *Fréchet subdifferential* of f at z which is the set

$$\partial^- f(z) := \left\{ z^* \in Z^* : \liminf_{\|u\| \to 0_+} \frac{1}{\|u\|} [f(z+u) - f(z) - \langle z^*, u \rangle] \geq 0 \right\}.$$

In finite dimensional spaces $\partial^- f(z)$ also coincides with the *Hadamard* (or contingent) *subdifferential* of f at z, given by

$$\partial f(z) := \left\{ z^* \in Z^* : \forall w \in Z, \liminf_{(t,v) \to (0_+, w)} \frac{1}{t} [f(z+tv) - f(z) - \langle z^*, tv \rangle] \geq 0 \right\}.$$

but, in the general case, $\partial f(z)$ contains $\partial^- f(z)$. Outside the class of spaces with a Fréchet smooth Lipschitz bump function, the notion of solution given with the Fréchet subdifferential is stronger than the corresponding one using the viscosity subdifferential and is weaker than the one with the Hadamard subdifferential.

Since each of the preceding subdifferentials (and others) can be used for a concept of solution, it may be convenient to use an unspecified subdifferential $\partial^?$ considered as a mapping $\partial^? : \mathcal{F}(X) \times X \to 2^{X^*}$, where $\mathcal{F}(X)$ is some function space on X. In principle, four concepts of solution could be considered, since an equality is replaced by an inequality and since in general $\partial(-f)(x) \neq -\partial f(x)$. Originally, only two such concepts have been selected.

Definition 1. Given a subdifferential $\partial^?$, a function $u : X \times \mathbb{R} \to \overline{\mathbb{R}}$ is a *supersolution* (for $\partial^?$) if for any $(x,t) \in X \times \mathbb{P}$ (where $\mathbb{P} :=]0, \infty[$ is the set of positive numbers) and any $(p,q) \in \partial^? u(x,t)$ one has $q + H(p, u(x,t)) \geq 0$. Similarly, u is called a (Crandall-Lions) *subsolution* (for $\partial^?$) if for $(x,t) \in X \times \mathbb{P}$ and any $(p,q) \in -\partial^?(-u)(x,t)$ one has $q + H(p, u(x,t)) \leq 0$. It is a (Crandall-Lions) *solution* if it is both a supersolution and a subsolution.

When $\partial^?$ is the Hadamard (or contingent) subdifferential ∂ we say that u is a *Hadamard* or *contingent* subsolution or supersolution. Observe that we do not impose that a solution is continuous or even l.s.c.

Although the preceding notions seem to be the most important concepts, it may be useful to consider other ones. If $q + H(p, u(x,t)) \leq 0$ for any $(x,t) \in X \times \mathbb{P}$ and any $(p,q) \in \partial u(x,t)$ we said in [53] that u is a *Hadamard lower solution*. Note that if u is a Hadamard lower solution and a Hadamard supersolution, then for any $(x,t) \in X \times \mathbb{P}$ and any $(p,q) \in \partial u(x,t)$ one has $q + H(p, u(x,t)) = 0$; in such a case, we say that u is a *subdifferential solution* or a *bilateral solution* ([5]) or a *Barron-Jensen solution,* in view of [9], where that notion emerged. This concept of solution is particularly attractive since it only involves subdifferentials. However, the next two examples show that a bilateral solution is not always l.s.c., so that we cannot use the terminology of [9] in such a case.

Example 2. Let $X = \mathbb{R}$ and let g and H be given by $H(p) = |p|$, $g(x) = \chi_{\mathbb{R}_+}(x)$ where $\chi_{\mathbb{R}_+}(x) = 0$ for $x < 0$, $\chi_{\mathbb{R}_+}(x) = 1$ for $x \geq 0$. Then u given by $u(x,t) = \chi_{\mathbb{R}_+}(x - t)$ is a Hadamard bilateral solution to (1) which is not l.s.c. □

Example 3. ([1]) Let g and H be given by $g(x) = -\frac{1}{2}\|x\|^2$, $H(p) = \frac{1}{2}\|p\|^2$. Then the Lax solution u given by $u(x,t) = \frac{1}{2(t-1)}\|x\|^2$ for $(x,t) \in X \times [0,1)$, $u(0,1) = 0$, $u(x,1) = -\infty$ for $x \neq 0$, $u(x,t) = -\infty$ for $(x,t) \in X \times (1,\infty)$ is a bilateral solution which is not l.s.c. Note that the relation $q + H(p) \geq 0$ is not satisfied for $(p,q) \in \{0\} \times (-\mathbb{P}) \subset -\partial(-u)(0,1)$.

Let us note that a bilateral solution can be a Crandall-Lions solution without satisfying the relation $q + H(p) \geq 0$ for any $(x,t) \in X \times \mathbb{P}$, $(p,q) \in -\partial(-u)(x,t)$.

Example 4. Let g and H be given by $g(x) = -\|x\|$, $H(p) = \|p\|$. Then the Lax solution u given by $u(x,t) = -\|x\| - t$ for $(x,t) \in X \times \mathbb{R}_+$ is a

Hadamard bilateral solution and a Crandall-Lions solution but for $(0,-1) \in -\partial(-u)(0,t)$ one has $q + H(p) < 0$. □

It may happen that a Crandall-Lions solution is not a Hadamard lower solution.

Example 5. Let g and H be given by $g(x) = \|x\|$, $H(p) = -\|p\|$. Then the Hopf solution v given by $v(x,t) := \|x\| + t$ is a Crandall-Lions solution; however v is not a Hadamard lower solution since $(0,1) \in \partial v(0,t)$ for each $t \geq 0$. □

Other subdifferentials can be used in our general definition; for instance, the Fréchet subdifferential or the *Fenchel-Moreau subdifferential* of convex analysis given by

$$\partial^c f(z) := \{z^* \in Z^* : \forall w \in Z \ \langle z^*, w - z\rangle \leq f(w) - f(z)\}$$

or the *lower subdifferential* of Plastria given by

$$\partial^< f(z) := \{z^* \in Z^* : \forall w \in [f < f(z)] \ \langle z^*, w - z\rangle \leq f(w) - f(z)\}$$

or the *infradifferential* $\partial^{\leq} f(z)$ in which $[f < f(z)]$ is replaced with $[f \leq f(z)]$. We will also use the following variant of the Greensberg-Pierskalla subdifferential

$$\partial^\nu f(z) = N^r([f \leq f(z)], z),$$

where $N^r(S,z)$ is the set of $z^* \in Z^*$ such that $\langle z^*, w\rangle \leq 0$ for any $w \in Z$ for which there exists a sequence $(t_n) \to 0_+$ with $z + t_n w \subset S$ for each $n \in \mathbb{N}$. Note that the radial normal cone $N^r(S,z)$ is larger than the usual normal cone $N(S,z) := \partial\iota_S(z)$ and coincides with it when S is convex. Therefore, for a quasiconvex function f, the *normal subdifferential* $\partial^\nu f(z)$ coincides with the subdifferential introduced in [46], [48]:

$$\partial^\nu f(z) = \left\{z^* \in Z^* : \ \langle z^*, w - z\rangle \leq 0 \ \forall w \in f^{-1}(f(z) - \mathbb{R}_+) \ \right\}.$$

It is easy to show that for any function f finite at z one has $\mathbb{R}_+ \partial f(z) \subset \partial^\nu f(z)$. As a consequence, the concept of *normal solution* which consists in replacing ∂ by ∂^ν is stronger than the concept of Hadamard solution.

Since our approach relies on a notion of subdifferential, a knowledge of the calculus rules for subdifferentials may be useful. The convex case is classical; for the quasiconvex case we refer to [48], [56].

4 The Hopf-Lax formulas

In order to instil some versatility in the search of solutions to (1), we introduce a function $G : X^* \to \mathbb{R} \cup \{+\infty\}$ which is meant to replace the conjugate of g. This idea appears in [54], where a function $F : X^* \to \mathbb{R} \cup \{+\infty\}$ which is comparable to H, if not H itself, is also used. Taking for F and G functions for which the explicit formulae are easier to compute and using comparison results, one can obtain bounds for the solutions. Here, for the sake of simplicity, we limit our study to the case $F = H$. We assume that

$$\mathrm{dom} H \cap \mathrm{dom} G \neq \emptyset. \tag{5}$$

The Hopf-Lax formula in the convex case can be adapted to our data as

$$v(x,t) := (G + tH)^*(x) := \sup_{p \in \mathrm{dom} H \cap \mathrm{dom} G} (p.x - G(p) - tH(p)) \quad \text{for } (x,t) \in X \times \mathbb{R}_+ \tag{6}$$

where $0H$ is interpreted as $\iota_{\mathrm{dom}\, H}$, the indicator function of $S := \mathrm{dom} H$. This formula is an incentive to use other dualities adapted to the quasiconvex case. In this case we will consider the following function as a possible candidate for a solution of (1):

$$v(x,t) := (G + tH)^\sharp(x) \ \text{ for } (x,t) \in X \times \mathbb{R}_+. \tag{7}$$

Note that this formula introduced in [11] in the case $G = g^\sharp$ and studied in [1], [8], [48], coincides with the one obtained by using the duality $\flat$, and G can stand for $g^\flat$ instead of $g^\sharp$.

Lemma 6. *For any functions G, H, the function v defined by (7) (resp. (6)) is quasiconvex (resp. convex) and lower semicontinuous.*

Note that if assumption (5) is not satisfied, the function v is identically $-\infty$ on $X \times \mathbb{R}_+$ so that the following statements are trivial. In the classical case $G = g^*$, g is supposed to be Lipschitz and convex and H is supposed to be finite and continuous ([30], [4], [38], [26], [27]...) or boundedness and uniform continuity assumptions are made; see [5] for instance for a comprehensive treatment. The fact that g and H are not supposed to be finite everywhere allows to deal with cases such as indicator functions or cases similar to the example $H(p) = -\sqrt{1-p^2}$ for $p \in [-1,1]$, $H(p) = +\infty$ for $p \in \mathbb{R} \backslash [-1,1]$.

Example 6. Let $g := \iota_{\{a\}}$. Then, with $G = g^*$, formula (6) gives $v(x,t) = tH^*((x-a)/t)$ and, with $G = g^\sharp$, formula (7) yields $v(x,t) = \max(H^\sharp((x-a)/t), 0)$. □

Example 7. Let $g := \min(\|\cdot\|, 1)$, $H(p,q) = c\,\|p\|$. Then, with $G = g^*$, formula (6) gives $v(x,t) = -tH(0)$ and, with $G = g^\sharp$, formula (7) yields $v(x,t) = \min((\|x\| - ct)_+, 1)$. □

Example 8. Let $g := \log\|\cdot\| - c$, $H(p,q) = e^q\,\|p\|$. Then, with $G = g^\sharp$, formula (7) yields $v(x,t) = \log\|x\| - \log(t + e^c)$. □

In our first result, no condition on H is required (but our standing assumption (5)) when one takes $G = g^*$.

Theorem 1. *The function v given by formula (6) is a Hadamard supersolution to (1) when one of the following assumptions holds:*

(a) $(G + \iota_{\mathrm{dom}\, H})^{**} \geq G$ (or, a fortiori, $G^{**} = G$, what occurs when $G = g^*$);

(b) $(G + tH)^{**} = G^{**} + tH^{**}$ for each $t \in \mathbb{P}$.

The following theorem is an analog for the quasiconvex case; see [1] and [8] for case b) and [54] for cases a), b) and c). These last two cases make use of Lemma 5.

Theorem 2. *The function v given by formula (7) is a supersolution to (1) for the normal subdifferential ∂^ν, hence for the Hadamard subdifferential, under each of the following assumptions:*

(a) $(G + \iota_{\mathrm{dom}\, H})^{\sharp\sharp} \geq G$, and for each $p \in X^*$ the functions $H(p,\cdot)$ and $G(p,\cdot)$ are u.s.c.

(b) $G = g^\sharp$ with g l.s.c., quasiconvex and H is nondecreasing in its second variable and jointly u.s.c.

(c) $G := g^\sharp$ where g is inf-compact for the weak topology and H is u.s.c. in its second variable.

A similar result holds when the hemi-duality $\sharp$ is replaced with the hemi-duality %, the normal subdifferential being replaced with the infradifferential $\partial^{\leq}$ (see [54]). Note that the upper semicontinuity assumption adopted here relaxes the continuity assumption made in [1] Th. 6.9 but is probably not the best possible assumption (see [49]).

Proof of case (a). Let $(x,t) \in X \times \mathbb{P}$ and $(p,q) \in \partial^\nu v(x,t)$, and let $r := v(x,t) = (G+tH)^\sharp(x)$. For any $w \in X$ such that $v(w,0) := (G+\iota_{\mathrm{dom}\, H})^\sharp(w) \leq r$, by the definition of ∂^ν and the quasiconvexity of v we have

$$p.(w-x) - qt \leq 0. \tag{8}$$

Using the assumption $(G + \iota_{\mathrm{dom}\, H})^{\sharp\sharp} \geq G$ and taking the supremum over $w \in [(G + \iota_{\mathrm{dom}\, H})^\sharp \leq r]$, we get

$$G(p,r) \leq (G + \iota_{\mathrm{dom}\, H})^{\sharp\sharp}(p,r) \leq p.x + qt. \tag{9}$$

Suppose $\varepsilon := -(q + H(p,r)) > 0$, so that

$$p.x - G(p,r) - tH(p,r) \geq -qt - tH(p,r) = t\varepsilon.$$

The semicontinuity assumption ensures that for each $s > r$ close enough to r one has

$$(G(\cdot,s) + tH(\cdot,s))^*(x) \geq p.x - G(p,s) - tH(p,s) > 0.$$

Therefore, one gets $v(x,t) \geq s > r$, a contradiction. □

Now let us give conditions ensuring that v given by (6) is a subsolution. Let us say that the conjugate of $G+tH$ is *exact* if for any $x \in X$ the supremum of $p \mapsto p.x - G(p) - tH(p)$ on X^* is attained.

Theorem 3. *If for each $t > 0$ the conjugate of $G + tH$ is exact, then (6) gives a subsolution to (1).*

Concrete conditions ensuring exactness are given in the following corollary.

Corollary 1. *([53]) Under assumption (A) or (A') or (B) below, (6) gives a subsolution to (1):*
(A) G and H are weak l.s.c. and* $\liminf_{\|p\|\to\infty} G(p)/\|p\| = +\infty$,
$\liminf_{\|p\|\to\infty} H(p)/\|p\| > -\infty$;
(A') $G = g^*$, g is majorized on bounded subsets of X, H is weak l.s.c. and* $\liminf_{\|p\|\to\infty} H(p)/\|p\| > -\infty$;
(B) $G = g^*$, H is weak l.s.c. and* $\liminf_{\|p\|\to\infty} H(p)/\|p\| = +\infty$.

Let us note that assumption (A') implies (A) and is satisfied whenever X is finite dimensional, $G = g^*$, g is finite, convex and H is l.s.c. proper convex. However, (A) and (B) do not involve an explicit convexity assumption. In particular, it is known that certain integral functionals are weakly lower semicontinuous without being convex ([15],...).

Now, let us turn to the quasiconvex formula (7).

Proposition 4. *([54]) Suppose $H^{\flat\flat} = H$. Then v given by $v(x,t) = (g^\flat + tH)^\flat(x)$ is a subsolution to (1) for the normal subdifferential ∂^ν.*

Finally, let us present assumptions ensuring that v given by (6) is a lower solution.

Proposition 5. *([53]) If for each $t \in \mathbb{P}$ the function $G + tH$ is a closed proper convex function, or if* $\operatorname{dom} G^* \neq \emptyset$, *$H^{**} = H$, then v given by (6) is a lower solution. In particular, if $G = g^*$, $H = H^{**}$, with our standing assumption (5), then v given by (6) is a Hadamard subdifferential solution to (1).*

5 The Lax-Oleinik formulas

In the convex case, the Lax-Oleinik formula is defined with the help of the classical infimal convolution operator $\Box$ given for $f, g \in \overline{\mathbb{R}}^X$, $x \in X$ by

$$(f \Box g)(x) := \inf\{f(u) + g(v) : u, v \in X,\ u + v = x\}. \qquad (10)$$

In the quasiconvex case we use instead the *sublevel convolution* operator $\Diamond$ on $\overline{\mathbb{R}}^X$ given by

$$(f \Diamond g)(x) = \inf\{f(u) \vee g(v) : u, v \in X,\ u + v = x\} \qquad f, g \in \overline{\mathbb{R}}^X,\ \ x \in X. \quad (11)$$

Again, we introduce some versatility by using for each $t \in \mathbb{P}$ a function h_t as a substitute to a "preconjugate" of tH, i.e. a function h_t on X whose conjugate can be compared with tH. The semigroup property $h_r \Box h_s = h_{r+s}$ (resp. $h_r \Diamond h_s = h_{r+s}$) plays an important role. It is easy to check when h_t is deduced from a single function h (which is a kind of preconjugate of H).

Lemma 7. *(a) If h is convex and $h_t := th(t^{-1}\cdot)$, then, for any $r, s > 0$, one has $h_r \Box h_s = h_{r+s}$.*

(b) If h is quasiconvex and $h_t := h(t^{-1}\cdot)$, then, for any $r, s > 0$, one has $h_r \Diamond h_s = h_{r+s}$.

Note that $(h_t)^* = tH$ when $h^* = H$ (resp. $(h_t)^\sharp = tH$ when $h^\sharp = H$). However, assuming the property $h = H^\sharp$ does not imply that $h^\sharp \leq H$, although this occurs when H is u.s.c. and nondecreasing in its second variable (with equality when moreover H is l.s.c. and sublinear in its first variable). The following formulas generalize the Lax-Oleinik representation:

$$u(\cdot, t) := g \Box h_t, \tag{12}$$

$$u(\cdot, t) = h_t \Diamond g. \tag{13}$$

We also assume that

$$\mathrm{dom} g \neq \emptyset, \quad \mathrm{dom}\, h \neq \emptyset. \tag{14}$$

When $h \leq H^*$, the assumption $\mathrm{dom}\, h \neq \emptyset$ is satisfied when $H(\cdot) \geq \langle \cdot, a \rangle - b$ for some $a \in X$, $b \in \mathbb{R}$; if $h = H^*$ it is equivalent to the later condition. In the classical case, H is supposed to be convex (or concave), $h = H^*$ and g is supposed to be Lipschitz ([30], [37], [26], [4]...) or H is supposed to be Lipschitz, convex and g is supposed to be continuous ([59]).

Lemma 8. *When g and h are convex (resp. quasiconvex), u given by (12) (resp. (13)) is convex (resp. quasiconvex), but not necessarily l.s.c.*

The proof relies on the fact that u is the performance function (with respect to the parameter (x,t)) of the function $(x,t,y) \mapsto g(x-y) + th(t^{-1}y)$ (resp. $(x,t,y) \mapsto g(x-y) \vee h(t^{-1}y)$) which is convex (resp. quasiconvex) in view of the following relations valid for $r', r'' \geq 0$, $r' + r'' = 1$, $t := r't' + r''t''$:

$$r't'h(\frac{x'}{t'}) + r''t''h(\frac{x''}{t''}) \geq th(\frac{1}{t}(r'x' + r''x''))$$

$$h(\frac{x'}{t'}) \vee h(\frac{x''}{t''}) \geq h(\frac{1}{t}(r'x' + r''x'')).$$

Example 9. With $g := \iota_{\{a\}}$ one has $u(x,t) = tH^*((x-a)/t)$ for u given by (12) (resp. $u(x,t) = (tH)^\sharp(x-a)$ for u given by (13)).
Example 10. Let $g := \min(\|\cdot\|, 1)$, $H(p,q) = c\|p\|$. Then, with $h = H^*$, formula (12) gives $u(x,t) = \min((\|x\| - ct)_+, 1)$ as does formula (13) for $h = H^\sharp$. □
Example 11. Let $g := \log\|\cdot\| - c$, $H(p,q) = e^q\|p\|$. Then, with $G = g^\sharp$, formula (13) yields $u(x,t) = \log\|x\| - \log(t + e^c)$. □

Theorem 4. *([53]) (a) If $h^* \leq H$, in particular if $h = H^*$, and if in (12) the inf-convolution is exact, then u is a Hadamard supersolution;*

(b) If $h^* \leq H$ and if g and h are convex, then u is a Hadamard supersolution;

(c) If $h^* \leq H$ and if the following condition holds, then u is a Fréchet supersolution:

$$(C) \qquad \liminf_{\|x\| \to \infty} g(x)/\|x\| > -\liminf_{\|x\| \to \infty} h(x)/\|x\|.$$

It is easy to see that this assumption is satisfied whenever the following condition holds:

(C') h^* is bounded above on the ball $B(0,r)$ of center 0 and radius r (a fortiori when $h \geq H^*$ with H bounded above on $B(0,r)$) and g is bounded below by $d - c\|\cdot\|$ for some $c < r$ and some $d \in \mathbb{R}$.

Corollary 2. *Suppose X is a dual space, g and h are weak* l.s.c. and assumption (C) holds. Then u given by (12) is a l.s.c. Hadamard supersolution.*

The following result and its corollary are the central facts about Lax solutions: when one takes $h = H^*$, the assumption that H is closed proper convex is a natural assumption since h only depends on H^{**}.

Theorem 5. *Suppose h is convex and $h^* \geq H$. Then u given by (12) is a Hadamard lower solution and a Hadamard subsolution.*

Corollary 3. *Suppose H is a closed proper convex function and $h = H^*$. Then u given by (12) is a Hadamard lower solution and a Hadamard subsolution. If moreover one of the following conditions holds, then u is a Hadamard subdifferential solution:*

(a) the inf-convolution in the definition of u is exact at (x,t);

(b) X is reflexive, condition (C) holds and g is weakly l.s.c.;

(c) X is finite dimensional and condition (C) holds;

(d) g is convex.

Now let us turn to the quasiconvex case, closely following [1], [8], [48] and [61], [62], albeit our assumptions and conclusions are slightly more general. Here g is not assumed to be convex or quasiconvex and H is not necessarily sublinear in its first variable.

Theorem 6. *(a) Suppose $h^\sharp \geq H$ and h is quasiconvex. Then u, as defined in (13), is a normal lower solution to (1).*

(b) If $h^\flat \geq H$ and h is quasiconvex, u is a normal subsolution and a normal lower solution to (1). In particular, these conclusions hold when $h = H^\flat$, H is l.s.c., is a proper (or identically $-\infty$) sublinear function of its first variable and is nondecreasing in its second variable.

Theorem 7. *Let u be defined by (13) with $h_t := h(\cdot/t)$ for some function h on X such that $h^\sharp \leq H$.*

(a) When the sublevel convolution is exact, u is a normal supersolution to (1).

(b) If moreover h is quasiconvex and $H = h^\sharp$, then u is such that $q + H(p, u(x,t)) = 0$ for each $(x,t) \in X \times \mathbb{P}$ and each $(p,q) \in \partial^\nu u(x,t)$.

(c) If $h^\sharp \leq H$ and if the following condition holds, then u is a Fréchet supersolution:

$$(C_\sharp) \qquad \liminf_{\|x\| \to \infty} g(x) \vee \liminf_{\|x\| \to \infty} h(x) = +\infty.$$

Corollary 4. *Suppose X is reflexive, g and h are weakly l.s.c., $H \geq h^\sharp$ and either g is coercive or $h := H^\sharp$, with $H^{\sharp\sharp}$ is finite, $H^{\sharp\sharp} \leq H$. Then u defined by (13) is a l.s.c. normal supersolution to (1).*

If moreover $H = h^\sharp$ then u is a normal subdifferential solution to (1).

Recall that the equality $H = h^\sharp$ occurs in particular when $h = H^\sharp$ and H is sublinear, l.s.c. in its first variable and nondecreasing, u.s.c. in its second variable.

6 Initial conditions

Unless one imposes stringent assumptions, one cannot expect that the initial condition (2) is satisfied in a simple, naive sense. In [48] Prop. 12.11-12.13 and in [53] section 4 we suggested it is satisfied in the generalized sense of epiconvergence without knowing that pioneering contributions in this line had been made by Barles and Perthame [7] and Barron and Jensen [9]. In fact, these early contributions do not take into account the full force of epiconvergence.

Let us briefly recall the basic definitions of epi-convergence (see [2], [57] for more information). Given a family $(f_t)_{t>0}$ of functions on X parametrized by $\mathbb{P} :=]0, \infty[$, its (sequential) weak epi-limit inferior and its strong epi-limit superior are given respectively by

$$\left(e_w - \liminf_{t \to 0_+} f_t\right)(x) = \inf_{(t_n) \to 0_+} \inf_{(x_n) \rightharpoonup x} \liminf_{n \to \infty} f_{t_n}(x_n)$$
$$= = \text{seq} - \text{weak} \liminf_{(w,t) \to (x, 0_+)} f_t(w),$$
$$\left(e - \limsup_{t \to 0_+} f_t\right)(x) = \sup_{(t_n) \to 0_+} \inf_{(x_n) \to x} \limsup_{n \to \infty} f_{t_n}(x_n).$$

The family $(f_t)_{t>0}$ is said to Mosco epi-converge to a function g on X if $e_w - \liminf_{t \to 0_+} f_t = g = e - \limsup_{t \to 0_+} f_t$; if the weak epi-limit inferior is replaced by the strong epi-limit inferior in this definition, one speaks of epi-convergence. These concepts can be given simple interpretations in terms of set-convergence of epigraphs; they are of great importance when dealing with duality questions. We refer to [60] and [57] for the related notion of level-convergence, an adapted concept when dealing with quasiconvex functions.

Proposition 6. *([53], [54]) Suppose v is given by formula (6): $v(x,t) := (G+tH)^*(x)$.*
(a) One has $v(\cdot,0) \le e - \liminf_{t\to 0_+} v(\cdot,t)$ and if $G^* \le g$ one has $v(\cdot,0) \le g$.
(b) If $\operatorname{dom} H^* \ne \emptyset$, then $v(\cdot,t)$ Mosco epi-converges: $v(\cdot,0) = e - \lim_{t\to 0_+} v(\cdot,t)$.
(c) If H is bounded below, then $v(\cdot,t)$ pointwise converges to $v(\cdot,0)$ as $t \to 0$.
(d) If $G^* \le g$ and if g or G^* is u.s.c. at x, then $\limsup_{(w,t)\to(x,0)} v(w,t) \le g(x)$.
(e) If $G^* = g$ and $\operatorname{dom} G \subset \operatorname{dom} H$ then $v(\cdot,0) = g$.
(f) If $G^* = g$ and if g is continuous at x, $\operatorname{dom} G \subset \operatorname{dom} H$, then v is continuous at $(x,0)$ on $X \times \mathbb{R}_+$ and $v(x,0) = g(x)$.

In the following proposition which deals with the Lax solution, we use the *asymptotic function* of h and the *incident asymptotic function* of h given by

$$h_\infty(z) := e_w - \liminf_{t\to 0_+} t^{-1} h(t\cdot)(z) = \liminf_{w\to z,\ t\to\infty} th(t^{-1}w),$$

$$h^i_\infty(z) := e - \limsup_{t\to\infty} th(t^{-1}\cdot)(z)$$

for $z \in X$, where the topology on X is the weak topology and the strong topology respectively. Let us note that when h is a closed proper convex function one has $h^i_\infty = h_\infty$.

Proposition 7. *([53], [54]) Suppose u is given by formula (12): $u(x,t) := (g\Box h_t)(x)$ and (14) holds.*
(a) One has

$$v(\cdot,0) \le e - \liminf_{t\to 0_+} u(\cdot,t) \le g\Box h_\infty \le g$$

and

$$e - \limsup_{t\to 0_+} u(\cdot,t) \le g\Box h^i_\infty \le g.$$

(b) Suppose condition (C):

$$\liminf_{\|w\|\to\infty} h(w)/\|w\| > -\liminf_{\|w\|\to\infty} g(w)/\|w\|.$$

Suppose also that X is reflexive and that the restriction of g to any bounded set is weakly l.s.c. Then $g\Box h_\infty = e - \liminf_{t\to 0_+} u(\cdot,t)$ and if $h^i_\infty = h_\infty$ then $u(\cdot,t)$ Mosco epiconverges to $g\Box h_\infty$ as $t \to 0_+$.
(c) Suppose X is finite dimensional, g is l.s.c., and $h_\infty(z) > -g_\infty(-z)$ for any $z \in X\backslash\{0\}$. Then $g\Box h_\infty = e - \liminf_{t\to 0} u(\cdot,t)$. If moreover $h_\infty = h^i_\infty$ then $u(\cdot,t)$ epi-converges to $g\Box h_\infty$ as $t \to 0_+$.
(d) If g is l.s.c. at x, if g and h are bounded below on bounded sets and are such that

$$\liminf_{\|w\|\to\infty} g(w)/\|w\| > -\infty, \qquad \liminf_{\|w\|\to\infty} h(w)/\|w\| = +\infty,$$

then $u(x,t) \to g(x)$ as $t \to 0_+$.

Assertion (d) is inspired by [1] Theorem 5.2.

Remark 1. It is observed in [32] that when g and h are closed proper convex functions the condition $g\Box h_\infty = g$ is equivalent to the assumption $\mathrm{cl}(\mathrm{dom}\, g^*) \subset \mathrm{cl}(\mathrm{dom}\, h)$. □

Combining the results of the preceding two propositions, we get new assertions.

Corollary 5. *Let u and v be given by $u(x,t) := (g\Box h_t)(x)$, $v(x,t) := (g^* + tH)^*(x)$. Suppose (14) holds and* $\mathrm{dom}\, g^* \cap \mathrm{dom}\, H \neq \emptyset$.
(a) If $g = g^{**}$ and $\mathrm{dom}\, g^* \subset \mathrm{dom}\, h^*$ *then* $e - \liminf_{t\to 0_+} u(\cdot,t) = e - \liminf_{t\to 0_+} v(\cdot,t) = g$.
(b) If moreover $h(0) < +\infty$, then $u(\cdot,t)$ and $v(\cdot,t)$ pointwise converge to g and epi-converge to g as $t \to 0_+$.

Studies of initial conditions in the quasiconvex case are given in [1], [48], [54], [61], [62]. Let us give a short direct proof of [48] Prop. 12.13 with a slight improvement in the assumptions (we assume $\inf g \geq \inf h$ instead of $\inf g \geq h(0)$).

Proposition 8. *Suppose g is weakly l.s.c., H is finite and* $\inf g \geq \inf h$ *with $h = H^\flat = H^\sharp$. Let $h_t := (tH)^\flat = (tH)^\sharp$. Then the Lax solution $u_t := g\Diamond h_t$ epiconverges to g as $t \to 0_+$.*

Proof. Let us first note that h is coercive, its sublevel sets being scalarly bounded (since $h^\flat = H^{\flat\flat} \leq H$), hence bounded. It follows that h has a bounded minimizing sequence (z_n). Given $(x,r) \in \mathrm{epi}\, g$ and a sequence $(t_n) \searrow 0$, for $s_n := g(x) \vee h(z_n)$, $r_n := s_n + r - g(x)$, $x_n := x + t_n z_n$ we have

$$r_n \geq s_n = g(x_n - t_n z_n) \vee h(z_n) \geq (g\Diamond h_{t_n})(x_n),$$

and since $g(x) \geq \lim h(z_n)$ we have $(s_n) \to g(x)$, $(x_n, r_n) \to (x, r)$ and $(x_n, r_n) \in \mathrm{epi}\, u(\cdot, t_n)$. Thus $g \geq e - \limsup_{t\searrow 0} u(\cdot,t)$.

In order to prove that $g \leq e - \liminf_{t\searrow 0} u(\cdot,t)$, let us consider sequences $(t_n) \searrow 0$, $(r_n) \to r$, $(x_n) \to x$ weakly with $r_n \geq u(x_n, t_n)$ for each n. Let us pick $y_n \in X$ such that $g(x_n - t_n y_n) \vee h(y_n) < r_n + t_n$. Since h is coercive, the sequence (y_n) is bounded and $(x_n - t_n y_n) \to x$ weakly. Then $g(x) \leq \liminf_n g(x_n - t_n y_n) \leq \liminf_n (r_n + t_n) = r$. □

7 Comparison and uniqueness

We first observe that when $h^* \leq H$, $g^* \leq G$, since $g^* + th^* \leq G + tH$, definitions (6) and (12) yield for $t > 0$

$$u(\cdot,t) := g\Box h_t \geq (g\Box h_t)^{**} = (g^* + th^*)^* \geq (G + tH)^* = v(\cdot,t). \qquad (15)$$

In fact

$$(u(\cdot,t))^{**} := (g\Box h_t)^{**} = (g^* + th^*)^* \geq (G + tH)^* = v(\cdot,t), \qquad (16)$$

with equality when $G = g^*$, $H = h^*$. Similarly, for $u := g \diamond h_t$, when $G \geq g^\flat$, $H \geq h^\flat$ we have

$$(u(\cdot,t))^{\flat\flat} := (g \diamond h_t)^{\flat\flat} = \left(g^\flat + th^\flat\right)^\flat \geq (G + tH)^\flat = v(\cdot,t), \qquad (17)$$

with equality when $G = g^\flat$, $H = h^\flat$.

In [48] and [53] some consequences of the coincidence of the Hopf and the Lax solutions are drawn. Criteria for such a coincidence are also presented.

Let us turn to the important question of uniqueness. There are several methods for such a question: partial differential equations techniques ([5], [6], [37]...), nonsmooth analysis results such as the striking Barron-Jensen touching theorem ([8], [9]), the fuzzy sum rule ([13], [24], [25]), multidirectional mean value inequalities ([31], [32], [53]). Let us note that the last two results are almost equivalent.

In the first result below, we use an "ordinary" mean value theorem for the Hadamard subdifferential ∂ (see [47] and its references). It is valid provided ∂ satisfies a kind of fuzzy sum rule in $\overline{\mathbb{R}}^X$ called reliability in [47], which is close to the notion of trustworthiness due to Ioffe (see [33] for a recent account). A space X is said to be *reliable* if for any l.s.c. function $f \in \overline{\mathbb{R}}^X$, for any Lipschitzian convex function g on X, for any $x \in \operatorname{dom} f$ at which $f + g$ attains its infimum and for any $\varepsilon > 0$ there exist $u, v \in B(x,\varepsilon)$ such that $|f(u) - f(x)| < \varepsilon$ and $0 \in \partial f(u) + \partial g(v) + \varepsilon B^*$, where B^* is the unit ball of X^*. This condition is satisfied when X is an Asplund space or when X has a smooth enough bump function.

Theorem 8. *([53] Theorem 6.2). Suppose X is reliable. Let $w : X \times \mathbb{R}_+ \to \mathbb{R}$ be a l.s.c. lower solution to (1) such that for each $x \in X$ one has*

$$\liminf_{(z,t)\to(x,0)} w(z,t) \leq g(x). \qquad (18)$$

Then $w \leq u$, the Lax solution.

Note that u itself satisfies condition (18) and that Corollary 2 presents conditions ensuring that u is l.s.c.

The following corollary has been obtained in [32] for X a Hilbert space under the condition that $\operatorname{dom} H^*$ is open. It is proved in [1] Theorem 2.1 under the additional assumptions that X is finite dimensional, H is finite everywhere and for the subclass of subdifferential (or bilateral) solutions which are l.s.c. and bounded below by a function of linear growth.

Corollary 6. *Suppose X is reliable, g and H are closed proper convex functions and $\operatorname{dom} g^* \subset \operatorname{dom} H$. Then the Hopf solution v given by (6) is the greatest lower solution w of (1) which is l.s.c. and satisfies (18).*

The next result uses a multi-directional mean value inequality, a tool more sophisticated than the ordinary mean value theorem (for a simple version under some convexity assumptions, see [18] p. 114, [53] Lemma 6.5 and Theorem 6.6).

Lemma 9. *([16]-[18], [33]) Let Z be a reliable Banach space for a subdifferential ∂ and let $f : Z \to \mathbb{R} \cup \{+\infty\}$ be a l.s.c. function. Given $\varepsilon > 0$, $z_0 \in \operatorname{dom} f$ and a bounded closed convex subset Y of Z, there exist $z \in \operatorname{co}(z_0, Y) + B(0, \varepsilon)$ and $z^* \in \partial f(z)$ such that*

$$\sup_{\delta > 0} \inf_{y' \in Y + B(0,\delta)} f(y') - f(z_0) \leq z^*.(y - z_0) \quad \forall y \in Y.$$

The use of this mean value inequality for a comparison result appears in [31], [32] Theorem 3.3 which assume that H is convex and globally Lipschitzian and that X is a Hilbert space. Our statement removes these restrictions and involves a kind of local uniform lower semicontinuity assumption which improves ([53] Theorem 6.8); the proof is similar.

Theorem 9. *Suppose $X \times \mathbb{R}$ is reliable for a subdifferential ∂, H is u.s.c. on $\operatorname{dom} g^* \neq \emptyset$ and such that $H(\cdot) \leq b + c\|\cdot\|$ for some $b, c \in \mathbb{R}$. Let $w : X \times \mathbb{R}_+ \to \mathbb{R}$ be a l.s.c. supersolution to (1) such that for any continuous affine function $a(\cdot)$ minorizing g, for any $\alpha > 0$ and any bounded subset M of X there exists $\tau > 0$ such that $w(x,t) \geq a(x) - \alpha$ for any $x \in M$, $t \in [0, \tau]$. Then $w \geq v$, the Hopf solution.*

We observe that v is a supersolution which satisfies the lower semicontinuity condition of the statement, hence is the lowest such supersolution when H fulfils the assumptions. It is shown below that the lower semicontinuity condition is satisfied when X is reflexive and w is weakly l.s.c.

Corollary 7. *Suppose X is reflexive and H is u.s.c. on $\operatorname{dom} g^*$ and such that $H(\cdot) \leq b + c\|\cdot\|$ for some $b, c \in \mathbb{R}$. Let w be a l.s.c. function on $X \times \mathbb{R}_+$ which is such that $w(\cdot, 0) = e_w - \liminf_{t \to 0_+} w(\cdot, t) = g$ and is a viscosity supersolution and a lower solution to (1). Then $v \leq w \leq u$.*

If moreover g and H are convex, then $w = v$, the Hopf solution.

Proof. Since X and $X \times \mathbb{R}$ are reflexive, they are reliable for the viscosity subdifferential (which coincides with the Fréchet subdifferential there). By Theorem 8, we have $w \leq u$, the Lax solution. Suppose the uniform lower semicontinuity of the preceding statement is not satisfied: there exist $(\ell, c) \in X^* \times \mathbb{R}$, a bounded subset M of X, $\alpha > 0$ and sequences $(t_n) \to 0_+$, (y_n) in M such that $g \geq \ell - c$, hence $g^*(\ell) \leq c$, and

$$w(y_n, t_n) < \ell(y_n) - c - \alpha$$

for each $n \in \mathbb{N}$. If x is limit of a subsequence $(y_k)_{k \in K}$ of the sequence (y_n), we get

$$w(x, 0) \leq \liminf_{k \in K} w(y_k, t_k) \leq \lim_{k \in K} \ell(y_k) - c - \alpha = \ell(x) - c - \alpha \leq g(x) - \alpha,$$

a contradiction with $w(\cdot, 0) = g$. Using the last theorem, we get $w \geq v$.

The last assertion follows from the fact ([53] Prop. 5.1) that when g and H are convex $v = \overline{u}$, the lower semicontinuous hull of u since then H is continuous. □

When X is finite dimensional, with quite different methods, it can be shown that the growth condition on H can be dropped. It would be interesting to know whether such a result can be obtained in infinite dimensions and if one can relax the continuity assumption.

Theorem 10. *([1] Thms 2.1 and 2.5) Suppose X is finite dimensional, g is a closed proper convex function and H is finite and continuous on X^*. Then the Hopf solution v given by (6) is the least supersolution which is l.s.c. and bounded below by a function of linear growth. If moreover H is convex, then v is the unique l.s.c. viscosity solution that is bounded below by a function of linear growth.*

A similar result is proved in [1] Thm. 6.9 in the quasiconvex case.

Theorem 11. *Suppose X is finite dimensional, g is a l.s.c. quasiconvex function and $H : X^* \times \mathbb{R} \to \mathbb{R}$ is continuous, sublinear in its first variable and nondecreasing in its second variable. Then the Hopf solution $v_\sharp$ given by (7) is the least supersolution which is l.s.c.*

Combining this result with Theorem 9 and using the facts that a l.s.c. quasiconvex function is weakly l.s.c. and that a continuous sublinear function satisfies the growth condition of Theorem 9, we obtain that the two Hopf formulas we studied give the same result when they are both applicable. A direct proof would be interesting.

Corollary 8. *Suppose X is finite dimensional, g is a closed proper convex function and H is finite, sublinear and continuous on X^*. Then the Hopf solution v given by (6) coincides with the Hopf solution $v_\sharp$ given by (7).*

8 Conclusion

In this short survey we endeavoured to present some recent developments of the study of Hamilton-Jacobi equations relying on duality theory. It was not our purpose to deal with all the advances in this field. In particular, we have not considered regularizing effects, sensitivity analysis, links with other fields such as optimal control theory, game theory...Even for what concerns the use of duality methods, we refrained to present all possibilities. Still we hope that the reader will be able to grasp the contribution of duality and generalized convexity to a field which is usually considered as relatively remote.

References

1. O. Alvarez, E.N. Barron and H. Ishii, Hopf-Lax formulas for semicontinuous data, Indiana Univ. Math. J. 48 (3) (1999), 993-1035.
2. H. Attouch, Variational convergence for functions and operators, Pitman, Boston, (1984).
3. H. Attouch and H. Brézis, Duality for the sum of convex functions in general Banach spaces. In: Aspects of Mathematics and its applications, J.A. Barroso ed., North Holland, Amsterdam (1986), 125-133.
4. M. Bardi and L.C. Evans, On Hopf's formulas for solutions of Hamilton-Jacobi equations, Nonlinear Anal. 8 (1984), 1373-1381.
5. M. Bardi and I. Capuzzo-Dolcetta, Optimal Control and Viscosity Solutions of Hamilton-Jacobi-Bellman Equations, Birkhäuser, Basel (1998).
6. G. Barles, Solutions de viscosité des équations de Hamilton-Jacobi, Math. et Appl. #17, Springer, Berlin (1994).
7. G. Barles and B. Perthame, Discontinuous solutions of deterministic optimal stopping time problems, Math. Modeling and Numer. Anal. 21 (1987), 557-579.
8. E.N. Barron, Viscosity solutions and analysis in L^∞, in Nonlinear Analysis, Differential Equations and Control, F.H. Clarke and R.J. Stern (eds.), Kluwer, Dordrecht (1999), 1-60.
9. E.N. Barron, R. Jensen, Semicontinuous viscosity solutions of Hamilton-Jacobi equations with convex Hamiltonians, Comm. Partial Diff. Eq. 15 (1990), 1713-1742.
10. E.N. Barron, R. Jensen and W. Liu, Hopf-Lax formula for $u_t + H(u, Du) = 0$, J. Differ. Eq. 126 (1996), 48-61.
11. E.N. Barron, R. Jensen and W. Liu, Hopf-Lax formula for $u_t + H(u, Du) = 0$. II, Comm. Partial Diff. Eq. 22 (1997), 1141-1160.
12. E.N. Barron and W. Liu, Calculus of variations in L^∞, Applied Math. Opt. 35 (1997), 237-243.
13. J.M. Borwein and Q.J. Zhu, Viscosity solutions and viscosity subderivatives in smooth Banach spaces with applications to metric regularity, SIAM J. Control Optim. 34 (1996), 1568-1591.
14. J.M. Borwein and Q.J. Zhu, A survey of subdifferential calculus with applications, Nonlinear Anal. Th. Methods Appl. 38 (1999), 687-773.
15. G. Buttazzo, Semicontinuity, relaxation and integral representations in the calculus of variations, Pitman 207, Longman, Harlow (1989).
16. F.H. Clarke, Yu.S. Ledyaev, Mean value inequalities, Proc. Amer. Math. Soc. 122 (1994), 1075-1083.
17. F.H. Clarke, Yu.S. Ledyaev, Mean value inequalities in Hilbert space, Trans. Amer. Math. Soc. 344 (1994), 307-324.
18. F.H. Clarke, Yu.S. Ledyaev, R.J. Stern and P.R. Wolenski, Nonsmooth analysis and control theory, Springer, New York 1998.
19. M.G. Crandall, L.C. Evans and P.-L. Lions, Some properties of viscosity solutions of Hamilton-Jacobi equations, Trans. Amer. Math. Soc. 27 (1992), 1-67.
20. M.G. Crandall, H. Ishii and P.-L. Lions, User's guide to viscosity solutions of second-order partial differential equations, Bull. Amer. Math. Soc. 282 (1984)
21. M.G. Crandall and P.-L. Lions, Viscosity solutions to Hamilton-Jacobi equations, Trans. Amer. Math. Soc. 277 (1983), 1-42.

22. M.G. Crandall and P.-L. Lions, Hamilton-Jacobi equations in infinite dimensions, Part I, Uniqueness of viscosity solutions, J. Funct. Anal. 62 (1985), 379-396, Part II, Existence of viscosity solutions 65 (1986) 368-405; Part III 68 (19) 214-247; Part IV Unbounded linear terms 90 (1990), 237-283; Part V B-continuous solutions, 97 (1991), 417-465.
23. J.-P. Crouzeix, Continuity and differentiability properties of quasiconvex functions on $\mathbb{R}^n$, in "Generalized concavity in optimization and economics", S. Schaible and W.T. Ziemba, eds. Academic Press, New York, (1981), 109-130.
24. R. Deville, Smooth variational principles and nonsmooth analysis in Banach spaces, in Nonlinear Analysis, Differential Equations and Control, F.H. Clarke and R.J. Stern (eds.), Kluwer, Dordrecht, (1999), 369-405.
25. E. El Haddad and R. Deville, The viscosity subdifferential of the sum of two functions in Banach spaces. I First order case, J. Convex Anal. 3 (1996), 295-308.
26. L.C. Evans, Some max-min methods for Hamilton-Jacobi equations, Indiana Univ. Math. J. 33 (1984), 31-50.
27. L.C. Evans, Partial differential equations, Graduate Studies in Math. #19, Amer. Math. Soc., Providence (1998).
28. H. Frankowska, Optimal trajectories associated with a solution of the contingent Hamilton-Jacobi equation, Applied Math. Optim. 19 (1989), 291-311.
29. H. Frankowska, Lower semicontinuous solutions of Hamilton-Jacobi-Bellman equations, SIAM J. Control Optim. 31 (1) (1993), 257-272.
30. E. Hopf, Generalized solutions to non linear equations of first order, J. Math. Mech. 14 (1965), 951-974.
31. C. Imbert, Convex analysis techniques for Hopf-Lax' formulae in Hamilton-Jacobi equations with lower semicontinuous initial data, preprint, Univ. P. Sabatier, Toulouse, May 1999.
32. C. Imbert and M. Volle, First order Hamilton-Jacobi equations with completely convex data, preprint, October 1999.
33. A.D. Ioffe, Fuzzy principles and characterization of trustworthiness, Set-Valued Anal. 6 (1998), 265-276.
34. R. Janin, On sensitivity in an optimal control problem, J. Math. Anal. Appl. 60 (1977), 631-657.
35. S.N. Kružkov, On the minmax representation of solutions of first order nonlinear equations, Functional Anal. Appl. 2 (1969), 128-136.
36. P.D. Lax, Hyperbolic systems of conservation laws II, Commun. Pure Appl. Math. 10 (1957), 537-566.
37. P.-L. Lions, Generalized Solutions of Hamilton-Jacobi Equations, Research Notes in Math. #69, Pitman, London (1982).
38. P.-L. Lions and C. Rochet, Hopf formula and multi-time Hamilton-Jacobi equations, Proc. Amer. Math. Soc. 90 (1980), 79-84.
39. P.-L. Lions and P.E. Souganidis, Differential games, optimal control and directional derivatives of viscosity solutions of Bellman's and Isaac's equations, SIAM J. Control and Optim. 23 (1985), 566-583.
40. Ph. Loewen, Optimal Control via Nonsmooth Analysis, CRM Proceedings and Lecture Notes # 2, Amer. Math. Soc., Providence, 1993.
41. J.-E. Martinez-Legaz, On lower subdifferentiable functions, in "Trends in Mathematical Optimization", K.H. Hoffmann et al. eds, Int. Series Numer. Math. 84 Birkhauser, Basel, 1988, 197-232.

42. J.-E. Martinez-Legaz, Quasiconvex duality theory by generalized conjugation methods, Optimization, 19 (1988) 603-652.
43. J.-E. Martinez-Legaz and I. Singer, Dualities between complete lattices, Optim. 21 (1990), 481-508.
44. J.-E. Martinez-Legaz and I. Singer, $\vee$-dualities and $\perp$-dualities, Optim. 22 (1991), 483-511.
45. J.-E. Martinez-Legaz and I. Singer, $*$-dualities, Optim. 30 (1994), 295-315.
46. J.-P. Penot, Are generalized derivatives useful for generalized convex functions? in "Generalized Convexity, Generalized Monotonicity", J.-P. Crouzeix, J-E. Martinez-Legaz and M. Volle (eds.) Kluwer, Dordrecht (1998), 3-60.
47. J.-P. Penot, Mean-value theorem with small subdifferentials, J. Opt. Th. Appl. 94 (1) (1997), 209-221.
48. J.-P. Penot, What is quasiconvex analysis? Optimization 47 (2000), 35-110.
49. J.-P. Penot, Questions and observations about Hamilton-Jacobi equations, in preparation.
50. J.-P. Penot and M. Volle, Dualité de Fenchel et quasi-convexité, C.R. Acad. Sc. Paris série I, 304 (13) (1987), 269-272.
51. J.-P. Penot and M. Volle, On quasi-convex duality, Math. Oper. Research 15 (4) (1990), 597-625.
52. J.-P. Penot and M. Volle, Another duality scheme for quasiconvex problems, in "Trends in Mathematical Optimization", K.H. Hoffmann et al. eds, Int. Series Numer. Math. 84 Birkhauser, Basel, 1988, 259-275.
53. J.-P. Penot and M. Volle, Hamilton-Jacobi equations under mild continuity and convexity assumptions, J. Nonlinear and Convex Anal. 1 (2000), 177-199.
54. J.-P. Penot and M. Volle, Duality methods for the study of Hamilton-Jacobi equations, preprint, December 1999.
55. J.-P. Penot and C. Zalinescu, Harmonic sum and duality, J. Convex Anal. 7 (1) (2000), 95-113.
56. J.-P. Penot and C. Zalinescu, Elements of quasiconvex analysis, J. Convex Anal. 7 (2) (2000), .
57. R.T. Rockafellar and R. J.-B. Wets, Variational Analysis, Springer-Verlag, Berlin, 1997.
58. P.E. Souganidis, Existence of viscosity solutions of Hamilton-Jacobi equations, J. Diff. Equations 56 (1985), 345-390.
59. A.I. Subbotin, Generalized solutions of first-order PDE's, Birkhäuser, Basel, 1995.
60. M. Volle, Convergence en niveaux et en épigraphes, C.R. Acad. Sci. Paris 299 (8) (1984), 295-298.
61. M. Volle, Duality for the level sum of quasiconvex functions and applications, ESAIM: Control, Optimisation and Calculus of Variations, 3 (1998), 329-343, http://www.emath.fr/cocv/
62. M. Volle, Conditions initiales quasiconvexes dans les équations de Hamilton-Jacobi, C.R. Acad. Sci. Paris série I, 325 (1997), 167-170.
63. I. Zang and M. Avriel, On functions whose local minima are global, J. Optim. Th. Appl. 16 (1975), 183-190.
64. I. Zang, E.V. Choo and M. Avriel, A note on functions whose local minima are global, J. Optim. Th. Appl. 18 (1976), 555-559.

Higher-order Monotone Functions and Probability Theory

Dinis D. Pestana[1] and Sandra Mendonça[2]

[1]DEIO, Faculdade de Ciências da Universidade de Lisboa
[2]Departamento de Matemática, Universidade da Madeira
[1,2]Centro de Estatística e Aplicações da Universidade de Lisboa

Abstract. From topological properties of the cone of real functions with higher order monotonicity properties, we derive an extreme point Choquet representation theorem; this, in fact, follows from Krein-Milman's theory, since there is a natural compact convex base. This fills the gap between convexity and complete monotonicity, in the sense that we obtain integral representations for functions such that $(-1)^k f^{(k)}(x) \geq 0, k = 0, 1, \ldots, n$ (in the form of special "beta transforms"), that in the case of completely monotone functions is Bernstein's classical result about Laplace transforms of non-decreasing functions. The integral representation is used to reobtain Khinchine's characterization of unimodal distribution functions and Pólya's characteristic functions. On the other hand, we obtain extended classes of unimodal distributions (different from the star-shaped unimodality of Olshen and Savage), and put in a proper light Mejzler's classes in extreme value theory, and special classes of characteristic functions such as Sakovič's class. Using fractional calculus, we have straightforward extensions of the previous results, and consider generalized beta transforms, naturally associated with order statistics with general parent distribution.

2000 Mathematics Subject Classification. Primary 60 E 05; Secondary 60 E 10.

Keywords and phrases. unimodality, generalized unimodality, Pólya's characteristic functions, Choquet integral representations, beta transforms, fractional integrals and derivatives

1 Introduction

The idea of model is the core concept of modern science. The best we can achieve is a useful lie about reality, a lie that should be as simple as possible — the ideal of Occam's razor still has its place in the modern paradigm of Science.

The usefulness of the model is certainly related to its mathematical tractability, and hence, when our approach is analytical, some kind of smoothness is desirable (on the other hand, in modern intensive computation approaches, it is advantageous to work out with a combination of simple functions, or to use wavelets to approximate less obvious functions).

The vague idea of smoothness may be expressed in many different ways. Monotonicity of the function f, existence of monotone derivatives, unimodality, are perhaps the most usual "regular conditions" needed. Many results do exist in what concerns monotone functions, functions with monotone derivatives, and convex functions (a concept easily expressed in terms of second order derivatives, when $f \in C^2$); on the other hand, functions $f \in C^\infty$ with monotone derivatives do have interesting properties — for instance, the class of Bernstein's completely monotone functions, i.e. functions f such that $(-1)^k f^{(k)}(x) \geq 0$ for $k = 0, 1, 2, \ldots$, $x \geq 0$, is the class of Laplace transforms of non-decreasing bounded functions: f is completely monotone on $[0, +\infty)$ iff there exists a non-decreasing bounded function G defined on $[0, +\infty)$ such that

$$f(x) = \int_0^{+\infty} \exp(-xt)\, dG(t). \tag{1}$$

We shall exploit the intermediate situation, when $(-1)^k f^{(k)}(x) \geq 0$ for $k = 0, 1, 2, \ldots, n$, for $x \geq 0$, and obtain an extreme point integral representation similar to (1). (The fact that (1) is an extreme point integral representation is established, for instance, by Phelps [17], and we shall show that this representation is obtained as the limit when $n \to +\infty$ of the general representation we establish.)

The inversion formula of the integral transform allows further generalizations, using "fractional calculus", or "differintegral calculus" — in other words, integrals and derivatives of non-integer order, an idea inspired by Cauchy's iterated integral (for a well-documented presentation of the early development of the idea of fractional integrals and derivatives, and a discussion and conciliation of Riemann's and Liouville's definitions, cf. Oldham and Spanier [13]).

Our inversion formula has been derived using standard methods, cf. Hirschman and Widder [7], and should be compared with previous work by Williamson [22]. It inspires the discussion leading to Definition 5, and the extension using fractional calculus, see also Williamson [22] and Lévy [10]. Our approach differs from their work in the consideration of generalized beta-transforms, more related to Olshen and Savage [14] star-shaped unimodality.

Integral representation of functions with monotone derivatives, and inversion theorems, have been studied by Williamson [22] and Lévy [10], who characterizes classes of distribution functions (see Definition 3) and of characteristic functions (see Definition 4) with monotone derivatives, and points out that his results generalize Khinchine's theorem on unimodality and Pólya's sufficient condition for φ to be a characteristic function. Their pioneering work solved the main problems of characterization of functions with monotone derivatives. In fact, monotonicity also plays an interesting role in the study of characteristic functions. Bochner's and Cramér's characterizations of characteristic functions are virtually impossible to use, and thus, in practical terms, only Pólya's sufficient condition for φ to be a characteristic function is really helpful. We show that this is, once again, an extreme point integral representation theorem, and a simple corollary of our results on extended monotonicity. Recent deep results by Gneiting [4] and Gneiting [5], broadened the scope of this important field, namely by presenting characteristic functions of random variables with finite moments (see our comment on Pólya's characteristic functions).

Our approach, using Choquet's theorem, provides further insight on the structure of higher order monotonicity, and has a simple probabilistic interpretation, in terms of products of independent random variables; in view of the analytical form of the extremal rays of the pointed convex cone, that we characterize, one of the factors is a beta random variable. This shows that Khinchine's theorem on unimodal distribution functions may be viewed as an extreme point integral representation theorem, and we thus generalize the concept of unimodality, and discuss general "beta-transformed" variables, broadening the scope of subject to encompass order statistics.

Unimodality has been characterized by Khinchine, whose result may be restated as: Y is unimodal (at 0) iff $Y = UX$, with U uniform in $(0,1)$, independent of X. U is both a member of the Beta$(a,1)$ family, used by Olshen and Savage [14] and of the Beta$(1,n)$ used by us. Our approach has the advantage of dealing both with probability density functions and with characteristic functions. Generalized beta transforms, a topic we shall discuss elsewhere further, have some bearing in stochastic evolution of populations.

We further discuss generalizations of log-concave distribution functions. When Mejzler [12] investigated a class similar to Khinchine's L class (that we shall call Mejzler's M class) in the scheme of maxima of independent, but not necessarily identically distributed, random variables, he has shown that the possible limiting distributions (or, in case of finite right endpoint R, a simple transformation of the limiting distribution function, $F(R - \exp(-x))$ are log-concave. Inspired by Urbanik's [21] refinements of Khinchine's L class, we shall investigate refinements of Mejzler's M class, and establish the potentialities of distribution functions of the form $F(x) = \exp[-K(x)]$, with K completely monotone, as the proper framework for modelling maxima. Keilson (cf. Keilson and Steutel [9] and references therein) has several results in log-concavity and log-concave sequences.

2 Definitions and preliminary results

In the sequel, a probability distribution function is a non-decreasing and right continuous function defined on $[-\infty, +\infty]$, such that $0 \leq \lim_{x\to-\infty} F(x) \leq \lim_{x\to+\infty} F(x) \leq 1$.

In other words, we do not discard the possibility of improper random variables, with atoms of probability at $-\infty$ and/or at $+\infty$. (The advantage lies on the fact that we shall thus obtain a compact set of extreme points, and this allows us to use Choquet's theory.)

We shall first consider the cone of nonnegative bounded functions f defined on $[0, +\infty]$ with n−th order monotone derivative, and then allow f to be unbounded at 0. We then focus our attention on the absolutely continuous probability distribution functions F with probability density function f such that $f(x)\mathbf{I}_{(0,+\infty)}(x)$ and $f(-x)\mathbf{I}_{(0,+\infty)}(x)$, where $\mathbf{I}$ is the indicator function, are in the previous class, in order to have a straightforward identification of the extremal rays of higher order monotone functions.

Definition 1. Let f be a nonnegative, bounded function defined in $[-\infty,+\infty]$, non-increasing in $[0,+\infty]$, such that $f(x)=0$ for all $x<0$. We shall say that such function f is monotone of order 0 on $[0,+\infty]$. Suppose, additionally, that f is convex in $A=\{x : f(x)>0\}$. We say that f is monotone of order n on $[0,+\infty]$, $n=1,2,\ldots$, if and only if

1. $f \in C^n(A)$, i.e., f possesses derivative of order n in the interior of the set A, $\mathrm{int}(A)$.
2. The derivative of order n of f, $f^{(n)}$, is monotone in $\mathrm{int}(A)$.

We shall denote by M_n^*, $n=0,1,\ldots$, the class of monotone functions of order n in $[0,+\infty]$; in the sequel, whenever necessary, we shall build our arguments in terms of functions whose support is $\mathbf{R}^+$.

Observe that M_n^* is a pointed convex cone. It is easy to establish that $M_n^* \supseteq M_{n+1}^*$ and that if $f \in M_n^*$ then $(-1)^k f^{(k)}$ is non-negative, non-increasing and convex for $k=0,1,\ldots,n-1$, and then $(-1)^n f^{(n)}$ is non-negative and non-increasing.

Hence $M_\infty^* = \bigcap_{n=1}^{\infty} M_n^*$ is the class of completely monotone functions.

Definition 2. We say that the function f is monotone of order n in $(0,+\infty]$ iff for all $a>0$ the function $f_a(x) = f(a+x)$ is monotone of order n in $[0,+\infty]$.

We denote M_n the class of monotone functions of order n in $(0,+\infty]$.

Definition 3. Let F be a probability distribution function,

$$\alpha = \inf\{x \in [-\infty,+\infty] : F(x)>0\},$$
$$\omega = \sup\{x \in [-\infty,+\infty] : F(x)<1\}$$

and $A=(\alpha,\omega)$. Let F be differentiable in $A-\{0\}$, $f(x)=\frac{d}{dx}F(x)$ if $x \in A-\{0\}$, $f(x)=0$ if $x \notin A$. We say that the probability distribution function F belongs to the class D_{n+1}, $n=0,1,\ldots$, if and only if the functions $f_1(x)=f(x)\mathbf{I}_{(0,+\infty)}(x)$ and $f_2(x)=f(-x)\mathbf{I}_{(0,+\infty)}(x)$ are monotone of order n in $(0,+\infty]$.

The class D_{n+1}^* is similarly defined, with the obvious transformations. The following result is easily established (cf. Williamson [22]):

Lemma 1. *If the function f is monotone of order n, then the function $x^{k-1}f^{(k)}(x)$ is integrable in $(0,+\infty)$ and $\lim_{x\to+\infty} x^k f^{(k)}(x)=0$ for $k=1,2,\ldots,n; n=1,2,\ldots$.*

This result has interesting applications in regularly varying functions in the sense of Karamata, cf. Bingham et al. [1].

3 Extreme points of D_n and integral representations

We are now going to identify the extreme points of D_n, which is obviously a convex set. Recall that given a convex set C we say that $a \in C$ is an extreme point iff $C - \{a\}$ is still convex. In other words, if a is an extreme point of C and $a = px + (1-p)y$ with $x,\ y \in C$, $p \in [0,1]$, either $x = y = a$, or p is either 0 or 1.

The next proposition is an immediate consequence of the following:

1. $F \in D_n$ if and only if both $F\mathbf{I}_{(0,+\infty]}$ and $F\mathbf{I}_{[-\infty,0)} + \mathbf{I}_{[0,+\infty]}$ belong to D_n.
2. If F is the distribution function of the random variable X, $F \in D_n$ if and only if $\tilde{F} \in D_n$, where $\tilde{F}$ is the distribution function of the random variable $-X$.

Proposition 1. *If F is an extreme point of the set D_n, $n = 1, 2, \ldots$, then $F(0) = 0$ or $F(0) = 1$. The degenerate distributions at $-\infty$, at 0 and at $+\infty$, which we shall denote by $F_{n,-\infty}$, $F_{n,0}$ and $F_{n,+\infty}$, are the only degenerate extreme points of D_n, $n = 1, 2, \ldots$.*

In what concerns non-degenerate extreme points of D_n, $n = 1, 2, \ldots$, we have:

Theorem 1. *The distribution functions $F_{n,a}$ defined by*

$$\frac{d}{dx} F_{n,a}(x) = \frac{n}{|a|^n} \left(|a| - x\ \mathrm{sign}(x)\right)^{n-1} \mathbf{I}_{(\min(0,a),\max(0,a))}(x), \qquad (2)$$

where $a \in \mathbf{R} - \{0\}$, are extreme points of D_n, $n = 1, 2, \ldots$.

Proof. It is immediate that if $F_{n-1,a}$ is an extreme point of D_{n-1}, then $F_{n,a}$ is an extreme point of D_n, $n = 2, 3, \ldots$. In view of this and the above remarks, it is sufficient to prove that $F_{1,a}$ is an extreme point of D_1 for every $a > 0$. Observe that, in this case $\frac{d}{dx}F_{1,a} = \frac{1}{a}\mathbf{I}_{(0,a)}$, i.e., $F_{1,a}$ is the distribution of a uniform variable in $(0,a)$. Let $G_1, G_2 \in D_1, \lambda \in [0,1]$ such that

$$F_{1,a} = \lambda G_1 + (1-\lambda) G_2.$$

If $G_1(a) = 0$ (resp. $G_2(a) = 0$), then $\lambda = 0$ and $G_2(a) = 1$ (resp. $\lambda = 1$ and $G_1(a) = 1$), and hence $F_{1,a} \equiv G_2$ (resp. $F_{1,a} \equiv G_1$).

If $0 < G_1(a) < 1$ (which, as $F_{1,a}(a) = 1$ implies that $G_2(a) > 0$) we have that

$$\frac{1}{a}\mathbf{I}_{(0,a)}(x) = \lambda \frac{d}{dx} G_1(x) + (1-\lambda) \frac{d}{dx} G_2(x),$$

where, since $G_i \in D_1$, $\frac{d}{dx}G_i$ are non-negative and non-increasing in $(0, a)$, for $i = 1, 2$. Hence it follows that $\frac{d}{dx}G_i$ is constant in $(0, a)$, i.e., $G_i = \mu_i F_{1,a}$, $i = 1, 2$, where $\mu_i \in (0, 1)$ and such that $1 = F_{1,a}(a) = \lambda\mu_1 + (1-\lambda)\mu_2$; $F_{1,a}$ is an extreme point of D_1.

□

We now state and prove the converse result:

Theorem 2. *If F is a non degenerate extreme point of D_n, $n = 1, 2, \ldots$, then $F = F_{n,a}$ for some $a \in \mathbf{R} - \{0\}$, where $F_{n,a}$ is the function defined by (2) .*

Proof. Clearly it is sufficient to consider the case $F(0) = 1$. Let us show that $F^{(n)}$ is constant in $(a, 0)$, where $a = \sup\{x \in \mathbf{R} : F(x) = 0\}$ (observe that the case $a = -\infty$ is a possible one and it corresponds to a degenerate extreme point).

Given that $F \in D_n$ we know that $F^{(1)}$ is monotone of order n and, consequently, $F^{(n-1)}$ is convex and non-decreasing. Hence there exists a non negative, non decreasing and integrable function f defined on $\mathbf{R}$ such that $f(x) = 0$ if $x > 0$ or $x < a$ and

$$F^{(n-1)}(x) = (-1)^n \int_a^x f(u)\, du.$$

Consider $\alpha \in (a, 0)$ and the auxiliary function

$$G(x) = \begin{cases} 0 & \text{if } x \le a \\ F^{(n-1)}(x) & \text{if } x \in (a, \alpha) \\ F^{(n-1)}(\alpha) + (x-\alpha) f(\alpha) & \text{if } x \in [\alpha, 0] \\ G(0) & \text{if } x > 0 \end{cases} .$$

Note that G is non-decreasing, non-negative and convex in $(-\infty, 0)$. We can then admit the existence of a function $G^* \in D_n$ such that $G^{*(n-1)} = G$. From $F = G^* + (F - G^*)$ and the fact that F is a non degenerate extreme point of D_n we have $G^* = \lambda F$ for some $\lambda \in [0, 1]$, and hence

$$G^{*^{(n)}} = \lambda F^{(n)} \Leftrightarrow \begin{cases} 0 & \text{if } x < a \text{ or } x > 0 \\ f(x) & \text{if } x \in (a, \alpha) \\ f(\alpha) & \text{if } x \in (\alpha, 0) \end{cases} = \lambda (-1)^n f(x),$$

almost everywhere. This equality implies that f is constant in $(\alpha, 0)$. Since α was arbitrarily taken in $(a, 0)$, f and, consequently, $F^{(n)}$, are constant in $(a, 0)$.

□

We are now going to give an integral representation of the elements of M_n^*, i.e., of the monotone functions of order n in $[0, +\infty]$.

Theorem 3. *A function f defined on $[-\infty,+\infty]$ is monotone of order n in $[0,+\infty]$ if and only if $f(x)=0$ for $x<0$ and*

$$f(x)=\int_0^{1/x}(1-xt)^n\,dG(t)\,,\quad x\geq 0$$

where G is bounded, right-continuous and non-decreasing.

Results on these and related classes of functions have been obtained by Schoenberg [19], Williamson [22], Lévy [10], Gneiting [4] and Gneiting [5].

Proof. It is obvious that M_n^*, $n=0,1,\ldots$, is a convex cone. Let E be the space of all real-valued functions with support in $[0,+\infty]$ whose restriction to $(0,+\infty)$ is n times derivable, and consider the topology induced by the countable family of semi-norms

$$p_{m,k}(f)=\sup\left\{\left|f^{(i)}(x)\right|:m^{-1}\leq x\leq m, 0\leq i\leq k\right\}$$

$(k=1,2,\ldots,n;m=1,2,\ldots)$. Let us define the subset of E,

$$k_n=\left\{f\in M_n^*:\int_0^{+\infty}f(x)\,dx=1\right\}.$$

The set k_n is a compact convex base for the cone M_n^* (see Pestana [16]).

From what we have seen above, it easily follows that the extreme points of k_n, $n=0,1,\ldots$, are

$$f_{n,0}(x)=\delta_0(x)\,;$$

$$f_{n,a}(x)=\frac{n}{a^n}\left(1-\frac{x}{a}\right)^n\mathbf{I}_{(0,a)}(x)\,;$$

$$f_{n,+\infty}(x)=\delta_\infty(x)\,.$$

Consider the map

$$\begin{array}{rcl}\varphi:[0,+\infty] & \longrightarrow & k_n\\ a & \longmapsto & \varphi(a)=f_{n,a}\end{array}.$$

It is obvious that φ is continuous, when we consider the induced topologies, and hence the set of extreme points of k_n is $\text{ext}(k_n)=\varphi([0,+\infty])$ which, being the image of a compact set, is also compact. Thus to each $f\in k_n$ there corresponds a regular Borel probability measure P supported by the set of extreme points of k_n (cf. Phelps [17]) such that

$$L(f)=\int_{\text{ext}(k_n)}L\,dP$$

for each continuous linear functional L on E. In the particular case that for $L_x(f) = f(x)$, $f \in E$, we obtain

$$f(x) = L_x(f) = \int_{\text{ext}(k_n)} L_x \, dP, \quad x \geq 0.$$

Define μ on each Borel subset B of $[0, +\infty]$ by $\mu(B) = P[\varphi(B)]$. Since $L_x[\varphi(a)] = f_{n,a}(x)$ we have

$$\begin{aligned} f(x) &= \int_{[0,+\infty]} f_{n,a}(x) \, d\mu(a) \\ &= \int_{(0,+\infty)} f_{n,a}(x) \, d\mu(a) + \int_{\{0\}} f_{n,a}(x) \, d\mu(a) \\ &= \int_{(0,+\infty)} \left(1 - \frac{x}{a}\right)^n \mathbf{I}_{[0,a)}(x) \, d\mu(a) + f_{n,0}(x) \, \mu(\{0\}) \\ &= \int_x^{+\infty} \left(1 - \frac{x}{a}\right)^n d\mu(a) + f_{n,0}(x) \, \mu(\{0\}), \quad x \geq 0. \end{aligned}$$

□

A monotone function of order n in $(0, +\infty]$ may be represented as

$$f(x) = \int_0^{1/x} (1 - xt)^n \, dG(t), \quad x > 0$$

where the representing measure G is non-decreasing.

There is also an integral representation for the elements of $M_\infty^* = \bigcap_{n=1}^{\infty} M_n^*$, the class of completely monotone functions.

Theorem 4. *A necessary and sufficient condition that f be completely monotone in $[0, +\infty)$ is that*

$$f(x) = \int_0^{+\infty} \exp(-xt) \, dG(t)$$

where G is bounded and non-decreasing and the integral converges for $0 \leq x < +\infty$, i.e., f is the Laplace transform of G (cf. Feller [3]).

Proof. This may be easily proven by observing that $\exp(-kx) - (1 - x)^{k-1}$ approaches zero uniformly in $[0, 1]$ as $k \to +\infty$, and having in mind Lemma 1.

The above Bernstein's theorem (cf. Feller [3]) on the integral representation of absolutely and completely monotone functions has been established along these lines by Pollard [15] and by Mattner [11].

□

4 Distribution functions in the class D_n and characteristic functions in class C_n

Based on the results of the previous section we have the following characterization of the elements of the class D_n.

Theorem 5. *A distribution F is an element of D_n if and only if F is the distribution function of a random variable $Y \stackrel{d}{=} \left(1 - U^{1/n}\right) Y_n$, where U denotes the uniform distribution in $[0,1]$ and Y_n is independent of U.*

Remark that $U^{1/n}$ is a beta random variable with parameters n and 1; hence $\left(1 - U^{1/n}\right)$ is a Be$(1,n)$ random variable.

From this theorem and the fact that D_1 is formed by all unimodal distributions with vertex 0, we have the following:

Corollary 1. *A distribution function F is unimodal with vertex 0 if and only if the corresponding characteristic function φ may be represented in the form*

$$\varphi(t) = \frac{1}{t}\int_0^t \theta(x)\, dx, \quad t \in \mathbf{R}$$

where θ is a characteristic function.

This celebrated result of Khinchine can be generalized in the following way:

Corollary 2. *A distribution F is an element of D_n if and only if the corresponding characteristic function φ is of the form*

$$\varphi(t) = \int_{-\infty}^{+\infty} M(1; n+1; itx)\, dG(x), \quad t \in \mathbf{R}$$

where G is a distribution function and $M(\alpha;\beta;z)$ denotes the confluent hypergeometric function with parameters α and β and argument z,

$$M(\alpha;\beta;z) = \sum_{k=0}^{\infty} \frac{(\alpha)_k}{(\beta)_k}\frac{z^k}{k!}.$$

As a special result we have Sakovič's theorem on characteristic functions of convex probability density functions

$$\varphi(t) = \frac{2}{t^2}\int_{-\infty}^{+\infty} \frac{1 + itx - \exp(itx)}{x^2}\, dG(x), \quad t \in \mathbf{R}$$

where G is a probability distribution function. Is is interesting to point out similarities with the representation of infinitely divisible random variables, cf. Steutel [20]. Observe as well, Johansen [8], that Lévy's representation of infinitely divisible characteristic functions is nothing but an extreme point Choquet integral representation.

We now define C_n and characterize its elements using an integral representation.

Definition 4. We say that a characteristic function $\varphi(t)$ belongs to the class C_n, $n = 0, 1, \ldots$, if and only if φ corresponds to a symmetric random variable and $\varphi \mathbf{I}_{[0,+\infty)}$ is monotone of order n in $[0, +\infty]$.

Observe that

1. If $\varphi \in C_n$, $n = 0, 1, \ldots$, φ is real and even.
2. For $n \geq 1$, $C_n \subseteq C_{n-1}$.
3. The characteristic function φ is in the class C_n, $n = 0, 1, \ldots$, if and only if it may be represented in the form

$$\varphi(t) = \varphi(-t) = \int_0^{1/t} (1 - xt)^n \, dG(x), \quad t > 0$$

where G is a probability distribution function such that $G(0) = 0$.

It is always troublesome to decide if a given function φ is or is not a characteristic function. Bochner and Cramér characterizations cannot be easily handled. Besides the condition $\varphi(0) = 1$, (i.e. a proper random variable), Bochner's theorem demands the function to be non-negative definite, and Cramér's criterion the integral $\int_0^A \int_0^A \varphi(t-u) \exp[ix(t-u)] \, dtdu$ to be real and non-negative for all real x and for all $A > 0$. Pólya's condition gives easy sufficient conditions for φ, real and even, to be a characteristic function, namely the convexity of $\varphi(t)$, $t > 0$.

If in the above definition we take $n = 1$ we obtain the Pólya type distributions. The well known integral representation for Pólya type characteristic function follows immediately: φ is a Pólya type characteristic function if, and only if,

$$\varphi(t) = \int_0^{1/|t|} (1 - x|t|) \, dG(x),$$

where G is a probability distribution function such that $G(0) = 0$.

This is not of an utmost interest, since characteristic functions in Pólya's class correspond to random variables without finite expectation. The usefulness of such models is rather limited. However, Gneiting [4] and Gneiting [5] has just established path breaking results giving criteria for characteristic function that corresponds to random variables with finite expectation and finite variance.

5 Inversion theory

Having in mind the results of Lemma 1 and using the Taylor's expansion of the elements of M_n^* it is not difficult to prove the following:

Theorem 6. *If $f(x) = \int_0^{1/x} (1-tx)^n \, dG(t), x \geq 0$, with G bounded and non-decreasing, $G(0) = 0$, then G is uniquely determined at its points of continuity and given at these points by*

$$G(t) = \sum_{k=0}^{n} \frac{(-1)^k}{k!} \left(\frac{1}{t}\right)^k f^{(k)}\left(\frac{1}{t}\right), \quad t > 0.$$

6 Alternative characterization of monotone functions of order n, and extensions to fractional order

We know that a monotone function of order n in $(0, +\infty]$ may be represented as

$$f(x) = \int_0^{1/x} (1-xt)^n \, dG(t), \quad x > 0$$

where G is non-negative and non-decreasing. The application G, given $G(0)$, can be easily obtained as a transformation of f for

$$x^n f\left(\frac{1}{x}\right) = \int_0^x (x-t)^n \, dG(t)$$

and

$$\frac{d^n}{dx^n}\left[x^n f\left(\frac{1}{x}\right)\right] = n! \left[G(x) - G(0)\right].$$

On the other hand if we assume that f is such that $\frac{d^n}{dx^n}\left[x^n f\left(\frac{1}{x}\right)\right]$ is non-negative and non-decreasing, and define G^* in $[0, +\infty]$ such that

$$G^*(x) = \frac{1}{n!} \frac{d^n}{dx^n}\left[x^n f\left(\frac{1}{x}\right)\right], \quad x > 0$$

and $0 \leq G^*(0) \leq \lim_{x \to 0^+} G^*(x)$, then

$$h(x) = \int_0^{1/x} (1-xt)^n \, dG^*(t), \quad x > 0$$

is a monotone function of order n and, assuming that $\lim_{x \to +\infty} f(x)$ and $\lim_{x \to +\infty} h(x)$ are both finite, $f = h$, provided that $G^*(0) = G(0)$.

We may then say that a function f is monotone of order n in $(0, +\infty]$ if and only if $0 \leq \lim_{x \to +\infty} f(x) < +\infty$, $\frac{d^n}{dx^n}\left[x^n f\left(\frac{1}{x}\right)\right]$ non-negative and non-decreasing in $(0, +\infty]$. This definition of monotone function of order n

will allow us to extend the concept by the use of fractional derivatives, D^ν, for the integer parameter n will be replaced by a non-negative parameter ν (instead of M_n, M_n^*, D_n and D_n^* we will have M_ν, M_ν^*, D_ν and D_ν^*). This derivative operator is defined as

$$D^\alpha [f(x)] = I^{-\alpha} [f(x)] = \frac{d^n}{dx^n} I^{n-\alpha} [f(x)]$$

where n is an integer greater than α, and, for $\beta > 0$

$$I^\beta [f(x)] = \frac{x^\beta}{\Gamma[\beta]} \int_0^1 (1-t)^{\beta-1} f(tx)\, dt.$$

Definition 5. A function f is said to be monotone of order ν ($\nu \geq 0$) in $(0, +\infty]$ if and only if

1. $0 \leq \lim_{x\to+\infty} f(x) < +\infty$;
2. $D^\nu \left[x^\nu f\left(\frac{1}{x}\right)\right]$ is non-negative and non-decreasing.

The class of monotone functions of order ν ($\nu \geq 0$) in $(0, +\infty]$ will be denoted M_ν.

The integral representation of the elements of M_ν can be easily guessed:

Theorem 7. *A function f is monotone of order ν ($\nu \geq 0$) in $(0, +\infty]$ if and only if*

$$f(x) = \int_0^{1/x} (1-xt)^\nu\, dG(t), \quad x > 0$$

with G non-negative and non-decreasing in $(0, +\infty]$.

As a consequence we have the following corollary:

Corollary 3. *$F \in D_\nu$, $\nu \geq 1$, if and only if F is the distribution function of a random variable*

$$X \stackrel{d}{=} \left(1 - U^{1/\nu}\right) Y_\nu,$$

where U denotes the uniform distribution in $[0,1]$ and Y_ν is independent of U.

And between classes we have the following inclusions:

Theorem 8. *$0 < \mu < \nu$ implies that $D_{\nu+1} \subseteq D_{\mu+1}$.*

Proof. We will use Mellin transforms in this proof. The Mellin transform of a non-negative random variable X is defined as

$$\mathcal{M}_X(s) = \int_0^{+\infty} x^s dF_X(x)$$

where, F_X is the associated distribution function. If X is a beta random variable with parameters p and q then

$$\mathcal{M}_X(s) = \mathcal{M}_{p,q}(s) = \frac{\mathrm{B}(p+s,q)}{\mathrm{B}(p,q)}, \quad \mathrm{Re}(s) > -p.$$

From this and the fact that the Mellin transform of product of independent, non-negative random variables is the product of the Mellin transform of each variable, the following identity is valid

$$\mathcal{M}_{1,\nu+1}(s) = \mathcal{M}_{1,\mu+1}(s)\,\mathcal{M}_{\mu+2,\nu-\mu}(s), \quad \mathrm{Re}(s) > -1. \tag{3}$$

From the previous corollary we know that $F \in D_{\nu+1}$ if and only if it is the distribution function of

$$X \stackrel{d}{=} \left(1 - U^{\frac{1}{\nu+1}}\right) Y_{\nu+1},$$

with $Y_{\nu+1}$ independent of U, the uniform distribution in $[0,1]$. As we mentioned before, the variable $1-U^{\frac{1}{\nu+1}}$ is a beta random variable with parameters 1 and $\nu+1$. Using (3) we obtain that F is the probability distribution function of the random variable

$$X \stackrel{d}{=} \left(1 - U^{\frac{1}{\mu+1}}\right) Y_{\mu+1},$$

where $Y_{\mu+1} = Z_{\mu+2,\nu-\mu} Y_{\nu+1}$, and hence $F \in D_{\mu+1}$. □

Corollary 4. *$0 < \mu < \nu$ implies that $M_{\nu+1} \subseteq M_{\mu+1}$.*

7 Mejzler's and Khinchine's classes

Higher order monotonicity also plays an important role in the known Mejzler's and Khinchine's classes. Consider a sequence $\{X_k\}_{k\in N}$ of independent (not necessarily identically distributed) random variables, $S_n = \sum_{k=1}^{n} X_k$ and $X_n^* = \max_{1\le k\le n} X_k$. These sequences adequately normalized may have non-degenerate weak limits:

$$\frac{S_n - b_n}{a_n} \stackrel{w}{\longrightarrow} Y, \quad \frac{X_n^* - b_n}{a_n} \stackrel{w}{\longrightarrow} Z,$$

when $n \to +\infty$.

The classes constituted by the limits distributions are know as the class L if sums are considered, and as the class M when maximums are considered. Consider a distribution function F, α and ω as defined before. Consider also the function

$$G(x) = \begin{cases} G_1(x) = F(\alpha + \exp(x)) & \alpha > -\infty \\ G_2(x) = F(\omega - \exp(-x)) & \omega < +\infty \\ G_3(x) = F(x) & \text{otherwise} \end{cases}.$$

Mejzler [12] proved that a distribution function F belongs to M iff G_i is log-concave for some (eventually all) $i = 1, 2, 3$. Note that $F \in M$ if and only if

$$G_i(x) = G_i(a + x) F_a(x) \tag{4}$$

for some $i = 1, 2, 3$, $\forall a < 0$ if $i = 1, \forall a > 0$ otherwise, being F_a a non-decreasing function.

Let $\mathcal{M}_{-1}$ the set of all distributions functions, $\mathcal{M}_0 = M$, and define iteratively $\mathcal{M}_r, r = 1, 2, \ldots$ by $F \in \mathcal{M}_r$ if and only if G_i, for some $i = 1, 2, 3$ satisfies (4) with F_a proportional to an $\mathcal{M}_{r-1}$ distribution function. Let $\mathcal{M}_\infty = \bigcap_{r=1}^{\infty} \mathcal{M}_r$. Distribution functions in the class $\mathcal{M}_r$ have nice monotonicity properties:

$$F \in \mathcal{M}_r \text{ if and only if } F(x) = \exp[-k(x)],$$

with k a monotone function of order $r + 1$, in particular

$$F \in \mathcal{M}_{+\infty} \text{ if and only if } F(x) = \exp[-k(x)]$$

with k a completely monotone function.

$\mathcal{M}_{+\infty}$ is the smallest closed class, under pointwise products and limits, containing all max-stable extreme value distribution functions.

Acknowledgments. The authors are deeply grateful to the referees for the very helpful comments, and namely for calling their attention to Williamson and Lévy previous work in the field, and to the recent deep developments achieved by Gneiting.
This research was partially supported by the Foundation of Science and Technology of Portugal, FCT – PRAXIS XXI, FEDER.

References

1. Bingham, N, Goldie, C. and Teugels, J. (1987) *Regular Variation.* Cambridge University Press.
2. Choquet, G. (1969) *Lectures on Analysis*, II. Benjamin, New York.
3. Feller, W. (1971) *An Introduction to Probability Theory and Its Applications*, II. John Wiley & Sons.
4. Gneiting, T. (1999) *Radial Positive Definite Functions Generated by Euclid's Hat.* J. of Multivar. An. **69**, 88-119.
5. Gneiting, T. (2000) *Kuttner's Problem and a Pólya Type Criterion for Characteristic Functions.* Proc. Amer. Math. Soc. **128**, 1721-1728.
6. Graça Martins, M. E. and Pestana, D. D. (1987) *Nonstable Limit Laws in Extreme Value Theory.* In Puri et al (eds.) New Perspectives in Theoretical and Applied Statistics, Wiley, 449-458.
7. Hirschman, I. I. and Widder, D. V. (1955) *The Convolution Transform.* Princeton University Press, Princeton.

8. Johansen, S. (1966) *An Application of Extreme Point Methods to the Representation of Infinitely Divisible Distributions.* Z. Wahrscheinlichkeitstheorie Verw. Geb. **5**, 304-316.
9. Keilson, J. and Steutel, W. (1974) *Mixtures of Distributions, Moment Inequalities and Measures of Exponentiality and Normality.* Ann. Prob. **2**, 112-130.
10. Lévy, P. (1962) *Extensions d'un Théorème de D. Dugué et M. Girault.* Z. Wahrscheinlichkeitstheorie Verw. Geb. **1**, 159-173.
11. Mattner, L. (1993) *Bernstein's Theorem, Inversion Formula of Post and Widder, and the Uniqueness Theorem for Laplace Transforms.* Expo. Math. **11**, 137-140.
12. Mejzler, D. (1956) *On the Problem of the Limit Distribution for the Maximal Term of a Variational Series.* L'vov Pol. Inst. Naucn Zp. **38**, 90-109.
13. Oldham, K. B. and Spanier, J. (1974) *The Fractional Calculus.* Academic Press, New York.
14. Olshen, R. A. and Savage, L. J. (1970) *A Generalized Unimodality.* J. Appl. Probab. **7**, 21-34.
15. Pollard, H. (1944) *The Bernstein-Widder Theorem on Completely Monotonic Functions.* Duke Math. J. **11**, 427-430.
16. Pestana, D. D. (1978) *Some Contributions to Unimodality, Infinite Divisibility, and Related Topics.* Ph.D. thesis, Univ. of Sheffield.
17. Phelps, R. R. (1966) *Lectures on Choquet's Theorem.* van Nostrand, New Jersey.
18. Sakovič, D. N. (1972) *On the Characteristic Functions of Concave Distributions.* Transl. in Theor. Prob. and Math. Statist. **6** (1975), 103-108.
19. Schoenberg, I. J. (1941) *On Integral Representations of Completely Monotone and Related Functions (abstract).* Bull. Amer. Math. Soc. **47**, 208.
20. Steutel, F. W. (1968) *A Class of Infinitely Divisible Mixtures.* Ann. Math. Statist. **39**, 1153-1157.
21. Urbanik, K. (1973) *Limit Laws for Sequences of Normed Sums Satisfying Some Stability Conditions.* In Krishnaiah (ed.) Multivariate Analysis, III, Academic Press, 225-237.
22. Williamson, R. E. (1956) *Multiply Monotone Functions and their Laplace Transforms.* Duke Math. J. **23**, 189-207.

Convexity and Decomposability in Multivalued Analysis

Adrian Petruşel[1] and Ghiocel Moţ[2]

[1] "Babeş-Bolyai" University
Department of Applied Mathematics
3400, Cluj-Napoca, Romania
[2] "Aurel Vlaicu" University
Department of Mathematics and Computer Science
2900, Arad, Romania.

Abstract. The main goal of this paper is to present some results in the field of multivalued analysis related to the following question: could convexity be replaced by decomposability ?

2000 Mathematics Subject Classification. 47H05, 54C55.

Keywords and phrases. decomposable set, multivalued operator, continuous selection.

1 Introduction

Throughout this paper $(T, \mathcal{A}, \mu)$ is a complete σ-finite nonatomic measure space and E is a Banach space. Let $L^1(T, E)$ be the Banach space of all measurable functions $u : T \to E$ which are Bochner μ-integrable. We call a set $K \subset L^1(T, E)$ *decomposable* if for all $u, v \in K$ and each $A \in \mathcal{A}$:

$$u\chi_A + v\chi_{T\setminus A} \in K, \qquad (1)$$

where χ_A stands for the characteristic function of the set A.

This notion is, somehow, similar to convexity, but there exist also major differences. However, in several cases the decomposability condition is a good substitute for convexity (see [3], [4], [6], [11], [12], [16], [17], [18]).

On the other hand, in 1977 (see [15]) Hiai and Umegaki proved that a closed subset Y of $L^1(T, E)$ is decomposable if and only if it is the set of all L^1-selections of a measurable multivalued operator (see also [19] for a characterization of decomposable subsets of L^∞). In fact, a decomposable set has been considered for the first time in the field of multivalued analysis by Antosiewicz and Cellina in 1975, as follows :

$$K(x) = \{u \in L^1([0,1], \mathbf{R}^n) |\ u(t) \in F(t, x(t))\ a.e.\}$$

in connection with the problem of the existence of a continuous selection for a continuous multifunction with not necessarily convex values.

The purpose of this paper is to present some results in the field of multivalued analysis related to this topic: convexity replaced by decomposability. Some open questions are also suggested.

2 Preliminaries

Throughout this paper we use the following notations and symbols:

$$\mathcal{P}(E) = \{Z | Z \subset E\}, \quad \mathcal{P}_0(E) = \{Z \in \mathcal{P}(E) | Z \text{ nonempty}\}$$

$$\mathcal{P}_p(E) = \{Z \in \mathcal{P}_0(E) |\ Z \text{ has the property "}p\text{"}\},$$

where "p" could be: cl = closed, b = bounded, cv = convex, cp = compact, dec = decomposable, etc.

If $x \in E$ and Z is a nonempty subset of E then:

$$D(x, Z) = \inf\{d(x, z) |\ z \in Z\}$$

and

$$V(Z, \varepsilon) = \{x \in E |\ D(x, Z) \leq \varepsilon\}$$

(where $\varepsilon > 0$). We denote by H the Hausdorff-Pompeiu metric on $\mathcal{P}_{b,cl}(E)$. Several basic notions are considered in the following definitions:

Definition 1. Let (X, d) be a metric space. Then $F : X \to \mathcal{P}_0(E)$ is called:
a) *lower semi-continuous* (briefly l.s.c.) iff for each closed $C \subset E$ the set $\{x \in X |\ F(x) \subset C\}$ is closed.
b) *upper semi-continuous* (briefly u.s.c.) iff for each closed $C \subset E$ the set $\{x \in X |\ F(x) \cap C\}$ is closed.
c) *continuous* iff is l.s.c. and u.s.c.
d) *H-u.s.c.* iff for each $\varepsilon > 0$ there is $\delta > 0$ such that $F(B(x_0;\delta)) \subset V(F(x_0);\varepsilon)$.
e) *locally selectionable* at $x_0 \in X$ iff for all $y_0 \in F(x_0)$ there exist a neighborhood $N(x_0)$ and a continuous function $f : N(x_0) \to E$ such that $f(x_0) = y_0$ and $f(x) \in F(x)$, for each $x \in N(x_0)$.
f) *measurable* iff $F^{-1}(C)$ belongs to the σ-algebra $\mathcal{B}$ of Borel subsets of X, for each closed subset C of E.

Definition 2. Let X be a nonempty set.
i) Let $F : X \to \mathcal{P}_0(E)$ be a multivalued operator. The set defined by $F^{-1}(y) = \{x \in X | y \in F(x)\}$ is said to be *the fibre* of F at the point $y \in E$. Also, a singlevalued operator $f : X \to E$ is a selection for F iff $f(x) \in F(x)$ for every $x \in X$.
ii) If $F : X \to \mathcal{P}_0(X)$, then x^* is a *fixed point* for F iff $x^* \in F(x^*)$. Denote by $FixF$ the fixed points set for F.

Definition 3 (Deguire-Lassonde [9]). Let X be a topological space and $\{Y_i | i \in I\}$ an arbitrary family of topological spaces.
i) We say that $\{f_i : X \to Y_i | i \in I\}$ is *a selecting family* for the family of multivalued operators $\{F_i : X \to \mathcal{P}(Y_i) | i \in I\}$ iff for each $x \in X$ there exists $i \in I$ such that $f_i(x) \in F_i(x)$.
ii) If $\{Y_i | \ i \in I\}$ is an arbitrary family of convex subsets of a Hausdorff topological vector space then the family $\{F_i : X \to \mathcal{P}(Y_i) | i \in I\}$ is said to be of *Ky Fan-type* iff each F_i has convex values and open fibres and for every $x \in X$ there is $i \in I$ such that $F_i(x) \neq \emptyset$.

Definition 4. The topological space X has *the compact fixed point property* if every continuous mapping $f : X \to X$ with relatively compact image has a fixed point.

We also recall the definition of *Kuratowski's index* α :

Definition 5. Let X be a metric space and S a bounded subset of X. We set
$$\alpha(S) = \inf\{\varepsilon > 0 | \exists m, \exists S_1, ..., S_m, S = \bigcup_{i \leq m} S_i \text{ and } diam(S_i) \leq \varepsilon\}.$$

Finally, we say that a metric space X is *an absolute retract* for metric spaces if, for any metric space Y and any nonempty closed subset Z of Y, every continuous function $f : Z \to X$ has a continuous extension $g : Y \to X$ over Y. It is well-known that an absolute retract is arcwise connected.

3 A review of known results

There are several results in the analysis of multivalued operators where in the assumptions convexity can be replaced by decomposability. Some of these theorems will be considered in what follows.

The first theorem is a "decomposable" version of the well-known Michael's selection theorem for l.s.c. multifunctions with convex values.

Theorem 1. *Let (X, d) be a separable metric space, E a separable Banach space and let $F : X \to \mathcal{P}_{cl,dec}(L^1(T, E))$ be a l.s.c. multivalued operator. Then F has a continuous selection.*

For the u.s.c. case we have the following:

Theorem 2 (Bressan-Colombo [3]). *Let (X, d) be a separable metric space and let $F : X \to \mathcal{P}_{dec}(L^1(T, E))$ be a H-u.s.c. multivalued operator. If either X or $L^1(T, E)$ is separable, then for each $\varepsilon > 0$ there is a continuous function $f_\varepsilon : X \to L^1(T, E)$ such that $Graph\, f_\varepsilon \subseteq V(Graph\, F, \varepsilon)$ and $f_\varepsilon(X) \in \mathcal{P}_{dec}(F(X))$.*

Remark 1. The main tool for the decomposable case is to consider instead of convex combinations some continuous interpolations between different elements of a decomposable set. More precisely, consider an increasing family $\{A_\lambda \mid \lambda \in [0,1]\}$ (where $A_\lambda \in \mathcal{A}$ with $\mu(A_\lambda) = \lambda\mu(T)$, for every $\lambda \in [0,1]$) and let $u_1, ..., u_n$ be elements of a decomposable set $K \subset L^1(T,E)$. Let λ_i be p nonnegative numbers such that $\sum_{i=1}^{p} \lambda_i = 1$. Setting $\eta_0 = 0$ and $\eta_i = \sum_{j=1}^{i} \lambda_i$, ($i \in \{1,2,...,p\}$ then the decomposable combination $\sum_{i=1}^{p} u_i \cdot \chi_{A_{\eta_i} \setminus A_{\eta_{i-1}}}$ lies inside K. For the compact case the construction below is given by Fryskowski in [11] and the extension for the paracompact case appear in Bresan-Colombo in [3]. (they consider continuous combinations of an infinite family of functions, taking advantage of the fact that at any given time only a finite number of u_i enter in a decomposable combination).

We state now a result about the Kuratowski index of a bounded and decomposable subset of $L^1(T,E)$.

Theorem 3 (Cellina-Mariconda [7]). *Let E be a separable Banach space and let K a nonempty, bounded, decomposable subset of $L^1(T,E)$.Then the Kuratowski's index α of K is the diameter of the set K.*

Finally, some fixed points results for multifunctions with decomposable values are pointed out.

Theorem 4 (Cellina-Colombo-Fonda [6]). *Let E be a separable Banach space and let K be a nonempty, closed subset of $L^1(T,E)$ and let $F : K \to \mathcal{P}_{dec}(K)$ be u.s.c. and with closed graph. If $F(K)$ is decomposable and totally bounded then $Fix\,F \neq \emptyset$.*

Theorem 5 (Bressan-Cellina-Fryszkowski [4]). *Let (X,d) be a separable metric space, E be a separable Banach and let $F : X \times L^1(T,E) \to \mathcal{P}_{b,cl,dec}(L^1(T,E))$ be a continuous multivalued operator satisfying the following condition: there exists $\alpha \in [0,1[$ such that $H(F(\lambda,u),F(\lambda,v)) \leq \alpha\|u-v\|$, for $\forall\ u,v \in L^1(T,E)$ and for all $\lambda \in X$. Then each set $Fix_\lambda F = \{u \mid u \in F(\lambda,u)\}$ is an absolute retract. Moreover a retraction can be chosen that depends continuously on λ.*

Theorem 6 (Fryszkowski [11], Bressan-Colombo [3]). *Every closed decomposable set $K \subset L^1(T,E)$ has the compact fixed point property.*

There are also other results of this type, see for example: [6], [7], [12], [14], [16], [17], [18].

On the other hand, the following theorems, known in the field of multivalued analysis, make use of some convexity assumptions.

Theorem 7 (Aubin-Cellina [2]). *Let F be a l.s.c. multivalued operator with closed, convex values from a paracompact space X to a Banach space E. Let $G : X \to \mathcal{P}_{cl}(E)$ be a multivalued operator with open graph. If $F(x) \cap G(x) \neq \emptyset$, for each $x \in X$, then there exists a continuous selection of $F \cap G$.*

Theorem 8 (Browder [5]). *Let X be a Hausdorff topological vector space and K be a nonempty compact convex subset of X. Let $F : K \to \mathcal{P}_{cv}(K)$ a multivalued operator with open fibres. Then $Fix\, F \neq \emptyset$.*

Theorem 9 (Deguire-Lassonde [9]). *Let X be a paracompact space and $\{Y_i | i \in I\}$ be a family of nonempty, convex subsets of a Hausdorff topological vector space. Then, each family $\{F_i : X \to \mathcal{P}(Y_i) | i \in I\}$ of Ky Fan-type has a selecting family.*

Using the property of decomposability as substitute for convexity, the analogues of the previous results are considered in the next section.

4 Main results

Our first result, concerning the existence of continuous selections for a locally selectionable multifunction, is as follows:

Lemma 1. *Let (X, d) be a separable metric space, $(T, \mathcal{A}, \mu)$ be a complete σ-finite and nonatomic measure space and E be a Banach space. Let $F : X \to \mathcal{P}_{dec}(L^1(T, E))$ be a locally selectionable multivalued operator. Then F has a continuous selection.*

Proof. We associate to any $y \in X$ and $z \in F(y)$ an open neighborhood $N(y)$ and a local continuous selection $f_y : N(y) \to L^1(T, E)$, satisfying $f_y(y) = z$ and $f_y(x) \in F(x)$ when $x \in N(y)$. We denote by $\{V_n\}_{n \in \mathbf{N}^*}$ a countable locally finite open refinement of the open covering $\{N(y) |\ y \in X\}$ and by $\{\psi_n\}_{n \in \mathbf{N}^*}$ a continuous partition of unity associated to $\{V_n\}_{n \in \mathbf{N}^*}$.

Then, for each $n \in \mathbf{N}^*$ there exist $y_n \in X$ such that $V_n \subset N(y_n)$ and a continuous function $f_{y_n} : N(y_n) \to L^1(T, E)$ with $f_{y_n}(y_n) = z_n$, $f_{y_n}(x) \in F(x)$, for all $x \in N(y_n)$. We define $\lambda_0(x) = 0$ and $\lambda_n(x) = \sum\limits_{m \leq n} \psi_m(x)$, $n \in \mathbf{N}^*$. Let $g_{m,n} \in L^1(T, \mathbf{R}_+)$ be the function defined by $g_{m,n}(t) = \|z_n(t) - z_m(t)\|$, for each $m, n \geq 1$.

We arrange these functions into a sequence $\{g_k\}_{k \in \mathbf{N}^*}$.

Consider the function $\tau(x) = \sum\limits_{m,n \geq 1} \psi_m(x)\psi_n(x)$. From Lemma 1 in [4], there exists a family $\{T(\tau, \lambda)\}$ of measurable subsets of T such that:

(a) $T(\tau, \lambda_1) \subseteq T(\tau, \lambda_2)$, if $\lambda_1 \leq \lambda_2$

(b) $\mu(T(\tau_1, \lambda_1) \Delta T(\tau_2, \lambda_2)) \leq |\lambda_1 - \lambda_2| + 2|\tau_1 - \tau_2|$

(c) $\int_{T(\tau,\lambda)} g_n d\mu = \lambda \int_T g_n d_\mu$, $\forall\ n \leq \tau_0$ for all $\lambda, \lambda_1, \lambda_2 \in [0, 1]$, and all $\tau_0, \tau_1, \tau_2 \geq 0$.

Define $f_n(x) = f_{y_n}(x)$ and $\chi_n(x) = \chi_{T(\tau(x),\lambda_n(x))\setminus T(\tau(x),\lambda_{n-1}(x))}$ for each $n \in \mathbf{N}^*$.

Let us consider the singlevalued operator $f : X \to L^1(T,E)$, defined by $f(x) = \sum_{n\geq 1} f_n(x)\chi_n(x)$, $x \in X$.Then, f is continuous because the functions τ and λ_n are continuous, the characteristic function of the set $T(\tau,\lambda)$ varies continuously in $L^1(T,E)$ with respect to the parameters τ and λ and because the sumation defining f is locally finite. On the other hand, from the properties of the sets $T(\tau,\lambda)$ (see Remark 1. and see also [3] p.70-71 and [16] p.199-203 for more details) and because F has decomposable values, it follows that f is a selection of F. □

The following result is similar to Theorem 7.

Theorem 10. *Let (X,d) be a separable metric space, E a separable Banach space, $F : X \to \mathcal{P}_{cl,dec}(L^1(T,E))$ be a l.s.c. multivalued operator and $G : X \to \mathcal{P}_{dec}(L^1(T,E))$ be with open graph. If $F(x) \cap G(x) \neq \emptyset$ for each $x \in X$ then there exists a continuous selection of $F \cap G$.*

Proof. Let $x_0 \in X$ and for each $y_0 \in F(x_0)$ we define the multifunction

$$F_0(x) = \begin{cases} \{y_0\}, & \text{if } x = x_0 \\ F(x), & \text{if } x \neq x_0. \end{cases}$$

Obviously $F_0 : X \to \mathcal{P}_{cl,dec}(L^1(T,E))$ is l.s.c. From Theorem 1 there exists a continuous selection f of F_0, i.e. $f_0(x_0) = y_0$ and $f_0(x) \in F(x)$, for each $x \in X$, $x \neq x_0$. Using Proposition 4, p.81 in [2] it follows that $F \cap G$ is locally selectionable at x_0 and has decomposable values. From Lemma 1 the conclusion follows. □

Let us remark that a continuous selection theorem for a finite intersection of l.s.c. multifunctions with decomposable values appear in [14].

An important result is the following Browder-type selection theorem:

Theorem 11. *Let E be a Banach space such that $L^1(T,E)$ is separable. Let K be a nonempty, paracompact, decomposable subset of $L^1(T,E)$ and let $F : K \to \mathcal{P}_{dec}(K)$ be a multivalued operator with open fibres. Then F has a continuous selection.*

Proof. For each $y \in K$, $F^{-1}(y)$ is an open subset of K. Since K is paracompact it follows that the open covering $\{F^{-1}(y)\}_{y\in K}$ admits a locally finite open refinement, let say $K = \bigcup_{j\in J} F^{-1}(y_j)$, with $y_j \in K$. Let $\{\psi_j\}_{j\in J}$ be a continuous partition of unity subordinate to $\{F^{-1}(y_j)\}_{j\in J}$. Using the same construction as in the proof of Lemma 1, one can construct a continuous function $f : K \to K$, $f(x) = \sum_{j\in J} f_j(x)\chi_j(x)$, where $f_j(x) \in F(x)$ for each $x \in K$. This function is a continuous selection for F. □

Theorem 12. *Let E be a Banach space such that $L^1(T,E)$ is separable. Let I be an arbitrary set of indices, $\{K_i | i \in I\}$ be a family of nonempty, decomposable subsets of $L^1(T,E)$ and X a paracompact space. Let us suppose that the family*
$\{F_i : X \to \mathcal{P}_{dec}(K_i) | i \in I\}$ is of Ky Fan-type. Then there exists a selecting family for $\{F_i\}_{i \in I}$.

Proof. Let $\{U_i\}_{i \in I}$ be the open covering of the paracompact space X given by $U_i = \{x \in X | \ F_i(x) \neq \emptyset\}$ for each $i \in I$. It follows that there exists a locally finite open cover $\{W_i\}_{i \in I}$ such that $\overline{W_i} \subset U_i$ for $i \in I$.(see [10] p.374, Remark 5.1.7.) Let $V_i = \overline{W}_i$. For each $i \in I$ let us consider the multivalued operator $G_i : X \to \mathcal{P}(K_i)$, defined by the relation

$$G_i(x) = \begin{cases} F_i(x), & \text{if } x \in V_i \\ K_i, & \text{if } x \notin V_i. \end{cases}$$

Then G_i is a multifunction with nonempty and decomposable values having open fibres (indeed, $G_i^{-1}(y) = F_i^{-1}(y) \cup (X \setminus V_i))$, for each $i \in I$.

Using Theorem 11 we have that there exists $f_i : X \to K_i$ continuous selection for G_i $(i \in I)$, for each $i \in I$. It follows that for each $x \in X$ there exists $i \in I$ such that $x \in V_i$ and hence $f_i(x) \in G_i(x) = F_i(x)$, proving that $\{f_i\}_{i \in I}$ is a selecting family for $\{F_i\}_{i \in I}$. □

Using a similar argument we can prove:

Theorem 13. *Let E be a separable Banach space and X a separable metric space. Let I be an arbitrary set of indices, $\{K_i | i \in I\}$ be a family of nonempty, closed, decomposable subsets of $L^1(T,E)$. Let $\{F_i : X \to \mathcal{P}_{cl,dec}(K_i) | i \in I\}$ be a family of l.s.c. multivalued operators such that for each $x \in X$ there is $i \in I$ such that $F_i(x) \neq \emptyset$. Then $\{F_i\}_{i \in I}$ has a selecting family.*

Proof. There are only minor modifications of the above arguments. More precisely, the proof runs exactly as in the previous theorem, but instead of using Theorem 11, the conclusion follows from Theorem 1. □

5 Open questions

Let us consider now some open questions on the same subject: can convexity be replaced by decomposability?

Problem 1. Let Y be a Banach space, $K \in \mathcal{P}_{cl}(Y)$ and $x \in K$. The contingent cone $T_K(x)$ is defined by:

$$T_K(x) = \left\{ y \in Y | \ \liminf_{h \to 0_+} \frac{D(x+hy, K)}{h} = 0 \right\}.$$

If $Y = L^1(T, E)$ and K is decomposable then is $T_K(x)$ also decomposable?

Remark 2. For the Clarke tangent cone the question has an affirmative answer (private communication of Professor Z.Pales). Also, a positive answer to this question is given in E. Giner's thesis (see [13]) for the incident tangent cone.

Problem 2. Let (X, d) be a separable metric space, $K \in \mathcal{P}_{dec}(L^1(T, E))$ and let $F : X \times K \to \mathcal{P}_{cl,dec}(L^1(T, E))$ be a continuous and α-contractive (in the second argument) multivalued operator. tDo these conditions on F imply that $\mathrm{Fix}_\lambda F$ is an absolute retract?

Remark 3. When $F : X \times K \to \mathcal{P}_{cl,dec}(K)$, Tolstonogov gives a positive answer to this question (see [21]).

Problem 3. Let (X, d) be a separable metric space and let $F : X \times L^1(T, E) \to \mathcal{P}_{cl,dec}(L^1(T, E))$ be a continuous and locally contractive multivalued operator. Find conditions ensuring that $\mathrm{Fix}_\lambda F$ is an absolute retract.

Problem 4. A "decomposable" version of the Browder's fixed point theorem is also an open problem. From Theorem 3 it follows that a compact, decomposable subset of $L^1(T, E)$ consists of only one point. On the other hand, each closed, decomposable subset of $L^1(T, E)$ has the compact fixed point property (see Definition 4 and Theorem 6). So, the problem is if there exists a continuous, compact selection for the multifunction F.

Acknowledgment. We wish to express sincere thanks to Professor N. Hadjisavvas for his kind support and to Professor J.P. Penot and Professor Zs. Páles for their suggestions and for useful discussions.
Also, the authors are grateful to the learned referees for their fruitful comments and suggestions toward the improvement of this paper.

References

1. H. A. Antosiewicz, A. Cellina, *Continuous selections and differential relations*, J. Diff. Eq., 19(1975), 386-398.
2. J.P. Aubin, A. Cellina, *Differential Inclusions*, Springer, Berlin, 1984.
3. A. Bressan, G. Colombo, *Extensions and selections of maps with decomposable values*, Studia Math., 90(1988), 69-86.
4. A. Bressan, A. Cellina, A. Fryszkowski, *A class of absolute retracts in spaces of integrable functions*, Proc. A.M.S., 12(1991), 413-418.
5. F.E. Browder, *The fixed point theory of multivalued mappings in topological spaces*, Math. Ann. 177(1968), 283-301.
6. A. Cellina, G. Colombo, A. Fonda, *Approximate selections and fixed points for upper semicontinuous maps with decomposable values*, Proc. A.M.S. 98(1986), 663-666.

7. A. Cellina, C. Mariconda, *Kuratowski's index of a decomposable set*, Bull. Pol. Acad. Sci. Math., 37(1989), 679-685.
8. P. Deguire, *Browder-Fan fixed point theorem and related results*, Discuss. Math. Differential Inclusions, 15(1995), 149-162.
9. P.Deguire, M. Lassonde, *Familles selectantes*, Topological Methods in Nonlinear Anal., 5(1995), 261-269.
10. R.Engelking, *General Topology*, Monografie Mat., vol.60, PWN, Warszawa, 1977.
11. A. Fryszkowski, *Continuous selections for a class of non-convex multivalued maps*, Studia Math., 76(1983), 163-174.
12. A. Fryszkowski, *The generalization of Cellina's fixed point theorem*, Studia Math., 78(1984), 213-215.
13. E. Giner, *Étude sur les fonctionnelles intigrales*, Ph. D. Thesis, Pau, 1985.
14. V. V. Goncharov, A. A. Tolstogonov, *Joint continuous selections of multivalued mappings with non-convex values and their applications*, Math. USSR Sbornik, 73(1992), 319-339.
15. F. Hiai, H. Umegaki, *Integrals, conditional expectations and martingales of multivalued functions*, J. Multiv. Anal., 7(1977), 149-182.
16. A. Kisielewicz, *Selection theorems for set-valued functions with decomposable values*, Ann. Soc. Math. Polonae, 34(1994), 123-135.
17. C. Olech, *Decomposabiliy as substitute for convexity*, Lect. Notes in Math., 1091, Springer, Berlin, 1984.
18. A. Ornelas, *Approximation of relaxed solutions for lower semicontinuous differential inclusions*, Ann. Pol. Math., 56(1991), 1-10.
19. Zs. Páles, V. Zeidan, *Characterization of L^1-closed decomposable sets in L^∞*, J. Math. Anal. Appl., 238(1999), 491-515.
20. A. Petruşel, *Continuous selections for multivalued operators with decomposable values*, Studia Univ. Babeş-Bolyai, Mathematica, 41(1996), 97-100.
21. A. Petruşel, A. Muntean, *On Browder's fixed point theorem*, Studia Univ. Babeş-Bolyai, Seria Mathematica, 4(1998), 103-106.
22. A. A. Tolstonogov, *Continuous selectors of fixed points set of multifunctions with decomposable values*, Set-Valued Analysis, 6(1998), 129-147.

Scalar Characterization of Generalized Quasiconvex Functions

Nicolae Popovici

Babeş-Bolyai University, Faculty of Mathematics and Computer Science,
3400 Cluj-Napoca, Romania.

Abstract. This paper provides a general scheme for constructing characteristic functions associated to certain binary relations, which can be used for obtaining scalar characterizations of a large class of generalized quasiconvex functions. In particular, it is shown that the smallest strictly monotonic functions, which were used by Dinh The Luc for characterizing the cone-quasiconvex vector-valued functions in terms of scalar quasiconvexity, can be simply derived from this general setting.

2000 Mathematics Subject Classification. Primary 26B25; Secondary 52A30, 90C29.

Keywords and phrases. Generalized quasiconvexity, scalarization, value functions.

1 Introduction

The notion of cone-quasiconvex vector-valued function, introduced by Dinh The Luc in [4], represents a natural generalization of the classical concept of scalar quasiconvexity, which was widely explored in nonconvex vector optimization [16]. Recall that a function $f : X \to E_2$ defined on a nonempty convex subset X of a vector space E_1 into a vector space E_2, partially ordered by a convex cone C, is called C-quasiconvex on X if for any points $x^1, x^2 \in X$ and $y \in E_2$, the following implication holds:

$$y \in [f(x^1) + C] \cap [f(x^2) + C] \Longrightarrow f(tx^1 + (1-t)x^2) \in y - C, \ \forall\, t \in [0,1]$$

which is equivalent to the fact that for any point $y \in E_2$, the level set

$$L_f(y) = \{x \in X \mid f(x) \in y - C\}$$

is convex. Note that among various generalizations of quasiconvexity which play an important role in vector optimization (see for instance [3], [6] or [1]) this one is of special interest thanks to this property of the level sets.

In particular, if $E_2 = \mathbb{R}^n$ is the Euclidean space, partially ordered by the cone $C = \mathbb{R}^n_+$, then a vector-valued function $f = (f_1, \ldots, f_n) : X \to \mathbb{R}^n$ is C-quasiconvex if and only if the scalar components $f_1, \ldots, f_n$ are quasiconvex in the usual sense. This property shows that, at least in this component-wise setting, there is a direct relationship between vector quasiconvexity and scalar quasiconvexity.

Certainly, if E_2 is an arbitrary vector space and the ordering cone C has no special structure, such a relationship is not obvious (especially because composition with a positive linear form does not preserve quasiconvexity). However, in [5] it was shown that if E_2 is a topological vector space, partially ordered by a closed convex cone C having a nonempty interior, then a function $f : X \to E_2$ is C-quasiconvex on a nonempty convex set $X \subset E_1$ if and only if for every point $y \in E_2$, the composite function $h_y \circ f : X \to \mathbb{R}$ is quasiconvex on X in the usual sense, where $h_y : E_2 \to \mathbb{R}$ denotes the smallest strictly monotonic function, defined for an arbitrary fixed point $e \in \operatorname{int} C$ by

$$h_y(x) = \min\{t \in \mathbb{R} \mid x \in y + te - C\}, \ \forall\, x \in E_2. \tag{1}$$

In [7] we have shown that even if the algebraic and topological structures of E_1 and E_2 are replaced by ordinal structures endowed with an abstract convexity (induced by a multifunction Γ) and a binary relation Ω, a similar characterization in terms of scalar quasiconvexity can be given for the class of (Γ, Ω)-quasiconvex functions by means of certain scalar functions, namely the properly characteristic functions associated to the binary relation Ω.

The aim of this paper is to show that such characteristic functions can be constructed by a simple algebraic approach whenever the binary relation Ω is compatible with a binary operation on E_2, in a certain sense which will be specified in the last section. Before doing this, in the next two sections, we shall recall some basic definitions and preliminary results.

2 Γ-convex sets and (Γ, Ω)-quasiconvex functions

Let E_1 be a nonempty set and let $\Gamma : E_1 \times E_1 \to 2^{E_1}$ be a set-valued mapping, which assigns to each pair of points from E_1 a subset of E_1. The following definition, proposed by us in [7], extends many known notions of generalized convexity (which generally consist in replacing the line segments by continuous arcs or discrete sets) such as: metric-convexity, polygonal-convexity, midpoint-convexity, etc. Moreover, it also permits to treat the case when Γ is not symmetric, which arise for instance in various optimization problems on oriented graphs.

Definition 1. *A subset X of E_1 is said to be:*
i) Γ-convex, if $\Gamma(x^1, x^2) \subset X$, $\forall\, x^1, x^2 \in X$;
ii) Γ-convex with respect to a point $x^0 \in X$, if $\Gamma(x, x^0) \subset X$, $\forall\, x \in X$.

By means of Γ-convexity, we have introduced in [7] an appropriate notion of generalized quasiconvexity for functions which map a Γ-convex set into a nonempty set E_2 which is just endowed with a binary relation $\Omega \subset E_2 \times E_2$ and has neither linear nor topological structure. In the sequel, this relation will be identified with the set-valued mapping $\Omega : E_2 \to 2^{E_2}$ defined by

$$\Omega y = \{y' \in E_2 \mid (y, y') \in \Omega\}, \forall\, y \in E_2.$$

We shall use the following notations:

$$\Omega^- y = \{y' \in E_2 \mid y \in \Omega y'\} \text{ and } \Omega^c y = E_2 \setminus (\Omega y), \ \forall\, y \in E_2;$$

$$\Omega Y = \cup\{\Omega y \mid y \in Y\} \text{ and } [\Omega]Y = \cap\{\Omega y \mid y \in Y\},\ \forall\, Y \subset E_2.$$

Definition 2. *Let $X \subset E_1$ be Γ-convex with respect to a point $x^0 \in X$. A function $f : X \to E_2$ is said to be (Γ, Ω)-quasiconvex at x^0 if*

$$\forall\, x \in X,\ f(x) \in \Omega f(x^0) \implies f(\Gamma(x, x^0)) \subset \Omega f(x^0).$$

Definition 3. *Let X be a nonempty and Γ-convex subset of E_1. A function $f : X \to E_2$ is said to be (Γ, Ω)-quasiconvex on X if*

$$\forall\, x^1, x^2 \in X,\ \forall y \in E_2,\ f(\{x^1, x^2\}) \subset \Omega y \implies f(\Gamma(x^1, x^2)) \subset \Omega y.$$

Remark that, in particular, if E_1 and E_2 are vector spaces and C is a convex cone in E_2, then taking $\Gamma(x^1, x^2) = \mathrm{co}\{x^1, x^2\}$ for all $x^1, x^2 \in E_1$ and $\Omega y = y - C$ for all $y \in E_2$, the (Γ, Ω)-quasiconvexity becomes exactly the C-quasiconvexity in the sense of Dinh The Luc [4].

Note also that Definition 3 can be formulated in a manner similar to the classical definition of quasiconvexity (where "$\max\{f(x^1), f(x^2)\}$" stands in the right-hand side) by means of the so-called "cyrtological closure" (cf. [2]) operator $\mathrm{cl}_{\Omega^-} : 2^{E_2} \to 2^{E_2}$, defined by $\mathrm{cl}_{\Omega^-} Y = [\Omega][\Omega^-]Y$ for all $Y \subset E_2$. Actually, a function $f : X \to E_2$ is (Γ, Ω)-quasiconvex on a nonempty and Γ-convex set $X \subset E_1$ if and only if

$$f(\Gamma(x^1, x^2)) \subset \mathrm{cl}_{\Omega^-}\{f(x^1), f(x^2)\},\ \forall\, x^1, x^2 \in X.$$

Moreover, if we define the level sets of a function f with respect to Ω by

$$L_f(y) = \{x \in X \mid f(x) \in \Omega y\},\ \forall\, y \in E_2$$

then (Γ, Ω)-quasiconvexity can be characterized as follows:

Theorem 1. ([9]) *Let $f : X \to E_2$ be defined on a nonempty set $X \subset E_1$.*

a) If X is Γ-convex with respect to a point $x^0 \in X$ and $f(x^0) \in \Omega f(x^0)$, then f is (Γ, Ω)-quasiconvex at x^0 if and only if $L_f(f(x^0))$ is Γ-convex with respect to x^0;

b) If X is Γ-convex, then f is (Γ, Ω)-quasiconvex on X if and only if $L_f(y)$ is Γ-convex, for every point $y \in E_2$.

Concerning the relationship between the global (Γ, Ω)-quasiconvexity and the (Γ, Ω)-quasiconvexity with respect to a point we have the following result.

Theorem 2. ([8]) *Suppose that Ω is transitive and complete and Γ is symmetric. Then, for any function $f : X \to E_2$ defined on a nonempty and Γ-convex set $X \subset E_1$, the following assertions are equivalent:*

i) f is (Γ, Ω)-quasiconvex on X;

ii) f is (Γ, Ω)-quasiconvex at any point of X.

Note that implication *i)* $\Rightarrow$ *ii)* is still valid whenever Ω is just reflexive, but each assumption on Γ and Ω is essential for the implication *ii)* $\Rightarrow$ *i)*.

3 Properly characteristic functions

The concept of scalar characteristic function associated to a binary relation $\Omega \subset E_2 \times E_2$ was introduced by us in [7] by observing that in the particular case where E_2 is a real topological vector space and $\Omega y = y - C$, $\forall y \in E_2$ is the partial order induced by a closed convex cone C with nonempty interior, the smallest strictly monotonic function h_y defined for a fixed $e \in \operatorname{int} C$ by (1) has the following property:

$$\forall\, x, y \in E_2,\ h_y(x) \leq 0 \Longleftrightarrow x \in y - C.$$

Now, by making use of the function $g : E_2 \times E_2 \to \mathbb{R}$ defined in [6] by

$$g(y^1, y^2) = h_{y^2}(y^1),\ \forall\, y^1, y^2 \in E_2 \tag{2}$$

we can observe that the above property can be rewritten as

$$\forall\, y^1, y^2 \in E_2,\ g(y^1, y^2) \leq 0 \Longleftrightarrow y^1 \in \Omega y^2,$$

which means that the function g is 0-characteristic for Ω in the sense of the definition below.

Definition 4. *Let $\Omega \subset E_2 \times E_2$ be a binary relation on a nonempty set E_2, and let λ be a real number. A function $g : E_2 \times E_2 \to \mathbb{R}$ is called:*

i) λ-characteristic for Ω, if

$$\forall\, y^1, y^2 \in E_2,\ g(y^1, y^2) \leq \lambda \Longleftrightarrow y^1 \in \Omega y^2;$$

ii) properly λ-characteristic if, in addition, there exists $y^0 \in E_2$ such that

$$g(y^1, y^0) \leq g(y^2, y^0) \leq \lambda < g(y^3, y^0),$$

whenever

$$g(y^1, y) \leq g(y^2, y) < g(y^3, y)$$

for some $y, y^1, y^2, y^3 \in E_2$.

Note that if $g : E_2 \times E_2 \to \mathbb{R}$ is (properly) λ-characteristic for Ω then, for any $k \in \mathbb{R}$, the translated function $g + k$ is (properly) $(\lambda + k)$-characteristic.

Remark 1. Denoting by $L_g(t) = \{(y^1, y^2) \in E_2 \times E_2 \mid g(y^1, y^2) \leq t\}$, $\forall t \in \mathbb{R}$ the level sets of $g : E_2 \times E_2 \to \mathbb{R}$, Definition 4 can be interpreted as follows:

i) g is λ-characteristic for Ω if and only if $L_g(\lambda) = \Omega^-$.

ii) g is properly λ-characteristic for Ω if, in addition, for any points $y, y^1, y^2, y^3 \in E_2$ such that

$$(y^1, y) \in L_g(g(y^2, y)) \text{ and } (y^3, y) \notin L_g(g(y^2, y))$$

there exists $y^0 \in E_2$ such that

$$(y^1, y^0),\ (y^2, y^0) \in L_g(\lambda) \text{ and } (y^3, y^0) \notin L_g(\lambda).$$

As we have already seen, the Dinh The Luc's function g defined by (2) is 0-characteristic for the partial order relation induced by a closed convex cone with nonempty interior in a topological vector space. Actually, we have proved in [7] that this function is properly characteristic. The following example shows that besides this vectorial setting, some other characteristic functions can be constructed.

Example 1. Let $\Omega \subset E_2 \times E_2$ be a binary relation on a nonempty set E_2. It is easy to see that for any $\lambda \in [0,1[$ the Heaviside-type function $h : E_2 \times E_2 \to \mathbb{R}$, defined by

$$h(y^1, y^2) = \begin{cases} 0 \text{ if } y^1 \in \Omega y^2 \\ 1 \text{ if } y^1 \in \Omega^c y^2 \end{cases} \tag{3}$$

is properly λ-characteristic, and the "sign"-type function $s : E_2 \times E_2 \to \mathbb{R}$, defined by

$$s(y^1, y^2) = \begin{cases} -1 \text{ if } y^1 \in [\Omega \cap (\Omega^-)^c] y^2 \\ \ \ 0 \text{ if } y^1 \in (\Omega \cap \Omega^-) y^2 \\ \ \ 1 \text{ if } y^1 \in \Omega^c y^2 \end{cases} \tag{4}$$

is λ-characteristic for Ω. Note that s is not properly characteristic in general.

The following two theorems extend some similar results given by Dinh The Luc in [5] and [6] in order to characterize the cone-quasiconvex functions in terms of scalar quasiconvex functions, by means of the smallest strictly monotonic function defined by (1) and the function g defined by (2). In fact, the theorems below allow us to characterize the (Γ, Ω)-quasiconvexity in terms of the $(\Gamma, \geq)$-quasiconvexity of some scalar functions with respect to the usual ordering "$\geq$" in $\mathbb{R}$ and the generalized convexity induced by the set-valued mapping $\Gamma : E_1 \times E_1 \to 2^{E_1}$.

Theorem 3. ([7]) *Let $X \subset E_1$ be Γ-convex with respect to a point $x^0 \in X$ and let $f : X \to E_2$. If the function $g : E_2 \times E_2 \to \mathbb{R}$ is λ_0-characteristic for Ω, with $\lambda_0 = g(f(x^o), f(x^0))$, then the following assertions are equivalent:*
 i) f is (Γ, Ω)-quasiconvex at x^0;
 ii) $g(f(\cdot), f(x^0)) : X \to \mathbb{R}$ is $(\Gamma, \geq)$-quasiconvex at x^0.

Theorem 4. ([7]) *Let $f : X \to E_2$ be a function defined on a nonempty and Γ-convex set $X \subset E_1$ and let $g : E_2 \times E_2 \to \mathbb{R}$ be a properly λ-characteristic function for Ω, with respect to an arbitrary $\lambda \in \mathbb{R}$. Then the following assertions are equivalent:*
 i) f is (Γ, Ω)-quasiconvex on X;
 ii) $g(f(\cdot), y) : X \to \mathbb{R}$ is $(\Gamma, \geq)$-quasiconvex on X, for any point $y \in E_2$.

Note that by applying these theorems for the (properly) characteristic functions g, h or s defined by (2), (3) and (4), we obtain practical criteria for testing (Γ, Ω)-quasiconvexity via scalar quasiconvexity.

4 A scheme for constructing characteristic functions

In this section we shall show that the characteristic functions can be constructed by a simple algebraic approach whenever Ω is compatible with some binary operation on E_2 in the sense of the following definition.

Definition 5. *Let $\omega : E_2 \times E_2 \to E_2$ be a binary operation on a nonempty set E_2. A binary relation $\Omega \subset E_2 \times E_2$ is said to be compatible with ω if there exists a point $\theta \in E_2$ such that $L_\omega(\theta) = \Omega^-$, where*

$$L_\omega(\theta) = \{(y^1, y^2) \in E_2 \times E_2 \mid \omega(y^1, y^2) \in \Omega\theta\}.$$

Example 2. If $(E_2, +)$ is a group and Ω is defined by $\Omega y = y - D$, $\forall\, y \in E_2$, where D is a nonempty subset of E_2, then Ω is compatible with the binary operation $\omega : E_2 \times E_2 \to E_2$, defined by

$$\omega(y^1, y^2) = -y^2 + y^1, \ \forall\, y^1, y^2 \in E_2.$$

Indeed, denoting by 0 the zero element of the group, we have

$$\begin{aligned} L_\omega(0) &= \{(y^1, y^2) \in E_2 \times E_2 \mid -y^2 + y^1 \in -D\} \\ &= \{(y^1, y^2) \in E_2 \times E_2 \mid y^1 \in y^2 - D\} \\ &= \{(y^1, y^2) \in E_2 \times E_2 \mid y^1 \in \Omega y^2\} = \Omega^-. \end{aligned}$$

Remark 2. Let E_2 be a nonempty set. If $\Omega \subset E_2 \times E_2$ is compatible with a binary operation $\omega : E_2 \times E_2 \to E_2$, which possesses a neutral element $0 \in E_2$ at the right, i.e. $\omega(y, 0) = y$, $\forall\, y \in E_2$, then $L_\omega(0) = \Omega^-$. Actually, if $\theta \in E_2$ is such that $L_\omega(\theta) = \Omega^-$ then for $y \in E_2$ we have

$$y \in \Omega\theta \Longleftrightarrow \omega(y, 0) \in \Omega\theta \Longleftrightarrow (y, 0) \in \Omega^- \Longleftrightarrow y \in \Omega 0$$

which shows that $\Omega\theta = \Omega 0$. Hence $L_\omega(0) = L_\omega(\theta) = \Omega^-$.

The following result shows that under the above compatibility assumption, the characteristic functions are easy to construct by making use of certain real-valued functions defined on E_2.

Theorem 5. *Let $\Omega \subset E_2 \times E_2$ be a binary relation which is compatible with a binary operation $\omega : E_2 \times E_2 \to E_2$ and let $\theta \in E_2$ be a point which satisfies $L_\omega(\theta) = \Omega^-$. If there exist a function $G : E_2 \to \mathbb{R}$ and a real number λ such that $L_G(\lambda) = \Omega\,\theta$, where*

$$L_G(\lambda) = \{y \in E_2 \mid G(y) \leq \lambda\},$$

then the function $g = G \circ \omega : E_2 \times E_2 \to \mathbb{R}$ is λ-characteristic for Ω.

Proof. Under the hypotheses of the theorem, we have

$$\begin{aligned} L_g(\lambda) &= \{(y^1, y^2) \in E_2 \times E_2 \mid G(\omega(y^1, y^2)) \leq \lambda\} \\ &= \{(y^1, y^2) \in E_2 \times E_2 \mid \omega(y^1, y^2) \in L_G(\lambda)\} \\ &= \{(y^1, y^2) \in E_2 \times E_2 \mid \omega(y^1, y^2) \in \Omega\theta\} \\ &= L_\omega(\theta) = \Omega^-. \end{aligned}$$

In view of Remark 1 the proof is complete. □

Remark 3. If a function $g : E_2 \times E_2 \to \mathbb{R}$ is λ-characteristic ($\lambda \in \mathbb{R}$) for a binary relation $\Omega \subset E_2 \times E_2$ then, for any point $y^0 \in E_2$, the function $G : E_2 \to \mathbb{R}$ defined by

$$G(y) = g(y, y^0), \ \forall\, y \in E_2$$

satisfies the condition $L_G(\lambda) = \Omega y^0$.

Example 3. Let E_2 be a real topological vector space partially ordered by a closed convex cone $C \subset E_2$ having a nonempty interior, and let Ω and ω be given by

$$\Omega y = y - C \quad \text{and} \quad \omega(y^1, y^2) = y^1 - y^2, \quad \forall\, y, y^1, y^2 \in E_2.$$

Then, for any $e \in \operatorname{int} C$, the function g defined by (2) can be expressed as $g = G \circ \omega$, where

$$G(y) = \min\{t \in \mathbb{R} \mid y \in te - C\}, \ \forall y \in E_2.$$

Similarly, the function h defined by (3) can be expressed as $h = H \circ \omega$, where

$$H(y) = \begin{cases} 0 \text{ if } y \in -C \\ 1 \text{ if } y \in E_2 \setminus (-C). \end{cases}$$

and the function s defined by (4) can be expressed as $s = S \circ \omega$, where

$$S(y) = \begin{cases} -1 \text{ if } y \in (-C) \setminus C \\ \ \ 0 \text{ if } y \in (-C) \cap C \\ \ \ 1 \text{ if } y \in E_2 \setminus (-C). \end{cases}$$

In order to conclude, let us remark that the scalar characteristic functions could be relevant to the decision making theory, since they are directly related to the concept of value function. In fact, if Ω is a preference relation on the criteria space E_2, which is representable by a value function $v : E_2 \to \mathbb{R}$, i.e. a function which satisfies the following condition:

$$\forall\, y^1, y^2 \in E_2, \ v(y^1) \leq v(y^2) \Longleftrightarrow y^1 \in \Omega y^2,$$

then it can be easily seen that the function $g : E_2 \times E_2 \to \mathbb{R}$ defined by

$$g(y^1, y^2) = v(y^1) - v(y^2), \ \forall \, y^1, y^2 \in E_2$$

is 0-characteristic for the preference relation Ω.

Obviously, the existence of a value function imposes a very restrictive assumption on Ω, namely its completeness in E_2. Thus, the representation of preference relations by using characteristic functions (which always exist) instead of value functions, could be a possible approach for decision making problems involving general (not necessary complete) preference relations.

References

1. Cambini, A., Martein, L. (1998) Generalized Concavity in Multiobjective Programming. In: Crouzeix, J.-P., Martinez-Legaz, J.-E., Volle, M. (Eds.) Generalized Convexity, Generalized Monotonicity: Recent Results. Nonconvex Optimization and Its Applications **27**. Kluwer Academic Publishers, Dordrecht
2. Dolecki, S., Malivert, C. (1988) Stability of Efficient Sets: Continuity of Mobile Polarities. Nonlinear Analysis, Theory, Methods & Appl. **12**, 12, 1461–1486
3. Jahn, J. (1986) Mathematical Vector Optimization in Partially Ordered Linear Spaces. Peter Lang Verlag, Frankfurt
4. Luc, D.T. (1987) Connectedness of the efficient point sets in quasiconcave vector maximization. J. Math. Analysis Appl. **122**, 346–354
5. Luc, D.T. (1989) Theory of Vector Optimization. Lecture Notes in Econ. and Math. Systems **319**. Springer-Verlag, Berlin
6. Luc, D.T. (1990) On Three Concepts of Quasiconvexity in Vector Optimization. Acta Mathematica Vietnamica **15**, 1, 3-9
7. Popovici, N. (1995) Contribution à l'optimisation vectorielle. Thèse de doctorat. Université de Limoges
8. Popovici, N. (1997) Sur une notion abstraite de quasiconvexité. Rev. d'Anal. Numérique et de Théorie de l'Approximation **26**, 1-2, 191–196
9. Popovici, N. (1998) On the level sets of (Γ, Ω)-quasiconvex functions. Studia Univ. Babes-Bolyai, Ser. Math. **43**, 1, 71–78

Optimality and Wolfe Duality for Multiobjective Programming Problems Involving n-set Functions

Vasile Preda[1] and Ion M. Stancu-Minasian[2]

[1] Faculty of Mathematics
University of Bucharest
[2] Centre for Mathematical Statistics
The Romanian Academy

Abstract. In this paper we establish some optimality and Wolfe duality results under generalized convexity assumptions for a multiobjective programming problem involving d-type-I and related n-set functions.

2000 Mathematics Subject Classification. Primary 90C29; Secondary 90 C 30.

Keywords and phrases. d-type-I n-set functions, multiobjective programming, optimality conditions, duality results.

1 Introduction

In this paper we consider the following multiobjective programming problem involving n-set functions:

$$(VP)\qquad \begin{array}{l} \text{Minimize } F(S) = (F_1(S), \ldots, F_p(S)) \\ \text{subject to} \\ \qquad G_j(S) \leqq 0, \ j \in M \\ \qquad S = (S_1, \ldots, S_n) \in \mathcal{S}^n \end{array}$$

where $\mathcal{S}^n$ is the n-fold product of a σ-algebra $\mathcal{S}$ of subsets of a given set X, $M = \{1, 2, .., m\}$, F_i, $i \in P = \{1, 2, \ldots, p\}$, and G_j, $j \in M$ are real-valued functions defined on $\mathcal{S}^n$. Let $\mathcal{S}_0 = \{S \mid S \in \mathcal{S}^n,\ G(S) \leqq 0\}$ be the set of all feasible solutions to (VP), where $G = (G_1, \ldots, G_m)$.

The analysis of optimization problems with set functions has been the subject of several papers ([1], [2], [5]-[8], [15]-[22], [25], [26] [29]-[35], [37], [41], [42]). These problems arise in various applications [3], [4], [9], [10], [28] and [38]. The general theory for optimizing set functions was first developed by Morris [26]. Many results of Morris [26] are confined only to set functions of a single set. Corley [8] started to give the concepts of partial derivatives and derivatives of real-valued n-set functions. In [1], [2], [5], [6], [15], [18]-[20], [29]-[32] and [37] are studied optimality conditions and duality results for vector-valued n-set functions. For some details, the readers may consult [1]-[5], [9], [13], [15]-[17], [22], [30].

Starting from the methods used by Jeyakumar and Mond [14] and Ye [40], Suneja and Srivastava [36] defined some new classes of scalar or vectorial functions called d-type-I, d-pseudo-type-I, d-quasi-type-I etc. for a multiobjective nondifferentiable programming problem and obtained necessary and sufficient optimality criteria. Also, they established duality between this problem and its Wolfe-type and Mond-Weir-type duals and obtained some duality results considering the concept of a weak minimum following the lines of Weir and Mond [39].

In this paper we consider some classes of functions of d-type-I for a multiobjective differentiable programming problem involving n-set functions and give optimality and Wolfe duality results for this problem.

2 Definitions and preliminaries

In this section we shall introduce some notation and definitions and recall a number of basic definitions which will be used frequently throughout the sequel.

For $x = (x_1, \ldots, x_m)$, $y = (y_1, \ldots, y_m) \in \mathbf{R}^m$ we put $x \leqq y$ iff $x_i \leqq y_i$ for each $i \in M$; $x \leq y$ iff $x_i \leqq y_i$ for each $i \in M$, with $x \neq y$; $x < y$ iff $x_i < y_i$ for each $i \in M$ and $x \nless y$ is the negation of $x < y$. We note that $x \in \mathbf{R}^m_+$ iff $x \geqq 0$.

Let $(X, \mathcal{S}, \mu)$ be a finite atomless measure space with $L_1(X, \mathcal{S}, \mu)$ separable, and let d be the pseudometric on $\mathcal{S}^n$ defined by:

$$d(S,T) = \left[\sum_{k=1}^{n} \mu^2 (S_k \bigtriangleup T_k)\right]^{1/2}$$

$S = (S_1, ..., S_n)$, $T = (T_1, ..., T_n) \in \mathcal{S}^n$, where $\bigtriangleup$ denotes the symmetric difference. Thus $(\mathcal{S}^n, d)$ is a pseudometric space, which will serve as the domain for most of the functions that will be used in this paper.

For $h \in L_1(X, \mathcal{S}, \mu)$ and $S \in \mathcal{S}$ with the indicator (characteristic) function $I_S \in L_\infty(X, \mathcal{S}, \mu)$, the integral $\int_S h d\mu$ will be denoted by $\langle h, I_S \rangle$.

Now we will introduce the notion of differentiability for n-set functions. This was originally introduced by Morris [26] for set functions; their n-set counterparts are discussed in Corley [8].

A function $\varphi : \mathcal{S} \to \mathbf{R}$ is differentiable at $S^0 \in \mathcal{S}$ if there exists $D\varphi(S^0) \in L_1(X, \mathcal{S}, \mu)$, called the derivative of φ at S^0, and there exists $\psi : \mathcal{S} \times \mathcal{S} \to \mathbf{R}$ such that for each $S \in \mathcal{S}$,

$$\varphi(S) = \varphi(S^0) + \left\langle D\varphi(S^0), I_S - I_{S^0} \right\rangle + \psi(S, S^0),$$

where $\psi(S, S^0)$ is $o(d(S, S^0))$, that is, $\lim\limits_{d(S,S^0) \to 0} \frac{\psi(S,S^0)}{d(S,S^0)} = 0.$

A function $F : \mathcal{S}^n \to \mathbf{R}$ is said to have a partial derivative at $S^0 = (S_1^0, \ldots, S_n^0)$ with respect to its k-th argument if the function

$$\varphi(S_k) = F(S_1^0, \ldots, S_{k-1}^0, S_k, S_{k+1}^0, \ldots, S_n^0)$$

has derivative $D\varphi(S_k^0)$ and we define $D_k F(S^0) = D\varphi(S_k^0)$. If there exists $D_k F(S^0)$, $1 \leqq k \leqq n$, we put $DF(S^0) = (D_1 F(S^0), \ldots, D_n F(S^0))$.

The function $F : \mathcal{S}^n \to \mathbf{R}$ is differentiable at S^0 if there exist $DF(S^0)$ and $\psi : \mathcal{S}^n \times \mathcal{S}^n \to \mathbf{R}$ such that

$$F(S) = F(S^0) + \sum_{k=1}^{n} \left\langle D_k F(S^0), I_{S_k} - I_{S_k^0} \right\rangle + \psi(S, S^0),$$

where $\psi(S, S^0)$ is $o(d(S, S^0))$.

A feasible solution S^0 of (VP) is said to be an efficient solution of (VP) if there exists no other feasible solution S of (VP) such that $F_i(S) \leqq F_i(S^0)$, for all $i \in P$, with strict inequality for at least one $i \in P$.

A feasible solution S^0 of (VP) is said to be a weak minimum (weakly efficient) solution of (VP) if there exists no other feasible solution S of (VP) such that $F_i(S) < F_i(S^0)$, for all $i \in P$.

Along the lines of Jeyakumar and Mond [14] and Suneja and Srivastava [36], we now define the following classes of n-set functions, called d-type-I, d-quasi type-I, d-pseudo type-I, d-quasi-pseudo type-I, d-pseudo-quasi type-I.

Definition 2.1. We say that (F, G) is of *d-type*-I at $S^0 \in \mathcal{S}^n$ if there exist functions $\alpha_i, \beta_j : \mathcal{S}^n \times \mathcal{S}^n \to \mathbf{R}_+ \backslash \{0\}$, $i \in P$, $j \in M$, such that for all $S \in \mathcal{S}_0$

$$F_i(S) - F_i(S^0) \geqq \alpha_i(S, S^0) \sum_{k=1}^{n} \left\langle D_k F_i(S^0), I_{S_k} - I_{S_k^0} \right\rangle, \quad i \in P \tag{2. 1}$$

and

$$-G_j(S^0) \geqq \beta_j(S, S^0) \sum_{k=1}^{n} \left\langle D_k G_j(S^0), I_{S_k} - I_{S_k^0} \right\rangle, \quad j \in M. \tag{2. 2}$$

We say that (F, G) is of d-semistrict type-I at S^0 if in the above definition we have $S \neq S^0$ and (2. 1) is a strict inequality.

Definition 2.2. We say that (F, G) is of d-quasi type-I at $S^0 \in \mathcal{S}^n$ if there exist functions $\gamma_i, \delta_j : \mathcal{S}^n \times \mathcal{S}^n \to \mathbf{R}_+ \backslash \{0\}$, $i \in P$, $j \in M$, such that for every $S \in \mathcal{S}_0$

$$\sum_{i=1}^{p} \gamma_i(S, S^0) F_i(S) \leqq 0 \sum_{i=1}^{p} \gamma_i(S, S^0) F_i(S^0) \Rightarrow$$

$$\sum_{i=1}^{p} \sum_{k=1}^{n} \left\langle D_k F_i(S^0), I_{S_k} - I_{S_k^0} \right\rangle \leqq 0 \tag{2. 3}$$

and

$$-\sum_{j=1}^{m}\delta_j(S,S^0)G_j(S^0) \leqq 0 \Rightarrow \sum_{j=1}^{m}\sum_{k=1}^{n}\left\langle D_kG_j(S^0), I_{S_k} - I_{S_k^0}\right\rangle \leqq 0 \quad (2.\ 4)$$

Definition 2.3. We say that (F,G) is of *d-pseudo type*-I at $S^0 \in \mathcal{S}^n$ if there exist the functions $\gamma_i, \delta_j : \mathcal{S}^n \times \mathcal{S}^n \to \mathbf{R}_+\backslash\{0\}$, $i \in P$, $j \in M$, such that for every $S \in \mathrm{S}_0$

$$\sum_{i=1}^{p}\sum_{k=1}^{n}\left\langle D_kF_i(S^0), I_{S_k} - I_{S_k^0}\right\rangle \geqq 0 \Rightarrow$$
$$\sum_{i=1}^{p}\gamma_i(S,S^0)F_i(S) \geqq \sum_{i=1}^{p}\gamma_i(S,S^0)F_i(S^0) \quad (2.\ 5)$$

and

$$\sum_{j=1}^{m}\sum_{k=1}^{n}\left\langle D_kG_j(S^0), I_{S_k} - I_{S_k^0}\right\rangle \geqq 0 \Rightarrow -\sum_{j=1}^{m}\delta_j(S,S^0)G_j(S^0) \geqq 0. \quad (2.\ 6)$$

Definition 2.4. We say that (F,G) is of *d-quasi-pseudo type*-I at $S^0 \in \mathcal{S}^n$ if there exist the functions $\gamma_i, \delta_j : \mathcal{S}^n \times \mathcal{S}^n \to \mathbf{R}_+\backslash\{0\}$, $i \in P$, $j \in M$, such that for every $S \in \mathrm{S}_0$

$$\sum_{i=1}^{p}\gamma_i(S,S^0)F_i(S) \leqq \sum_{i=1}^{p}\gamma_i(S,S^0)F_i(S^0) \Rightarrow$$
$$\sum_{i=1}^{p}\sum_{k=1}^{n}\left\langle D_kF_i(S^0), I_{S_k} - I_{S_k^0}\right\rangle \leqq 0 \quad (2.\ 7)$$

and

$$\sum_{j=1}^{m}\sum_{k=1}^{n}\left\langle D_kG_j(S^0), I_{S_k} - I_{S_k^0}\right\rangle \geqq 0 \Rightarrow -\sum_{j=1}^{m}\delta_j(S,S^0)G_j(S^0) \geqq 0. \quad (2.\ 8)$$

Definition 2.5. We say that(F,G) is of *d-pseudo-quasi type*-I at $S^0 \in \mathcal{S}^n$ if there exist the functions $\gamma_i, \delta_j : \mathcal{S}^n \times \mathcal{S}^n \to \mathbf{R}_+\backslash\{0\}$, $i \in P$, $j \in M$, such that for every $S \in \mathrm{S}_0$

$$\sum_{i=1}^{p}\sum_{k=1}^{n}\left\langle D_kF_i(S^0), I_{S_k} - I_{S_k^0}\right\rangle \geqq 0 \Rightarrow$$
$$\sum_{i=1}^{p}\gamma_i(S,S^0)F_i(S) \geqq \sum_{i=1}^{p}\gamma_i(S,S^0)F_i(S^0) \quad (2.\ 9)$$

and

$$-\sum_{j=1}^{m}\delta_j(S,S^0)G_j(S^0) \leqq 0 \Rightarrow \sum_{j=1}^{m}\sum_{k=1}^{n}\left\langle D_kG_j(S^0), I_{S_k} - I_{S_k^0}\right\rangle \leqq 0.$$

Remark 2.1. If we fix $S^0 \in \mathcal{S}^n$ then it suffices to define the functions $\alpha_i, \beta_j, \gamma_i, \delta_j$ on $\mathcal{S}^n$ instead on $\mathcal{S}^n \times \mathcal{S}^n$.

Some relationships between these classes of functions can be established on the lines of Suneja and Srivastava [36] and Jeyakumar and Mond [14].

3 Optimality conditions

In this section, we derive necessary and sufficient optimality conditions for a feasible solution to be a weak minimum for (VP).

Theorem 3.1. *Let S^0 be a feasible solution for (VP). We suppose that:*
(i1) There exist scalars $\lambda_i^0 \geqq 0$, $i \in P$, $\sum_{i=1}^{p} \lambda_i^0 = 1$, $\mu_j^0 \geqq 0$, $j \in M_0 = M(S^0) = \{j \in M \mid G_j(S^0) = 0\}$, such that for all $S \in \mathcal{S}^n$

$$\sum_{i=1}^{p}\sum_{k=1}^{n} \lambda_i^0 \left\langle D_k F_i(S^0), I_{S_k} - I_{S_k^0} \right\rangle + \sum_{j \in M_0} \mu_j^0 \sum_{k=1}^{n} \left\langle D_k G_j(S^0), I_{S_k} - I_{S_k^0} \right\rangle \geqq 0; \tag{3. 1}$$

(i2) (F, G_0) is of d-type-I at S^0 where $G_0 = (G_j)_{j \in M_0}$.
Then S^0 is a weak minimum for (VP).

Proof. Suppose, on the contrary, that S^0 is not a weak minimum for (VP). Then there exists a feasible solution $S (S \neq S^0)$ for (VP) such that

$$F_i(S) < F_i(S^0), \quad i \in P. \tag{3. 2}$$

Using (i2), we have for all $S \in \mathcal{S}_0$

$$F_i(S) - F_i(S^0) \geqq \alpha_i(S, S^0) \sum_{k=1}^{n} \left\langle D_k F_i(S^0), I_{S_k} - I_{S_k^0} \right\rangle, i \in P$$

and

$$0 = -G_j(S^0) \geqq \beta_j(S, S^0) \sum_{k=1}^{n} \left\langle D_k G_j(S^0), I_{S_k} - I_{S_k^0} \right\rangle, j \in M_0.$$

Since $\alpha_i(S, S^0) > 0$, $i \in P$ and $\beta_j(S, S^0) > 0$, $j \in M_0$, dividing by $\alpha_i(S, S^0)$ and $\beta_j(S, S^0)$ respectively, the above inequalities reduce to the following:

$$\frac{1}{\alpha_i(S, S^0)} F_i(S) - \frac{1}{\alpha_i(S, S^0)} F_i(S^0) \geqq \sum_{k=1}^{n} \left\langle D_k F_i(S^0), I_{S_k} - I_{S_k^0} \right\rangle, \ i \in P \tag{3. 3}$$

and

$$0 \geqq \sum_{k=1}^{n} \left\langle D_k G_j(S^0), I_{S_k} - I_{S_k^0} \right\rangle, \quad j \in M_0. \tag{3. 4}$$

As $\lambda_i^0 \geqq 0$, $i \in P$, $\mu_j^0 \geqq 0$, $j \in M_0$, using (i1), (3.3) and (3.4), we obtain:

$$\sum_{i=1}^{p} \frac{\lambda_i^0}{\alpha_i(S,S^0)} F_i(S) - \sum_{i=1}^{p} \frac{\lambda_i^0}{\alpha_i(S,S^0)} F_i(S^0) \geqq \sum_{i=1}^{p} \sum_{k=1}^{n} \lambda_i^0 \left\langle D_k F_i(S^0), I_{S_k} - I_{S_k^0} \right\rangle +$$
$$+ \sum_{j \in M_0} \mu_j^0 \sum_{k=1}^{n} \left\langle D_k G_j(S^0), I_{S_k} - I_{S_k^0} \right\rangle \geqq 0.$$

Thus

$$\sum_{i=1}^{p} \frac{\lambda_i^0}{\alpha_i(S,S^0)} F_i(S) \geqq \sum_{i=1}^{p} \frac{\lambda_i^0}{\alpha_i(S,S^0)} F_i(S^0) \tag{3. 5}$$

Since $\frac{\lambda_i^0}{\alpha_i(S,S^0)} \geqq 0$ for any $i \in P$ and $\sum_{i=1}^{p} \lambda_i^0 = 1$, by (3.2) we get:

$$\sum_{i=1}^{p} \frac{\lambda_i^0}{\alpha_i(S,S^0)} F_i(S) < \sum_{i=1}^{p} \frac{\lambda_i^0}{\alpha_i(S,S^0)} F_i(S^0)$$

which contradicts (3.5.). Hence S^0 is a weak minimum for (VP) and the theorem is proved. □

Theorem 3.2. *Let S^0 be a feasible solution for (VP). We suppose that:*
(j1) There exist the scalars $\lambda_i^0 \geqq 0$, $i \in P$, $\sum_{i=1}^{p} \lambda_i^0 = 1$ and $\mu_j^0 \geqq 0$, $j \in M_0$, such that for all $S \in \mathcal{S}^n$

$$\sum_{i=1}^{p} \sum_{k=1}^{n} \lambda_i^0 \left\langle D_k F_i(S^0), I_{S_k} - I_{S_k^0} \right\rangle + \sum_{j \in M_0} \sum_{k=1}^{n} \mu_j^0 \left\langle D_k G_j(S^0), I_{S_k} - I_{S_k^0} \right\rangle \geqq 0$$

(j2) $\left((\lambda_i^0 F_i)_{i \in P}, (\mu_j^0 G_j)_{j \in M_0}\right)$ is of d-pseudo-quasi type-I at S^0.
Then S^0 is a weak minimum for (VP).

Proof. Assume that S^0 is not a weak minimum for (VP). Hence, another feasible solution S ($S \neq S^0$) for (VP) exists with this property, i.e.

$$F_i(S) < F_i(S^0), \quad \text{forany } i \in P. \tag{3. 6}$$

According to (j2) there exist $\gamma_i, \delta_j : \mathcal{S}^n \times \mathcal{S}^n \to \mathbf{R}_+ \backslash \{0\}$, $i \in P$, $j \in M_0$, such that for all $S \in \mathcal{S}_0$

$$\begin{aligned} &\sum_{i=1}^{p} \gamma_i(S,S^0) \lambda_i^0 F_i(S) < \sum_{i=1}^{p} \gamma_i(S,S^0) \lambda_i^0 F_i(S^0) \Rightarrow \\ &\Rightarrow \sum_{i=1}^{p} \sum_{k=1}^{n} \lambda_i^0 \left\langle D_k F_i(S^0), I_{S_k} - I_{S_k^0} \right\rangle < 0 \end{aligned} \tag{3. 7}$$

and

$$-\sum_{j\in M_0}\delta_j(S,S^0)\mu_j^0G_j(S^0)\leqq 0\Rightarrow\sum_{j\in M_0}\sum_{k=1}^{n}\mu_j^0\left\langle D_kG_j(S^0),I_{S_k}-I_{S_k^0}\right\rangle\leqq 0 \tag{3. 8}$$

From (3.6) and $\lambda_i^0\geqq 0$, $i\in P$, $\sum_{i=1}^{p}\lambda_i^0=1,\gamma_i>0$ for any $i\in P$ we get

$$\sum_{i=1}^{p}\gamma_i(S,S^0)\lambda_i^0F_i(S)<\sum_{i=1}^{p}\gamma_i(S,S^0)\lambda_i^0F_i(S^0).$$

Now, using (3.7) we get

$$\sum_{i=1}^{p}\sum_{k=1}^{n}\lambda_i^0\left\langle D_kF_i(S^0),I_{S_k}-I_{S_k^0}\right\rangle<0 \tag{3. 9}$$

Since S^0 is a feasible solution for (VP) and $G_j(S^0)=0$ for $j\in M_0$ we obtain:

$$-\sum_{j\in M_0}\delta_j(S,S^0)\mu_j^0G_j(S^0)\leqq 0$$

This last relation, together with relation (3. 8), implies:

$$\sum_{j\in M_0}\sum_{k=1}^{n}\mu_j^0\left\langle D_kG_j(S^0),I_{S_k}-I_{S_k^0}\right\rangle\leqq 0 \tag{3. 10}$$

Summing the two relations (3. 9) and (3. 10) we obtain:

$$\sum_{i=1}^{p}\sum_{k=1}^{n}\lambda_i^0\left\langle D_kF_i(S^0),I_{S_k}-I_{S_k^0}\right\rangle+\sum_{j\in M_0}\sum_{k=1}^{n}\mu_j^0\left\langle D_kG_j(S^0),I_{S_k}-I_{S_k^0}\right\rangle<0$$

which contradicts (j1).

Hence S^0 is a weak minimum for (VP) and the proof is complete. □

The following optimality result will be needed later in our discussions of duality for (VP).

Theorem 3.3. *[42] Let S^0 be an efficient (or weakly efficient) solution for (VP) and assume that F_i, $i\in P$, and G_j, $j\in M$ are differentiable at S^0.*

Then there exist real numbers $\lambda=(\lambda_1,\ldots\lambda_p)\in\mathbf{R}_+^p$, $\mu=(\mu_1,\ldots,\mu_m)\in\mathbf{R}_+^m$, $(\lambda,\mu)\neq 0$ such that

$$\sum_{k=1}^{n}\left\langle\sum_{i=1}^{p}\lambda_iD_kF_i(S^0)+\sum_{j=1}^{m}\mu_jD_kG_j(S^0),I_{S_k}-I_{S_k^0}\right\rangle\geqq 0,\ \text{for all } S\in\mathcal{S}^n$$

$$\mu_jG_j(S^0)=0,\ j\in M.$$

Now we define the following constraint qualification:

Definition 3.1. [42] A feasible solution S^0 of (VP) is said to be a regular feasible solution if there exists $\widehat{S} \in \mathcal{S}^n$ such that

$$G_j(S^0) + \langle DG_j(S^0), I_{\widehat{S}} - I_{S^0} \rangle < 0, \; j \in M_0.$$

It is easy to see that in the presence of the above constraint qualification, the vector λ in Theorem 3.3 cannot be zero. Thus, incorporating the above constraint qualification in Theorem 3.3, normalizing λ such that $\sum_{i=1}^{p} \lambda_i = 1$ and redefining $(\mu_1, \ldots, \mu_m)$ we obtain the following optimality result for (VP).

Theorem 3.4. *[42] Let S^0 be an efficient (or weakly efficient) solution for (VP) and assume that F_i, $i \in P$, and G_j, $j \in M$ are differentiable at S^0.*

Then there exist $\lambda = (\lambda_1, \ldots \lambda_p) \in \mathbf{R}^p_+$, $\sum_{i=1}^{p} \lambda_i = 1$, $\mu = (\mu_1, \ldots, \mu_m) \in \mathbf{R}^m_+$, such that

$$\sum_{k=1}^{n} \left\langle \sum_{i=1}^{p} \lambda_i D_k F_i(S^0) + \sum_{j \in M_0} \mu_j D_k G_j(S^0), I_{S_k} - I_{S_k^0} \right\rangle \geqq 0, \text{ for all } S \in \mathcal{S}^n$$

$$\mu_j G_j(S^0) = 0, \; j \in M.$$

4 Wolfe duality

In this section we shall consider a Wolfe-type dual for (VP) and we shall prove some duality results assuming the functions to be of the types defined in Section 2.

$$\text{(WD)} \quad \begin{aligned} &\text{Maximize } \psi(T, \lambda, \mu) \equiv F(T) + \sum_{j=1}^{m} \mu_j G_j(T) e \\ &\text{subject to} \\ &\sum_{i=1}^{p} \sum_{k=1}^{n} \lambda_i \langle D_k F_i(T), I_{S_k} - I_{T_k} \rangle + \sum_{j=1}^{m} \sum_{k=1}^{n} \mu_j \langle D_k G_j(T), I_{S_k} - I_{T_k} \rangle \geqq 0, \\ &\qquad\qquad \text{for all } S \in \mathcal{S}^n \\ &\lambda_i \geqq 0, \; i \in P, \; \sum_{i=1}^{p} \lambda_i = 1 \\ &\mu_j \geqq 0, \; j \in M \\ &T \in \mathcal{S}^n \end{aligned}$$

where $e = (1, \cdots, 1) \in \mathbf{R}^p$.

Let $\mathcal{D}$ be the set of all feasible solutions for (WD).

Theorem 4.1. *(Weak Duality) We suppose that*
(a1) $S \in \mathcal{S}_0$ *and* $(T, \lambda, \mu) \in \mathcal{D}$*;*
(a2) (F, G) *is of d-type-I at* T*;*
(a3) $\beta_j = 1$ *for all* $j \in M$*;* $\sum_{i=1}^{p} \frac{\lambda_i}{\alpha_i(S,T)} = 1$.

Then

$$F(S) \nless F(T) + \sum_{j=1}^{m} \mu_j G_j(T)e.$$

Proof. We proceed by contradiction. Then there exist $S \in \mathcal{S}_0$ and $(T, \lambda, \mu) \in \mathcal{D}$ feasible solutions for (VP) and (WD) respectively, such that

$$F(S) < F(T) + \sum_{j=1}^{m} \mu_j G_j(T)e. \tag{4. 1}$$

According to (a2) for α_i $(i \in P), \beta_j$ $(j \in M)$ we have:

$$F_i(S) - F_i(T) \geqq \alpha_i(S,T) \sum_{k=1}^{n} \langle D_k F_i(T), I_{S_k} - I_{T_k} \rangle, \quad i \in P \tag{4. 2}$$

and

$$-G_j(T) \geqq \beta_j(S,T) \sum_{k=1}^{n} \langle D_k G_j(T), I_{S_k} - I_{T_k} \rangle, \quad j \in M. \tag{4. 3}$$

Multiplying inequality (4. 2) by $\frac{\lambda_i}{\alpha_i(S,T)} \geqq 0$, $\forall i \in P$, and inequality (4. 3) by $\mu_j \geqq 0$, $\forall j \in M$, and summing after all i and j respectively, yields

$$\sum_{i=1}^{p} \frac{\lambda_i}{\alpha_i(S,T)} (F_i(S) - F_i(T)) \geqq \sum_{i=1}^{p} \sum_{k=1}^{n} \lambda_i \langle D_k F_i(T), I_{S_k} - I_{T_k} \rangle$$

and

$$-\sum_{j=1}^{m} \mu_j G_j(T) \geqq \sum_{j=1}^{m} \sum_{k=1}^{n} \mu_j \langle D_k G_j(T), I_{S_k} - I_{T_k} \rangle.$$

Adding the last two inequalities and using the feasibility of (T, λ, μ) for (WD) we get:

$$\sum_{i=1}^{p} \frac{\lambda_i}{\alpha_i(S,T)} (F_i(S) - F_i(T)) - \sum_{j=1}^{m} \mu_j G_j(T) \geqq$$
$$\geqq \sum_{i=1}^{p} \sum_{k=1}^{n} \lambda_i \langle D_k F_i(T), I_{S_k} - I_{T_k} \rangle + \sum_{j=1}^{m} \sum_{k=1}^{n} \mu_j \langle D_k G_j(T), I_{S_k} - I_{T_k} \rangle \geqq 0.$$

i.e.

$$\sum_{i=1}^{p} \frac{\lambda_i}{\alpha_i(S,T)} (F_i(S) - F_i(T)) - \sum_{j=1}^{m} \mu_j G_j(T) \geqq 0 \tag{4. 4}$$

Using $\lambda_i \geqq 0,\ i \in P,\ \sum_{i=1}^{p} \lambda_i = 1,\ \alpha_i(S,T) > 0,\ i \in P$, by (4. 1) we get:

$$\sum_{i=1}^{p} \frac{\lambda_i}{\alpha_i(S,T)} (F_i(S) - F_i(T)) - \sum_{i=1}^{p} \frac{\lambda_i}{\alpha_i(S,T)} \sum_{j=1}^{m} \mu_j G_j(T) < 0.$$

Then, by (a3) we get:

$$\sum_{i=1}^{p} \frac{\lambda_i}{\alpha_i(S,T)} (F_i(S) - F_i(T)) - \sum_{j=1}^{m} \mu_j G_j(T) < 0.$$

This inequality contradicts (4. 4). Thus the theorem is proved. □

Theorem 4.2. *(Strong Duality) Let $S^0 \in \mathcal{S}_0$ be such that:*
(b1) S^0 is a weak minimum for (VP);
(b2) S^0 is a regular solution for (VP);
Then there exist $\lambda^0 \in \mathbf{R}^p$ and $\mu^0 \in \mathbf{R}^m$ such that (S^0, λ^0, μ^0) is a feasible solution for (WD) and the values of the objective functions of (VP) and (WD) are equal at these points.

Further, if the conditions of the weak duality theorem 4.1 hold for each feasible solution (T, λ, μ) of (WD), then (S^0, λ^0, μ^0) is a weak maximum for (WD).

Proof. Using Theorem 3.5 we obtain that there exist $\lambda_i^0 \geqq 0,\ i \in P$, with $\sum_{i=1}^{p} \lambda_i^0 = 1$ and $\mu_i^0 \geqq 0,\ i \in M_0$ such that

$$\begin{aligned}&\sum_{i=1}^{p}\sum_{k=1}^{n} \lambda_i^0 \left\langle D_k F_i(S^0), I_{S_k} - I_{S_k^0} \right\rangle + \\ &+ \sum_{j \in M_0}\sum_{k=1}^{n} \mu_j^0 \left\langle D_k G_j(S^0), I_{S_k} - I_{S_k^0} \right\rangle \geqq 0 \quad \text{for all } S \in \mathcal{S}^n.\end{aligned}$$

Now, if we put $\mu_j^0 = 0$ for $j \in M \backslash M_0$, we obtain

$$\begin{aligned}&\sum_{i=1}^{p}\sum_{k=1}^{n} \lambda_i^0 \left\langle D_k F_i(S^0), I_{S_k} - I_{S_k^0} \right\rangle + \\ &+ \sum_{j=1}^{m}\sum_{k=1}^{n} \mu_j^0 \left\langle D_k G_j(S^0), I_{S_k} - I_{S_k^0} \right\rangle \geqq 0 \quad \text{for all } S \in \mathcal{S}^n\end{aligned}$$

and

$$\mu_j^0 G_j(S^0) = 0, \quad \text{for all } j \in M. \tag{4. 5}$$

It follows that (S^0, λ^0, μ^0) is a feasible solution for (WD) and further $\psi(S^0, \lambda^0, \mu^0) = F(S^0)$, i.e. the values of the objective functions of (VP) and (WD) are equal.

For the last part of this theorem we suppose that (S^0, λ^0, μ^0) is not a weak maximum for (WD). Then, there exists a feasible solution (T, λ, μ) for

(WD) such that $\psi(S^0, \lambda^0, \mu^0) < \psi(T, \lambda, \mu)$. From this inequality and (4.5) we have:

$$F(S^0) < \psi(T, \lambda, \mu)$$

which is a contradiction to the weak duality theorem 4.1 for the feasible solutions S^0 of (VP) and (T, λ, μ) of (WD).

Hence (S^0, λ^0, μ^0) is a weak maximum solution for (WD) and the theorem is proved. □

Now we establish a Mangasarian-type [22] strict converse duality theorem for (VP) and (WD).

Theorem 4.3. *(Strict Converse Duality) Let $S^0 \in \mathcal{S}_0$ and $(T^0, \lambda^0, \mu^0) \in \mathcal{D}$ such that:*
(c1) S^0 is a weak minimum for (VP);
(c2) (T^0, λ^0, μ^0) is a weak maximum for (WD);
(c3) $\sum_{i=1}^{p} \lambda_i^0 F_i(S^0) \leqq \sum_{i=1}^{p} \lambda_i^0 F_i(T^0) + \sum_{j=1}^{m} \mu_j^0 G_j(T^0)$;
(c4) $((F_i)_{i\in P}, (G_j)_{j\in M})$ is d-semistrict type-I at T^0 with $\alpha_i(S^0, T^0) = 1,\ i \in P,\ \beta_j(S^0, T^0) = 1,\ j \in M$.

Then $S^0 = T^0$ i.e. T^0 solves (VP) and

$$F(S^0) = F(T^0) + \sum_{j=1}^{m} \mu_j^0 G_j(T^0) e. \tag{4. 6}$$

Proof. We suppose $S^0 \neq T^0$ and we obtain a contradiction. Using (c4) we obtain

$$\sum_{i=1}^{p} \lambda_i^0 \left(F_i(S^0) - F_i(T^0)\right) > \sum_{i=1}^{p}\sum_{k=1}^{n} \lambda_i^0 \left\langle D_k F_i(T^0), I_{S_k^0} - I_{T_k^0} \right\rangle$$

and

$$-\sum_{j=1}^{m} \mu_j^0 G_j(T^0) \geqq \sum_{j=1}^{m}\sum_{k=1}^{n} \mu_j^0 \left\langle D_k G_j(T^0), I_{S_k^0} - I_{T_k^0} \right\rangle.$$

Adding these inequalities and using the feasibility of (T^0, λ^0, μ^0) for (WD) we obtain

$$\sum_{i=1}^{p} \lambda_i^0 F_i(S^0) > \sum_{i=1}^{p} \lambda_i^0 F_i(T^0) + \sum_{j=1}^{m} \mu_j^0 G_j(T^0)$$

which is a contradiction to (c3). Hence $S^0 = T^0$.

Now the last part of this theorem results by (c1) and (c2). Thus, the proof is complete. □

References

1. C.R. BECTOR, D. BHATIA and S. PANDEY, *Duality for multiobjective fractional programming involving n-set functions*, J. Math. Anal. Appl. **186**(1994) 3, 747-768.
2. C.R. BECTOR and M. SINGH, *Duality for multiobjective B-vex programming involving n-set functions*, J. Math. Anal. Appl. **202** (1996) 3, 701-726.
3. D. BEGIS and R. GLOWINSKI, *Application de la méthode des éléments finis à l'approximation d'une probléme de domaine optimal: Méthodes de résolution de problémes approchés*, Appl. Math. Optim., **2** (1975) 2, 130-169.
4. J. CEA, A. GIOAN and J. MICHEL, *Quelques résultats sur l'identification de domaines*, Calcolo, **10**(1973), 133-145.
5. J.H. CHOU, W.S. HSIA and T.Y. LEE, *On multiple objective programming problems with set functions*, J. Math. Anal. Appl., **105** (1985) 2, 383-394.
6. J.H. CHOU, W.S. HSIA and T.Y. LEE, *Second order optimality conditions for mathematical programming with set functions*, J. Austral. Math. Soc. Ser.B, **26**(1985), 284-292.
7. J.H. CHOU, W.S. HSIA and T.Y. LEE, *Epigraphs of convex set functions*, J. Math. Anal. Appl., **118**(1986) 1, 247-254.
8. H.W. CORLEY, *Optimization theory for n-set functions*, J. Math. Anal. Appl., **127**(1987) 1, 193-205.
9. H.W. CORLEY and S.D. ROBERTS, *A partitioning problem with applications in regional design*, Oper. Res., **20**(1972), 1010-1019.
10. G. DANTZIG and A. WALD, *On the fundamental lemma of Neyman and Pearson*, Ann. Math. Stat., **22**(1951), 87-93.
11. R.R. EGUDO *Multiobjective fractional duality*, Bull. Austral. Math. Soc., **37**(1988) 3, 367-378.
12. R.V. HOGG and A.T. CRAIG, *Introduction to Mathematical Statistics*, Macmillan Co., New Jork, 1978.
13. W.S. HSIA and T.Y. LEE, *Proper D-solutions of multiobjective programming problems with set functions*, J. Optim. Theory Appl., **53**(1987), 247-258.
14. V. JEYAKUMAR and B. MOND, *On generalised convex mathematical programming*, J. Austral. Math. Soc., Ser. B, **34**(1992) 1, 43-53.
15. D.S. KIM, C.L. JO and G.M. LEE, *Optimality and duality for multiobjective fractional programming involving n-set functions*, J. Math. Anal. Appl., **224**(1998) 1 , 1-13.
16. H.C. LAI and S.S. YANG, *Saddlepoint and duality in the optimization theory of convex set functions*, J. Austral. Math. Soc. Ser. B, **24** (1982), 130-137.
17. H.C. LAI , S.S. YANG and G.R. HWANG, *Duality in mathematical programming of set functions: on Fenchel duality theorem*, J. Math. Anal. Appl., **95**(1983) 1, 223-234.
18. L.J. LIN, *Optimality of differentiable, vector-valued n-set functions*, J. Math. Anal. Appl., **149** (1990) 1, 255-270.
19. L.J. LIN, *On the optimality conditions of vector-valued n-set functions*, J. Math. Anal. Appl., **161**(1991) 2, 367-387.
20. L.J. LIN, *Duality theorems of vector valued n-set functions*, Comput. Math. Appl., **21**(1991), 165-175.
21. L.J. LIN, *On optimality of differentiable nonconvex n-set functions*, J. Math. Anal. Appl., **168** (1992), 351-366.

22. O.L. MANGASARIAN, *Nonlinear Programming*, McGraw-Hill, New York, 1969.
23. P. MAZZOLENI, *On constrained optimization for convex set functions,* In: Survey of Mathematical Programming, Vol.1, A.PREKOPA (ed.), North-Holland, Amsterdam, 1979, pp.273-290.
24. B. MOND and T. WEIR, *Generalized concavity and duality,* In: Generalized Concavity in Optimization and Economics, S. Schaible and W.T. Ziemba (eds.), Academic Press, New York 1981, 263-279.
25. R.J.T. MORRIS, *Optimization Problem Involving Set Functions*, Ph. D. dissertation, University of California, Los Angeles, 1978.
26. R.J.T. MORRIS, *Optimal constrained selection of a measurable subset*, J. Math. Anal. Appl. **70**(1979) 2, 546-562.
27. R.N. MUKHERJEE, *Generalized convex duality for multiobjective fractional programs*, J. Math. Anal. Appl. **162**(1991), 309-316.
28. J. NEYMANN and E.S. PEARSON, *On the problem of the most efficient tests of statistical hypotheses*, Philos. Trans. Roy. Soc. London, Ser.A, **231**(1933), 289-337.
29. V. PREDA, *On minmax programming problems containing n-set functions*, Optimization, **22**(1991) 4, 527-537.
30. V. PREDA, *On duality of multiobjective fractional measurable subset selection problems*, J. Math. Anal. Appl.,**196** (1995), 514-525.
31. V. PREDA and I.M. STANCU-MINASIAN, *Mond-Weir duality for multiobjective mathematical programming with n-set functions*, Analele Universitatii Bucuresti, Matematica-Informatica **46**(1997), 89-97.
32. V. PREDA and I.M. STANCU-MINASIAN, *Mond-Weir duality for multiobjective mathematical programming with n-set functions*, Rev. Roumaine Math. Pures Appl. **44**(1999) 4, 629-644.
33. J. ROSENMULLER, *Some properties of convex set functions*, Arch. Math., **22**(1971), 420-430.
34. J. ROSENMULLER and H.G. WEIDNER, *A class of extreme convex set functions with finite carrier*, Adv. Math., **10**(1973), 1-38.
35. J. ROSENMULLER and H.G. WEIDNER, *Extreme convex set functions with finite carrier: General theory*, Discrete Math, **10** (1974), 343-382.
36. S.K. SUNEJA and M.K. SRIVASTAVA, *Optimality and duality in nondifferentiable multiobjective optimization involving d-type I and related functions*, J. Math. Anal. Appl., **206**(1997) 2, 465-479.
37. K. TANAKA and Y. MARUYAMA, *The multiobjective optimization problem of set function*, J. Inform. Optim. Sci., **5** (1984) 3, 293-306.
38. P.K.C. WANG, *On a class of optimization problems involving domain variations*. Lectures Notes in Control and Information Sciences, Vol.2., Springer-Verlag, Berlin, 1977, 49-60.
39. T. WEIR and B.MOND, *Generalized convexity and duality in multiple objective programming,* Bull. Austral. Math. Soc., **39**(1989), 287-299.
40. Y.L. YE, *D-invexity and optimality conditions*, J. Math. Anal. Appl., **162**(1991), 242-249.
41. G.J. ZALMAI, *Optimality conditions and duality for constrained measurable subset selection problems with minmax objective functions*, Optimization, **20**(1989) 4, 377-395.
42. G.J. ZALMAI, *Optimality conditions and duality for multiobjective measurable subset selection problems*, Optimization, **22** (1991)2, 221-238.

Vector Stochastic Optimization Problems

Giovanna Redaelli

Istituto di Econometria e Matematica per le Decisioni Economiche Universita Cattolica di Milano
Largo Gemelli 1, 20123 Milano
ITALY

Abstract. A vector optimization problem is studied, whose objective function is a vector of distribution functions depending on a vector of decision variables. Properties of the model are investigated and a scalar representation in terms of the joint distribution function is proposed. Furthermore, assuming to have only empirical estimates of the true distribution functions, we get a sequence of approximate problems and convergence results of the level sets and optimal solution set are proved.

2000 Mathematics Subject Classification. Primary 90C15; Secondary 90C29

Keywords and phrases. Vector sochastic optimization, probability distribution functions, almost sure convergence.

1 Introduction

Many economic and financial problems can be formulated through a decision model with multiple and (eventually) conflicting objectives. Let us think, for example, to portfolio managers who should deal with and keep under control several kinds of financial risks (market risk, credit risk, interest rate risk, exchange risk, and so on) and are requested, more and more often, to produce an integrated management of such different risks. Another typical example is given by investment managers who must distribute resources among different desks and have to do it in such a way to minimize the overall risk of the firm (cfr. [1]).

The uncertainty of the economic and financial environment induces stocasticity in the definition of the objective function and the constraints. In such a context, it is often useful to describe the problem at hand in terms of a stochastic programming model where the functions involved are random variables defined on a given probability space. The objective functions and the constraints may be defined in many different ways according to the problem. In this paper, we consider stochastic programming problems with a convex compact feasible set and assume, as vector objective function, what is known as *probability function.* Let a probability space (Ω, Σ, P), a vector $\varphi = [\varphi_1, \varphi_2, ..., \varphi_l] \in \mathbf{R}^l$ $(l \geq 1)$, a convex compact set $X \subseteq \mathbf{R}^n$ be given and define the vector valued function $\mathbf{g} : X \times \Omega \to \mathbf{R}^l$, $\mathbf{g}(x, \omega) = [g_1(x, \omega), g_2(x, \omega), ..., g_l(x, \omega)]$ to measure the losses from the investment strategy $\mathbf{x}$ when the state of the world is ω. We call *probability function* the vector valued function:

$$\mathbf{\Psi}_{\varphi}(\mathbf{x}) = \begin{bmatrix} F_1(\varphi_1, \mathbf{x}) \\ F_2(\varphi_2, \mathbf{x}) \\ .. \\ .. \\ F_l(\varphi_l, \mathbf{x}) \end{bmatrix} = \begin{bmatrix} P(\omega \in \Omega : g_1(\mathbf{x}, \omega) \leq \varphi_1) \\ P(\omega \in \Omega : g_2(\mathbf{x}, \omega) \leq \varphi_2) \\ .. \\ .. \\ P(\omega \in \Omega : g_l(\mathbf{x}, \omega) \leq \varphi_l) \end{bmatrix}$$

i.e. the probabilty that losses don't exceed a prespecified level φ. If we think each g_i as representing the loss with respect to a specific kind of risk, the vector $[\mathbf{1} - \Psi_{\varphi}(\mathbf{x})]$ can be interpreted as a measure of the risks related to each feasible strategy, i.e. the probability that our losses will exceed the admissible level φ.

What usually really matters is to limit losses and, given the randomness of the problem, we can deal with it from two different but strictly related perspectives. From one hand, we can look for the investment strategy maximizing the amount of capital φ such that the risk $[\mathbf{1} - \Psi_{\varphi}(\mathbf{x})]$ of the strategy is less than or equal to a given level (see, for example, the VaR methodology in finance or the capital at risk computation in insurance, cfr. [8], [15], [7]); otherwise, we can fix the maximum acceptable loss φ and look for minimizing the probability that losses exceed that given value (cfr.[3], [13], [19]).

In this paper we consider this second approach and study the following vector optimization problem:

$$\inf_{\mathbf{x} \in X} \mathbf{\Phi}_{\varphi}(\mathbf{x}) = \inf_{x \in X} [-\mathbf{\Psi}_{\varphi}(\mathbf{x})] \tag{1}$$

As pointed out by an anonymous referee, we may also look for minimizing the average losses, eventually employing non linear utility functions. We have not considered this case explicitely, since formulation (1) seemed the one which better describe the modern approach to risk management. Nevertheless, the expected utility approach remains more natural and practical for many decision problems; in these cases, we can still prove convergence properties by using some results from [21] and [22], but under stronger assumptions than those given in section 2 (see the Remark at the end of section 5.2).

The paper is organized as follows: in section 2 we state the assumptions on the objective function and the feasible set of problem (1); these assumptions are quite general and usually satisfied in financial applications. In particular, we don't care for the way in which the feasible set is determined and concentrate mainly on the properties and characterizations of the objective function. In section 3, properties of (1) are deduced from the assumptions and results given in section 2. We then turn to scalarization in section 4; in particular, in section 4.1 we consider the so-called *complete scalarizations*, i.e. families of scalar functions such that every optimal solution of (1) can be obtained as optimal solution of a scalar problem written in terms of the scalar functions belonging to these families; then, in section 4.2 we consider a scalar optimization problem where the objective function is expressed in terms of the joint probability distribution of the loss function and study under what condition such a model is a scalar representation of problem (1).

Finally, in section 5 we use the results obtained in the previous sections to deal with convergence problems. Indeed, very often in the applications we don't know the true probability distribution of the random variables involved in model (1) but we only have empirical estimates; we then assume to have a sequence of empirical probability distributions, define a sequence of approximate versions of model (1) and give convergence conditions for the approximate optimal solution sets to the true one.

Section 6 gives our conclusions and comments.

2 Assumptions and preliminary results

Let us introduce the following notations and assumptions:

1. $\xi : \Omega \to \mathbf{R}^m$ is a random vector on the probability space (Ω, Σ, P) with absolutely continuous probability distribution function (with respect to the Lebesgue measure on $\mathbf{R}^m$). We denote with P_ξ its probability measure.
2. Let $g : \mathbf{R}^{n+m} \to \mathbf{R}^l$ be a vector valued function:
$$g(\mathbf{x}, \xi) = [g_1(\mathbf{x}, \xi), g_2(\mathbf{x}, \xi), ..., g_l(\mathbf{x}, \xi)]$$
such that the g_i $(i = 1, 2, ..., l)$ are quasi-convex and $g_i(\cdot, \xi)$ are lower semicontinuous for every $\xi \in \mathbf{R}^m$.
3. For every $x \in \mathbf{R}^n$, let $F_i(\cdot, x)$ be the distribution function of the random variable $g_i(x, \xi(\omega))$ defined on the probability space (Ω, Σ, P), i.e.:
$$F_i(\varphi, x) = P\{\omega \in \Omega : g_i(x, \xi(\omega)) \leq \varphi\} \tag{2}$$
for every $\varphi \in \mathbf{R}$. Similarly, let $F_g(\cdot, x)$ be the distribution function of the random vector
$$\mathbf{g}(x, \xi(\omega)) = [g_1(x, \xi(\omega)), g_2(x, \xi(\omega)), ..., g_l(x, \xi(\omega))]$$
defined on the probability space (Ω, Σ, P) and with values in $\mathbf{R}^l$, i.e.:
$$\begin{aligned} F_g(\varphi, x) &= P\{\omega \in \Omega : \mathbf{g}(x, \xi(\omega)) \leq \varphi\} \\ &= P\{\omega \in \Omega : g_i(x, \xi(\omega)) \leq \varphi_i \ i = 1, 2, ..., l\} \end{aligned} \tag{3}$$
for every $\varphi \in \mathbf{R}^l$.
4. Let X be a convex and compact subset of $\mathbf{R}^n$.

We recall now some properties concerning the marginal and the joint distribution functions ($F_i(\varphi_i, \cdot)$ and $F_g(\varphi, \cdot)$, respectively) as functions of the decision variables x. These results hold under the assumptions stated above; for the sake of simplicity, we will not explicitely recall them every time in the statement of the theorems.

Theorem 1. *Let $\varphi \in \mathbf{R}$ be given. The functions $F_i(\varphi, \cdot)$, defined in (2) are upper semicontinuous on $\mathbf{R}^n$.*

Proof. See [21], Theorem 5, p.154 with $l = 1$. □

Remark 1. Theorem 1 was proved assuming that the $g_i(\cdot, \xi)$ $(i = 1, 2, ..., l)$ were lower semicontinuos for every $\xi \in \mathbf{R}^m$. It is straightforward to state the same result under the weaker assumption that the $g_i(\cdot, \xi(\omega))$ are lower semicontinuous for every $\omega \in \Omega$.

Theorem 2. *Let $\varphi \in \mathbf{R}$ be given. The functions $F_i(\varphi, \cdot)$, defined in (2) are quasi-concave on $\mathbf{R}^n$.*

Proof. Let $\alpha \in [0, 1]$; then, for every x_1, $x_2 \in X$ and $\xi \in \mathbf{R}^m$, it is:

$$g_i(\alpha x_1 + (1 - \alpha) x_2, \xi) \leq \max[g_i(x_1, \xi), g_i(x_2, \xi)]$$

therefore:

$$\begin{gathered}\{\xi \in \mathbf{R}^m : g_i(\alpha x_1 + (1 - \alpha) x_2, \xi) \leq \varphi\} \supseteq \\ \supseteq \{\xi \in \mathbf{R}^m : \max[g_i(x_1, \xi), g_i(x_2, \xi)] \leq \varphi\}\end{gathered}$$

$$\Downarrow$$

$$F_i(\varphi, \alpha x_1 + (1 - \alpha) x_2) = P_\xi \{\xi \in \mathbf{R}^m : g_i(\alpha x_1 + (1 - \alpha) x_2, \xi) \leq \varphi\} \geq$$

$$\geq P_\xi \{\xi \in \mathbf{R}^m : \max[g_i(x_1, \xi), g_i(x_2, \xi)] \leq \varphi\} = \min[F_i(\varphi, x_1), F_i(\varphi, x_2)]$$

□

Remark 2. If g_i is strictly quasi-convex we get the strict quasi-concavity of $F_i(\varphi, \cdot)$ if the open set:

$$\{\xi \in \mathbf{R}^m : g_i(\alpha x_1 + (1 - \alpha) x_2, \xi) < \varphi_i < \max[g_i(x_1, \xi), g_i(x_2, \xi)]\}$$

has strictly positive measure.

Remark 3. Theorem 2 was proved assuming that the $g_i(\cdot, \xi)$ $(i = 1, 2, ..., l)$ were quasi-convex for every $\xi \in \mathbf{R}^m$. It is straightforward to state the same result under the weaker assumption that the $g_i(\cdot, \xi(\omega))$ are quasi-convex for every $\omega \in \Omega$.

Remark 4. To prove Theorem 2 in the case $l > 1$, it is necessary to assume that ξ has a log-concave distribution (see [20], Theorem 10.2.1, p.303). Under this further assumption, we get the quasi-concavity also of $F_g(\varphi, \cdot)$ as defined in (3).

3 The optimization problem

We consider the following vector stochastic programming problem:

$$\inf_{\mathbf{x}\in X} \mathbf{\Phi}_\varphi(\mathbf{x}) = -\begin{bmatrix} F_1(\varphi_1, x) \\ F_2(\varphi_2, x) \\ \cdot\cdot \\ \cdot\cdot \\ F_2(\varphi_l, x) \end{bmatrix} = -\begin{bmatrix} P(\omega \in \Omega : g_1(\mathbf{x}, \xi(\omega)) \leq \varphi_1) \\ P(\omega \in \Omega : g_2(\mathbf{x}, \xi(\omega)) \leq \varphi_2) \\ \cdot\cdot \\ \cdot\cdot \\ P(\omega \in \Omega : g_l(\mathbf{x}, \xi(\omega)) \leq \varphi_l) \end{bmatrix} \quad (4)$$

where φ is a fixed vector in $\mathbf{R}^l$ and X is a compact subset of $\mathbf{R}^n$.

From now on, let us assume $C = \mathbf{R}^l_+$ as the ordering cone for problem (4).

As an immediate consequence of the assumptions stated in the previous section, we can characterize problem (4) as follows.

Theorem 3. *Let $\varphi \in \mathbf{R}^l$ be given. The vector function $\mathbf{\Phi}_\varphi$ is $C-$ quasiconvex and $C-$ continuous on $\mathbf{R}^n$. Furthermore, it is level-closed and epi-closed.*

Proof. The quasi-convexity is a direct consequence of [17], Corollary 6.6 p.31, since in our case it is $E_2 = \mathbf{R}^l$ and the functions $-F_i(\varphi_i, \cdot)$ are quasi-convex on $\mathbf{R}^n$ (see Theorem 2). Similarly, we get $C-$ continuity of $\mathbf{\Phi}_\varphi$ from [17], Theorem 5.5, p.23 and Corollary 5.6 p.24, since in our context: (i) the functions $-F_i(\varphi_i, \cdot)$ are lower-semicontinuous on $\mathbf{R}^n$ (see Theorem 1), (ii) the topological polar cone (see [17],p.7) is $C' = \mathbf{R}^l_+$ which is a polyhedral cone and (iii) it is

$$C' = cone\left(conv\left\{e_i \in \mathbf{R}^l; i = 1, 2, ..., l\right\}\right)$$

e_i being the ith component of the canonical basis of $\mathbf{R}^l$.

Let us now turn to the level set and the epigraph of $\mathbf{\Phi}_\varphi$. By definition (see [17], p.18) it is:

$$lev(y) = \{x \in \mathbf{R}^n : \mathbf{\Phi}_\varphi(x) \in y - C\}$$

which is equivalent to

$$lev(y) = \bigcap_{i=1}^{l} \{x \in \mathbf{R}^n : -F_i(\varphi_i, x) \leq y_i\}$$

The sets $\{x \in \mathbf{R}^n : -F_i(\varphi_i, x) \leq y_i\}$ $(i = 1, 2, ..., l)$ are the lower level sets of $-F_i(\varphi_i, \cdot)$, which are closed since $-F_i(\varphi_i, \cdot)$ $(i = 1, 2, ..., l)$ are lower semicontinuous. Therefore, $lev(y)$ is closed since it is the intersection of closed sets. Furthermore, being $intC = int\mathbf{R}^l_+ \neq \emptyset$, from [17], Theorem 5.8, p.25, we get that also the set

$$epi(\varPhi_\varphi) = \left\{(x, y) \in \mathbf{R}^{n+l} : y \in \varPhi_\varphi + C\right\}$$

is closed. □

Let us denote with $WS(X;\mathbf{\Phi}_\varphi)$ the set of weakly optimal solutions, i.e.:

$$WS(X;\mathbf{\Phi}_\varphi) = \{x \in X : \mathbf{\Phi}_\varphi(x) \in WMin(\mathbf{\Phi}_\varphi(X)\,|C)\}$$

being $\mathbf{\Phi}_\varphi(X)$ the image of X under $\mathbf{\Phi}_\varphi$ and

$$WMin(\mathbf{\Phi}_\varphi(X)\,|C) = \{y \in \mathbf{\Phi}_\varphi(X) : \nexists y^* \in \mathbf{\Phi}_\varphi(X), y^* \neq y \text{ such that } y \geq_C y^*\}$$

where, in our context, $y \geq_C y^*, y \neq y^*$ means $y_i \geq y_i^*$ $(i = 1, 2, ..., l)$ with at least one strict inequality.

Lemma 1. *Let $y \in \mathbf{R}^l$ such that $lev(y) \neq \emptyset$. Then, the set*

$$WS(lev(y);\mathbf{\Phi}_\varphi) = \{x \in lev(y) : \mathbf{\Phi}_\varphi(x) \in WMin(\mathbf{\Phi}_\varphi(lev(y))\,|C)\}$$

is compact.

Proof. Let $\{x_n\}_{n\geq 1}$ be a sequence in $WS(lev(y);\mathbf{\Phi}_\varphi)$ convergent to a point $x^* \in X$. Let us suppose $x^* \notin WS(lev(y);\mathbf{\Phi}_\varphi)$; therefore, there exists $y^* \in \mathbf{\Phi}_\varphi(lev(y))$ such that $\mathbf{\Phi}_\varphi(x^*) \geq_C y^*$, $\mathbf{\Phi}_\varphi(x^*) \neq y^*$, i.e.: $-F_i(\varphi, x^*) \geq y_i^*$ with at least one strict inequality; without loss of generality, suppose $-F_1(\varphi, x^*) > y_1^*$; since $-F_i(\varphi,\cdot)$ are lower-semicontinuous, we have

$$\liminf(-F_1(\varphi, x_n)) \geq -F_1(\varphi, x^*) > y_1^*$$

$$\liminf(-F_i(\varphi, x_n)) \geq -F_i(\varphi, x^*) \geq y_i^* \quad (i = 2, ..., l)$$

Hence, there exists $\tilde{n}$ such that $-F_1(\varphi, x_n) > y_1^*$ and $-F_i(\varphi, x_n) \geq y_i^*$ $(i = 2, ..., l)$ for every $n \geq \tilde{n}$. But this is a contradiction, since we assumed $x_n \in WS(lev(y);\Phi_\varphi)$ for every n. Then, $WS(lev(y);\Phi_\varphi)$ is closed; therefore, since it is a closed subset of the compact set X, it is compact. □

Theorem 4. *The set $WS(X;\Phi_\varphi)$ of weakly optimal solutions is a closed connected set.*

Proof. The result immediately follows from [17], Theorem 4.6, p.150 which proves that $WS(X;\Phi_\varphi)$ is a closed connected set whenever Φ_φ is $C-$ continuous, $C-$ quasiconvex on a closed set X and $WS(lev(y);\Phi_\varphi)$ is compact for every $y \in \mathbf{R}^l$ such that $lev(y) \neq \emptyset$. In our context, these requirements are fullfilled (see Theorem 3 and Lemma 1). □

4 Scalarization

In the previous section, we studied some properties of problem (4) as consequences of the assumptions stated on the functions involved. In order to obtain convergence results, we now turn to scalarization methods. In particular, we first recall some very simple *weak complete scalarizations*, i.e. families of scalar optimization problems which completely characterize (in the sense of [17], Definition 3.1, p.95) the optimal solutions of the vector problem (4).

We then show in section 4.2 that under suitable conditions the scalar problem, whose objective function is the joint probability distribution (defined in (3)) of the loss functions g_i, is indeed a scalar representation of problem (4).

The scalar representations, described in the present section, will enable us to obtain some interesting convergence results in section 5.2.

4.1 Weak complete scalarization families

From [17], Definition 3.1, p.95, a family $\mathcal{H}$ of real functions from $\mathbf{R}^l$ to $\mathbf{R}$ is a complete scalarization of problem (4) if for every optimal solution x of (4), there exists $h \in \mathcal{H}$ such that x is an optimal solution of the scalar problem $\min(h \circ \Phi)$ and $S(X; h \circ \Phi) \subseteq S(X; \Phi)$.

Let $e \in intC$ be given. From [17], Proposition 3.2, p.95 the family $\mathcal{G} = \{h_{e,a}\}_{a \in \mathbf{R}^l}$ of the smallest monotonic functions:

$$h_{e,a}(\Phi(x)) = \min\{t \in \mathbf{R} : \Phi(x) \in a + te - C\} \tag{5}$$

is a complete weak scalarization. If we take $a \in WMin(X; \Phi)$ we get that $h_{e,a}$ is continuous and strictly increasing (see [17], Theorem 1.6, p.83 and Corollary 1.7, p.85). Furthermore, since the components of Φ, for every x, are distribution functions (changed in sign), $\{h_{e,a}\}$ is a family of bounded real functions.

Another well known family of scalar representations[1] of problem (4) is the following:

$$\begin{aligned} &\inf_{x \in X} \left[-F_i(\varphi_i, x)\right] \\ s.t.:\ & F_j(\varphi_j, x) \geq y_j \qquad j \neq i \end{aligned} \tag{6}$$

It consists of scalar problems whose objective function is one of the components of the objective $\mathbf{\Phi}_\varphi$ in the vector problem (4), while the other functions are considered as constraints with appropriately chosen bounds[2] y_j.

The following result is an immediate consequence of the properties of the functions $F_i(\varphi_i, \cdot)$.

Lemma 2. *The feasible and optimal solution sets of problem (6) are compact and convex subsets of* $\mathbf{R}^n$.

[1] This scalarization method is known as constraint method.

[2] Since $F_j(\cdot, x)$ are distribution functions, the constraints are meaningfull as long as y_j are chosen in $(0, 1]$.

4.2 The scalar problem with the joint distribution function

Let us consider the joint probability distribution of the loss functions g_i:

$$F_g(\varphi, x) = P\{\omega \in \Omega : \mathbf{g}(x, \xi(\omega)) \leq \varphi\}$$

$$= P\{\omega \in \Omega : g_i(x, \xi(\omega)) \leq \varphi_i \; i = 1, 2, ..., l\}$$

and define the following optimization problem:

$$\inf_{x \in X} [-F_g(\varphi, x)] \tag{7}$$

Since the objective function is lower semicontinuous over X, we get:

$$S(X; -F) = \arg\min_{x \in X} [-F_g(\varphi, x)] \neq \emptyset$$

The following result is immediate.

Theorem 5. *The set of optimal solutions $S(X; -F)$ is a compact convex set.*

If we assume the random vector ξ independent of x, we can interpret each

$$\{g_i(x); x \in X\} = \{g_i(x, \xi(\omega)); x \in X\}$$

$(i = 1, 2, .., l)$ as a family of random variables indexed by x varying in X. Recall the following definition (see [10], p.44): two random variables Y and Z are of the same type if[3] $Y =^d (Z - b)/a$ with $a > 0$ and $b \in \mathbf{R}$, i.e. they differ only by location parameters.

Theorem 6. *The following statements are true:*

1. *If $\mathbf{g}(\cdot, \xi(\omega))$ is a vector of mutually independent random variables for every given $x \in X$, then problem (7) is a scalar representation of (4).*
2. *If $\{g_i(x); x \in X\}$ $(i = 1, 2, .., l)$ is a family of random variables of the same type, then (7) is a scalar representation of (4).*

Proof. From [17] Definition 2.1 p.86, (7) is a scalar representation of (4) with respect to the ordering cone C if for every x_1, $x_2 \in X$, we get:

$$\begin{array}{ll} \Phi_\varphi(x_1) \geq_C \Phi_\varphi(x_2) & \text{implies } -F_g(\varphi, x_1) \geq -F_g(\varphi, x_2) \\ \Phi_\varphi(x_1) >_C \Phi_\varphi(x_2) & \text{implies } -F_g(\varphi, x_1) > -F_g(\varphi, x_2) \end{array} \tag{8}$$

1. The result is immediate, since in case of g_i's independent random variables it is:

$$F_g(\varphi, x) = \prod_{i=1}^{l} F_i(\varphi_i, x)$$

3 "$=^d$" means equal in distribution.

2. Suppose $\Phi_\varphi(x_1) \geq_C \Phi_\varphi(x_2)$, i.e.: $-F_i(\varphi, x_1) \geq -F_i(\varphi, x_2)$ for every i, which is equivalent to

$$P_i\{g_i(x_1) \leq \varphi_i\} \leq P_i\{g_i(x_2) \leq \varphi_i\}$$

for every i, being $P_i\{g_i(x) \leq \varphi_i\} = P\{\omega \in \Omega : g_i(x, \xi(\omega)) \leq \varphi_i\}$. Since, by assumption, $g_i(x_1)$ and $g_i(x_2)$ are random variables of the same type, there exist constants a_i and b_i ($a_i > 0$ and $b_i \in \mathbf{R}$) such that $g_i(x_1) = \frac{g_i(x_2) - b_i}{a_i}$. Therefore, we get:

$$P_i\{g_i(x_1) \leq \varphi_i\} = P_i\left\{\frac{g_i(x_2) - b_i}{a_i} \leq \varphi_i\right\}$$

$$= P_i\{g_i(x_2) \leq a_i\varphi_i + b_i\} \leq P_i\{g_i(x_2) \leq \varphi_i\}$$

Since the distribution function of any random variable is non-decreasing, it must be: $a_i\varphi_i + b_i \leq \varphi_i$ for every i. When we consider the distribution function of the random vector g, i.e.:

$$F_g(\varphi, x) = P_g\{g(x) \leq \varphi\} = P_g\{g_i(x) \leq \varphi_i, \, i = 1, 2, ..., l\}$$

we get:

$$\begin{aligned} F_g(\varphi, x_1) = P_g\{g_i(x_1) \leq \varphi_i, \, i = 1, 2, ..., l\} = \\ = P_g\{g_i(x_2) \leq a_i\varphi_i + b_i, \, i = 1, 2, ..., l\} \leq \\ \leq P_g\{g_i(x_2) \leq \varphi_i, \, i = 1, 2, ..., l\} = F_g(\varphi, x_2) \end{aligned}$$

which is the desired result. By reasoning similarly we can prove the implication in (8) with the strict inequality.

□

As an immediate consequence of the previous theorem, we get that the optimal solution set of problem (4), $WS(X; \Phi_\varphi)$ is non empty, since it holds (see [17], Proposition 2.2, p.87):

$$S(X; -F) \subseteq WS(X; \Phi_\varphi)$$

We proved Theorem 6 in two simple cases: when the random variables are mutually independent or linear trasforms of each other. In particular, this last assumption is frequently used in financial modelling (see, for example, the VaR methodology based on normal random variables described in *J.P. Morgan, Risk metrics: technical document*, or [9] where elliptical distributions are considered. We are looking for generalizing Theorem 6 to families of random variables which differ not only for location parameters but also for skewness; in particular, we are considering $\alpha - stable$ distributions. The problem of dealing with stochastically dependent random variables is an open question.

5 Convergence problems

Let us now turn to the problem of approximating the true probability measure, which usually is not known when dealing with applications; very often, we only have empirical estimates so that we can formulate approximate versions of the true model. The optimal solution and optimal value sets are then estimates of the true ones. In the sequel, we study if and how the empirical estimates of optimal solutions almost surely converge to the true ones.

Let $\{\xi^k\}_{k\geq 1}$ be a sequence of independent random vectors such that, for every k, ξ^k has the same probability distribution of the random variable ξ. Let F_ξ^n be the empirical distribution function determined by the first n random variables ξ^k; i.e. for any $\omega \in \Omega$ it is:

$$F_\xi^n(z,\omega) = \frac{1}{n}\sum_{k=1}^{n} \Upsilon_{(\xi^k \leq z)}(\omega) \quad (z \in \mathbf{R}^m)$$

(where $\Upsilon_{(A)}$ is the indicator function of A). Furthermore, given $x \in \mathbf{R}^n$, for any $\omega \in \Omega$ let[4] $F_i^n(\cdot,x) = F_i^n(\cdot,\omega,x)$ and $F_g^n(\cdot,x) = F_g^n(\cdot,\omega,x)$ denote the empirical distribution functions (determined by the first n random variables ξ^k) of, respectively, g_i $(i = 1,2,...,l)$ and $\mathbf{g} = [g_1,g_2,...,g_l]$. Finally, let us define:

$$\sup_{\mathbf{x}\in X} \mathbf{\Psi}_\varphi^n(\mathbf{x}) = \begin{bmatrix} F_1^n(\varphi,x) \\ F_2^n(\varphi,x) \\ \cdot\cdot \\ \cdot\cdot \\ F_l^n(\varphi,x) \end{bmatrix} \tag{9}$$

Under the given assumptions, we will study under what conditions it holds:

$$\arg\max_{\mathbf{x}\in X} \mathbf{\Psi}_\varphi^n(\mathbf{x}) \to \arg\max_{\mathbf{x}\in X} \mathbf{\Psi}_\varphi(\mathbf{x}) \ \text{ a.s.}$$

Since we consider sequences of stochastic programming problems like model (9) and we wish to determine whether or under what conditions the optimal solution sets converge to the corresponding ones of the true problem, notions of convergence for sequences of sets and functions are needed. To this end, in Section 5.1 we introduce the definitions of hypo-convergent sequences of functions and convergent (in the sense of Kuratowski) sequences of sets; we also recall some useful properties necessary to prove the main convergence results in section 5.2.

In order to get convergence results, we will refer to the scalarizations described in section 4.

[4] In order to simplify the notation, in the sequel we will omit the dependance of F_i^n on ω.

5.1 Preliminary definition and results

Let us remind first the definition of convergence in the sense of Kuratowski[5] for sequences of sets. Let $\mathbf{N}$ be the set of positive integers.

Definition 1. (cfr. [26], p.19) A sequence $\{K_n\}_{n\in\mathbf{N}}$ of subsets of a normed finite-dimensional linear space Z converges in the sense of Kuratowsky to the set $K \subseteq Z$ and we write $K_n \to K$ if:

$$\limsup K_n = K = \liminf K_n$$

where:

$$\limsup K_n = \left\{ \begin{array}{c} x \in Z : x = \lim_m x_m \ , \ x_m \in K_{n_m} \text{ for } \{K_{n_m}\}_{m\geq 1} \\ \text{subsequence of } \{K_n\} \end{array} \right\}$$

$$\liminf K_n = \left\{ x \in Z : x = \lim_{n\in\mathbf{N}} x_n \ , \ x_n \in K_n \text{ for all but finitely many } n \right\}$$

Let us now introduce a particular type of convergence for sequences of real functions; we first give the definition and then state some results which will be useful in the study of problem (9).

Definition 2. (cfr. [2]) Given a sequence of functions $\{f_v : \mathbf{R}^n \to \mathbf{R}\}_{v\geq 1}$ and a function f, the sequence $\{f_v\}_{v\geq 1}$ *hypo-converges* to f ($f_v \to^h f$) if for every $x \in \mathbf{R}^n$ it holds:

$$f(x) \leq \lim_{v\to\infty} \inf f_v(x_v) \tag{10}$$

for some sequence $\{x_v\}_{v\geq 1}$ such that $x_v \to x$ and

$$f(x) \geq \lim_{v\to\infty} \sup f_v(x_v) \tag{11}$$

for every sequence $\{x_v\}_{v\geq 1}$ such that $x_v \to x$.

Moreover, the sequence $\{f_v\}_{v\geq 1}$ *epi-converges* to f ($f_v \to^e f$) if for every $x \in \mathbf{R}^n$ it holds:

$$f(x) \geq \lim_{v\to\infty} \inf f_v(x_v)$$

for every sequence $\{x_v\}_{v\geq 1}$ such that $x_v \to x$ and

$$f(x) \leq \lim_{v\to\infty} \sup f_v(x_v)$$

for some sequence $\{x_v\}_{v\geq 1}$ such that $x_v \to x$.

[5] The notion of convergence in the sense of Kuratowski for sequences of sets is, in general, weaker than other types of convergence. In [26] various types of convergence for sequences of sets are given (convergence in the Hausdorff distance, r-convergence and *-convergence) and the implications between such concepts and the convergence in the sense of Kuratowski are proved.

It is easy to prove that $\{f_v\}_{v\geq 1}$ hypo-converges to f iff $\{-f_v\}_{v\geq 1}$ epi-converges to $(-f)$. In the sequel, we consider hypo-convergent sequences of functions.

We recall two theorems proved in [22] for epi-convergent sequences of lower semicontinuous functions. We reformulate them, in an equivalent way, for hypo-convergent sequences of upper semicontinuous functions[6] over compact sets.

Theorem 7. *([22], Proposition 7.15, (a), p.252) Let $\{f, f_v : \mathbf{R}^n \to \mathbf{R}\}_{v\geq 1}$ be a family of upper semicontinuous functions over a set $X \subseteq \mathbf{R}^n$. If the sequence $\{f_v\}_{v\geq 1}$ converges uniformly to f on X ($f_v \to^u f$), then it also hypo-converges to f over X.*

Theorem 8. *([22], Proposition 7.33, p.266) Let $\{f, f_v : \mathbf{R}^n \to \mathbf{R}\}_{v\geq 1}$ be a family of eventually level-bounded upper semicontinuous functions over a compact set $X \subseteq \mathbf{R}^n$. If the sequence $\{f_v\}_{v\geq 1}$ hypo-converges to f over X, then*

$$\max_X f_v \to \max_X f$$

$$\limsup_v \left(\arg\max_X f_v\right) \subset \arg\max_X f$$

Therefore, despite the notion of hypo-convergence for sequences of functions is, in general, less strong than other types of convergence, in case of upper semicontinuous functions it is sufficient to guarantee the convergence of the sequence of the maxima.

Finally, it is easy to prove the following result.

Lemma 3. *Let $\{f, f_v : \mathbf{R}^n \to \mathbf{R}\}_{v\geq 1}$ be a family of upper semicontinuous functions over a set $X \subseteq \mathbf{R}^n$. If $f_v \to^u f$ over X, then[7] $f(x) \leq \liminf f_v(x)$.*

5.2 Convergence results

We now turn to convergence properties of problem (9) under the assumptions stated in section 2. To this end, we will refer to the scalarizations described in section 4. We firts prove convergence results for the scalar problem (7) whose objective function is the joint probability distribution of the loss functions; then we turn to problem (6): we deduce the a.s. convergence of the approximate feasible sets to the the true one and use it to state Theorem 11 which is the analogue, in the present context, of Theorem 8. Finally, we refer to the weak scalarization family (5) and use it to study the convergence for the vector optimization problems of the empirical optimal solution sets to the true one.

[6] Indeed, $f_n \to^h f$ iff $-f_n \to^e -f$ and f is upper semicontinuous iff $-f$ is lower semicontinuous.

[7] $f(x) \leq \liminf f_v(x)$ is equivalent to condition (10) in Definition 2 with $x_v = x$ for every v.

Theorem 9. *([21], Theorem 7, p.157) Let $\varphi \in \mathbf{R}^l$ be given. The sequence $\{F_g^n(\varphi,\cdot)\}_{n\geq 1}$ almost surely uniformly converges to $F(\varphi,\cdot)$ ($F_g^n(\varphi,\cdot) \to^u F_g(\varphi,\cdot)$) on $\mathbf{R}^n$.*

By using Theorem 9 we can restate Theorem 8 for problem (7).

Theorem 10. *Almost surely, it holds:*

$$\max_{x\in X}\left[F_g^n(\varphi,x)\right] \to \max_{x\in X}\left[F_g(\varphi,x)\right] \tag{12}$$

$$\limsup_n \left(\arg\max_X F_g^n(\varphi,x)\right) \subset \arg\max_X F_g(\varphi,x) \tag{13}$$

Proof. Functions F_g and F_g^n ($n \geq 1$) are bounded since they are distribution functions. By assumption the feasible set X is compact, hence bounded; then we have F_g and F_g^n level bounded and upper semicontinuous[8] functions over X. As immediate consequence of Theorems 7 and 9, it is $F_g^n(\varphi,\cdot) \to^h F_g(\varphi,\cdot)$ over[9] X. Then, (12) and (13) immediately follow from Theorem 8. □

Let us now turn to the scalar representation obtained by using the constraint method. We first claim that Theorems 1 and 2 hold for the emprical distribution functions $F_i^n(\varphi,\cdot)$ in place of $F_i(\varphi,\cdot)$; i.e.: $F_i^n(\varphi,\cdot)$ are upper semicontinuous and quasi-concave on $\mathbf{R}^n$; furthermore, Theorem 9 can be restated for the marginal distribution functions[10] in place of the joint one; i.e.: $F_i^n(\varphi,\cdot) \to^u F_i(\varphi,\cdot)$ a.s. on $\mathbf{R}^n$ for every i.

Let $y = [y_2,..,y_l]$, $y_i \in (0,1]$ and $\varphi \in \mathbf{R}^l$ be given. Define:

$$lev(y) = \bigcap_{i=2}^{l} \{x \in X : F_i(\varphi_i,x) \geq y_i\}$$

$$lev^n(y) = \bigcap_{i=2}^{l} \{x \in X : F_i^n(\varphi_i,x) \geq y_i\}$$

Without loss of generality, we consider the following scalar problem:

$$\max_{x\in lev(y)} F_1(\varphi_1,x) \tag{14}$$

and the corresponding empirical versions:

$$\max_{x\in lev^n(y)} F_1^n(\varphi_1,x) \tag{15}$$

We first deduce the almost sure convergence (in the sense of Kuratowski) of the empirical level sets to the true one; then we turn to the convergence of the optimal solution sets.

[8] See [21], Theorem 5, p.154.

[9] The following implications trivially hold: $F_g^n(\varphi,\cdot) \to^u F_g(\varphi,\cdot)$ over $\mathbf{R}^n$ implies $F_g^n(\varphi,\cdot) \to^u F_g(\varphi,\cdot)$ over X, which implies (see Theorem 7) $F_g^n(\varphi,\cdot) \to^h F_g(\varphi,\cdot)$ over X.

[10] See [21], Theorem 7, p.157 with g real valued.

Lemma 4. *We get:*

$$\limsup lev^n(y) \subseteq lev(y) \quad a.s.$$

Proof. Let $x \in \limsup lev^n(y)$; then, there exists a sequence $\{x_m\}$ such that $x_m \to x$ and $x_m \in lev^m(y)$, $\{lev^m(y)\}$ being a subsequence of $\{lev^n(y)\}$. Therefore, it is: $F_i^m(\varphi_i, x_m) \geq y_i$ for every m and for every i, which implies: $\limsup F_i^m(\varphi_i, x_m) \geq y_i$ for every i. Furthermore, $F_i(\varphi_i, \cdot)$ and, for every m, $F_i^m(\varphi_i, \cdot)$ are upper semicontinuous; moreover, $\{F_i^m(\varphi_i, \cdot)\}$ a.s. uniformly converges to $F_i(\varphi_i, \cdot)$ $(i = 2, ..., l)$. Therefore, as an immediate consequence of Theorem 7 we get:

$$F_i(\varphi_i, x) \geq \limsup_m F_i^m(\varphi_i, x_m) \geq y_i \quad \text{a.s.}$$

which holds for every i. Then, $x \in \bigcap_{i=1}^{l} \{x \in X : F_i(\varphi_i, x) \geq y_i\} = lev(y)$. □

Lemma 5. *We get:*

$$lev^*(y) \subseteq \liminf lev^n(y) \quad a.s.$$

where $lev^*(y) = \bigcap_{i=1}^{l} \{x \in X : F_i(\varphi, x) > y_i\}$

Proof. Let $x \in lev^*(y)$. Since by Theorem 1 $F_i(\varphi, \cdot)$ and, for every n, $F_i^n(\varphi, \cdot)$ are upper semicontinuous functions and the sequence $\{F_i^n(\varphi_i, \cdot)\}_{n \in \mathbf{N}}$ a.s. uniformly converges to $F_i(\varphi_i, \cdot)$, from Lemma 3 it is:

$$y_i < F_i(\varphi_i, x) \leq \liminf_{n \to \infty} F_i^n(\varphi_i, x) \tag{16}$$

Condition (16) implies that, at least definitively, a.s. $F_i^n(\varphi_i, x) > y_i$; therefore, at least definitively, a.s. $x \in lev^n(y)$. Then a.s. $x \in \liminf lev^n(y)$. □

If sequences $\{F_i^n(\varphi_i, \cdot)\}_{n \geq 1}$ monotonically decreasing to $F_i(\varphi_i, \cdot)$ $(i = 2, ..., l)$ are considered, the result in Lemma 5 can be stated with $lev(y)$ in place of $lev^*(y)$. In particular, we can prove the following Lemma.

Lemma 6. *Let* $\{F_i^n(\varphi_i, \cdot)\}_{n \geq 1}$ *monotonically decrease to* $F_i(\varphi_i, \cdot)$ $(i = 2, .., l)$. *Then, we have:*

$$lev^n(y) \to lev(y) \quad a.s. \tag{17}$$

where the convergence is in the sense of Kuratowski.

Proof. Condition (17) is equivalent to ask that a.s. it is:

$$lev(y) \subseteq \liminf lev^n(y) \subseteq \limsup lev^n(y) \subseteq lev(y) \tag{18}$$

From Lemma 4 it is $\limsup lev^n(y) \subseteq lev(y)$ a.s.. Furthermore, if the sequence $\{F_i^n(\varphi_i, \cdot)\}_{n \in \mathbf{N}}$ monotonically decreases to $F_i(\varphi_i, \cdot)$ we have $lev(y) \subseteq lev^n(y)$ for every n. But this implies: $lev(y) \subseteq \liminf lev^n(y)$. Then, we get (17). □

Let us now come to the convergence of the empirical optimal values and optimal solution sets of problems (15) to the corresponding optimal values and optimal solution sets of problem (14). We can prove the following theorem.

Theorem 11. *Let* $y = [y_2, .., y_l]$, $y_i \in (0,1]$ *and* $\varphi \in \mathbf{R}^l$ *be given. Suppose* $\{lev^n(y)\}_{n\geq 1}$ *a.s. converges (in the sense of Kuratowski) to* $lev(y)$, *it holds:*

$$\max_{x\in lev^n(y)} [F_1^n(\varphi_1, x)] \to \max_{x\in lev(y)} [F_1(\varphi_1, x)] \qquad a.s. \qquad (19)$$

$$\limsup_n \left(\arg \max_{x\in lev^n(y)} [F_1^n(\varphi_1, x)] \right) \subset \arg \max_{x\in lev(y)} [F_1(\varphi_1, x)] \qquad a.s.$$

Proof. Let $\varepsilon > 0$. From the definition of upper level sets it follows: $lev(y) \subseteq lev(y-\varepsilon)$; furthermore, since $F_i^n(\varphi, \cdot) \to^u F_i(\varphi, \cdot)$ a.s. on $\mathbf{R}^n$, for every i there exists n_i^* such that:

$$F_i(\varphi_i, x) - \varepsilon < F_i^n(\varphi_i, x) < F_i(\varphi_i, x) + \varepsilon \qquad (20)$$

whenever $n \geq n_i^*$. If we choose $n^* = \max(n_i^*)$ condition (20) holds for every i; hence we easily get:

$$lev^n(y) \subseteq lev(y-\varepsilon)$$

whenever $n \geq n^*$. Let us now define new auxiliary objective functions:

$$\bar{F}_1(\varphi_1, x) = \begin{cases} F_1(\varphi_1, x) & \text{if } x \in lev(y) \cap lev(y-\varepsilon) \\ 0 & \text{otherwise} \end{cases}$$

$$\bar{F}_1^n(\varphi_1, x) = \begin{cases} F_1^n(\varphi_1, x) & \text{if } x \in lev^n(y) \cap lev(y-\varepsilon) \\ 0 & \text{otherwise} \end{cases}$$

Clearly, it is:

$$\max_{lev(y-\varepsilon)} \bar{F}_1(\varphi_1, x) = \max_{lev(y)} F_1(\varphi_1, x) \qquad (21)$$

$$\max_{lev(y-\varepsilon)} \bar{F}_1^n(\varphi_1, x) = \max_{lev^n(y)} F_1^n(\varphi_1, x)$$

$$\arg \max_{lev(y-\varepsilon)} \bar{F}_1(\varphi_1, x) = \arg \max_{lev(y)} F_1(\varphi_1, x)$$

$$\arg \max_{lev(y-\varepsilon)} \bar{F}_1^n(\varphi_1, x) = \arg \max_{lev^n(y)} F_1^n(\varphi_1, x)$$

We claim $\bar{F}_1^n(\varphi_1, \cdot) \to^h \bar{F}_1(\varphi_1, \cdot)$ on $lev(y-\varepsilon)$; indeed, it is:

1. Let $x^* \in lev(y) \cap lev(y-\varepsilon)$. Since, by assumption, $lev^n(y) \to lev(y)$ (in the sense of Kuratowski) a.s., there exists a sequence $\{x_n\}_{n\geq 1}$ such that $x_n \to x^*$ and $x_n \in lev^n(y)$; then, a.s. $x_n \in lev^n(y) \cap lev(y-\varepsilon)$ for $n \geq n^*$. Hence, we get:

   $$\bar{F}_1(\varphi_1, x^*) = F_1(\varphi_1, x^*) \leq \liminf F_1^n(\varphi_1, x_n) = \liminf \bar{F}_1^n(\varphi_1, x)$$

   since $F_1^n(\varphi_1, \cdot) \to^h F_1(\varphi_1, \cdot)$.

2. Let $x^* \in lev(y) \backslash lev(y-\varepsilon)$. It is $F_1(\varphi_1, x^*) = 0$ by definition. Then, $\bar{F}_1(\varphi_1, x^*) \le \bar{F}_1^n(\varphi_1, x^*)$ where $\bar{F}_1^n(\varphi_1, x^*) = F_1^n(\varphi_1, x)$ if $x^* \in lev^n(y)$ $\cap lev(y-\varepsilon)$ and $\bar{F}_1^n(\varphi_1, x^*) = 0$ if $x^* \in lev(y-\varepsilon) \backslash lev^n(y)$.
3. Let $x^* \in lev(y-\varepsilon)$ and $\{x_n\}_{n\ge1}$ be a sequence in $lev(y-\varepsilon)$ such that $x_n \to x^*$. If it exists at least a subsequence $\{x_{n_m}\}$ of $\{x_n\}$ such that at least definitively $x_{n_m} \in lev^{n_m}(y)$, we get $x^* \in lev(y) \cap lev(y-\varepsilon)$ since $lev^n(y) \to lev(y)$ (in the sense of Kuratowski). Therefore,

$$\limsup \bar{F}_1^n(\varphi_1, x_n) \le \limsup F_1^n(\varphi_1, x_n) \le F_1(\varphi_1, x^*) = \bar{F}_1(\varphi_1, x^*)$$

since $F_1^n(\varphi_1, \cdot) \to^h F_1(\varphi_1, \cdot)$. If no subsequence$\{x_{n_m}\}$ exists such that at least definitively $x_{n_m} \in lev^{n_m}(y)$, we get:

$$0 = \limsup F_1^n(\varphi_1, x_n) \le \bar{F}_1(\varphi_1, x^*)$$

where $\bar{F}_1(\varphi_1, x^*) = F_1(\varphi_1, x)$ if $x^* \in lev(y) \cap lev(y-\varepsilon)$ and $\bar{F}_1(\varphi_1, x^*)$ $= 0$ if $x^* \in lev(y-\varepsilon) \backslash lev(y)$.

From points (1) and (2) we get: for every $x^* \in lev(y-\varepsilon)$ there exists at least a sequence $\{x_n\}$, $\{x_n\} \subset lev(y)$ and $x_n \to x^*$ such that $\bar{F}_1(\varphi_1, x^*) \le \liminf \bar{F}_1^n(\varphi_1, x_n)$; from point (3) we get: for every sequence $\{x_n\}$, $\{x_n\} \subset lev(y)$, $x_n \to x^*$, it is $\limsup \bar{F}_1^n(\varphi_1, x_n) \le \bar{F}_1(\varphi_1, x^*)$. Hence, $\bar{F}_1^n(\varphi_1, \cdot) \to^h \bar{F}_1(\varphi_1, \cdot)$ a.s. on $lev(y-\varepsilon)$. Therefore, Theorem 8 holds for $\{\bar{F}_1^n(\varphi_1, \cdot)\}$ and $\bar{F}_1(\varphi_1, \cdot)$. Then, we get (19) from (21). □

Finally, let us consider the optimal solution sets for the true problem $(WS(X; \Psi_\varphi))$ and for the approximate ones $(WS(X; \Psi_\varphi^n))$. The following result hold.

Theorem 12. *It holds:*

$$\limsup WS(X; \Psi_\varphi^n) \subseteq WS(X; \Psi_\varphi)$$

Proof. Let $\Phi_\varphi = -\Psi_\varphi$, $\Phi_\varphi^n = -\Psi_\varphi^n$ and $\mathcal{H}$ be the family of the smallest monotonic functions introduced in section 4.1 which is a complete scalarization for the vector problem $\min_{x \in X} \Phi_\varphi(\mathbf{x})$. Since each $h \in \mathcal{H}$ is a continuous and increasing function and it is $\Phi_\varphi^n \to^u \Phi_\varphi$ a.s., we have $(h \circ \Phi_\varphi^n) \to^h (h \circ \Phi_\varphi)$ a.s.. As an immediate consequence of Theorem 8, we get:

$$\limsup (\arg\min (h \circ \Phi_\varphi^n)) \subseteq \limsup (\arg\min (h \circ \Phi_\varphi)) \subseteq WS(X; \Phi_\varphi)$$

where the last inclusion follows from [17], Proposition 2.2, p.87.
Since $WS(X; \Phi_\varphi^n)$ is the union, over $h \in \mathcal{H}$, of $\arg\min (h \circ \Phi_\varphi^n)$ the result follows. □

Remark 5. As pointed out by an anonymous referee, sometimes it may be more natural to consider the problem of "minimizing the average losses":

$$\min_{x\in X} \Xi(\mathbf{x}) = \begin{bmatrix} E[g_1(\mathbf{x},\xi)] \\ E[g_2(\mathbf{x},\xi)] \\ \cdot\cdot \\ \cdot\cdot \\ E[g_l(\mathbf{x},\xi)] \end{bmatrix} = -\max_{x\in X} \begin{bmatrix} E[-g_1(\mathbf{x},\xi)] \\ E[-g_2(\mathbf{x},\xi)] \\ \cdot\cdot \\ \cdot\cdot \\ E[-g_l(\mathbf{x},\xi)] \end{bmatrix}$$

In this case, to get convergence results we make the following assumptions[11]: each g_i $(i = 1, 2, ..., l)$ is a real functions defined on $\mathbb{R}^{n+m}$ such that $g_i(x, \cdot)$ is bounded and continuous on $\mathbb{R}^m$ for every $x \in \mathbb{R}^n$ and $g_i(\cdot, \xi)$ is lower semicontinuous for every $\xi \in \mathbb{R}^m$; furthermore, given the empirical distribution function introduced at the beginning of section 5, let each sequence $\{E^n[g_i(\cdot,\xi)]\}_{n\geq 1}$ $(i = 1, 2, ..., l)$ be a.s. asympotically equi-lower semicontinuous[12] in x (E^n being the expectation with respect to the empirical probability measure).

Let us first apply Theorem 11, p.160 in [21] (replacing g_0 with $-g_i$) thus obtaining $E^n[-g_i(\cdot,\xi)] \to^h E[-g_i(\cdot,\xi)]$ $(i = 1, 2, ..., l)$ on the feasible set X (which is compact by assumption). Then, we can plainly prove Lemmas 4 trough 6 and Theorems 11 and 12 replacing $E[-g_i(\cdot,\xi)]$ with F_i and $E^n[-g_i(\cdot,\xi)]$ with F_i^n, respectively. In particular, condition (16) in Lemma 5 and condition (20) in Theorem 11 still hold because $\{E^n[-g_i(\cdot,\xi)]\}$ pointwise converges[13] to $E[-g_i(\cdot,\xi)]$.

6 Conclusion

We studied a vector optimization problem whose objective function was expressed as a vector of distribution functions depending on a vector of decision variables. Under quite general assumptions, usually satisfied in financial applications, the problem was identified as a $C-$continuous and $C-$quasiconvex vector problem (being C the ordering cone on the image space); furthermore, we proved the optimal solution set to be compact and connected. We then dealt with scalarization; in particular, we considered a scalar optimization

[11] For the sake of simplicity, we state only the assumptions which are different from those given in section 2. We assume verified all the other ones not explicitely mentioned here.

[12] For the definition of asymptotically equi-lower semicontinuity, see [22], p.248-249. The lower semicontinuity of E^n in x is a consequence of the assumptions on the g_i. Then what we really require is the almost sure *equi*-lower semicontinuity of the sequence $\{E^n\}_{n\geq 1}$ with respect to x.

[13] From Theorem 7.10, p.249 in [22], we have that hypo-convergence is equivalent to pointwise convergence for sequences of *equi*-upper semicontinuous functions.

problem where the objective function is expressed in terms of the joint probability distribution and proved (see Theorem 6) that, if each loss function g_i (depending on the decision variables) describes a family of random variables of the same type such a model is a scalar representation of the original vector problem. A question remains open, which is object of our present research: if it is possible to generalize Theorem 6 to families of random variables which differ not only for location parameters but also for skewness; this is a more realistic situation and we are studying it in the context of *mean-variance-skewness efficient* random variables (see [11]).

Finally, in section 5 we dealt with convergence problems: we assumed to have a sequence of empirical probability distributions, defined a sequence of approximate versions of the original vector model and gave convergence conditions for the approximate optimal solution sets to the true one.

References

1. Artzner P., F. Delbaen, J-M. Eber, D. Heath, "Thinking coherently", Market Risk, Vol.10, N.11, November 1997
2. Avriel M., W.E.Diewert, S.Schaible, I.Zang, *Generalized Concavity*, Plenum Press, New York, 1988
3. Bawa V.S., "Safety-first stochastic dominance and optimal portfolio choice", Journal of Financial and Quantitative Analysis (1978), p.255-271
4. Billingsley P., *Probability and Measure*, Wiley Series in Probability and Mathematical Statistics, John Wiley & Sons Inc. 1986
5. Borovkov A.A., *Mathematical Statistics*, Gordon and Breach Science Publishers,1998
6. Cirelli R., *Elementi di Topologia*, Cooperativa libraria universitaria del Politecnico, Milano, 1992
7. Daykin C.D., T. Pentikäinen, M. Pesonen, *Practical Risk Theory for Actuaries*, Chapman and Hall, 1994
8. Embrechts P., C. Klüppelberg, T. Mikosch, *Modelling extremal events for insurance and finance*, Springer-Verlag, 1997
9. Embrechts P., A. McNeil, D. Straumann, "Correlation and dependency in risk management: properties and pitfalls", Working Paper on Internet, November 1998
10. Feller W., *An Introduction to probability theory and its applications*, Vol. I e II, John Wiley & Sons Inc., 1966
11. Gamba A., F.Rossi,"A three moment based portfolio selection model", Rivista di matematica per le scienze economiche e sociali, 21 (1998), p.25-48
12. Goovaerts M.J., "Net stop-loss ordering and related orderings", in De Vydler F.at al. (eds), *Premium calculation in insurance*, D. reidel Publishing Company (1984), p.195-234
13. Goovaerts M.J., F. De Vylder, J. Haezendonck, "Ordering of risks: a review", Insurance: Mathematics and Economics 1 (1982), p.131-163
14. Huang C., R. Litzenberger, *Foundations for financial economics*, Elsevier Science London, 1988
15. Kast R., E. Luciano, L. Peccati, "VaR and optimization", Working Paper, 1999

16. Kolmogorov A.N. - S.V. Fomin, *Introductory Real Analysis*, Dover Publications Inc., New York, 1970.
17. Luc D.T., *Theory of Vector Optimization*, Springer Verlag,1989
18. Luenberger D.G., *Optimization by vector space methods*, John Wiley & Sons Inc.
19. Pedersen C., Satchell S.E., "An extended family of financial risk measures"; The Geneva Papers on Risk and Insurance Theory, 23 (1998), p.89-117
20. Prekòpa v, *Stochastic Programming*, Kluwer Academic Publishers, Dordrecht, 1995
21. Redaelli G., "Convergence problems in stochastic programming models with probabilistic constraints", Rivista di matematica per le scienze economiche e sociali, 21 (1998), p.147-164
22. Rockafellar R.T., R.J-B.Wets, Variational Analysis, Springer Verlag, 1998
23. Rudin W., *Principles of Mathematical Analysis*, McGraw-Hill International Editions, 1976
24. G.Salinetti, "Convergence for Measurable Multifunctions: an Application to Stochastic Optimization Problems", SIAM Review,
25. G.Salinetti, R.J.B.Wets, "On the Relations between two Types of Convergence for Convex Functions", Journal of Mathematical Analysis and applications, Vol.60, 1977
26. G.Salinetti, R.J.B.Wets, "On the Convergence of Sequences of Convex Sets in Finite Dimensions", SIAM Review, Vol.21, N.1, 1979
27. G.Salinetti, R.J.B.Wets, "Glivenko-Cantelli Type Theorems: an Application of the Convergence Theory of Stochastic Suprema", Annals of Operations Research Vol.30, 1991

On Suprema of Abstract Convex and Quasi-convex Hulls

Ivan Singer

Institute of Mathematics,
P. O. Box 1-764,
RO-70700 Bucharest, Romania.

Abstract. We show that the (pointwise) supremum of two arbitrary hull operators on a complete lattice is again a hull operator, which can be expressed in a natural way in terms of the two initial hull operators. Our results yield, in particular, very simple proofs of some results of Attéia and ElQortobi on projective bipolars of sets and functions. Also, we give a new way of recuperating abstract convex hulls of sets from abstract quasi-convex hulls of their indicator functions.

2000 Mathematics Subject Classification. Primary 52A01; Secondary 52A05, 52A41, 06B23.

Keywords and phrases. convex hull, quasi-convex hull, abstract convexity.

1 Introduction

For a locally convex space X, with conjugate space X^*, Attéia and ElQortobi ([2], [4]) have introduced, based on the notion of projective polarity studied by Attéia [1], the *projective bipolar* $G^{\#\#}$ *of a set* $G \subseteq X$ and the *projective bipolar* $f^{\#\#}$ *of a function* $f : X \to \overline{R} = [-\infty, +\infty]$, as follows:

$$G^{\#\#} := G^{00} \cap G^{\nabla\nabla}, \tag{1. 1}$$

where

$$G^{00} := \cap_{\{\Phi \in X^* \mid \sup \Phi(G) \leq 1\}} \{x \in X \mid \Phi(x) \leq 1\}, \tag{1. 2}$$

$$G^{\nabla\nabla} := \cap_{\{\Phi \in X^* \mid \inf \Phi(G) \geq 1\}} \{x \in X \mid \Phi(x) \geq 1\}, \tag{1. 3}$$

and, respectively,

$$f^{\#\#} := \max(f^{00}, f^{\nabla\nabla}), \tag{1. 4}$$

where

$$f^{00}(x) := \sup_{\substack{\Phi \in X^* \\ \Phi(x) > 1}} \inf_{\substack{y \in X \\ \Phi(y) > 1}} f(y) \qquad (x \in X), \tag{1. 5}$$

$$f^{\nabla\nabla}(x) := \sup_{\substack{\Phi \in X^* \\ \Phi(x) < 1}} \inf_{\substack{y \in X \\ \Phi(y) < 1}} f(y) \qquad (x \in X). \tag{1. 6}$$

Among other results, Attéia and ElQortobi have shown ([2], [4]) that

$$G^{\#\#} = \overline{\text{co}}\, G, \tag{1. 7}$$

$$f^{\#\#} = f_{\overline{q}}, \tag{1. 8}$$

where $\overline{\text{co}}\, G$ and $f_{\overline{q}}$ denote the closed convex hull of G and the lower semi-continuous quasi-convex hull (i.e., the greatest lower semi-continuous quasi-convex minorant) of f, respectively. From the point of view of abstract convex analysis, G^{00} and $G^{\nabla\nabla}$ of (1. 2), (1. 3) are abstract convex hulls of G, so (1. 7) says that the intersection of these two hulls is again an abstract convex hull of G, namely $\overline{\text{co}}\, G$, and, similarly, f^{00} and $f^{\nabla\nabla}$ of (1. 5) and (1. 6) are abstract quasi-convex hulls of f, so (1. 8) says that the maximum of these two hulls is again an abstract quasi-convex hull of f, namely $f_{\overline{q}}$.

In the present paper we shall study such phenomena in the framework of abstract convex analysis. Namely, we shall show that, more generally, the supremum of two arbitrary hull operators on a complete lattice, and hence, in particular, on the complete lattice of all subsets of a set and on the complete lattice of all (extended-real valued) functions on a set, is again a hull operator, which can be expressed in a natural way in terms of the two initial hull operators.Our results will yield, in particular, very simple proofs of (1. 7) and (1. 8). Also, we shall give a new way of recuperating abstract convex hulls of sets from abstract quasi-convex hulls of their indicator functions.

Now we shall recall some concepts of abstract convex analysis (see [5] or [6]), and we shall give a brief description of the sections of the present paper.

We recall that a *hull operator* on a complete lattice $E = (E, \leq)$ is an isotone, contractive and idempotent mapping $u : E \to E$, i.e., such that for any $x, y \in E$ we have

$$x \leq y \Rightarrow u(x) \leq u(y), \tag{1. 9}$$

$$u(x) \leq x, \tag{1. 10}$$

$$u(u(x)) = u(x). \tag{1. 11}$$

For any complete lattice $E = (E, \leq)$ we shall denote the lattice operations in E by $\vee$ or sup, respectively by $\wedge$ or inf.

Definition 1.1. The supremum of two hull operators $u_1, u_2 : E \to E$ on a complete lattice E is the mapping $u_1 \vee u_2 : E \to E$ defined by

$$(u_1 \vee u_2)(x) := u_1(x) \vee u_2(x) \qquad (x \in E). \tag{1. 12}$$

We recall that if E is a complete lattice and $\mathcal{M} \subseteq E$, an element $x \in E$ is said to be $\mathcal{M}$-*convex*, in symbols, $x \in C(\mathcal{M})$, if

$$x = \sup\{m \in M | \, m \leq x\}. \tag{1. 13}$$

For the $\mathcal{M}$-*convex hull* $\text{co}_{\mathcal{M}} x$ of any element $x \in E$, i.e., for the greatest $\mathcal{M}$-convex minorant of x, we have (see [6], p.39, theorem 1.2)

$$\text{co}_{\mathcal{M}} x = \sup\{m \in \mathcal{M} | \, m \leq x\}. \tag{1. 14}$$

It is well-known (see e.g. [6], p.38, theorem 1.1) that $\mathrm{co}_{\mathcal{M}}$ is a hull operator on E.

In section 2 we shall show that the supremum of two hull operators u_1, u_2 on a complete lattice E is a hull operator on E and we shall give a characterization of the $u_1 \vee u_2$-convex elements of E, that is, of the elements x satisfying $x = (u_1 \vee u_2)(x)$. Also, we shall give a formula for the supremum of two abstract convex hull operators $u_1 = \mathrm{co}_{\mathcal{M}}, u_2 = \mathrm{co}_{\mathcal{N}}$ on E, where $\mathcal{M}, \mathcal{N} \subseteq E$, and a characterization of the elements x of E satisfying $x = (\mathrm{co}_{\mathcal{M}} \vee \mathrm{co}_{\mathcal{N}})(x)$.

In section 3 we shall consider the complete lattice $E = (2^X, \supseteq)$ of all subsets of a set X, ordered by containment (i.e., for $G_1, G_2 \in 2^X, G_1 \leq G_2$ if and only if $G_1 \supseteq G_2$). We recall that if X is a set and $\mathcal{M}$ is a family of subsets of X, then, by the above definition applied in the complete lattice $E = (2^X, \supseteq)$, a set $G \subseteq X$ is said to be *$\mathcal{M}$-convex*, in symbols, $G \in \mathcal{C}(\mathcal{M})$, if

$$G = \cap \{M \in \mathcal{M} | G \subseteq M\}. \tag{1. 15}$$

For the *$\mathcal{M}$-convex hull* $\mathrm{co}_{\mathcal{M}} G$ of any set $G \subseteq X$, i.e., the intersection of all $\mathcal{M}$-convex sets containing G, by (1. 14) we have

$$\mathrm{co}_{\mathcal{M}} G = \cap \{M \in \mathcal{M} | G \subseteq M\}. \tag{1. 16}$$

By the above, the mapping $u : G \to \mathrm{co}_{\mathcal{M}} G$ is a hull operator on $E = (2^X, \supseteq)$. Since $G_1 \vee G_2 = G_1 \cap G_2$ in $E = (2^X, \supseteq)$, by definition 1.1 applied to the hull operators $u_1 = \mathrm{co}_{\mathcal{M}}, u_2 = \mathrm{co}_{\mathcal{N}}$ on $E = (2^X, \supseteq)$, where $\mathcal{M}, \mathcal{N} \subseteq 2^X$, we have

$$(\mathrm{co}_{\mathcal{M}} \vee \mathrm{co}_{\mathcal{N}})(G) = \mathrm{co}_{\mathcal{M}} G \cap \mathrm{co}_{\mathcal{N}} G \qquad (G \subseteq X). \tag{1. 17}$$

In section 3 we shall deduce from the results of section 2 a formula for the supremum of two abstract convex hull operators $u_1 := \mathrm{co}_{\mathcal{M}}, u_2 := \mathrm{co}_{\mathcal{N}}$ on $E = (2^X, \supseteq)$, where $\mathcal{M}, \mathcal{N} \subseteq 2^X$, and a characterization of the subsets G of X satisfying $G = (\mathrm{co}_{\mathcal{M}} \vee \mathrm{co}_{\mathcal{N}})(G)$.

Let X be a set. In section 4 we shall consider the complete lattice $E = (\overline{R}^X, \leq)$ of all functions $f : X \to \overline{R} = [-\infty, +\infty]$, with the usual pointwise order (i.e., for $f_1, f_2 \in \overline{R}^X, f_1 \leq f_2$ if and only if $f_1(x) \leq f_2(x)$ for all $x \in X$). We recall that if X is a set and W is a subset of $\overline{R}^X$, then, by the above definition applied in the complete lattice $E = (\overline{R}^X, \leq)$, a function $f : X \to \overline{R}$ is said to be *W-convex*, in symbols, $f \in \mathcal{C}(W)$, if

$$f = \sup_{\substack{w \in W \\ w \leq f}} w. \tag{1. 18}$$

Following the notations of [6], we shall denote an arbitrary subset of $\overline{R}^X$ by W (instead of $\mathcal{M}$), and by $f_{\mathrm{co}(W)}$ (instead of $\mathrm{co}_W f$) the W-convex hull of f, i.e., the greatest W-convex function majorized by f. By (1. 14), for any function $f : X \to \overline{R}$ we have

$$f_{\mathrm{co}(W)} = \sup_{\substack{w \in W \\ w \leq f}} w, \tag{1. 19}$$

the sup being taken pointwise (i.e., $(\sup_{w\in A} w)(x) := \sup_{w\in A} w(x)$, for each $x \in X$). By the above, the mapping $u = \mathrm{co}(W) : f \to f_{\mathrm{co}(W)}$ is a hull operator on $E = (\overline{R}^X, \leq)$. By definition 1.1 applied to the hull operators $u_1 = \mathrm{co}(W_1)$, $u_2 = \mathrm{co}(W_2)$ on $E = (\overline{R}^X, \leq)$, where $W_1, W_2 \subseteq \overline{R}^X$, we have

$$f_{\mathrm{co}(W_1)\vee \mathrm{co}(W_2)} = \max\{f_{\mathrm{co}(W_1)}, f_{\mathrm{co}(W_2)}\} \qquad (f \in \overline{R}^X). \tag{1. 20}$$

In section 4 we shall deduce from the results of section 2 a formula for the maximum of two abstract convex hull operators $u_1 := \mathrm{co}(W_1), u_2 := \mathrm{co}(W_2)$ on $E = (\overline{R}^X, \leq)$, where $W_1, W_2 \subseteq \overline{R}^X$, and a characterization of the functions $f : X \to \overline{R}$ satisfying $f = \max\{f_{\mathrm{co}(W_1)}, f_{\mathrm{co}(W_2)}\}$.

In section 5 we shall consider abstract quasi-convex hulls of functions. We recall that if X is a set and $\mathcal{M}$ is a family of subsets of X, a function $f : X \to \overline{R}$ is said to be $\mathcal{M}$-*quasi-convex*, in symbols, $f \in Q(\mathcal{M})$, if all level sets

$$S_d(f) := \{x \in X \mid f(x) \leq d\} \qquad (d \in R) \tag{1. 21}$$

are $\mathcal{M}$-convex. For any function $f : X \to \overline{R}$, the $\mathcal{M}$-*quasi-convex hull* $f_{\mathrm{q}(\mathcal{M})} : X \to \overline{R}$ *of* f is the greatest $\mathcal{M}$-quasi-convex function majorized by f. It is well-known (see [6], p. 131, theorem 4.1) that the mapping $u = \mathrm{q}(\mathcal{M})$: $f \to f_{\mathrm{q}(\mathcal{M})}$ is a hull operator on $E = (\overline{R}^X, \leq)$. In section 5 we shall give a formula for the maximum of two quasi-convex hull operators $u_1 := \mathrm{q}(\mathcal{M})$, $u_2 := \mathrm{q}(\mathcal{N})$ on $E = (\overline{R}^X, \leq)$, where $\mathcal{M}, \mathcal{N} \subseteq 2^X$, and a characterization of the functions $f : X \to \overline{R}$ satisfying $f = \max\{f_{\mathrm{q}(\mathcal{M})}, f_{\mathrm{q}(\mathcal{N})}\}$.

In section 6 we shall give a new formula for $(\chi_G)_{\mathrm{q}(\mathcal{M})}$, where G is an arbitrary subset of X and χ_G denotes the *indicator function* of the set G, defined by

$$\chi_G(x) = \begin{cases} 0 & \text{if } x \in G \\ +\infty & \text{if } x \in X \setminus G; \end{cases} \tag{1. 22}$$

this result will show a new way of recuperating (uniquely) $\mathrm{co}_{\mathcal{M}}$ from $\mathrm{q}(\mathcal{M})$. Applying this result, we shall give another proof of the main formula of section 5 on the maximum of two quasi-convex hull operators $u_1 := \mathrm{q}(\mathcal{M})$, $u_2 := \mathrm{q}(\mathcal{N})$, which is also revealing.

We wish to thank A. M. Rubinov for useful discussions, especially about part of Section 6.

2 The supremum of two hull operators on a complete lattice

Proposition 2.1. *If u_1 and u_2 are two hull operators on a complete lattice E, then the mapping $u_1 \vee u_2$ is a hull operator.*

Proof. Since u_1 and u_2 are hull operators, for any $x, y \in E$ with $x \leq y$ we have

$$(u_1 \vee u_2)(x) = u_1(x) \vee u_2(x) \leq x,$$
$$(u_1 \vee u_2)(x) = u_1(x) \vee u_2(x) \leq u_1(y) \vee u_2(y) = (u_1 \vee u_2)(y).$$

Finally, by the above,

$$(u_1 \vee u_2)(u_1 \vee u_2)(x) \leq (u_1 \vee u_2)(x),$$

and, on the other hand,

$$\begin{aligned}(u_1 \vee u_2)(u_1 \vee u_2)(x) &= u_1(u_1 \vee u_2)(x) \vee u_2(u_1 \vee u_2)(x) \\ &= u_1(u_1(x) \vee u_2(x)) \vee u_2(u_1(x) \vee u_2(x)) \\ &\geq u_1(u_1(x)) \vee u_2(u_2(x)) = u_1(x) \vee u_2(x) = (u_1 \vee u_2)(x).\end{aligned}$$

□

Proposition 2.2. *If u_1 and u_2 are two hull operators on a complete lattice E, then for any $x \in E$ we have the equivalence*

$$x = (u_1 \vee u_2)(x) \Leftrightarrow x = y \vee z, \ y = u_1(y), \ z = u_2(z). \tag{2. 1}$$

Proof. The implication $\Rightarrow$ is obvious, taking $y = u_1(x), z = u_2(x)$.

Conversely, assume now that we have $x = y \vee z, \ y = u_1(y), \ z = u_2(z)$. Then

$$\begin{aligned}(u_1 \vee u_2)(x) &= (u_1 \vee u_2)(y \vee z) = u_1(y \vee z) \vee u_2(y \vee z) \\ &\geq u_1(y) \vee u_2(z) = y \vee z = x,\end{aligned}$$

and the opposite inequality holds by (1. 10) (applied to the hull operator $u_1 \vee u_2$). Thus, $(u_1 \vee u_2)(x) = x$. □

Theorem 2.1. *Let E be a complete lattice and let $\mathcal{M}, \mathcal{N} \subseteq E$. Then*

$$\mathrm{co}_{\mathcal{M}} \vee \mathrm{co}_{\mathcal{N}} = \mathrm{co}_{\mathcal{M} \cup \mathcal{N}} = \mathrm{co}_{\mathcal{P}}, \tag{2. 2}$$

where

$$\mathcal{P} = \{m \vee n | \, m \in \mathcal{M}, n \in \mathcal{N}\}. \tag{2. 3}$$

Furthermore, for any $x \in E$ we have the equivalence

$$x = (\mathrm{co}_{\mathcal{M}} \vee \mathrm{co}_{\mathcal{N}})x \Leftrightarrow x = x_1 \vee x_2, \, x_1 \in \mathcal{C}(\mathcal{M}), \, x_2 \in \mathcal{C}(\mathcal{N}). \tag{2. 4}$$

Proof. By (1. 12), (1. 13) and the complete distributivity of sup, we obtain, for any $x \in E$,

$$\begin{aligned}(\mathrm{co}_{\mathcal{M}} \vee \mathrm{co}_{\mathcal{N}})x &= \mathrm{co}_{\mathcal{M}}x \vee \mathrm{co}_{\mathcal{N}}x \\ &= \sup\{m \in \mathcal{M} | \, m \leq x\} \vee \sup\{n \in \mathcal{N} | \, n \leq x\} = \mathrm{co}_{\mathcal{M} \cup \mathcal{N}}x \\ &= \sup\{m \vee n | \, m \in \mathcal{M}, n \in \mathcal{N}, m \vee n \leq x\} = \mathrm{co}_{\mathcal{P}}x.\end{aligned}$$

Finally, the equivalence (2. 4) follows from proposition 2.2. □

Remark 2.1. Since in general $\mathcal{P} \neq \mathcal{M} \cup \mathcal{N}$, theorem 2.1 shows again the well-known fact that $\mathcal{C}(\mathcal{P})$ does not determine uniquely the set $\mathcal{P}$.

3 The intersection of two abstract convex hulls of a set

From (1. 17) and theorem 2.1, we obtain

Theorem 3.1. *Let X be a set and $\mathcal{M}, \mathcal{N}$ two families of subsets of X. Then*

$$\mathrm{co}_{\mathcal{M}} G \cap \mathrm{co}_{\mathcal{N}} G = \mathrm{co}_{\mathcal{M}\cup\mathcal{N}} G = \mathrm{co}_{\mathcal{P}} G \qquad (G \subseteq X), \tag{3. 1}$$

where

$$\mathcal{P} = \{M \cap N \mid M \in \mathcal{M}, N \in \mathcal{N}\}. \tag{3. 2}$$

Furthermore, for any set $G \subseteq X$ we have the equivalence

$$G = \mathrm{co}_{\mathcal{M}} G \cap \mathrm{co}_{\mathcal{N}} G \Leftrightarrow G = G_1 \cap G_2,\ G_1 \in \mathcal{C}(\mathcal{M}), G_2 \in \mathcal{C}(\mathcal{N}). \tag{3. 3}$$

Remark 3.1. a) In particular, for a locally convex space X, with conjugate space X^*, theorem 3.1 yields the following simple proof of formula (1. 7): Let $\mathcal{M}$ be the family of all subsets of X of the form

$$M_\Phi := \{x \in X \mid \Phi(x) \leq 1\} \qquad (\Phi \in X^*), \tag{3. 4}$$

and let $\mathcal{N}$ be the family of all subsets of X of the form

$$N_\Phi := \{x \in X \mid \Phi(x) \geq 1\} \qquad (\Phi \in X^*). \tag{3. 5}$$

Then, by (1. 16), (3. 4) and (1. 2), for any set $G \subseteq X$ we have

$$\mathrm{co}_{\mathcal{M}} G = \cap\{M_\Phi \in \mathcal{M} \mid G \subseteq M_\Phi\} = G^{00}, \tag{3. 6}$$

and, similarly,

$$\mathrm{co}_{\mathcal{N}} G = \cap\{N_\Phi \in \mathcal{N} \mid G \subseteq N_\Phi\} = G^{\nabla\nabla}. \tag{3. 7}$$

Hence, by (1. 1), theorem 3.1 and (1. 16), we obtain

$$\begin{aligned} G^{\#\#} := G^{00} \cap G^{\nabla\nabla} &= \mathrm{co}_{\mathcal{M}} G \cap \mathrm{co}_{\mathcal{N}} G = \mathrm{co}_{\mathcal{M}\cup\mathcal{N}} G \\ &= \cap\{H \in \mathcal{M} \cup \mathcal{N} \mid G \subseteq H\}. \end{aligned} \tag{3. 8}$$

But, by (3. 4), $\mathcal{M}$ is nothing else than the family of all closed half-spaces in X containing 0 and, by (3. 5), $\mathcal{N}$ is the family of all closed half-spaces in X which do not contain 0, so $\mathcal{M} \cup \mathcal{N}$ is the family of closed-half-spaces in X. Hence, by the well-known fact that $\overline{\mathrm{co}}\, G$ is the intersection of all closed half-spaces containing G, we have

$$\mathrm{co}_{\mathcal{M}\cup\mathcal{N}} G = \cap\{H \in \mathcal{M} \cup \mathcal{N} \mid G \subseteq H\} = \overline{\mathrm{co}}\, G. \tag{3. 9}$$

Consequently, by (3. 8) and (3. 9), we obtain (1. 7). Moreover, the last equality of (3. 1), applied to the families $\mathcal{M}, \mathcal{N}$ of (3. 4) and (3. 5), yields a new expression for $G^{\#\#}$.

b) Formula (1. 7) and the argument of part a) above can be extended to more general situations in which, instead of the pair (X, X^*) one considers a pair (X, W), where X is a set and W is a set of functions $w : X \to \overline{R}$, satisfying certain additional assumptions, and in which $\overline{\mathrm{co}}\, G$ is replaced by the so-called "W-convex hull" of G. We omit the details.

4 The maximum of two abstract convex hulls of a function

From (1. 20) and theorem 1 we obtain

Theorem 4.1. *Let X be a set and W_1, W_2 two sets of functions from X into $\overline{R}$. Then*

$$\max\{f_{\text{co}(W_1)}, f_{\text{co}(W_2)}\} = f_{\text{co}(W_1 \cup W_2)} = f_{\text{co}(\mathcal{P}')}, \qquad (4.\ 1)$$

where

$$\mathcal{P}' := \{\max(w_1, w_2) | \, w_1 \in W_1, w_2 \in W_2\} \qquad (4.\ 2)$$

Furthermore, for any function $f : X \to \overline{R}$ we have the equivalence

$$f = \max\{f_{\text{co}(W_1)}, f_{\text{co}(W_2)}\} \Leftrightarrow f = \max\{f_1, f_2\}, \; f_1 \in \mathcal{C}(W_1), f_2 \in \mathcal{C}(W_2). \qquad (4.\ 3)$$

5 The maximum of two abstract quasi-convex hulls of a function

The main result of this section is

Theorem 5.1. *Let X be a set and $\mathcal{M}, \mathcal{N}$ two families of subsets of X. Then for any function $f : X \to \overline{R}$ we have*

$$\max\{f_{\text{q}(\mathcal{M})}, f_{\text{q}(\mathcal{N})}\} = f_{\text{q}(\mathcal{P})}, \qquad (5.\ 1)$$

with $\mathcal{P}$ of (3. 2).

Furthermore, for any function $f : X \to \overline{R}$ we have the equivalence

$$f = \max\{f_{\text{q}(\mathcal{M})}, f_{\text{q}(\mathcal{N})}\} \Leftrightarrow f = \max\{f_1, f_2\}, \; f_1 \in Q(\mathcal{M}), f_2 \in Q(\mathcal{N}). \qquad (5.\ 2)$$

Proof. We recall that, by [5], theorem 2.4 (or [6], p.138, theorem 4.6), for any family of subsets $\mathcal{M}$ of X and any function $f : X \to \overline{R}$ we have

$$f_{\text{q}(\mathcal{M})} = f_{\text{co}(J_{\mathcal{M}} + R)}, \qquad (5.\ 3)$$

where

$$J_{\mathcal{M}} + R := \{-\chi_{X \setminus M} + d | \, M \in \mathcal{M}, d \in R\} \; (\subseteq \overline{R}^X). \qquad (5.\ 4)$$

Now let

$$W_1 := J_{\mathcal{M}} + R, W_2 := J_{\mathcal{N}} + R \; (\subseteq \overline{R}^X). \qquad (5.\ 5)$$

Then, by (5. 3) and theorem 4.1, for any function $f : X \to \overline{R}$ we have

$$\max\{f_{\text{q}(\mathcal{M})}, f_{\text{q}(\mathcal{N})}\} = \max\{f_{\text{co}(J_{\mathcal{M}}+R)}, f_{\text{co}(J_{\mathcal{N}}+R)}\} = f_{\text{co}(\mathcal{P}')}, \qquad (5.\ 6)$$

where

$$\mathcal{P}' = \{\max(-\chi_{X\setminus M} + d_1, -\chi_{X\setminus N} + d_2) \mid M \in \mathcal{M}, N \in \mathcal{N}, d_1, d_2 \in R\} \subseteq \overline{R}^X. \tag{5. 7}$$

But, by the definition (1. 22) of χ,

$$\max(-\chi_{X\setminus M} + d_1, -\chi_{X\setminus N} + d_2)(x) = \begin{cases} -\infty & \text{if } x \in M \cap N \\ d_1 & \text{if } x \in M \cap (X\setminus N) \\ d_2 & \text{if } x \in (X\setminus M) \cap N \\ \max(d_1, d_2) & \text{if } x \in (X\setminus M) \cap (X\setminus N), \end{cases} \tag{5. 8}$$

whence

$$\mathcal{P}' = \{-\chi_{X\setminus(M\cap N)} + d \mid M \in \mathcal{M}, N \in \mathcal{N}, d \in R\} = J_{\mathcal{P}} + R, \tag{5. 9}$$

with $\mathcal{P}$ of (3. 2). Consequently, by (5. 9) and (5. 3), we obtain, for any function $f : X \to \overline{R}$,

$$f_{\mathrm{co}(\mathcal{P}')} = f_{\mathrm{co}(J_{\mathcal{P}}+R)} = f_{\mathrm{q}(\mathcal{P})}, \tag{5. 10}$$

which, together with (5. 6), yields (5. 1).

Finally, since $f_1 \vee f_2 = \max\{f_1, f_2\}$ in $E = (\overline{R}^X, \leq)$, the first equality in formula (5. 2) means, by definition 1.1 applied to the hull operators $u_1 :=\mathrm{q}(\mathcal{M})$ and $u_2 :=\mathrm{q}(\mathcal{N})$ in $E = (\overline{R}^X, \leq)$, that

$$f = f_{\mathrm{q}(\mathcal{M})} \vee f_{\mathrm{q}(\mathcal{N})}, \tag{5. 11}$$

and hence, by proposition 2.2, we obtain the equivalence (5. 2). □

Corollary 5.1. *Let X be a set and $\mathcal{M}, \mathcal{N}$ two families of subsets of X. Then for any function $f : X \to \overline{R}$ we have*

$$\begin{aligned} \max\{f_{\mathrm{q}(\mathcal{M})}, f_{\mathrm{q}(\mathcal{N})}\}(x) &= \inf_{\substack{d\in R \\ x\in \mathrm{co}_{\mathcal{M}} S_d(f) \cap \mathrm{co}_{\mathcal{N}} S_d(f)}} d = \inf_{\substack{d\in R \\ x\in \mathrm{co}_{\mathcal{M}\cup\mathcal{N}} S_d(f)}} d \\ &= f_{\mathrm{q}(\mathcal{M}\cup\mathcal{N})}(x) \qquad (x \in X). \end{aligned} \tag{5. 12}$$

Proof. Let us first recall that, by [5], theorem 2.1 (or [6], p.133, theorem 4.2), for any family of subsets $\mathcal{M}$ of X and any function $f : X \to \overline{R}$ we have

$$f_{\mathrm{q}(\mathcal{M})}(x) = \inf_{\substack{d\in R \\ x\in \mathrm{co}_{\mathcal{M}} S_d(f)}} d \qquad (x \in X). \tag{5. 13}$$

Now, by (5. 1), (5. 13) and (3. 1), for $\mathcal{P}$ of (3. 2) we have

$$\max\{f_{\mathrm{q}(\mathcal{M})}, f_{\mathrm{q}(\mathcal{N})}\} = f_{\mathrm{q}(\mathcal{P})}(x) = \inf_{\substack{d\in R \\ x\in \mathrm{co}_{\mathcal{P}} S_d(f)}} d = \inf_{\substack{d\in R \\ x\in \mathrm{co}_{\mathcal{M}} S_d(f) \cap \mathrm{co}_{\mathcal{N}} S_d(f)}} d \ (x \in X),$$

which proves the first equality of (5. 12). Finally, by theorem 3.1 and (5. 13), there hold also the second and third equalities of (5. 12). □

Remark 5.1. One can also proceed in the converse direction, to prove directly the first equality of corollary 5.1 and then to deduce from it formula (5. 1). To this end, let us first prove

Lemma 5.1. *Let $\{U_d\}_{d\in R}$ and $\{V_d\}_{d\in R}$ be two families of subsets of a set X, satisfying*

$$d_1 < d_2 \Rightarrow U_{d_1} \subseteq U_{d_2}, \quad V_{d_1} \subseteq V_{d_2}. \tag{5. 14}$$

Then

$$\inf_{\substack{d\in R \\ x\in U_d\cap V_d}} d = \max\left\{\inf_{\substack{d\in R \\ x\in U_d}} d, \inf_{\substack{d\in R \\ x\in V_d}} d\right\} \quad (x \in X). \tag{5. 15}$$

Proof. Given $x \in X$, let us denote

$$d_1 := \inf_{\substack{d\in R \\ x\in U_d}} d, \quad d_2 := \inf_{\substack{d\in R \\ x\in V_d}} d, \tag{5. 16}$$

$$d_3 := \inf_{\substack{d\in R \\ x\in U_d\cap V_d}} d. \tag{5. 17}$$

Then, by (5. 16), we have the implications

$$x \in U_d \Rightarrow d \geq d_1, \quad x \in V_d \Rightarrow d \geq d_2,$$

whence also the implication

$$x \in U_d \cap V_d \Rightarrow d \geq \max(d_1, d_2).$$

Thus, by (5. 17),

$$d_3 = \inf_{\substack{d\in R \\ x\in U_d\cap V_d}} d \geq \max(d_1, d_2). \tag{5. 18}$$

Assume now, a contrario, that this inequality is strict, i.e., that

$$d_3 > \max(d_1, d_2). \tag{5. 19}$$

Then, by (5. 16), there exist $d \in R$ with $d_1 \leq d < d_3$ and $d' \in R$ with $d_2 \leq d' < d_3$ such that $x \in U_d$ and $x \in V_{d'}$. Let $d'' := \max(d, d')$. Then $d'' < d_3$, and, on the other hand, by (5. 14), we have $x \in U_d \subset U_{d''}$ and $x \in V_{d'} \subseteq V_{d''}$, so $x \in U_{d''} \cap V_{d''}$, whence, by (5. 17), $d'' \geq d_3$, a contradiction. □

Now, by lemma 5.1, for the families of subsets of X defined by

$$U_d := \mathrm{co}_{\mathcal{M}} S_d(f), \quad V_d := \mathrm{co}_{\mathcal{N}} S_d(f) \tag{5. 20}$$

(which satisfy (5. 14)), and by (5. 13), we obtain

$$\inf_{\substack{d\in R \\ x\in \mathrm{co}_{\mathcal{M}} S_d(f)\cap \mathrm{co}_{\mathcal{N}} S_d(f)}} d = \max\left\{\inf_{\substack{d\in R \\ x\in \mathrm{co}_{\mathcal{M}} S_d(f)}} d \;, \; \inf_{\substack{d\in R \\ x\in \mathrm{co}_{\mathcal{N}} S_d(f)}} d\right\}$$
$$= \max\{f_{q(\mathcal{M})}, f_{q(\mathcal{N})}\} \quad (x \in X),$$

which proves the first equality of (5. 12).

Using this equality, one can give the following proof of formula (5. 1): By (5. 13) and (3. 1), for any function $f : X \to \overline{R}$ there holds

$$f_{q(\mathcal{P})}(x) = \inf_{\substack{d \in R \\ x \in \mathrm{co}_{\mathcal{P}} S_d(f)}} d = \inf_{\substack{d \in R \\ x \in \mathrm{co}_{\mathcal{M}} S_d(f) \cap \mathrm{co}_{\mathcal{N}} S_d(f)}} d \quad (x \in X), \tag{5. 21}$$

whence, by the first equality of (5. 12), we obtain (5. 1).

Corollary 5.2. *Let X be a set and $\mathcal{M}, \mathcal{N}$ two families of subsets of X. Then for any function $f : X \to \overline{R}$ we have*

$$\max\{f_{q(\mathcal{M})}(x), f_{q(\mathcal{N})}(x)\} = \sup_{\substack{M \in \mathcal{M}, N \in \mathcal{N} \\ x \in X \setminus (M \cap N)}} \inf f(X \setminus (M \cap N)) \quad (x \in X). \tag{5. 22}$$

Proof. We recall that, by [5], theorem 2.2 (or [6], p. 134, theorem 4.3), for any function $f : X \to \overline{R}$ we have

$$f_{q(\mathcal{M})}(x) = \sup_{\substack{M \in \mathcal{M} \\ x \in X \setminus M}} \inf f(X \setminus M) \quad (x \in X). \tag{5. 23}$$

Hence, by (5. 1) and (5. 23), for $\mathcal{P}$ of (3. 2) there holds

$$\max\{f_{q(\mathcal{M})}(x), f_{q(\mathcal{N})}(x)\} = f_{q(\mathcal{P})}(x) = \sup_{\substack{P \in \mathcal{P} \\ x \in X \setminus P}} \inf f(X \setminus P)$$

$$= \sup_{\substack{M \in \mathcal{M}, N \in \mathcal{N} \\ x \in X \setminus (M \cap N)}} \inf f(X \setminus (M \cap N)) \quad (x \in X).$$

□

Remark 5.2. a) Similarly to the case of remark 5.1, one can also proceed in the converse direction, to prove directly corollary 5.2 and then to deduce from it formula (5. 1). We omit the details.

b) In particular, for a locally convex space X, with conjugate space X^*, theorem 5.1 yields the following proof of formula (1. 8): Let $\mathcal{M}$ and $\mathcal{N}$ be the families of subsets of X considered in remark 3.1a), i.e., defined by (3. 4) and (3. 5) respectively. Then, by (5. 23) and (1. 5), (1. 6), we have

$$f_{q(\mathcal{M})}(x) = \sup_{\substack{M_\Phi \in \mathcal{M} \\ x \in X \setminus M_\Phi}} \inf f(X \setminus M_\Phi) = \sup_{\substack{\Phi \in X^* \\ \Phi(x) > 1}} \inf_{\substack{y \in X \\ \Phi(y) > 1}} f(y) = f^{00}(x) \quad (x \in X), \tag{5. 24}$$

$$f_{q(\mathcal{N})}(x) = \sup_{\substack{N_\Phi \in \mathcal{N} \\ x \in X \setminus N_\Phi}} \inf f(X \setminus M_\Phi) = \sup_{\substack{\Phi \in X^* \\ \Phi(x) < 1}} \inf_{\substack{y \in X \\ \Phi(y) < 1}} f(y) = f^{\nabla\nabla}(x) \quad (x \in X). \tag{5. 25}$$

Hence, by (1. 4), (5. 12), and (3. 9), we obtain

$$f^{\#\#}(x) = \max\{f_{q(\mathcal{M})}(x), f_{q(\mathcal{N})}(x)\} = \inf_{\substack{d \in R \\ x \in \mathrm{co}_{\mathcal{M} \cup \mathcal{N}} S_d(f)}} d = \inf_{\substack{d \in R \\ x \in \overline{\mathrm{co}} S_d(f)}} d \quad (x \in X), \tag{5. 26}$$

which, together with the well-known equality

$$\inf_{\substack{d\in R \\ x\in \overline{\mathrm{co}} S_d(f)}} d = f_{\overline{\mathrm{q}}}, \tag{5. 27}$$

due to Crouzeix (see [3], p.11, proposition 5 or [6], p. 143, formula (4.67)), yields (1. 8).

c) Formula (1. 8) and the argument of part b) above can be extended to more general situations in which, instead of the pair (X, X^*) one considers a pair (X, W), where X is a set and W is a set of functions $w : X \to \overline{R}$, satisfying certain additional assumptions, and in which $f_{\overline{\mathrm{q}}}$ is replaced by the so-called "W-quasi-convex hull" of f. We omit the details.

6 Abstract convex and quasi-convex hulls revisited

It will be relevant to give also another proof of formula (5. 1), based on an analysis of the operator $f \to \max\{f_{\mathrm{q}(\mathcal{M})}, f_{\mathrm{q}(\mathcal{N})}\}$. Let us first recall that, by [5], proposition 2.5b) and [6], p. 136, theorem 4.5(b), *there holds* $\emptyset \in \mathcal{C}(\mathcal{M})$ *if and only if for each set* $G \subseteq X$ *we have*

$$(\chi_G)_{\mathrm{q}(\mathcal{M})} = \chi_{\mathrm{co}_{\mathcal{M}} G}. \tag{6. 1}$$

Now we shall give an extension of this result, which we shall need in the sequel, and which may have some interest also for other applications, since it does not involve the condition $\emptyset \in \mathcal{C}(\mathcal{M})$.

Theorem 6.1. *Let X be a set and $\mathcal{M}$ a family of subsets of X. Then*

a) *For any set $G \subseteq X$ we have*

$$(\chi_G)_{\mathrm{q}(\mathcal{M})}(x) = \begin{cases} -\infty & \text{if} \quad x \in \mathrm{co}_{\mathcal{M}}\emptyset \\ 0 & \text{if} \quad x \in \mathrm{co}_{\mathcal{M}}G \backslash \mathrm{co}_{\mathcal{M}}\emptyset \\ +\infty & \text{if} \quad x \in X \backslash \mathrm{co}_{\mathcal{M}}G. \end{cases} \tag{6. 2}$$

b) *We have*

$$\mathrm{co}_{\mathcal{M}}G = \{x \in X | (\chi_G)_{\mathrm{q}(\mathcal{M})}(x) \leq 0\} \qquad (G \subseteq X). \tag{6. 3}$$

Proof. a) Let $G \subseteq X$. By (5. 13), there holds

$$(\chi_G)_{\mathrm{q}(\mathcal{M})}(x) = \inf_{\substack{d\in R \\ x\in \mathrm{co}_{\mathcal{M}} S_d(\chi_G)}} d \qquad (x \in X). \tag{6. 4}$$

But, by (1. 21) and (1. 22), we have

$$S_d(\chi_G) = \begin{cases} \emptyset & \text{if} \quad d < 0 \\ G & \text{if} \quad 0 \leq d < +\infty, \end{cases} \tag{6. 5}$$

whence

$$\mathrm{co}_{\mathcal{M}} S_d(\chi_G) = \begin{cases} \mathrm{co}_{\mathcal{M}} \emptyset & \text{if} \quad d < 0 \\ \mathrm{co}_{\mathcal{M}} G & \text{if} \quad 0 \leq d < +\infty, \end{cases} \tag{6. 6}$$

which, together with (6. 4), yields (6. 2) (note that by $\emptyset \subseteq G$ we have $\mathrm{co}_{\mathcal{M}}\emptyset \subseteq \mathrm{co}_{\mathcal{M}}G$).

b) Let $G \subseteq X$ and $x \in X$. Then, by (6. 2), we have $(\chi_G)_{\mathrm{q}(\mathcal{M})}(x) \leq 0$ if and only if $x \in \mathrm{co}_{\mathcal{M}}G$, which proves (6. 3). □

Remark 6.1. a) We have $\emptyset \in \mathcal{C}(\mathcal{M})$ if and only if $\mathrm{co}_{\mathcal{M}}\emptyset = \emptyset$, which, by (6. 2), happens if and only if for any set $G \subseteq X$ there holds (6. 1). Thus, theorem 6.1 is indeed an extension of the result mentioned before its statement.

b) Formula (5. 13) shows that the $\mathcal{M}$-convex hull operator $\mathrm{co}_{\mathcal{M}}$ on 2^X determines uniquely the $\mathcal{M}$-quasi-convex hull operator $\mathrm{q}(\mathcal{M})$ on $\overline{R}^X$ (but they do not determine uniquely the family $\mathcal{M}$–see e.g. [6], p. 51, remark 2.1). Theorem 6.1b) gives a (new) formula in the converse direction, showing that one can recuperate uniquely $\mathrm{co}_{\mathcal{M}}G$ from $(\chi_G)_{\mathrm{q}(\mathcal{M})}$, for any set $G \subseteq X$. Let us also mention the following related formulas for recuperating uniquely $\mathrm{co}_{\mathcal{M}}$ from $\mathrm{q}(\mathcal{M})$, which follow from [5], proposition 2.5a) (or [6], p. 136, theorem 4.5(a)) and [6], p. 137, remark 4.5(b) respectively:

$$\mathrm{co}_{\mathcal{M}}G = \{x \in X \mid (-\chi_{X \setminus G})_{\mathrm{q}(\mathcal{M})}(x) = -\infty\} \qquad (G \subseteq X), \tag{6. 7}$$

$$\mathrm{co}_{\mathcal{M}}G = \{x \in X \mid (\eta_G)_{\mathrm{q}(\mathcal{M})}(x) = -\infty\} \qquad (G \subseteq X), \tag{6. 8}$$

where η_G is the representation function of G, defined by

$$\eta_G(x) = \begin{cases} -\infty & \text{if} \quad x \in G \\ +\infty & \text{if} \quad x \in X \setminus G. \end{cases} \tag{6. 9}$$

Now we can give

Another proof of formula (5. 1). We claim that the mapping

$$v : f \to f_v := \max\{f_{\mathrm{q}(\mathcal{M})}, f_{\mathrm{q}(\mathcal{N})}\} \tag{6. 10}$$

is a quasi-convex hull operator on $\overline{R}^X$, i.e., that there exists a hull operator $u : 2^X \to 2^X$ such that

$$f_v = f_{\mathrm{q}(u)} \qquad (f \in \overline{R}^X) \tag{6. 11}$$

(we recall that for a hull operator $u : 2^X \to 2^X$, a function $f : X \to \overline{R}$ is said to be *u-quasi-convex*, if all level sets (1. 21) of f are *u-convex*, i.e., $u(S_d(f)) = S_d(f)$ for all $d \in R$, and that for any function $f : X \to \overline{R}$, its *u-quasi-convex hull* $f_{\mathrm{q}(u)}$ is the greatest u-quasi-convex function majorized by f). We know from proposition 2.1 applied to the hull operators $u_1 := \mathrm{q}(\mathcal{M})$ and $u_2 := \mathrm{q}(\mathcal{N})$ in the complete lattice $E = (\overline{R}^X, \leq)$, that the mapping v is a hull operator on $\overline{R}^X$ (since there $f_1 \vee f_2 = \max\{f_1, f_2\}$). We shall show that for any function $k \in C_\infty$, where C_∞ denotes the set of all functions $k : \overline{R} \to \overline{R}$ satisfying

$$k(\sup_{i\in I} d_i) = \sup_{i\in I} k(d_i) \qquad (I \neq \emptyset, \{d_i\}_{i\in I} \subseteq R), \tag{6. 12}$$

$$k(\inf_{i\in I} d_i) = \inf_{i\in I} k(d_i) \qquad (I \neq \emptyset, \{d_i\}_{i\in I} \subseteq R), \tag{6. 13}$$

$$k(-\infty) = -\infty, \tag{6. 14}$$

we have

$$(k \circ f)_v = k \circ f_v \quad (f \in \overline{R}^X), \tag{6. 15}$$

which, by a result of Volle (see [7], p. 108, theorem 7 or [6], p.167, theorem 4.15) on characterizations of quasi-convex hull operators among hull operators, will imply that v is a quasi-convex hull operator on $\overline{R}^X$. Indeed, since $u_1 :=\mathrm{q}(\mathcal{M})$ and $u_2 :=\mathrm{q}(\mathcal{N})$ are quasi-convex hull operators on $\overline{R}^X$, by the same theorem of Volle we have, for any $f \in \overline{R}^X$ and $k \in C_\infty$,

$$(k \circ f)_{\mathrm{q}(\mathcal{M})} = (k \circ f_{\mathrm{q}(\mathcal{M})}), \tag{6. 16}$$

$$(k \circ f)_{\mathrm{q}(\mathcal{N})} = (k \circ f_{\mathrm{q}(\mathcal{N})}). \tag{6. 17}$$

Consequently, by (6. 10), (6. 16), (6. 17) and (6. 12) we obtain, for any $f \in \overline{R}^X$ and $k \in C_\infty$,

$$\begin{aligned}(k \circ f)_v &= \max\{(k \circ f)_{\mathrm{q}(\mathcal{M})}, (k \circ f)_{\mathrm{q}(\mathcal{N})}\} = \max\{k \circ f_{\mathrm{q}(\mathcal{M})}, k \circ f_{\mathrm{q}(\mathcal{N})}\} \\ &= k \circ \max\{f_{\mathrm{q}(\mathcal{M})}, f_{\mathrm{q}(\mathcal{N})}\} = k \circ f_v.\end{aligned}$$

Thus, there holds (6. 15), which proves the claim that the mapping v is a quasi-convex hull operator on $\overline{R}^X$. Namely, by the proof of [6], p.167, theorem 4.15, if we define

$$u(G) := \{x \in X | (\chi_G)_v(x) \leq 0\} \qquad (G \subseteq X), \tag{6. 18}$$

then $u : 2^X \to 2^X$ is a hull operator satisfying (6. 11).

Let $x \in X$. Then, by (6. 18), (6. 10), and theorem 6.1b),

$$\begin{aligned} x \in u(G) &\Leftrightarrow 0 \geq (\chi_G)_v(x) = \max\{(\chi_G)_{\mathrm{q}(\mathcal{M})}(x), (\chi_G)_{\mathrm{q}(\mathcal{N})}(x)\} \\ &\Leftrightarrow x \in \mathrm{co}_{\mathcal{M}} G \cap \mathrm{co}_{\mathcal{N}} G, \end{aligned}$$

whence, by (3. 1),

$$u(G) = \mathrm{co}_{\mathcal{M}} G \cap \mathrm{co}_{\mathcal{N}} G = \mathrm{co}_{\mathcal{P}} G \qquad (G \subseteq X), \tag{6. 19}$$

with $\mathcal{P}$ of (3. 2). Thus,

$$f_v = f_{\mathrm{q}(\mathrm{co}_{\mathcal{P}})} = f_{\mathrm{q}(\mathcal{P})} \qquad (f \in \overline{R}^X), \tag{6. 20}$$

which, together with (6. 10), proves (5. 1) (in order to see the last equality of (6. 20), it is enough to observe that $\mathcal{C}(\mathrm{co}_{\mathcal{P}}) = \{G \in 2^X | \mathrm{co}_{\mathcal{P}} G = G\} = \mathcal{C}(\mathcal{P})$, whence $Q(\mathrm{co}_{\mathcal{P}}) = Q(\mathcal{P})$, and thus $\mathrm{q}(\mathrm{co}_{\mathcal{P}}) = \mathrm{q}(\mathcal{P})$.

Remark 6.2. Since being a quasi-convex hull operator $\mathrm{q}(u)$ on $\overline{R}^X$ for some hull operator u on 2^X is equivalent to being a quasi-convex hull operator $\mathrm{q}(\mathcal{L})$ on $\overline{R}^X$ for some family of subsets $\mathcal{L}$ of X (see e.g. [6], p. 56, remark 2.5), from remark 6.1b) above it follows that, once we know that v is a quasi-convex hull operator $\mathrm{q}(u)$, i.e., that we have (6. 11), the convex hull operator u on 2^X satisfying (6. 11) is uniquely determined by $\mathrm{q}(u)$, namely it is necessarily the operator given by formula (6. 18).

Acknowledgment. Research supported in part by the National Agency for Science, Technology and Innovation, grant nr. 4073/1998.

References

1. M. Attéia, Analyse convexe projective. Comptes rendus Ac. Sci. Paris 276 (1973), 855-858.
2. M. Attéia and A. ElQortobi, Quasi-convex duality. In: Optimization and Optimal Control (A. Auslender, W. Oettli and J.Stoer, eds.), Lecture Notes Control Inform. Sciences 30, Springer-Verlag, Berlin, Heidelberg (1981), 3-8.
3. J.-P. Crouzeix, Contributions à l'étude des fonctions quasi-convexes. Thèse. Univ. de Clermont (1977).
4. A. ElQortobi, Contributions à la théorie de la dualité pour les fonctionnelles quasi-convexes. Thèse de 3-ème cycle, Univ. de Toulouse (1980).
5. I. Singer, Generalized convexity, functional hulls and applications to conjugate duality in optimization. In: Selected Topics in Operations Research and Mathematical Economics (G. Hammer and D. Pallaschke, eds.), Lecture Notes Econ. Math. Systems 226, Springer-Verlag, Berlin, Heidelberg (1984), 49-79.
6. I. Singer, Abstract convex analysis. Wiley-Interscience, New York, 1997.
7. M. Volle, Contributions à la dualité en optimisation et à l'épi-convergence. Thèse. Univ. de Pau (1986).

Specific Numerical Methods for Solving Some Special Max-min Programming Problems Involving Generalized Convex Functions

Ştefan Ţigan[1], Ion M. Stancu-Minasian[2], and Iliana Ţigan[3]

[1] University of Medicine and Pharmacy, Cluj-Napoca, Romania
[2] Centre for Mathematical Statistics of the Romanian Academy, Bucharest, Romania
[3] National College "G. Coşbuc", Cluj-Napoca, Romania

Abstract. In this paper, we deal with several classes of max-min programming problems. The first class refers to some max-min bicriterion fractional programming problems with separate polyhedral constraints. We show that these problems can be transformed into bicriterion fractional maximization problems. The second class involves some polynomial max-min problems with separate polyhedral constraints. The third class comprises some max-min problems with joint linear constraints.

2000 Mathematics Subject Classification. Primary 90C26; Secondary 90C29.

Keywords and phrases. fractional maximin, bicriterion programming, polynomial maximin, pseudo-monotonic maximin.

1 Preliminaries

The max-min problem under consideration is of the form:
MP. Find $V = \max_{x \in X} \quad \min_{y \in T(x)} G(x, y)$, where $X \subseteq R^n$ and $Y \subseteq R^m$ are two given sets, T is a given point to set function from X to Y and $G : X \times Y \to R$ is a real-valued function.

A point $(x^*, y^*) \in X \times Y$ is said to be an optimal solution for problem **MP**, if: i) $G(x^*, y^*) = V$; ii) $\min_{y \in T(x)} G(x^*, y) = G(x^*, y^*)$.

Max-min problem **MP** can be reformulated as a min-max problem if $G(x, y)$ is replaced by $-G(x, y)$. These problems have important applications in modeling various conflict situations (see, [46]).

Min-max fractional problems arise in the formulation of discrete and continuous rational approximation problems with respect to the Chebyshev norm [3], in continuous rational games [27], in multi-objective programming [29], in engineering design and in some parameter estimation problems as well as in some portfolio selection problems discussed by Bajona-Xandri [1]. One can add to these the application of fractional max-min problems to some minimum-risk problems studied by Stancu-Minasian and Tigan [32], [33], [34], [44]. Some applications of the discrete minimax problems can be found in Du and Pardalos [15], Katoh [21], Pardalos and Du [23] and Peteanu and Tigan [25].

Recently, min-max fractional problems have been studied by Gwinner and Jeyakumar [18], who have established duality and stability results, and Bajona-Xandri and Martinez-Legaz [2], who have obtained necessary and sufficient Kuhn-Tucker conditions assuming quasiconvexity of the objective function, under lower subdifferentiability hypothesis. A transformation variable method and some parametric procedures for solving a large class of fractional max-min problems have been presented by Tigan [37], [40].

The goal of this paper is to present specific finite procedures for solving some special max-min problems, involving generalized convex objective functions.

We recall that on a convex set $\widetilde{A} \subset R^k$ the function $f : \widetilde{A} \to R$ is said to be :

(i) pseudo-concave if $\nabla f(x')(x'' - x') \leq 0 \Rightarrow f(x'') \leq f(x')$, for every $x', x'' \in \widetilde{A}$; (ii) pseudo-convex if $-f$ is pseudo-concave; (iii) quasi-concave if $f(x'') \leq f(x') \Rightarrow f(x'') \leq f(tx' + (1-t)x'')$, for every $x', x'' \in \widetilde{A}$ and $t \in]0,1[$, or equivalently, when f is a differentiable function, $f(x'') \leq f(x') \Rightarrow \nabla f(x')(x'' - x') \leq 0$, for every $x', x'' \in \widetilde{A}$; (iv) quasi-convex if $-f$ is quasi-concave; (v) quasi-monotonic if f is both quasi-concave and quasi-convex; (vi) pseudo-monotonic if f is both pseudo-concave and pseudo-convex; (vii) strictly quasi-convex if $f(x'') < f(x') \Rightarrow f(tx' + (1-t)x'') < f(x')$, for every $x', x'' \in \widetilde{A}$ and $t \in]0,1[$.

The paper is organized as follows. In Section 2, we will consider the case of max-min problem **MP**, when the objective function G is of the form $G(x,y) = h(F(x), Q(x,y))$, where $h : D \to R$ $(D \subseteq R^2)$ is a real-valued function, non-decreasing and continuous on a given set D such that $D \supseteq \{(F(x), Q(x,y)) : x \in X, y \in Y\}$, $X \subseteq R^n$ and $Y \subseteq R^m$ are two compact convex polyhedral given sets, $F : X \to R$, $Q : X \times Y \to R$ are given real functions, while $T(x) = Y$ for any $x \in X$.

Section 3 is dedicated to a procedure of solving a polynomial max-min problem with non-joint linear constraints, that is a max-min problem **MP** in which $G(x,y) = P(f(x) + g(y))$, where P is a polynomial, $f : X \to R$, $g : Y \to R$ are appropriate continuous functions, and X and Y are polyhedral bounded nonempty sets, while $T(x) = Y$ for any $x \in X$.

In section 4, we will consider a generalized linear-fractional max-min problem with joint linear constraints.

For each class of these max-min problems we suggest a numerical procedure for its solving.

2 Bicriterion Max Min fractional problem with non joint linear constraints

2.1 Problem formulation

In this section, we consider a max-min bicriterion fractional problem with separate polyhedral constraints. We show that this problem can be transformed into a bicriterion fractional maximization problem. For this type of nonlinear program, Warburton [45] described a finite procedure, which consists in solving a one-parameter linear program and a series of one-dimensional maximization subproblems.

Our algorithm for max-min bicriterion fractional problem consists in applying Warburton's procedure [45] to solve a certain bicriterion fractional maximization problem and then the solution of a linear minimization program.

Applying some results from [17] and [10], Warburton showed how the bicriterion problem could be reduced to a one-parameter linear program and a finite set of one-dimensional sub-problems.

For another parametric algorithms for solving bi and multicriterion programming problems we refer to [14], [22], [28].

The max-min problem under consideration is of the form:

P. Find $V = \max_{x\in X} \min_{y\in Y} h(F(x), Q(x,y))$, where $X \subseteq R^n$ and $Y \subseteq R^m$ are two given compact convex polyhedral sets, $h : D \to R \quad (D \subseteq R^2)$ is a real-valued function, non-decreasing and continuous on a given set $D \supseteq \{(F(x), Q(x,y)) | \, x \in X, y \in Y\}$, and $F : X \to R, Q : X \times Y \to R$.

A particular case of the functions F and Q is that when they are fractional functions of the form:

$$F(x) = \frac{cx + p}{fx + q}, x \in X, \tag{1}$$

$$Q(x,y) = \frac{xCy + ey + gx + \alpha}{\max_{i\in K}(d^i x + r^i)}, x \in X, y \in Y, \tag{2}$$

where $K = \{1, 2, ..., s\}$.

In (1) and (2), C is a given $n \times m$ real matrix, $e \in R^m, g, f, c, d^i \in R^n$ and $p, q, \alpha, r^i (i \in K)$ are given real numbers.

Without loss of generality, in what follows, we consider the feasible sets X and Y of the form:

$$X = \{x \in R^n : Ax \leq a, x \geq 0\}, \tag{3}$$

$$Y = \{y \in R^m : By \leq b, y \geq 0\}, \tag{4}$$

where A is a $s \times n$ real matrix, B is a $k \times m$ real matrix, $a \in R^s$ and $b \in R^k$.

On functions F and Q, we make the following assumptions:

$$fx + q > 0, \text{for all } x \in X, \tag{5}$$

$$d^i x + r^i > 0, \text{for all } x \in X \text{ and } i \in K. \tag{6}$$

The problem **P** represents a generalization of the bicriterion linear fractional problem considered by Warburton [45], that is, the case when

$$C = 0, e = 0 \text{ and } s = 1.$$

The particular case of problem **P**, when

$$h(F(x), Q(x, y)) = max\{F(x), Q(x, y)\},$$

was studied by Belenkii [5]. Other approaches of max-min problems related to problem **P** have been obtained in references [7], [24], [31], [35].

2.2 Transformation of the Max Min problem P

Under the hypothesis of the following lemma, the max-min problem **P** can be transformed into a bicriterion maximization problem (problem **P1** given below).

Lemma 1. *([43]) If h is continuous and non-decreasing on D, then*

$$\max_{x \in X} \min_{y \in Y} h(F(x), Q(x, y)) = \max_{x \in X} h\left(F(x), \min_{y \in Y} Q(x, y)\right).$$

Consequently, by Lemma 1, the problem **P** can be reduced to the following bicriterion maximization problem
P1. Find

$$V1 = \max_{x \in X} h(F(x), F_1(x)), \text{ where } F_1(x) = \min_{y \in Y} Q(x, y). \tag{7}$$

Obviously, $V = V1$ and if (x^*, y^*) is an optimal solution of problem **P**, then x^* is an optimal solution of problem **P1**.

In dealing with problem **P1**, we denote by E the solution set for the bicriterion Pareto maximization problem over the feasible set X, with the objective functions F and F_1 .

The problem **P1** is closely related to the following problem
P2. Find $V2 = \max_{(z,t) \in Z} h\left(\frac{cz+pt}{fz+qt}, \min_{y \in Y}(zCy + tey + gz + \alpha t)\right)$, where

$$Z = \{(z, t) \in R^{n+1} : Az - at \leq 0, d^i z + r^i t \leq 1, z \geq 0, t \geq 0, i \in K\}.$$

Theorem 1. *If X is a bounded and non-empty set and (5), (6) hold then: (i) If $(z, t) \in Z$, then $t > 0$. (ii) If (z^*, t^*) is an optimal solution for **P2**, then $x^* = z^*/t^*$ is an optimal solution for **P1**.*

Proof. The assertion (i) can be easily obtained by a similar argument as those used by Charnes-Cooper [9] for usual linear fractional programming. The property (ii) can be easily proved, by a similar reason like Schaible [26], by performing in problem **P1**, the variable transformation $z = tx$. We mention that by this variable transformation, the problem **P1** is converted into the problem **P2**. □

Consider the following problem related to problem **P2**
P3. Find

$$V3 = \max_{z,t,u} h\left(\frac{cz+pt}{fz+qt}, gz+\alpha t - bu\right) \tag{8}$$

subject to

$$Az - at \leq 0, \tag{9}$$

$$d^i z + r^i t \leq 1, i \in K, \tag{10}$$

$$-uB \leq zC + te, \tag{11}$$

$$z \geq 0, t \geq 0, u \geq 0. \tag{12}$$

Denote by T the feasible set of problem **P3**, i.e., the set of points $(z,t,u) \in R^{n+k+1}$, which satisfy the constraints (9)-(12).

Theorem 2. *If h is continuous and non-decreasing on D, then: (i) problems **P2** and **P3** have the same optimal values, that is $V2 = V3$; (ii) if (z^*, t^*, u^*) is an optimal solution for **P3**, then (z^*, t^*) is an optimal solution for **P2**; (iii) if (z^*, t^*) is an optimal solution for **P2**, then there exists $u^* \geq 0$, such that (z^*, t^*, u^*) is an optimal solution for **P3**.*

Proof. Indeed, for every $(z,t) \in Z$, the problem

$$M(z,t) = \min_{y \in Y}(zCy + tey + gz + \alpha t)$$

is a linear program. Since Y is a nonempty, compact set of the form (4), by the duality property of linear programming, it follows that

$$M(z,t) = max\{gz + \alpha t - bu : -uB \leq zC + te, u \geq 0\}.$$

Consequently, for every $(z,t) \in Z$, we have

$$h(\frac{cz+pt}{fz+qt}, \min_{y \in Y}(zCy+tey+gz+\alpha t) = h(\frac{cz+pt}{fz+qt}, \max_{u \in S(z,t)}(gz+\alpha t - bu)) \tag{13}$$

where $S(z,t) = \{u \in R^k : -uB \leq zC + te, u \geq 0\}$, for all $(z,t) \in Z$.

Since h is continuous and non-decreasing, it result that

$$h(\frac{cz+pt}{fz+qt}, \max_{u \in S(z,t)}(gz+\alpha t - bu)) = \max_{u \in S(z,t)} h\left(\frac{cz+pt}{fz+qt}, gz+\alpha t - bu\right), \tag{14}$$

for all $(z,t) \in Z$.

From (13) and (14), it follows immediately that the properties (i), (ii) and (iii) of the theorem hold. □

Therefore, by Lemma 1 and Theorems 1 and 2, it follows that the max-min problem **P** can be reduced to a bicriterion linear fractional problem **P3** with linear constraints, to which Warburton's parametrical procedure [45] can be applied.

2.3 The parametric algorithm

In this section, we describe an algorithm for solving the problem **P**.

Under assumptions (5) and (6), because X and Y are nonempty compact sets, the following values are well-defined:

$$f_1 = \max\{\frac{cz + pt}{fz + qt} : (z,t) \in Z\}, \tag{15}$$

$$f_2 = \max\{gz + \alpha t - bu : (z,t,u) \in T\}, \tag{16}$$

$$f_1^* = \max\{\frac{cz + pt}{fz + qt} : (z,t) \in Z, gz + \alpha t - bu \geq f_2\}, \tag{17}$$

$$f_2^* = \max\{gz + \alpha t - bu : (z,t,u) \in T, \frac{cz + pt}{fz + qt} \geq f_1\}. \tag{18}$$

Denote by E^* the Pareto optimal solutions over T associated to problem **P3**. From equations (15) - (17), it is obvious that $f_1^* \leq \frac{cz+pt}{fz+qt} \leq f_1$,whenever $(z,t,u) \in E^*$.

For each $w \in [f_1^*, f_1]$ consider the problem

P(w). Find $V(w) = \max\{gz + \alpha t - bu : (z,t,u) \in T, \frac{cz+pt}{fz+qt} \geq w\}$.

Since (5) holds, the problem P(w) may be rewritten:

P(w). Find $V(w) = max\{gz + \alpha t - bu : (z,t,u) \in T, cz + pt \geq w(fz + qt)\}$.

The solution of **P**(w) for $w \in [f_1^*, f_1]$ according to Warburton's procedure [45], leads to the following algorithm for solving problem **P**.

Step 1. Solve problems (15), (16) and (17) to obtain f_1, f_2 and f_1^*.

Step 2. Apply row parametric procedure to the linear program **P**(w) for $w \in [f_1^*, f_1]$ to obtain critical values $f_1^* = w_0 \leq w_1 \leq ... \leq w_{r^*} = f_1$, and the corresponding points $Z_i = (z^i, t^i, u^i)$ solving **P**(w_i).

Step 3. Solve r^* one-dimensional sub-problems defined by:

Q$_i$. Find $W_i = max\{h(H(\zeta)) : \zeta \in [Z_{i-1}, Z_i]\}, for 1 \leq i \leq r^*$, where

$$[Z_{i-1}, Z_i] = \{\theta Z_{i-1} + (1-\theta) Z_i : 0 \leq \theta \leq 1\},$$

and by $\zeta = (z, u, t)$ we denote a current feasible solution of problem **P3**, while by $H : T \to R^2$ the function

$$H(z,t,u) = (\tfrac{cz+pt}{fz+qt}, gz + \alpha t - bu), \text{ for all } (z,t,u) \in T.$$

Let Z_i^* for $1 \leq i \leq r^*$, be the corresponding optimal solution of sub-problem $\mathbf{Q}_i$.

Step 4. Find $j \in \{1, 2, ..., r^*\}$ such that:

$$h(H(Z_j^*)) = max\{W_i : 1 \leq i \leq r^*\}.$$

Then $Z^* = Z_j^*$ is an optimal solution for **P3**. Let $Z^* = (z^*, t^*, u^*)$ denote this optimal solution.

Step 5. Take $x^* = z^*/t^*$, which, according to Theorem 1, is an optimal solution for problem **P1**.

Step 6. Solve the following linear program

$$\min_{y \in Y} \frac{x^*Cy + ey + gx^* + \alpha}{\min_{i \in K}} (d^i x^* + r^i).$$

Let y^* be an optimal solution for this linear program. Then (x^*, y^*) is an optimal solution for problem **P**, and the algorithm stops.

As it is known, by Charnes-Cooper transformation [9], problems (15) and (17) in Step 1 are equivalent to certain linear programs. In Step 2, w_i occurs at the i-th basis change in the parametric solution of $\mathbf{P}(w)$. The Step 3 is justified by Theorem 1 from [45], while Step 5 is warranted by Theorem 1.

Since h is a non-decreasing function, it results that for every (z^*, t^*, u^*) in E^*, we have $x^* = z^*/t^*$ in E.

3 Polynomial Max Min problem

3.1 Introduction

In the papers [19] and [30] a minimization problem of a polynomial fractional function on a polyhedral set in R^n is considered. The objective function of this problem is a composition of a polynomial P with a linear-fractional function defined on a subset in R^n.

Cimoca and Tigan [11] presented a method for solving a max-min problem having as objective function the composition of a polynomial with a sum of two linear-fractional functions defined on two separable polyhedral sets. Iacob and Tigan in [20] give an extension of the method from [11] for solving a polynomial pseudo-monotonic max-min problem.

In this section we will show that the method considered in [20] can be employed for solving a polynomial generalized linear-fractional max-min problem, which generally is not pseudo-monotonic.

The polynomial max-min problem can be stated as follows:

PMF. Find $V = \max_{x \in X} \min_{y \in Y} P(f(x) + g(y))$,
where: (i) P is a polynomial of degree k' with real coefficients; (ii) $f : X \to R, g : Y \to R$ are continuous functions; (iii) X and Y are polyhedral bounded nonempty sets defined by (3) and (4).

3.2 The auxiliary Max Min problem

In what follows we associate to problem **PMF** a polynomial max-min problem having only two variables. This problem states as follows:

PA. Find $V^* = \max_{u\in[c,c^*]} \min_{v\in[d,d^*]} P(u+v)$, where

$$c = min\{f(x) : x \in X\}, \tag{19}$$

$$c^* = max\{f(x) : x \in X\}, \tag{20}$$

$$d = min\{g(y) : y \in Y\}, \tag{21}$$

$$d^* = max\{g(y) : y \in Y\}. \tag{22}$$

In what follows we suppose that $c < c^*$ and $d < d^*$. If $c = c^*$ or $d = d^*$, then **PMF** is a minimization or a maximization polynomial problem, so that a simpler method for its solving can be employed.

We have the following result concerning problems **PA** and **PMF**.

Theorem 3. *([20]) If f and g are continuous functions and X, Y are bounded nonempty sets given by (3), (4), then $V = V^*$ and $(x^*, y^*) \in X \times Y$ is an optimal solution of problem* ***PMF*** *if and only if $(f(x^*), g(y^*))$ is an optimal solution of problem* ***PA***.

From Theorem 3, we can conclude the following procedure for solving max-min problem **PA**.

Step 1. Find c, c^*, d and d^* by solving the programming problems (19)-(22), and let x', x'', y' and y'' be the optimal solutions of these optimization problems.

Step 2. Find the local minimum points $e_1, ..., e_r$ and the local maximum points $z_1, ..., z_{r^*}$ of the polynomial P situated in the open interval $(c+c^*, d+d^*)$. Denote $Z = \{z_1, ..., z_{r^*}\}$.

Step 3. Let $D = \{e_i - d^* : i \in I\} \cup \{e_i - d : i \in I\} \cup \{c, c^*\}$, where $I = \{1, 2, ..., r\}$. Denote by $N = \{t_1, ..., t_{r^*}\}$ the set of the points of D belonging to the interval $[c, c^*]$. We suppose that the elements of the set N are in the following order: $t_1 < t_2 < ... < t_{r^*}$.

Let $D_j = \{i^* \in I : d + t_{j+1} \leq e_{i^*} \leq d^* + t_j\}$.

Find, for any $j \in J = \{1, 2, ..., r^* - 1\}$,

$$f_j = \begin{cases} min\{P(e_i) : i \in D_j\}, \text{ when } D_j \text{ is a nonempty set,} \\ W, \text{ when } D_j \text{ is an empty set,} \end{cases}$$

W being a positive real number greater than maximum of P over $[c+c^*, d+d^*]$.

Step 4. For every $j \in J$, find the set M_j of all real solutions in closed interval $[t_j, t_{j+1}]$ of the equations:

$$P(u+d) = P(u+d^*), P(u+d) = f_j, P(u+d^*) = f_j.$$

Let denote $E_j = (Z \cap [t_j, t_{j+1}]) \cup M_j$.

Step 5. Find the values

$$V_j = max\{min\{P(u+d), P(u+d^*), f_j\} : u \in E_j\},$$

for every $j \in J$, and let u_j be a point in which this maximum is achieved.
Step 6. Find $u^* = u_{j^*}$, such that $V_{j^*} = max\{V_j : j \in J\}$.
Step 7. Find $v^* \in [d, d^*]$ such that

$$P(u^* + v^*) = min\{P(u^* + v) : v \in [d, d^*]\}.$$

If f and g are pseudo-monotonic functions, in Step 1 it must be solved four pseudo-monotonic programming problems. Algorithms to solve optimization problems with a pseudo-monotonic objective function are discussed, for instance, by Bector and Jolly [4], Bhatt [8] and Tigan [38], [42]. These approaches use linearization properties of the pseudo-monotonic functions to end up at linear programs that can be solved employing simplicial methods. In the next paragraph, we will consider a case when the function f and g are not generally pseudo-monotonic functions.

3.3 Generalized linear fractional programming

The solving of the problem **PMF** can be effectively performed under some supplementary assumptions concerning the functions f and g. In the particular cases, when the functions f and g are linear, linear-fractional, and pseudo-monotonic ref. [20] provide some approaches to obtain an optimal solution of problem **PMF**, having as starting point an optimal solution of problem **PA**. Now we consider the case when f and g are supposed to be generalized linear-fractional functions, that is:

$$f(x) = \min_{i \in H_0} \frac{c_i x + c_0}{d_i x + d_0},\ g(x) = \min_{i \in H_1} \frac{e_i y + e_0}{q_i y + q_0},$$

where $H_0 = \{1, 2, ..., p_0\}, H_1 = \{1, 2, ..., p_1\}, c_i, d_i \in R^n, e_i, q_i \in R^m$ and $c_0, d_0, e_0, q_0 \in R$.

Furthermore, we suppose that:
H1) $d_i x + d_0 > 0$, for any $x \in X$ and any $i \in H_0$,
H2) $q_j y + q_0 > 0$, for any $y \in Y$ and any $j \in H_1$.

Theorem 4. *Let f and g be generalized linear-fractional functions and let x', x'', y', y'' and (u'', v'') be optimal solutions of pseudo-monotonic programming problems (19)-(22) and max-min problem **PA**, respectively. Then the pair (x^*, y^*) given by*

$$x^* = \alpha^* x' + (1 - \alpha^*) x'',\ y^* = \beta^* y' + (1 - \beta^*) y'', \tag{23}$$

*is an optimal solution for **PMF** provided that α^* and β^* to be some solutions in $[0, 1]$ of the equations:*

$$f_1(\alpha) = f(\alpha x' + (1 - \alpha) x'') = u'', g_1(\beta) = g(\beta y' + (1 - \beta) y'') = v''. \tag{24}$$

Proof. By Theorem 3, x^* and y^* may be generally obtained by solving the equations: $f(x) = u'', g(y) = v''$.

But due to the assumption of convexity of the sets X and Y, we can solve instead of multivariable equations (24) the following special non-linear equations with a single variable (α or β):

$$f(\alpha x' + (1-\alpha)x'') = u'', g(\beta y' + (1-\beta)y'') = v''.$$

□

Theorem 4 suggests the following procedure for solving problem **PMF**.

Procedure for solving problem PMF

Step 1. In order to find c, c^*, d and d^*, under the assumptions H1, H2, problems (19) and (21) can be reduced to p_0 and p_1 usual linear-fractional minimization programs, respectively, while for solving problems (20) and (22), which are generalized linear-fractional programs, a parametrical procedure of Dinkelbach type can be employed (see, e.g. [13], [36], [40]). Let x', x'', y' and y'' be the optimal solutions of these optimization problems.

Step 2. Find an optimal solution (u'', v'') for max-min auxiliary problem **PA**.

Step 3. Find some solutions α^* and β^* in the interval $[0, 1]$ of the equations (24).

Step 4. Find x^* and y^* by using (23).

4 Max Min problem with joint linear constrains

The fourth section of this paper is concerned to a max-min problem having a special generalized linear-fractional objective function and joint linear constrains. A finite algorithm for solving this problem is suggested.

4.1 Problem formulation

We consider the following max-min problem

QM. Find

$$V = \max_{x} \min_{y} f(x, y) \tag{25}$$

subject to

$$Ax + By \leq b, x \geq 0, y \geq 0 \tag{26}$$

where $x \in R^n$, $y \in R^m$ are decision variable vectors, while A, B are $q \times n$, $q \times m$ given matrices respectively, $b \in R^q$ is a given right hand side vector and f is a real function defined on the feasible set $S \subset R^n \times R^m$,

$$S = \{(x, y) \in R^n \times R^m : Ax + By \leq b, x \geq 0, y \geq 0\}.$$

We denote the projections of S on R^n by

$$S_X = \{x \in R^n : \text{there exists } y \geq 0, \text{ such that } Ax + By \leq b, x \geq 0\}.$$

The problem **QM** is an extension of the linear max-min problem considered by Falk [16] or of the linear-fractional max-min problem studied by Cook, Kirby and Mehndiratta [12].

4.2 The quasi monotonic case

We shall assume throughout the remainder of this paragraph that: i1) S is a nonempty and compact set; i2) f is a quasi-monotonic function on the polyhedral convex set S.

Under the assumptions i1) and i2), we can define a function $F : S_X \to R$, by

$$F(x) = \min\{f(x,y) : y \in T(x)\}, \text{ for all } x \in S_X,$$

where $T(x)$ is given by

$$T(x) = \{y \in R^m : By \leq b - Ax, y \geq 0\}.$$

The following theorems characterize problem **QM** as a programming problem involving the maximization of a quasi-convex function over a linear polyhedron.

Theorem 5. *([39]) If the assumptions i1) and i2) hold, then the function F is quasi-convex on S_X.*

Theorem 6. *([41]) If the assumption i1) holds and f is a quasi-monotonic and strictly quasi-convex function on S, then F is strictly quasi-convex on S.*

Theorem 7. *([39]) If the assumptions i1), i2) hold, then there exists an optimal solution (x'', y'') of **QM** such that x'' is a vertex of the polyhedral set S_X.*

Theorem 8. *([39]) If the assumptions i1), i2) hold, then there exists an optimal solution (x'', y'') of **QM** that is a vertex of the polyhedral set S.*

Theorem 9. *([41]) If the problem **QM** has separate constraints (that is, $S = X \times Y$) and the assumptions i1) and i2) hold, then the function F is quasi-monotonic on $S_X = X$.*

From Theorem 5, it results that the quasi-monotonic problem **QM** can be reduced to a quasi-convex programming problem, for which a finite procedure similar to that of Falk [16] can be given (see, Tigan [39], [41]).

By Theorem 9, the problem **QM** with separate constraints is equivalent with a quasi-monotonic programming problem with linear constraints, for which the algorithm proposed by Belenkii [6] could be employed.

4.3 Generalized fractional Max Min problem

We denote by **FQM** the particular case of problem **QM** when f is a special generalized linear-fractional function

$$f(x,y) = \frac{a_0 x + e_0 y + \alpha_0}{\max_{i \in K}\ (a_i x + e_i y + \alpha_i)},$$

where $K = \{1,2,...,s\}$, and $a_i \in R^n$, $e_i \in R^m$, $\alpha_i \in R$ are given, for any $i \in K \cup \{0\}$. Next we suppose that

$$\max_{i \in K}\ (a_i x + e_i y + \alpha_i) > 0, \text{forany} (x,y) \in S. \tag{27}$$

Next we will show that problem **FQM** can be reduced to a linear max-min problem with joint linear constraints. We mention that the quasi-monotonic max-min problem with joint constraints, presented in 4.3, don't generally extend the max-min problem **FQM**.

In order to give a solution to **FQM**, we will associate to this fractional max-min problem the following linear max-min problem
AQM. Find $V_1 = \max_u\ \min_{v,t}(a_0 u + e_0 v + \alpha_0 t)$, subject to

$$Au + Bv - bt \leq 0, \tag{28}$$

$$a_i u + e_i v + \alpha_i t \leq 1, i \in K, \tag{29}$$

$$u \geq 0, v \geq 0, t \geq 0. \tag{30}$$

Let U be the feasible set of problem **AQM**.

Theorem 10. *If S is a bounded and nonempty set and (27) holds then: (i) If $(u,v,t) \in\ U$, then $t > 0$. (ii) If (u^*, v^*, t^*) is an optimal solution for* ***AQM****, then $(x^*, y^*) = (u^*/t^*, v^*/t^*)$ is an optimal solution for* ***QM****.*

Proof. The assertion (i) can be easily obtained by a similar argument as those used by Charnes-Cooper [9] for usual linear fractional programming. The property (ii) can be easily proved, by a similar reason like that of Schaible [26] (see, also [37]), by performing in problem **FQM**, the variable transformation $u = tx$ and $v = ty$. □

We mention that by this variable transformation, the problem **QM** is converted into the problem **AQM**, which can be solved, for instance, by the finite procedure given by Falk [16].

5 Conclusions

Although problem **P**, in Section 2, is nonlinear and non-convex, by applying Warburton's parametric algorithm, it is possible to obtain an optimal solution, and hence a Pareto maximal point for the corresponding bicriterion problem. This algorithm uses only one-dimensional parametric linear programming techniques and one-dimensional nonlinear maximization procedures. Moreover, in the initial and final steps only linear fractional or linear programs must be solved. A similar approach can be used, when in (2) we take

$$Q(x,y) = \frac{xCy + ey + gx + \alpha}{\max_{i\in K}\ (d^i x + r^i)}.$$

In this case, the auxiliary maximization problem **P3**, having linear constrains, must be replaced with a similar problem but with a special nonlinear constraint

$$\max_{i\in K}\ (d^i x + r^i) \leq 1.$$

In order to solve the polynomial generalized linear-fractional max-min problem **PMF**, in Section 3, a polynomial max-min problem having only two variables is associated to it, for which a numerical algorithm we have proposed.

In Section 4, it is shown that a generalized fractional max-min problem can be reduced to a linear max-min problem having at least an optimal solution, which is a vertex of the feasible set. The finite algorithm due to Falk for solving this problem can be applied. We mention that a similar procedure can be employed, when we take an objective function

$$f(x,y) = \frac{a_0 x + e_0 y + \alpha_0}{\min_{i\in K}(a_i x + e_i y + \alpha_i)}.$$

However, in this case, the corresponding auxiliary problem **AQM** becomes more complex, having the nonlinear constraint

$$\min_{i\in K}\ (a_i x + e_i y + \alpha_i) \leq 1,$$

instead of the constraint (29).

The main characteristic of the proposed methods for solving these special class of max-min problems is that these procedures are linear programming technique based. It would be useful to compare these special numerical procedures with another more general existing algorithms.

Acknowledgment. The authors are grateful to the editor and referees for helpful comments and suggestions.

References

1. Bajona-Xandri, C.: On fractional programming, Master thesis, Universitat Autonoma de Barcelona, 1993.
2. Bajona-Xandri, C.; Martinez-Legaz, J.E.: Lower subdifferentiability in minimax fractional programming, Optimization, 45 (1999),1-4, 1-12.

3. Barrodale, I.: Best rational approximation and strict quasi-convexity, SIAM J. Numer. Anal., 10(1973), 8-12.
4. Bector, C.R.; Jolly, P.L.: Programming problems with pseudomonotonic objectives, Math.Operationsforsch. und Statist., ser. Optimization, 15 (1984), 2, 217-219.
5. Belenkii, A.S.: Minimax scheduling problems with linear constraint and methods of their solutions, Avt. i Telemeh., 10 (1981), 157-170.
6. Belenkii, A.S.: Minimization of a monotone function on a polyhedral set, Avt. i Telemeh., 9(1982), 112-121.
7. Belenkii, A.S.: On minimax problems with monotone functions on polyhedral sets, Avt. i Telemeh.,10(1982), 94-106.
8. Bhatt, S.L.: Linearization technique for linear fractional and pseudomonotonic programs revisited, Cahiers du CERO, 23(1981), 53-56.
9. Charnes, A.; Cooper, W.W.: Programming with linear fractional functionals, Naval Res. Logist. Quart., 9 (1962), 181-186.
10. Choo, E.U.; Atkins, D.R.: Bicriteria linear fractional programming, J. Optim. Theory Appl., 36 (1982),2, 203-220.
11. Cimoca, G.; Tigan, S.: O problemă polinomială de max-min cu restricţii liniare, Seminarul "T. Angheluţă", Inst. Politeh. Cluj-Napoca, 1983, 35-38.
12. Cook, W.D.; Kirby, M.J.L.; Mehndiratta, S.L: A linear fractional max-min problem, Oper. Res., 23 (1975),3, 511-521.
13. Crouzeix, J.P.; Ferland, J.A.; Schaible, S., An algorithm for generalized fractional programs, J. Optim. Theory Appl., 47(1985),1, 35-49.
14. Dragomirescu, M.; Maliţa, M.: Programare neliniară, Editura Stiinţifică, Bucureşti, 1972.
15. Du, D.-Z.; Pardalos, P.M.(Eds): Minimax and Applications, Kluwer Academic Publishers, 1995.
16. Falk, J.E.: A linear max-min problem, Math. Programming, 5(1973), 169-188.
17. Geoffrion, A.M.: Solving bicriterion mathematical programs, Oper. Res., 15 (1967), 39-54.
18. Gwinner, J.; Jeyakumar, V.: A solvability theorem and minimax fractional programming, Z. Oper. Res., 37 (1993),1-12.
19. Hirche, J.: A note on minimizing a polynomial of a fractional function, Rev. Roumaine Math. Pures Appl., 26 (1981), 1193-1195.
20. Iacob, E.; Tigan, S.: On a polynomial pseudomonotonic max-min problem, Lucrările Sesiunii de Comunicări Ştiinţifice, mai 1992, Univ. "Aurel Vlaicu" Arad, vol. I, Matematica Informatica, 24-31.
21. Katoh, N.: An ε-approximation scheme for combinatorial optimization problems with minimum variance criterion, Discrete Applied Mathematics, 35 (1992), 131-141.
22. Katoh, N.; Ibaraki, T.: A parametric characterization and an ε-approximation scheme for the minimization of quasiconcave program, Discrete Applied Mathematics, 17 (1987), 39-66.
23. Pardalos, P.M.; Du, D.-Z.(Eds): Network Design: Connectivity and Facilities Location, DIMACS Series vol. 40, AMS, 1998.
24. Patkar, V.N.; Stancu-Minasian, I.M.: A disjunctive linear-fractional max-min problem, Portugal Math., 48 (1991),1, 67-73.
25. Peteanu, V.; Tigan, S.: On some discrete fractional max-min problem. Application to max-min problems in graphs, Analyse Numer. Theor. Approx., 13 (1984),2, 167-173.

26. Schaible, S.: Nonlinear fractional programming, Oper. Res. Verfahren, 19 (1974), 109-115.
27. Schroeder, R. G.: Linear programming solutions to ratio games, Oper. Res., 18 (1970), 300-305.
28. Sniedovich, M.: C-programming problems: a class of nonlinear optimization problems, Discrete Applied Mathematics, 9 (1984), 301-305.
29. Soyster, A. L.; Lev, B.; Di Toof: Conservative linear programming with mixed multiple objectives, Omega, 5 (1977),2, 193-205.
30. Stancu-Minasian, I.M.: On a class of nonlinear fractional programming problems, Rev. Roumaine Math. Pures Appl., 23 (1978), 285-290.
31. Stancu-Minasian, I.M.; Patkar, V.N.: A note on nonlinear fractional max-min problem, Nat. Acad. Sci. Letters, 8 (1985),2, 39-41.
32. Stancu-Minasian, I.M.; Tigan, S.: The stochastic max-min problem, Oper. Res. Verfahren, 51 (1984), 119-126.
33. Stancu-Minasian, I.M.; Tigan, S.: The minimum risk approach to max-min bilinear programming, An Stiint. Univ. "Alexandru I. Cuza", Iaşi, N. ser., Sect. la, 31 (1985), 205-209.
34. Stancu-Minasian, I.M.; Tigan, S.: On some fractional programming models occurring in minimum risk problems, Proc. of the International Workshop on Generalized Concavity, Fractional programming and Economic Applications, Pisa Univ.,1988, Generalized Concavity, Fractional programming and Economic Applications, Cambini A., Castagnoli E., Martein L., Mazzoleni P., Schaible S. (eds), Lecture Notes in Economics and Mathematical Systems 345, Springer Verlag, Berlin, 1990, 295-324.
35. Stancu-Minasian, I.M.; Tigan, S.; Lozovanu, D.D.; Solomon, D.I.: On a parametrical method for solving bicriterion max-min fractional problem, Buletinul Acad.Stiinte Rep. Moldova, Matematica, 3(9) (1992),56-61.
36. Tigan, S.: Sur une méthode pour la résolution d'un problème d'optimisation fractionnaire par segments, Analyse Numér. Théor. Approx., 4 (1975),1, 87-97.
37. Tigan, S.: On the max-min nonlinear fractional problem, Analyse Numér. Théor. Approx., 9 (1980),2, 283-288.
38. Tigan, S., On the linearization technique for quasimonotonic optimization problems, Analyse Num. Theor. Approx., 12 (1983),1, 89-96.
39. Tigan, S.: A quasimonotonic max-min programming problem with linked constraints, Itinerant Sem. Functional Eq., Approx. and Convexity, Cluj-Napoca Univ. (Romania), 1986, 279-284.
40. Tigan, S.: On some procedures for solving fractional max-min problems, Rev. Anal. Numcr. Thcor. Approx., 17 (1988),1, 73-91
41. Tigan, S.: On a quasimonotonic max-min problem, Rev. Anal. Numer. Theor. Approx., 19 (1990),4, 85-91.
42. Tigan, S.: Linearization procedure and Kortanek-Evans optimality conditions for symmetric pseudo-concave programming, Rev. Anal. Numer. Theor. Approx., 22 (1993),1, 113-120
43. Tigan, S.; Stancu-Minasian, I.M.: On a bicriterion max-min fractional problem, Rev. Anal. Numer. Theor. Approx., 20 (1991),1-2, 117-125.
44. Tigan, S.; Stancu-Minasian, I.M.: Methods for solving stochastic bilinear fractional max-min problems, Recherche Operationnelle/ Operations Research, 30 (1996),1, 81-98.
45. Warburton, A.R.: Parametric solution of bicriterion linear fractional programs, Opns. Res., 33 (1985),1, 74-84.

46. Wolfe, P.: Determinateness of polyhedral games, in "Linear Inequalities and Related Systems", Kuhn H.W., Tucker A. W.,(Eds.) Princeton University Press, Princeton, 1956, 195-198.

Lecture Notes in Economics and Mathematical Systems

For information about Vols. 1–315
please contact your bookseller or Springer-Verlag

Vol. 316: R.L. Basmann, D.J. Slottje, K. Hayes, J.D. Johnson, D.J. Molina, The Generalized Fechner-Thurstone Direct Utility Function and Some of its Uses. VIII, 159 pages. 1988.

Vol. 317: L. Bianco, A. La Bella (Eds.), Freight Transport Planning and Logistics. Proceedings, 1987. X, 568 pages. 1988.

Vol. 318: T. Doup, Simplicial Algorithms on the Simplotope. VIII, 262 pages. 1988.

Vol. 319: D.T. Luc, Theory of Vector Optimization. VIII, 173 pages. 1989.

Vol. 320: D. van der Wijst, Financial Structure in Small Business. VII, 181 pages. 1989.

Vol. 321: M. Di Matteo, R.M. Goodwin, A. Vercelli (Eds.), Technological and Social Factors in Long Term Fluctuations. Proceedings. IX, 442 pages. 1989.

Vol. 322: T. Kollintzas (Ed.), The Rational Expectations Equilibrium Inventory Model. XI, 269 pages. 1989.

Vol. 323: M.B.M. de Koster, Capacity Oriented Analysis and Design of Production Systems. XII, 245 pages. 1989.

Vol. 324: I.M. Bomze, B.M. Pötscher, Game Theoretical Foundations of Evolutionary Stability. VI, 145 pages. 1989.

Vol. 325: P. Ferri, E. Greenberg, The Labor Market and Business Cycle Theories. X, 183 pages. 1989.

Vol. 326: Ch. Sauer, Alternative Theories of Output, Unemployment, and Inflation in Germany: 1960–1985. XIII, 206 pages. 1989.

Vol. 327: M. Tawada, Production Structure and International Trade. V, 132 pages. 1989.

Vol. 328: W. Güth, B. Kalkofen, Unique Solutions for Strategic Games. VII, 200 pages. 1989.

Vol. 329: G. Tillmann, Equity, Incentives, and Taxation. VI, 132 pages. 1989.

Vol. 330: P.M. Kort, Optimal Dynamic Investment Policies of a Value Maximizing Firm. VII, 185 pages. 1989.

Vol. 331: A. Lewandowski, A.P. Wierzbicki (Eds.), Aspiration Based Decision Support Systems. X, 400 pages. 1989.

Vol. 332: T.R. Gulledge, Jr., L.A. Litteral (Eds.), Cost Analysis Applications of Economics and Operations Research. Proceedings. VII, 422 pages. 1989.

Vol. 333: N. Dellaert, Production to Order. VII, 158 pages. 1989.

Vol. 334: H.-W. Lorenz, Nonlinear Dynamical Economics and Chaotic Motion. XI, 248 pages. 1989.

Vol. 335: A.G. Lockett, G. Islei (Eds.), Improving Decision Making in Organisations. Proceedings. IX, 606 pages. 1989.

Vol. 336: T. Puu, Nonlinear Economic Dynamics. VII, 119 pages. 1989.

Vol. 337: A. Lewandowski, I. Stanchev (Eds.), Methodology and Software for Interactive Decision Support. VIII, 309 pages. 1989.

Vol. 338: J.K. Ho, R.P. Sundarraj, DECOMP: An Implementation of Dantzig-Wolfe Decomposition for Linear Programming. VI, 206 pages.

Vol. 339: J. Terceiro Lomba, Estimation of Dynamic Econometric Models with Errors in Variables. VIII, 116 pages. 1990.

Vol. 340: T. Vasko, R. Ayres, L. Fontvieille (Eds.), Life Cycles and Long Waves. XIV, 293 pages. 1990.

Vol. 341: G.R. Uhlich, Descriptive Theories of Bargaining. IX, 165 pages. 1990.

Vol. 342: K. Okuguchi, F. Szidarovszky, The Theory of Oligopoly with Multi-Product Firms. V, 167 pages. 1990.

Vol. 343: C. Chiarella, The Elements of a Nonlinear Theory of Economic Dynamics. IX, 149 pages. 1990.

Vol. 344: K. Neumann, Stochastic Project Networks. XI, 237 pages. 1990.

Vol. 345: A. Cambini, E. Castagnoli, L. Martein, P Mazzoleni, S. Schaible (Eds.), Generalized Convexity and Fractional Programming with Economic Applications. Proceedings, 1988. VII, 361 pages. 1990.

Vol. 346: R. von Randow (Ed.), Integer Programming and Related Areas. A Classified Bibliography 1984–1987. XIII, 514 pages. 1990.

Vol. 347: D. Ríos Insua, Sensitivity Analysis in Multi-objective Decision Making. XI, 193 pages. 1990.

Vol. 348: H. Störmer, Binary Functions and their Applications. VIII, 151 pages. 1990.

Vol. 349: G.A. Pfann, Dynamic Modelling of Stochastic Demand for Manufacturing Employment. VI, 158 pages. 1990.

Vol. 350: W. B. Zhang, Economic Dynamics. X, 232 pages. 1990.

Vol. 351: A. Lewandowski, V. Volkovich (Eds.), Multiobjective Problems of Mathematical Programming. Proceedings, 1988. VII, 315 pages. 1991.

Vol. 352: O. van Hilten, Optimal Firm Behaviour in the Context of Technological Progress and a Business Cycle. XII, 229 pages. 1991.

Vol. 353: G. Ricci (Ed.), Decision Processes in Economics. Proceedings, 1989. III, 209 pages 1991.

Vol. 354: M. Ivaldi, A Structural Analysis of Expectation Formation. XII, 230 pages. 1991.

Vol. 355: M. Salomon. Deterministic Lotsizing Models for Production Planning. VII, 158 pages. 1991.

Vol. 356: P. Korhonen, A. Lewandowski, J . Wallenius (Eds.), Multiple Criteria Decision Support. Proceedings, 1989. XII, 393 pages. 1991.

Vol. 357: P. Zörnig, Degeneracy Graphs and Simplex Cycling. XV, 194 pages. 1991.

Vol. 358: P. Knottnerus, Linear Models with Correlated Disturbances. VIII, 196 pages. 1991.

Vol. 359: E. de Jong, Exchange Rate Determination and Optimal Economic Policy Under Various Exchange Rate Regimes. VII, 270 pages. 1991.

Vol. 360: P. Stalder, Regime Translations, Spillovers and Buffer Stocks. VI, 193 pages . 1991.

Vol. 361: C. F. Daganzo, Logistics Systems Analysis. X, 321 pages. 1991.

Vol. 362: F. Gehrels, Essays in Macroeconomics of an Open Economy. VII, 183 pages. 1991.

Vol. 363: C. Puppe, Distorted Probabilities and Choice under Risk. VIII, 100 pages . 1991

Vol. 364: B. Horvath, Are Policy Variables Exogenous? XII, 162 pages. 1991.

Vol. 365: G. A. Heuer, U. Leopold-Wildburger. Balanced Silverman Games on General Discrete Sets. V, 140 pages. 1991.

Vol. 366: J. Gruber (Ed.), Econometric Decision Models. Proceedings, 1989. VIII, 636 pages. 1991.

Vol. 367: M. Grauer, D. B. Pressmar (Eds.), Parallel Computing and Mathematical Optimization. Proceedings. V, 208 pages. 1991.

Vol. 368: M. Fedrizzi, J. Kacprzyk, M. Roubens (Eds.), Interactive Fuzzy Optimization. VII, 216 pages. 1991.

Vol. 369: R. Koblo, The Visible Hand. VIII, 131 pages.1991.

Vol. 370: M. J. Beckmann, M. N. Gopalan, R. Subramanian (Eds.), Stochastic Processes and their Applications. Proceedings, 1990. XLI, 292 pages. 1991.

Vol. 371: A. Schmutzler, Flexibility and Adjustment to Information in Sequential Decision Problems. VIII, 198 pages. 1991.

Vol. 372: J. Esteban, The Social Viability of Money. X, 202 pages. 1991.

Vol. 373: A. Billot, Economic Theory of Fuzzy Equilibria. XIII, 164 pages. 1992.

Vol. 374: G. Pflug, U. Dieter (Eds.), Simulation and Optimization. Proceedings, 1990. X, 162 pages. 1992.

Vol. 375: S.-J. Chen, Ch.-L. Hwang, Fuzzy Multiple Attribute Decision Making. XII, 536 pages. 1992.

Vol. 376: K.-H. Jöckel, G. Rothe, W. Sendler (Eds.), Bootstrapping and Related Techniques. Proceedings, 1990. VIII, 247 pages. 1992.

Vol. 377: A. Villar, Operator Theorems with Applications to Distributive Problems and Equilibrium Models. XVI, 160 pages. 1992.

Vol. 378: W. Krabs, J. Zowe (Eds.), Modern Methods of Optimization. Proceedings, 1990. VIII, 348 pages. 1992.

Vol. 379: K. Marti (Ed.), Stochastic Optimization. Proceedings, 1990. VII, 182 pages. 1992.

Vol. 380: J. Odelstad, Invariance and Structural Dependence. XII, 245 pages. 1992.

Vol. 381: C. Giannini, Topics in Structural VAR Econometrics. XI, 131 pages. 1992.

Vol. 382: W. Oettli, D. Pallaschke (Eds.), Advances in Optimization. Proceedings, 1991. X, 527 pages. 1992.

Vol. 383: J. Vartiainen, Capital Accumulation in a Corporatist Economy. VII, 177 pages. 1992.

Vol. 384: A. Martina, Lectures on the Economic Theory of Taxation. XII, 313 pages. 1992.

Vol. 385: J. Gardeazabal, M. Regúlez, The Monetary Model of Exchange Rates and Cointegration. X, 194 pages. 1992.

Vol. 386: M. Desrochers, J.-M. Rousseau (Eds.), Computer-Aided Transit Scheduling. Proceedings. 1990. XIII, 432 pages. 1992.

Vol. 387: W. Gaertner, M. Klemisch-Ahlert, Social Choice and Bargaining Perspectives on Distributive Justice. VIII, 131 pages. 1992.

Vol. 388: D. Bartmann, M. J. Beckmann, Inventory Control. XV, 252 pages. 1992.

Vol. 389: B. Dutta, D. Mookherjee, T. Parthasarathy, T. Raghavan, D. Ray, S. Tijs (Eds.), Game Theory and Economic Applications. Proceedings, 1990. IX, 454 pages. 1992.

Vol. 390: G. Sorger, Minimum Impatience Theorem for Recursive Economic Models. X, 162 pages. 1992.

Vol. 391: C. Keser, Experimental Duopoly Markets with Demand Inertia. X, 150 pages. 1992.

Vol. 392: K. Frauendorfer, Stochastic Two-Stage Programming. VIII, 228 pages. 1992.

Vol. 393: B. Lucke, Price Stabilization on World Agricultural Markets. XI, 274 pages. 1992.

Vol. 394: Y.-J. Lai, C.-L. Hwang, Fuzzy Mathematical Programming. XIII, 301 pages. 1992.

Vol. 395: G. Haag, U. Mueller, K. G. Troitzsch (Eds.), Economic Evolution and Demographic Change. XVI, 409 pages. 1992.

Vol. 396: R. V. V. Vidal (Ed.), Applied Simulated Annealing. VIII, 358 pages. 1992.

Vol. 397: J. Wessels, A. P. Wierzbicki (Eds.), User-Oriented Methodology and Techniques of Decision Analysis and Support. Proceedings, 1991. XII, 295 pages. 1993.

Vol. 398: J.-P. Urbain, Exogeneity in Error Correction Models. XI, 189 pages. 1993.

Vol. 399: F. Gori, L. Geronazzo, M. Galeotti (Eds.), Nonlinear Dynamics in Economics and Social Sciences. Proceedings, 1991. VIII, 367 pages. 1993.

Vol. 400: H. Tanizaki, Nonlinear Filters. XII, 203 pages. 1993.

Vol. 401: K. Mosler, M. Scarsini, Stochastic Orders and Applications. V, 379 pages. 1993.

Vol. 402: A. van den Elzen, Adjustment Processes for Exchange Economies and Noncooperative Games. VII, 146 pages. 1993.

Vol. 403: G. Brennscheidt, Predictive Behavior. VI, 227 pages. 1993.

Vol. 404: Y.-J. Lai, Ch.-L. Hwang, Fuzzy Multiple Objective Decision Making. XIV, 475 pages. 1994.

Vol. 405: S. Komlósi, T. Rapcsák, S. Schaible (Eds.), Generalized Convexity. Proceedings, 1992. VIII, 404 pages. 1994.

Vol. 406: N. M. Hung, N. V. Quyen, Dynamic Timing Decisions Under Uncertainty. X, 194 pages. 1994.

Vol. 407: M. Ooms, Empirical Vector Autoregressive Modeling. XIII, 380 pages. 1994.

Vol. 408: K. Haase, Lotsizing and Scheduling for Production Planning. VIII, 118 pages. 1994.

Vol. 409: A. Sprecher, Resource-Constrained Project Scheduling. XII, 142 pages. 1994.

Vol. 410: R. Winkelmann, Count Data Models. XI, 213 pages. 1994.

Vol. 411: S. Dauzère-Péres, J.-B. Lasserre, An Integrated Approach in Production Planning and Scheduling. XVI, 137 pages. 1994.

Vol. 412: B. Kuon, Two-Person Bargaining Experiments with Incomplete Information. IX, 293 pages. 1994.

Vol. 413: R. Fiorito (Ed.), Inventory, Business Cycles and Monetary Transmission. VI, 287 pages. 1994.

Vol. 414: Y. Crama, A. Oerlemans, F. Spieksma, Production Planning in Automated Manufacturing. X, 210 pages. 1994.

Vol. 415: P. C. Nicola, Imperfect General Equilibrium. XI, 167 pages. 1994.

Vol. 416: H. S. J. Cesar, Control and Game Models of the Greenhouse Effect. XI, 225 pages. 1994.

Vol. 417: B. Ran, D. E. Boyce, Dynamic Urban Transportation Network Models. XV, 391 pages. 1994.

Vol. 418: P. Bogetoft, Non-Cooperative Planning Theory. XI, 309 pages. 1994.

Vol. 419: T. Maruyama, W. Takahashi (Eds.), Nonlinear and Convex Analysis in Economic Theory. VIII, 306 pages. 1995.

Vol. 420: M. Peeters, Time-To-Build. Interrelated Investment and Labour Demand Modelling. With Applications to Six OECD Countries. IX, 204 pages. 1995.

Vol. 421: C. Dang, Triangulations and Simplicial Methods. IX, 196 pages. 1995.

Vol. 422: D. S. Bridges, G. B. Mehta, Representations of Preference Orderings. X, 165 pages. 1995.

Vol. 423: K. Marti, P. Kall (Eds.), Stochastic Programming. Numerical Techniques and Engineering Applications. VIII, 351 pages. 1995.

Vol. 424: G. A. Heuer, U. Leopold-Wildburger, Silverman's Game. X, 283 pages. 1995.

Vol. 425: J. Kohlas, P.-A. Monney, A Mathematical Theory of Hints. XIII, 419 pages, 1995.

Vol. 426: B. Finkenstädt, Nonlinear Dynamics in Economics. IX, 156 pages. 1995.

Vol. 427: F. W. van Tongeren, Microsimulation Modelling of the Corporate Firm. XVII, 275 pages. 1995.

Vol. 428: A. A. Powell, Ch. W. Murphy, Inside a Modern Macroeconometric Model. XVIII, 424 pages. 1995.

Vol. 429: R. Durier, C. Michelot, Recent Developments in Optimization. VIII, 356 pages. 1995.

Vol. 430: J. R. Daduna, I. Branco, J. M. Pinto Paixão (Eds.), Computer-Aided Transit Scheduling. XIV, 374 pages. 1995.

Vol. 431: A. Aulin, Causal and Stochastic Elements in Business Cycles. XI, 116 pages. 1996.

Vol. 432: M. Tamiz (Ed.), Multi-Objective Programming and Goal Programming. VI, 359 pages. 1996.

Vol. 433: J. Menon, Exchange Rates and Prices. XIV, 313 pages. 1996.

Vol. 434: M. W. J. Blok, Dynamic Models of the Firm. VII, 193 pages. 1996.

Vol. 435: L. Chen, Interest Rate Dynamics, Derivatives Pricing, and Risk Management. XII, 149 pages. 1996.

Vol. 436: M. Klemisch-Ahlert, Bargaining in Economic and Ethical Environments. IX, 155 pages. 1996.

Vol. 437: C. Jordan, Batching and Scheduling. IX, 178 pages. 1996.

Vol. 438: A. Villar, General Equilibrium with Increasing Returns. XIII, 164 pages. 1996.

Vol. 439: M. Zenner, Learning to Become Rational. VII, 201 pages. 1996.

Vol. 440: W. Ryll, Litigation and Settlement in a Game with Incomplete Information. VIII, 174 pages. 1996.

Vol. 441: H. Dawid, Adaptive Learning by Genetic Algorithms. IX, 166 pages.1996.

Vol. 442: L. Corchón, Theories of Imperfectly Competitive Markets. XIII, 163 pages. 1996.

Vol. 443: G. Lang, On Overlapping Generations Models with Productive Capital. X, 98 pages. 1996.

Vol. 444: S. Jørgensen, G. Zaccour (Eds.), Dynamic Competitive Analysis in Marketing. X, 285 pages. 1996.

Vol. 445: A. H. Christer, S. Osaki, L. C. Thomas (Eds.), Stochastic Modelling in Innovative Manufactoring. X, 361 pages. 1997.

Vol. 446: G. Dhaene, Encompassing. X, 160 pages. 1997.

Vol. 447: A. Artale, Rings in Auctions. X, 172 pages. 1997.

Vol. 448: G. Fandel, T. Gal (Eds.), Multiple Criteria Decision Making. XII, 678 pages. 1997.

Vol. 449: F. Fang, M. Sanglier (Eds.), Complexity and Self-Organization in Social and Economic Systems. IX, 317 pages, 1997.

Vol. 450: P. M. Pardalos, D. W. Hearn, W. W. Hager, (Eds.), Network Optimization. VIII, 485 pages, 1997.

Vol. 451: M. Salge, Rational Bubbles. Theoretical Basis, Economic Relevance, and Empirical Evidence with a Special Emphasis on the German Stock Market.IX, 265 pages. 1997.

Vol. 452: P. Gritzmann, R. Horst, E. Sachs, R. Tichatschke (Eds.), Recent Advances in Optimization. VIII, 379 pages. 1997.

Vol. 453: A. S. Tangian, J. Gruber (Eds.), Constructing Scalar-Valued Objective Functions. VIII, 298 pages. 1997.

Vol. 454: H.-M. Krolzig, Markov-Switching Vector Autoregressions. XIV, 358 pages. 1997.

Vol. 455: R. Caballero, F. Ruiz, R. E. Steuer (Eds.), Advances in Multiple Objective and Goal Programming. VIII, 391 pages. 1997.

Vol. 456: R. Conte, R. Hegselmann, P. Terna (Eds.), Simulating Social Phenomena. VIII, 536 pages. 1997.

Vol. 457: C. Hsu, Volume and the Nonlinear Dynamics of Stock Returns. VIII, 133 pages. 1998.

Vol. 458: K. Marti, P. Kall (Eds.), Stochastic Programming Methods and Technical Applications. X, 437 pages. 1998.

Vol. 459: H. K. Ryu, D. J. Slottje, Measuring Trends in U.S. Income Inequality. XI, 195 pages. 1998.

Vol. 460: B. Fleischmann, J. A. E. E. van Nunen, M. G. Speranza, P. Stähly, Advances in Distribution Logistic. XI, 535 pages. 1998.

Vol. 461: U. Schmidt, Axiomatic Utility Theory under Risk. XV, 201 pages. 1998.

Vol. 462: L. von Auer, Dynamic Preferences, Choice Mechanisms, and Welfare. XII, 226 pages. 1998.

Vol. 463: G. Abraham-Frois (Ed.), Non-Linear Dynamics and Endogenous Cycles. VI, 204 pages. 1998.

Vol. 464: A. Aulin, The Impact of Science on Economic Growth and its Cycles. IX, 204 pages. 1998.

Vol. 465: T. J. Stewart, R. C. van den Honert (Eds.), Trends in Multicriteria Decision Making. X, 448 pages. 1998.

Vol. 466: A. Sadrieh, The Alternating Double Auction Market. VII, 350 pages. 1998.

Vol. 467: H. Hennig-Schmidt, Bargaining in a Video Experiment. Determinants of Boundedly Rational Behavior. XII, 221 pages. 1999.

Vol. 468: A. Ziegler, A Game Theory Analysis of Options. XIV, 145 pages. 1999.

Vol. 469: M. P. Vogel, Environmental Kuznets Curves. XIII, 197 pages. 1999.

Vol. 470: M. Ammann, Pricing Derivative Credit Risk. XII, 228 pages. 1999.

Vol. 471: N. H. M. Wilson (Ed.), Computer-Aided Transit Scheduling. XI, 444 pages. 1999.

Vol. 472: J.-R. Tyran, Money Illusion and Strategic Complementarity as Causes of Monetary Non-Neutrality. X, 228 pages. 1999.

Vol. 473: S. Helber, Performance Analysis of Flow Lines with Non-Linear Flow of Material. IX, 280 pages. 1999.

Vol. 474: U. Schwalbe, The Core of Economies with Asymmetric Information. IX, 141 pages. 1999.

Vol. 475: L. Kaas, Dynamic Macroeconomics with Imperfect Competition. XI, 155 pages. 1999.

Vol. 476: R. Demel, Fiscal Policy, Public Debt and the Term Structure of Interest Rates. X, 279 pages. 1999.

Vol. 477: M. Théra, R. Tichatschke (Eds.), Ill-posed Variational Problems and Regularization Techniques. VIII, 274 pages. 1999.

Vol. 478: S. Hartmann, Project Scheduling under Limited Resources. XII, 221 pages. 1999.

Vol. 479: L. v. Thadden, Money, Inflation, and Capital Formation. IX, 192 pages. 1999.

Vol. 480: M. Grazia Speranza, P. Stähly (Eds.), New Trends in Distribution Logistics. X, 336 pages. 1999.

Vol. 481: V. H. Nguyen, J. J. Strodiot, P. Tossings (Eds.). Optimation. IX, 498 pages. 2000.

Vol. 482: W. B. Zhang, A Theory of International Trade. XI. 192 pages. 2000.

Vol. 483: M. Königstein, Equity, Efficiency and Evolutionary Stability in Bargaining Games with Joint Production. XII, 197 pages. 2000.

Vol. 484: D. D. Gatti, M. Gallegati, A. Kirman, Interaction and Market Structure. VI, 298 pages. 2000.

Vol. 485: A. Garnaev, Search Games and Other Applications of Game Theory. VIII, 145 pages. 2000.

Vol. 486: M. Neugart, Nonlinear Labor Market Dynamics. X, 175 pages. 2000.

Vol. 487: Y. Y. Haimes, R. E. Steuer (Eds.), Research and Practice in Multiple Criteria Decision Making. XVII, 553 pages. 2000.

Vol. 488: B. Schmolck, Ommitted Variable Tests and Dynamic Specification. X, 144 pages. 2000.

Vol. 489: T. Steger, Transitional Dynamics and Economic Growth in Developing Countries. VIII, 151 pages. 2000.

Vol. 490: S. Minner, Strategic Safety Stocks in Supply Chains. XI, 214 pages. 2000.

Vol. 491: M. Ehrgott, Multicriteria Optimization. VIII, 242 pages. 2000.

Vol. 492: T. Phan Huy, Constraint Propagation in Flexible Manufacturing. IX. 258 pages. 2000.

Vol. 493: J. Zhu, Modular Pricing of Options. X, 170 pages. 2000.

Vol. 494: D. Franzen, Design of Master Agreements for OTC Derivatives. VIII, 175 pages. 2001.

Vol. 495: I Konnov, Combined Relaxation Methods for Variational Inequalities. XI, 181 pages. 2001.

Vol. 496: P. Weiß, Unemployment in Open Economies. XII, 226 pages. 2001.

Vol. 497: J. Inkmann, Conditional Moment Estimation of Nonlinear Equation Systems. VIII, 214 pages. 2001.

Vol. 498: M. Reutter, A Macroeconomic Model of West German Unemployment. X, 125 pages. 2001.

Vol. 499: A. Casajus, Focal Points in Framed Games. XI, 131 pages. 2001.

Vol. 500: F. Nardini, Technical Progress and Economic Growth. XVII, 191 pages. 2001.

Vol. 501: M. Fleischmann. Quantitative Models for Reverse Logistics. XI, 181 pages. 2001.

Vol. 502: N. Hadjisavvas, J. E. Martínez-Legaz, J.-P. Penot (Eds.), Generalized Convexity and Generalized Monotonicity. IX, 410 pages. 2001.

GPSR Compliance

The European Union's (EU) General Product Safety Regulation (GPSR) is a set of rules that requires consumer products to be safe and our obligations to ensure this.

If you have any concerns about our products, you can contact us on ProductSafety@springernature.com

In case Publisher is established outside the EU, the EU authorized representative is:

Springer Nature Customer Service Center GmbH
Europaplatz 3
69115 Heidelberg, Germany

Batch number: 10844669

Printed by Printforce, the Netherlands